KEY ENVIRONMENTS

General Editor: J. E. Treherne

WESTERN MEDITERRANEAN

The International Union for Conservation of Nature and Natural Resources (IUCN), founded in 1948, is the leading independent international organization concerned with conservation. It is a network of governments, non-governmental organizations, scientists and other specialists dedicated to the conservation and sustainable use of living resources.

The unique role of IUCN is based on its 502 member organizations in 114 countries. The membership includes 57 States, 121 government agencies and virtually all major national and international non-governmental conservation organizations.

Some 2000 experts support the work of IUCN's six Commissions: ecology; education; environmental planning; environmental policy, law and administration; national parks and protected areas; and the survival of species.

The IUCN Secretariat conducts or facilitates IUCN's major functions: monitoring the status of ecosystems and species around the world; developing plans (such as the World Conservation Strategy) for dealing with conservation problems, supporting action arising from these plans by governments or other appropriate organizations, and finding ways and means to implement them. The Secretariat co-ordinates the development, selection and management of the World Wildlife Fund's international conservation projects. IUCN provides the Secretariat for the Ramsar Convention (Convention on Wetlands of International Importance especially as Waterfowl Habitat). It services the CITES convention on trade in endangered species and the World Heritage Site programme of UNESCO.

IUCN, through its network of specialists, is collaborating in the Key Environments Series by providing information, advice on the selection of critical environments, and experts to discuss the relevant issues.

KEY ENVIRONMENTS
WESTERN MEDITERRANEAN

Edited by
RAMON MARGALEF

University of Barcelona, Spain

Foreword by

HRH THE DUKE OF EDINBURGH

Published in collaboration with the

INTERNATIONAL UNION FOR CONSERVATION OF
NATURE AND NATURAL RESOURCES

by

PERGAMON PRESS

OXFORD · NEW YORK · TORONTO · SYDNEY · FRANKFURT

U.K.	Pergamon Press Ltd., Headington Hill Hall, Oxford OX3 0BW, England.
U.S.A.	Pergamon Press Inc., Maxwell House, Fairview Park, Elmsford, New York 10523, U.S.A.
CANADA	Pergamon Press Canada Ltd., Suite 104, 150 Consumers Road, Willowdale, Ontario M2J 1P9, Canada
AUSTRALIA	Pergamon Press (Aust.) Pty. Ltd., PO Box 544, Potts Point, N.S.W. 2011, Australia.
FEDERAL REPUBLIC OF GERMANY	Pergamon Press GmbH, Hammerweg 6, D-6242 Kronberg-Taunus, Federal Republic of Germany

Copyright © 1985 Pergamon Press Ltd.

All Rights Reserved. No part of this publication may be reproduced, stored in a retrieval system or transmitted in any form or by any means: electronic, electrostatic, magnetic tape, mechanical, photocopying, recording or otherwise, without permission in writing from the publishers.

First edition 1985

Library of Congress Cataloging in Publication Data
Main entry under title:
Key environments: Western Mediterranean.
(Key environments)
Includes index.
1. Marine ecology: Mediterranean Sea.
I. Margalef, Ramón.
II. International Union for Conservation of Nature and Natural Resources.
III. Series.
QH93.K46   1984   574.5'2636'091638   84-10993.

British Library Cataloguing in Publication Data

Western Mediterranean—(Key environments)
1. Natural history—Western Mediterranean
I. Margalef, Ramón   II. Series
508.3182'2   QH150

ISBN 0-08-028870-7

Printed in Great Britain by A. Wheaton & Co. Ltd., Exeter

BUCKINGHAM PALACE.

The general problems of conservation are understood by most people who take an intelligent interest in the state of the natural environment. But if adequate measures are to be taken, there is an urgent need for the problems to be spelled out in accurate detail.

This series of volumes on "Key Environments" concentrates attention on those areas of the world of nature that are under the most severe threat of disturbance and destruction. The authors expose the stark reality of the situation without rhetoric or prejudice.

The value of this project is that it provides specialists, as well as those who have an interest in the conservation of nature as a whole, with the essential facts without which it is quite impossible to develop any practical and effective conservation action.

Philip

1984

General Preface

The increasing rates of exploitation and pollution are producing unprecedented environmental changes in all parts of the world. In many cases it is not possible to predict the ultimate consequences of such changes, while in some, environmental destruction has already resulted in ecological disasters.

A major obstacle, which hinders the formulation of rational strategies of conservation and management, is the difficulty in obtaining reliable information. At the present time the results of scientific research in many threatened environments are scattered in various specialist journals, in the reports of expeditions and scientific commissions and in a variety of conference proceedings. It is, thus, frequently difficult even for professional biologists to locate important information. There is consequently an urgent need for scientifically accurate, concise and well-illustrated accounts of major environments which are now or soon will be, under threat. It is this need which these volumes attempt to meet.

The series is produced in collaboration with the International Union for the Conservation of Nature. It aims to identify environments of international ecological importance, to summarize the present knowledge of the flora and fauna, to relate this to recent environmental changes and to suggest where possible, effective management and conservation strategies for the future. The selected environments will be re-examined in subsequent editions to indicate the extent and characteristics of significant changes.

The volume editors and authors are all acknowledged experts who have contributed significantly to the knowledge of their particular environments.

The volumes are aimed at a wide readership, including: academic biologists, environmentalists, conservationists, professional ecologists, some geographers as well as graduate students and informed lay people.

John Treherne

Contents

1. Introduction to the Mediterranean
   RAMON MARGALEF — 1

2. Evolution of the Mediterranean Basins and a Detailed Reconstruction of the Cenozoic Paleoceanography
   ANDRES MALDONADO — 17

3. The Driving Machine
   JORDI FLOS — 60

4. Physics of the Sea
   TOM SAWYER HOPKINS — 100

5. Chemistry of Mediterranean Waters
   ANTONIO CRUZADO — 126

6. Life and the Productivity of the Open Sea
   MARTA ESTRADA, FRANCISCO VIVES and MIGUEL ALCARAZ — 148

7. History of the Mediterranean Biota and the Colonization of the Depths
   J. M. PERES — 198

8. Diving in Blue Water. The Benthos
   JOAN DOMENEC ROS, JAVIER ROMERO, ENRIC BALLESTEROS and JOSEP MARIA GILI — 233

9. Fishes and Fishermen. The Exploitable Trophic Levels
   CHARLES BAS, ENRIQUE MACPHERSON and FRANCESC SARDA — 296

10. The Footprints of Life and of Man
    JOAN ALBAIGES, M. AUBERT and J. AUBERT — 317

Index — 353

CHAPTER 1

Introduction to the Mediterranean

RAMON MARGALEF

Department of Ecology, Facultad de Biologia, Universidad de Barcelona, avenida Diagonal, 645, Barcelona, 28, Spain

CONTENTS

1.1. A sea made to human scale	1
1.2. Two millennia of curiosity and research	2
1.3. Description and understanding of nature	4
1.4. A small scale model of a real ocean	6
1.5. The Mediterranean, a living machine	9
1.6. The edge of the sea. Coasts and marshes	12
1.6.1. The brackish environments	12
1.6.2. Salinity, a key factor	13
1.6.3. Economical use and importance	15

1.1. A SEA MADE TO HUMAN SCALE

To the inhabitants of its shores, the Mediterranean is a sea of blue water that only exceptionally becomes really rough, more important as a way of communication and commerce of ideas and goods than as a direct source of riches. The size of the sea is of a human scale. No long days of navigation are spent between harbour and harbour, and the sailors can wait in safety for acceptable sea conditions. The trips of the fishermen rarely take them away for more than one day. The sea keeps them together, as neighbours around the backfence. Beliefs, folklore, methods of cooking, become common property, without altogether losing their local flavour.

Attention was focused on the coasts and accurate navigation was not so important as in large oceans. Yet, since ancient times, real sailors were bred and spawned from the Mediterranean to sail past Gibraltar, to circumnavigate Africa and to explore the western expanse of ocean, but they were exceptional. Even today, much research is done in Summer (often with the excuse of academic holidays), under the undeniable influence of the philosophy that nice weather is necessary to do things at sea. However, the Gulf of Lions can be rough in winter, when the water mixes vertically to great depths. This process, so important for the biology of our sea, was better appreciated by extra-Mediterranean researchers, trapped on their ships at sea, than by the Mediterranean oceanographers, sitting in harbours, waiting for clearer skies.

Oceanographers have extended the name Mediterranean to other marine areas (the 'Arctic Mediterranean', the 'American Mediterranean'), in which the ratio between shore development and sea surface is high and in human terms could have made of the sea a link rather than a hindrance. But the Mediterranean remains unique, also because the Mediterranean countries have been the cradle of classical culture and can be truly regarded as harbingers of civilization.

1.2. TWO MILLENNIA OF CURIOSITY AND RESEARCH

If some theorists are to be believed, our ancestors passed through a semi-aquatic stage in their evolution, finding a living along the shore. The thought was half-facetiously advanced by Alastair Hardy, and has been unconvincingly elaborated by at least two writers. If early Man was a littoral animal then it would be a safe bet that the first ones climbed out of the Mediterranean and asked many of the questions that still tantalize us. It was not only the birth of Aphrodite, seen in the light playing on the tops of the waves, but also the observation and the record of plain facts about the form and the life of myriads of less glamourous but no less marvellous beings.

Aristotle (384 – 322 B.C.) is credited with an exceptional knowledge about the creatures of the Mediterranean, a knowledge that, without doubt, incorporated the wisdom of his forerunners and of fishermen. Aristotle personally made many observations (probably on the shores of the island of Lesbos). He had a good knowledge of the marine animals, in general, and generated a proper scientific interest in the products of the sea. The classic Renaissance writers on Natural History, such as the Swiss Konrad Gesner (1516 – 1565) and the Frenchman Rondelet (1507 – 1556) wrote about and published drawings on fishes and other marine creatures, based on specimens obtained at markets or directly from fishermen. Later E. Forbes sounded the Aegean in 1840, and recognized different zones according to the depth. At that time, it was widely believed that life could not exist in very deep water. The developing technologies of the time helped to solve the problem: in 1860 a submarine cable near Sardinia was recovered from 2100 m of water and found to be overgrown with all sorts of animals.

The 19th century witnessed a dramatic development in Marine Science. Several countries commissioned naval ships to make observations and collections. Even the Papal Navy joined in the fray: the Vatican Pirocorvette, *Immacolata Concezione*, introduced the Secchi disc named after Father Angelo Secchi (astronomer and translator of the work of the American oceanographer, Matthew F. Maury, 1806 – 1837), and a good scientist himself — the first recorded readings fell between 16 and 42.5 m (Cialdi, 1866). Coastal research laboratories were opened at Banyuls-sur-mer in 1863 – 82, Naples in 1872, Villefranche-sur-mer in 1880, Monaco in 1899 and then in many other coastal places with the purpose of facilitating the study of all forms of marine life. After the publication of *The Origin of Species* by Charles Darwin in 1856, the study of larval forms and of the organization of marine animals, became fashionable, providing clues as well as potential answers to evolutionary enigmas, resulting in an incredible rush to research. The Mediterranean was particularly favoured by this fashion. The Zoological Station of Naples is at once typical of the trend and exceptional for its results. Anton Dohrn, (a German from Stettin, born in 1840), established, after considerable efforts, a research laboratory and aquarium in Naples. Very soon it became the model for similar centres to be established in many countries. Dohrn's father, a 'late son of romanticism', had been acquainted with Humbolt, and Anton, as with many other scientists from northern countries, was struck by the beauty of the Mediterranean organisms, in the no less beautiful setting of the local landscape. It was instant love, an example of the Goethe syndrome, which overcomes not only Germans, for the Russians settled in Villefranche-sur-mer, where hydrographical factors favour the accumulation in the bay of the most marvellous pelagic creatures.

Anton Dohrn, a confidant of T. H. Huxley, was encouraged by Darwin in his project. The names of the guests of the Naples Station reads like a *Who's Who* of the biology of the times: Balfour and Ray Lankester of England; Haeckel, Selenka, von Ihering, Hensen, Chun, Driesch, Boveri, from Germany; Claus and Schulze from Austria; de Man from Holland and many more. Anton put to good use the sensibility and talent of fishermen and local people. Salvatore Lo Bianco (1860–1910), the son of a porter of a Neapolitan palace, became a scientist of great distinction. His grandson, Enrique Rioja Lo Bianco, was a pioneer of marine science in Spain. Comingio C. Merculiano (1845–1915) produced beautiful illustrations for the Station's publications.

Anton created the laboratory, his son Reinhold Dohrn consolidated it through the difficulties of Fascist Italy, and his grandson, Peter, could not sail easily over the turmoil of postwar Italy. But the Stazione had probably outlived its times. Each time has its challenge and each elicits appropriate responses.

The main purpose of the Naples Station, and many of the other early stations, was more the study of the biology of marine organisms, than a global understanding of the ecology of the ocean. Meanwhile, the interest in pelagic life and the physical and chemical oceanography grew. The relations between marine circulation and the distribution of biomass were recognized and studied theoretically in 1900. Knudsen established the empirical relations between chlorinity, salinity and density. The relevance for plankton, a relatively close community dependent on the chemistry of sea-water, became obvious, after the observations of Victor Hensen from Kiel, towards the end of the last century. The *Challenger*, in her most famous cruise around the world ocean (1827–1876), did not sail through the Mediterranean. Perhaps because it did not qualify as a real sea. But the stimulus of the times was sufficient and it was the Danes who accepted the challenge. The Danish cruises of the *Michael Sars* (just at the entrance to the Mediterranean) and of the *Dana* and the *Thor*, in 1908 and 1910, were highlights in the scientific study of the Mediterranean.

The Prince of Monaco, Albert, born in 1848, was of the same generation as Anton Dohrn. An enlightened amateur, he was a sailor at heart and organized several expeditions that led him to explore vaster seas than the Mediterranean. But he also worked in the Mediterranean: testing and introducing ingenious pieces of equipment, studying the pelagic and deep fauna, with the help of a number of distinguished scientists (among them R. J. Richard). The Prince was the creator of the Musée Océanographique de Monaco, dedicated in 1910.

The Mediterranean is not a very productive sea, but long tradition gave fishermen excellent skills and many of them made a sufficient living. Overfishing was only local and usually resulted from ill-advised developments. Mediterranean fisheries provided the stimulus for good theory. During the First World War, the fish populations in the Adriatic reacted noticeably to the modified activity and effort of the fishermen. The resulting population changes prompted the Italian zoologist, Umberto d'Ancona, to set the problem to his Father-in-law, the eminent mathematician Vito Volterra (1860–1940), who, in response, created what is to this day, the unsurpassed basis of much ecological theory.

After the Second World War, Marine Science received an immense impetus. Much sophisticated equipment was available, to measure environmental conditions (temperature, salinity, light, etc.) and to process numerical data. The exploration of deep water and, in particular, of the solid substrate of the oceans revolutionized marine science. Knowledge increased exponentially and the Mediterranean was not left behind. It emerged that the Mediterranean is an extremely complex residue of a very dynamic past. This included a virtual drying-up, before five million years ago, which set a limit to the age of colonization by living organisms. We know more and more about the quality and movements of water. The many coastal laboratories (usually working on fisheries) have published descriptions of yearly cycles of plankton in the local waters, and often engage in much more sophisticated research. Mediterranean science is now very active and it is on this living research that the different chapters of this book are based.

1.3. DESCRIPTION AND UNDERSTANDING OF NATURE

In enterprises such as this book, some compromise must be struck between detailed description and the outline of basic processes. A description of the geography, of the organisms, their life histories, their distribution, could go on and on into myopic detail. Perhaps this would be welcomed by the observer and traveller, but its sheer volume would be inconvenient and as an account of the available information would be uneven and incomplete. Science attempts to provide abridged descriptions of reality and recognizes the natural laws behind the variety of appearances. Instantaneous appearances are the result of dynamic processes. This is the approach that will be adopted. It is more congenial with the present spirit of science and may have the advantage of providing some insight into the workings of Nature and, in favourable situations, allow a certain measure of prediction.

Of course, in reality, things are much more involved. What has been said is barely an excuse for not going into extreme detail in this book. It is obvious that such phenomena as waves and plankton populations, as well as the clouds, impress us as processes. On the other hand, we are led to detailed description (giving up any attempt of an advanced causal understanding) when dealing with such things as the vegetation of a forest, the shape of a coast or the populations of a rocky bottom.

The principles of the physics of fluids are relevant to the understanding of the atmosphere and the ocean, and to their mutual interface. Basic properties of air and water (density, compressibility, heat capacity) the input of radiant energy from the Sun, the composition of the resulting movements with the rotation of Earth, led to the understanding of the relations between movements of water and distribution of mass. But the process happens in a frame that is by no means simple: the distribution of oceans and land masses, the topographic configuration of coasts and depths. These introduce much complexity, and the cascade of uncertainty increases down to the local event. The difficulties associated with the prediction of weather are but one example of the problems to be expected in understanding the behaviour of the sea.

The chemical composition of sea-water depends on a number of processes: weathering of the continents, segregation of substances in the sediments (metalliferous nodules) or trapping in evaporated basins (like the salts that remain in the bottom of the Mediterranean, as witnesses of its drying up some six million years ago), chemical equilibria in the water and changes caused by the activity of organisms. We can thus explain the main features of the distribution of the concentration of salt in the waters of the world ocean. However, local distributions of salinity cannot be totally explained and, in consequence, require further scientific inquiry. It is typical that in trying to understand the workings of a limited sea, the Mediterranean, we have to constantly refer to the world ocean. Important events in Mediterranean life have to be explained outside its boundaries, in past times, even worldwide: for instance, the large scale movements of continental plates, or in shifts of the climatic zones. This is also obvious in the shape of the Mediterranean basin. The Earth's crust consists of large discontinuous plates. These are never at rest; new material is added to oceanic plates along divergences and the old material drifts apart; old plates are overlapped and lie below other, mostly continental ones, along volcanic lines. The shapes of the oceans are altering continuously and the properties of their coasts is accordingly diversified. The continental margins are either abrupt and steep (with less apparent sediment accumulation) or are progressive and passive (formed by a thick wedge of sediments, with lateral and seaward continuity, that tracks the edge of the continent) and sinking or subsident, allowing for the accumulation of a complete stratigraphic record — including sometimes oil deposits. Geologically, the Mediterranean area has always been in turmoil, with upheavals and collisions. New oceanic crust has been added to old materials, reconstructed and patched. So many elements of stress and friction have accumulated, that we can foresee that the Mediterranean will remain active for a long time, both geologically and humanly.

The rules of life are the same everywhere. Organic molecules are synthetized by organisms referred to as primary producers (plants), using simpler chemicals and the energy of the Sun. Of course, this can happen only where there is light and it is pertinent to remember that the sea is sufficiently illuminated only in a layer 50 to 100 m thick, compared with the 4000 m of the average depth of the oceans. Organic matter travels through a food chain, from plants to phytophagous animals and then to carnivores, and so on. At each step a sizable part goes to bacteria, the smallest heterotrophs, that play an important and often undervalued role in the biology of the sea. In the end, the chemical elements return to the water in the form of relatively simple and usually oxidized compounds, to be taken again into the cycle of life. The thinness of the illuminated or photic layer is the main reason why, on average, primary production per unit of surface is much less in the sea than on land. But everywhere, the discontinuity of living beings — in individuals and in species — their seemingly infinite variety, the trend to slow down chemical cycles, the vertical organization of the ecosystems along the axis defined by light and gravity, and many other regularities provide a common link in the understanding of all the living cover of the earth.

Going into more detail, significant differences become obvious. The pelagic world consists of the populations suspended in open water; the small organisms are components of plankton, they are relatively short-lived, fugacious, and turnover of their populations is fast; the larger pelagic exploiters, such as fish and squid, are more in command of their own movement and are commonly referred to as nekton. Plankton organisms shift their positions almost continuously and a considerable component of randomness is introduced into their distributions, often through the turbulence of water.

Benthic populations, whether fixed or motile, are much more structured, usually extending in two dimensions, and related to a substrate. Each organism has neighbours that may last as such. These engage in much more complex and steady relations of competition and varied biological interactions. Turnover is usually slower than in plankton and not only because of the accumulation of supporting structures which are made to last (witness the coral reefs and the coralligenous communities in the Mediterranean). Communities continue to build on preceding history. No wonder that naturalists speak of organismic assemblages (communities and associations) to describe the regularities that may exist above the individual or the species level. Problems of scale are unavoidable, but it is undeniable that certain physical boundaries are reinforced at the organismic level, as in the zonation or distribution of organisms in distinguishable belts along the coasts of all the seas. The boundary between land and sea (or between river and sea) is especially interesting for the recognition of the factors and processes responsible for the distribution of organisms, but less so in the Mediterranean than in other seas, where tides are really important.

Marine populations change constantly, sometimes irregularly, as in plankton, and sometimes more regularly, as in the case of benthos, here in a way which is more comparable to the development of forests. The seasonal growth of plankton may be quite predictable: after initial enrichment of the water, diatoms develop, and as the turbulence declines, dinoflagellates continue to increase. The development of animals fits into the succession. Sometimes certain species follow other species, in waves, with the imposition of other, long-lived ones, that add another scale to the overall pattern. Comparable sequences occur almost everywhere, although different species may be present in homologous parts of the chain. Such changes can arise from a variety of causes: mixing, local upwelling and inflow of enriched water from elsewhere. Usually such unpredictable changes give rise to more-or-less regular successions. Sediments may accumulate regularly, with a progressive reduction of grain size, until some event discharges more coarse sediment and creates another discontinuity. This is an example of succession in a non-living system.

Short periods of local history are punctuated and restarted by major events that are less predictable. These are the boundary conditions for our deterministic models, conditions that may be the most important unknowns. Many phenomena cannot be explained on a local basis and have their origins in the whole Mediterranean, or in the world ocean. Local Mediterranean events may have world-wide implications. The Messinian crisis that dried up the Mediterranean arose from outside causes, but in its turn, influenced the whole ocean. The rapid storage of salt in the depths of the Mediterranean basin lowered the salinity of the world ocean, by as much as 2 parts per thousand, and had climatic (through changing the freezing point of water) and biological consequences.

This is an example of the illusory nature of our impressions of balance and persistence. Tremendous changes happen over a scale of only millions of years. Short term changes in the flow and properties of water also occur. The exchanges between the Mediterranean and Atlantic are subjected to accelerations and decelerations. Not many thousands of years ago, the low salinity water of the Black Sea extended into the Eastern Mediterranean, increasing fertility and creating anoxic bottom conditions. Even the deep Atlantic waters can show relatively large changes in temperature and salinity during a decade. Cold winters result in higher primary production, not only in surrounding freshwater lakes, but also in parts of the Mediterranean itself. Similar changes occur in the distribution of species and in the waxing and waning of their numbers. Life in the Mediterranean arose from organisms which were introduced after the Messinian crisis. However, discussion is still in progress on these origins, and their possible biogeographical meaning. We are so used to complaining about catastrophic human perturbations of natural balance, that the concept of dramatic changes, in a set of naturally interlocking processes, may strike us as a novelty. It should not diminish our sense of responsibility, for example, in relation to pollution, but should redirect it more efficiently.

1.4. A SMALL SCALE MODEL OF A REAL OCEAN

My purpose is to present a synthetic view of the Western Mediterranean. This chapter is a summary, but also includes some points of view that would be lost in the more detailed approach of the other sections. We can start by looking at the Mediterranean as a scale model of any large ocean or as a machine that produces organic matter. These do not exclude further approaches. A physical scientist wrote that the surface of the sea is so complex that only a poet can describe it. The beauty and mutual dependence of marine life is overwhelming. Poetic and scientific descriptions are complementary rather than opposite. Underwater films display coral reefs and schools of swimming fishes, but the iridescent beauty of pelagic life, and the silent complexity of the living cover of the submerged rocks, does not trail behind as a source of aesthetic wonder.

The Mediterranean is too small to develop important tidal waves; this element of the coastal scenery of large oceans is alien to our shores. No receding tides expose large surfaces covered by thousands of organisms. No wonder that scuba diving fast became popular among Mediterranean biologists, and helped to recognize what is the equivalent of the tropical coral reefs. In fact, transplanted coral reefs would probably grow in our present sea.

The shores of the Mediterranean are well-populated and even crowded. The historical changes in sea level have, therefore, been of much consequence. In the last 2 to 3 millenia, the sea level has risen by 5 to 6 m, engulfing all kinds of constructions, harbours, villas, towns, to which the remnants of a thousand shipwrecks along the coasts and in old harbours have to be added. The coasts of the Mediterranean are a treasure trove for archaeologists. Submarine caves are equally exciting for geologists and biologists, bearing traces of ancient sea levels and testimony of old colonizations.

Everything tends to attract the attention more to the shore than to the expanse of open water. Today, pollution is a menace especially along the coasts. So much has been written about the coastal waters of the Mediterranean that a change in this attitude, with more attention to the blue offshore waters is an unusual point of view that the authors of this book could offer. In practical terms, tourist trade and fisheries appear mutually exclusive. The Mediterranean is relatively poor, not in variety but in the quantity of organisms produced, and its transparent waters are attractive to holiday-makers. Pollution could mean increased productivity, but also unattractive shores. Somehow, in the future, Mediterraneans will have to learn how to use several inputs in some sort of advanced, large-scale, aquaculture. Meantime, perhaps it has to be considered a blessing that pollution is, as a matter of course, especially visible and noxious on the shores, the interface where man acts on the sea. This may evoke a rectifying feedback, before excessive damage is done. The alert can help to keep the sea relatively clean, but overconfidence would be dangerous, and it is not wise to rely too much on the relative safety provided by natural mechanisms.

Geographical appearance has to be harmonized with the underlying processes. One must be aware of the human scale, both in space and in time. Sometimes change looks dramatic, but usually the sea is the very image of eternity. The open sea appears more dynamic and looks more uniform than the bottom waters and along the shore. But even the offshore, blue water, is structured transiently — affecting the behaviour of large masses of water and of their inhabitants. Moreover, the configuration of the edges of the sea (the coasts and the depths) is the result of an uninterrupted and still very active history of crustal change. The system appears stationary only to the eyes of beings that enjoy life for a limited number of days.

The Mediterranean is situated between Europe, Asia and Africa, and, excluding the Black Sea, covers 2.542 millions of square kilometres, with an average depth close to 1,500 m. The Black Sea, the Eastern Mediterranean, and the Western Mediterranean form a string of basins which finally open into the Atlantic. The Western Mediterranean covers 860,000 km^2. It is effectively limited by the sills of Sicily — Tunis (close to 400 m depth) and of Gibraltar — Morocco (extending down to 320 m). The maximal depth of the Western Mediterranean is 3700 m. The salinity is close to 38.5 g of solid components per kg of deep water and a little less closer to the surface. The deep water temperature is near 13°C and is relatively constant; the temperature of the superficial water varies between approximately 13°C in winter, and about 26°C in summer. The water is well oxygenated throughout.

The Mediterranean lies in an area where evaporation exceeds both direct rainfall and the water carried by rivers which, in general, drain rather arid countries. Only the Black Sea receives relatively more freshwater. This balance explains the main properties of the liquid mass filling the Mediterranean. Water of relatively high salinity is formed by evaporative concentration and, being more dense, sinks to the bottom and cascades across the Gibraltar sill into the Atlantic, at a level where the Atlantic water has a lower density. The Mediterranean water pouring into the Atlantic sinks to the level where it finds equal density and is carried northwards, along the Portuguese coast. It interferes with the upwelling of deep water along the western coast of the Iberian Peninsula. The loss of deep Mediterranean water is over-compensated by the Atlantic water that enters with the superficial current of Gibraltar. Since the *Challenger* expedition it has been known that water of Mediterranean origin could be traced in the North Atlantic. William B. Carpenter (making observations from the *Porcupine* in 1870) observed that the two overlaying currents flowed in opposite directions in the Straits, and, a few years later, Helland-Hansen made more detailed measurements, including estimates of their variability.

We can now consider the circulation through Gibraltar from the point of view of general science. The excess of evaporation in the Mediterranean basin not only draws from the Atlantic the necessary water to compensate it, but also generates a much more important horizontal exchange. Similar mechanisms are at work at the mouth of rivers where the flow of freshwater drives, horizontally, a much larger amount of marine water which mixes slowly with the water contributed from land.

The Black Sea functions as an estuary, since an excess of freshwater flows into the Eastern Mediterranean through the straits, over deep layers of highly saline water. This deep water is slowly renewed and much of the surface organic matter is decomposed and leads to anoxic conditions. The deep water of the Black Sea accumulates hydrogen sulphide and lacks both oxygen and animal life below 190 m. It reacts easily with metals (for instance in paint coverings) to produce black sulphides — hence the Black Sea.

The Baltic Sea, 20 degrees further north, is comparable with the Black Sea. It receives large amounts of fresh water and sections of the Baltic contain freshwater life. However, the renewal of deep water occurs more easily in the Baltic, since sea-water from the North Sea creeps over the bottom, through the straits around Denmark, and is diluted in the Baltic proper with freshwater. The mixed water then flows back into the North Sea. It is the same model of estuarine circulation and, again, another example of how a delicate balance of salt and water sets in horizontal motion very large masses of water.

The comparison between the Mediterranean (minus the Black Sea) and the Baltic can be very instructive concerning the risks of eutrophication and pollution (*eutrophication* is any enrichment with nutrients that enhances biological production). This may be good for certain fishing grounds, or for aquaculture, but usually is considered undesirable, because water turns green, loses transparency, and, at depth, often develops oxygen lack, with potential serious danger to deep water fauna. Often, the results of mild organic and agricultural pollution appear only as eutrophication. However, people are ambivalent concerning this way of influencing nature: they want both fertile waters (able to produce rapidly-growing food) and, also, clear and transparent water.

The Mediterranean is well-adapted to avoid excessive eutrophication. It loses deep water, relatively rich with mineralized or recycled nutrients, and receives surface Atlantic water, in which most nutrients have been used before entering Gibraltar. The situation is exactly the opposite of the one in the Baltic, where ecological mechanisms tend to recycle and accumulate large amounts of nutrients. People around the Baltic might care more about conservation of Nature than people around the Mediterranean (or at least they believe so), but the Gods have always been fond of the Mediterraneans — that is all.

The Mediterranean, especially its western part, is a small sea. Almost a lake. It is not large enough to develop important tidal waves, except in the Adriatic, and here only by reason of its long and narrow shape. One difference between lakes and oceans concerns salinity. Yet the amount of sodium chloride and accompanying salts is not particularly important in relation to the adaptive capacity of life and its evolution. A more important difference is that temperate lakes mix, or circulate, vertically in the same place (usually once or twice in the year). In the oceans, on the other hand, such local mixing is rare, except in coastal areas. Usually there are regions of sinking water, and other distant areas of upwelling. These upwellings are most productive, because the ascending water brings towards the light water that is rich in necessary nutrients. The Mediterranean shows no appreciable upwelling. Vertical mixing can, however, influence adjacent areas by a mechanism which is more similar to that of a lake than to that in large oceans. In other ways the formation of cold and dense water in the Mediterranean is similar to that occurring in oceans at high latitude.

An important oceanographic process is the dynamic relations between currents and distributions of mass. The main circulation is driven by the wind. Water, as having a larger inertia than air, averages the effects of wind. The combination of the water transport with the rotation of the Earth is balanced against the distribution of mass. At the turn of the century V. Bjerknes, B. Helland-Hansen and J. Sandstrom developed the theoretical basis for the study of marine circulation. The surface of the sea is dependent on the distribution of mass and the movement of water and because of gravity and the difference of density between water and rock reflects also the topography of the bottom. The mass distributions (calculated from density measurements) along oceanographic profiles, taking into account the rotation of the Earth enables the marine currents that are needed to maintain a steady state to be computed.

The computation of geopotential topography (as the procedure is called) has been very successful and the predicted circulation, excepting transients, confirms the soundness of this method. However, this approach has been much less successful in the Mediterranean, where an equilibrium with the rotation of the Earth is less easily achieved. Changes in atmospheric pressure over the whole Mediterranean basin act as a piston, accelerating or decelerating flows through the Strait of Gibraltar and impairing the conditions required for a steady state analysis. Mediterranean oceanographers, with good reason, care less than their colleagues working in other areas about dynamic topography. But the principle remains useful, namely: that the persistence of surface water of lower salinity along the coasts requires a coastal current directed towards the right (as seen from land).

The sea dilutes, averages and transports energy as well as nutrients and pollutants. The oceans are the large flywheel of the planet. The specific heat of water is more than 3000 times higher than specific heat of air per unit volume. An atmospheric column of 1 m^2 section contains about 1 kg of carbon as CO_2. In contrast a column of sea-water of the same section, down to the average depth of the oceans (4000 m), contains about 100 kg of dissolved inorganic carbon. No wonder that, in the sea, apparently random inputs result in more regular fluctuations. The marine organisms also contribute to the process of averaging and regularizing. For this reason, fluctuations in populations are slower and narrower than fluctuations in the environment. This is particularly true in the Mediterranean, where biological activity is low and individual life usually longer. The minor importance of tides and of upwelling contribute to this apparent stability. In this respect the Mediterranean is comparable to coral reefs and other tropical ecosystems, where the general structure and appearance remains the same for years, despite rapid dynamics and with possible shifts in the biological populations.

1.5. THE MEDITERRANEAN, A LIVING MACHINE

The Mediterranean is a machine that uses solar energy and works to recycle chemical elements into biological production. Part of this production is used by man, in the form of sea food, of other products, and as coral or sponges.

In a surface layer of water, about 100 m deep, tiny organisms (which can be compared in their function to the leaves of a forest) change inorganic ions into organic matter. The cells are small, their relative surface area is high. Consequently, phytoplankton reacts rapidly to any increase in available nutrients. Planktonic growth is limited by the concentration of nutrients, the most scarce usually being phosphorus. Often phosphate cannot be detected at all in surface waters with conventional analytical methods. For most of the year, rain water contains more phosphate than the top layers of offshore sea-water.

Any atom that passes from solution to a particulate form (i.e. in the body of an organism) gets a ticket that allows it to travel downwards. At the end of the journey, everything is mineralized again, the average level at which the nutrients are returned to the water being below the average level at which the nutrients were assimilated. The final situation is that where there is light, there are no nutrients, and at the deeper level, where the used nutrients accumulate, there is no light.

The continuity of life requires the replenishment of nutrients on the top layers. A certain amount of nutrients may come from terrestrial run-off, but most become available through vertical mixing or upwelling.

The processes of fertilization require external energy (i.e. the energy of waves and marine circulation). The Mediterranean is, however, only occasionally rough, currents are not violent and real upwelling is rare (except, perhaps in the Alboran Sea, and in limited areas under the influence of strong wind). Consequently, light is used ineffectively in the Mediterranean. Characteristically, the level of maximum chlorophyll content, and often of maximum production, is about 100 m depth in summer — just at the limit where the dim light is matched by concentrations of nutrients close to the limit of utilization.

The surface of the Mediterranean receives an average solar radiation of 1.5 millions of kilocalories per square metre per year. Average primary production, in the western basin, corresponds to an assimilation of 60 g of carbon (per square metre per year). This is approximately equal to 660 kcal (i.e. only $660/1,500 = 0.44$ per mille of the input). The water is full of light, but the organisms can barely use it.

In general, light can be used only to a depth equal to two and a half times the depth of Secchi's disk visibility, about 110 m in the Mediterranean (44 × 2.5 = 110).

Fertility can, however, be unusually high at the mouths of rivers and along coasts in winter time, at the arrival of layers of water produced by mixing in the Gulf of Lions, and in large eddies (as in the Alboran Sea) where deep water comes close to the surface. Visible blooms and red tides are of limited occurrence. They occur in fertile conditions associated with low diffusion (in and around harbours) or in areas of high phosphate inflow (as in the bay of Tunis). Pink tides of *Noctiluca* (beautifully luminescent at night and physiologically a true animal) appear at Spring along some shores. Summer is a particularly difficult time for biological production in Mediterranean waters. Only organisms with very special adaptations thrive at the surface or accumulate at depth, in thin layers below the thermocline, protected from dispersal by the lack of vertical mixing.

The distribution of planktonic organisms is difficult to predict by conventional statistical models. Random effects on pre-existing distributions are further modified by specific interactions and movements. What is not uniform is qualified as 'patchy' and biologists are always looking for patches (perhaps with memories of schools of fishes in the back of their minds). When we search for patches they usually vanish, or we find that we have patches of patches of patches — truly reflecting the fractal aspect of Nature, seen also in clouds and trees and almost everything else.

Phytoplankton sinks and many of the cells are grazed by animals. The remainder die and decompose and together with faeces, moults, dead animals and materials from land, contribute to the detritus of the sea. If plankton is collected using a net with holes of 0.05 to 0.1 mm, the smaller organisms and most of the detritus pass through. The plankton obtained in this way appears very clean. It is composed of large and resplendent organisms, quite different from those of muddy bottom communities. However, the difference is not so great when the whole suspended material is separated, an essential operation for understanding the functioning of the pelagic community. In the Mediterranean there is, in fact, ten times more junk or garbage in the water than plankton (typical figures are 25 mg/l of dissolved inorganic carbon, 0.5 to 1.5 mg of dissolved organic carbon, close to 1 mg of particulated suspended carbon, only 0.1 mg of living carbon per litre). Much organic dissolved matter comes from the excretions of phytoplankton. Algae usually give off from 10 to 30 per cent of the assimilated carbon in soluble form. Part of this is used by bacteria. Everything contributes to, and draws from, the pool of organic matter at sea, and the material is constantly transformed (the final form being hard organic matter, which is difficult to decompose and akin to the humic substances of the soil). Besides, definite compounds are especially active from the biological point of view. In summary, sea-water can be visualized as a living plasma linking all the organisms in a potential web of interactions.

Zooplankton, basically, eats phytoplankton. The animals are able to sweep up diluted food, consisting of discrete particles suspended in a volume of water. This usually involves some form of automatic filtering or screening of water, but active searching and capture also occurs — more often than it is generally thought. These animals utilize between 20 and 90 per cent of the ingested food. The unused food, together with faeces, moults and excretory products, contribute to secondary bacterial foodchains. Up to one third of the ingested food is used in growth and reproduction, the rest is respired. Simply swimming around takes a considerable amount of energy. It is tempting to quantify these energy relations in plankton. Such exercises in modelling have a long history. A century ago Victor Hensen had clearly in mind the importance of drawing up a budget of the transfer of matter and energy in marine plankton.

The active zooplankton poses many problems of distribution, motility and behaviour. Its main components, copepods and euphausids, tend to migrate vertically with a 24 hour periodicity: they swim upwards and feed in the surface layers during the night; before daybreak they are down again in the dark water. It has been suspected that by avoiding the illuminated water they escape predation, but most predators rely on mechanical rather than optical sensors. Perhaps it is easier to catch algae when they are not assimilating. Whatever the reasons, this migratory behaviour is contagious, probably as a result of natural selection. The zooplankton is followed by its predators. At a given level, when the predators are coming from the depths, the first animals to go up are saved; sluggish ones fall as prey. The whole circus is alive with action.

Migration redistributes food, for planktonic animals move or drift horizontally between vertical movements. This may be 'good for the system' but it is difficult to explain in terms of individual selection. Even the advantage for the 'system' is difficult to visualize, since migrating animals feed near the surface and defecate at depth. Copepods and euphausids accelerate the sinking of materials because they compress and package their faecal pellets, which sink at speeds close to 200 m per day. This brings some food to the inhabitants of very deep water, but slows the recycling in the illuminated or photic layers. The behaviour is surprisingly similar to our own when we accumulate garbage and slow down the recycling of the materials.

Perhaps migration evolved in regions of upwelling as a mechanism to secure a countercurrent mechanism of production. But whatever its function it is a fact of life. Quite often, local hydrography introduce considerable perturbations and even abolish vertical migrations. Besides sinking and migration, there are other ways of carrying particulate materials towards the bottom. For example, salps are powerful filter feeders. They produce dense faecal pellets and, also, themselves often sink to the bottom where they decompose. They contribute much material to the benthic environment, but fishermen dislike them, when they fill their nets with gelatinous and messy stuff.

The top layers of the ocean can be a separate world, with separate populations of organisms. However, exchanges with the atmosphere can occur, and many cells become airborne with the spray. The surface layers of the Mediterranean are populated by interesting species, often deep blue in the blue water. Blue copepods gather and jump against the surface film, producing the effect of falling raindrops. Long stretches of beaches are covered by the remnants of the blue sail, *Velella*, when they are driven ashore by the anticylonic gyres.

There are few genuine pelagic fishes. Those that occur are either small fishes (such as sardines or anchovies) at the end of rather short food chains, living on local fertilization processes, or else large ones, at the end of long food chains. These have special abilities for swimming and detecting prey (such as several species of tunas, the dolphin or llampuga and a less active swimmer, the swordfish). In addition, a number of species of squids are very important in pelagic biology.

For the terrestrial observer, the life at the bottom of the sea looks more conventional. Its relatively slow turnover ensures that a given primary production can support a large and impressive looking standing crop. Typical Mediterranean populations (such as sea-grasses and encrusting algae and corals) have very low rates of turnover, rather like the wood in a grove of oaks. Benthic algae provide a primary production, that is probably always small. With the exception of some areas of *Laminaria* in the region of Alboran, no large, fast-growing seaweeds are abundant in the Mediterranean. However, the benthic vegetation provides conditions for the exciting diversification of life in coastal communities. Several coastal organisms produce planktonic larvae which when dispersed constitute another pelagic foodlink. A fraction of the material produced in the pelagic world finds its way to the bottom as can be shown with sediment traps placed at different depths. The benthos can easily extract from 3 to 6 g of organic carbon per square metre each year from the open waters. The pelagic community grows fast and is kept cropped. The sea bottom thus receives a bit of everything: it is a rather well balanced menu.

Benthic communities differ according to the nature of the bottom (i.e. soft or hard). The communities found on, or within, shifting sediment have to be repeatedly regenerated. Thus food is easily exported, or lost in old sediment. On a coherent substrate, the benthos grows slowly and maintains an important biomass: more impressive than the primary production or trapping of food elsewhere. Such benthic communities are, in fact, comparable to terrestrial ones. Mediterranean research has profited from the development of phytosociology which had led to careful floristic and faunistic descriptions. Most benthic fishes are associated with long food chains, their contribution to production and transfer of energy is low. However, the very nature of benthos adjusts fish biomass and makes their food extraction relatively efficient. The use of fishing gear for demersal fishes (trawls) disturbs the environment more than purse seining and long lining.

Benthic communities are dealt with in Chapters 7 and 8. Chapter 7, besides historical and biogeographical aspects, emphasizes deep water communities, that are not discussed in Chapter 8.

The annual yield of marine animals, both pelagic and demersal, obtained by Mediterranean fishermen amounts to 0.3 or 0.4 tonnes of fish per km^2 of sea. This is equivalent to about 40 mg of organic carbon per m^2 per year. It is, however, a long way down from the primary production and corresponds to less than 0.1% of it. On the basis of the conventional loss of 10:1 for each food link, this means an average of three food links. On a world basis the same proportion holds: a primary production estimated at 36.10^9 tonnes of carbon, yields an amount of fishes and other sea food equivalent to 8.10^6 tonnes of organic carbon. Mediterranean fishermen are as efficient as the best of their brethren; perhaps because the relative length of coasts and a long tradition in the craft helps them. Anyway, it is sobering to compare the catch of fishes (40 mg C/m^2 per year) with the decomposable organic carbon that the Mediterranean gets from the neighbouring lands, with a generous helping from man and towns (close to 1000 mg C/m^2 per year).

The Mediterranean has had a very turbulent history, as is told in the next chapter. But today, at least in human terms, life appears to be remarkably stable. Except in small coastal areas, pollution has not noticeably changed the Mediterranean as an ecological machine. Year after year, similar sequences of events are repeated, with only small departures from the average. The understanding of such a notable mechanism has to involve extrapolation and prediction. As in meteorology, the best prediction is perhaps that tomorrow will be like today. But the power and influence of Man is increasing and we still do not know how these will effect the existing ecosystem. A global view would perhaps provide constructive ideas leading to better management and conservation.

1.6. THE EDGE OF THE SEA. COASTS AND MARSHES

1.6.1. The brackish environments

In estuaries and deltas, the mixing of freshwater with marine water is not instantaneous, but gradual. This creates a number of environments, of intermediate and fluctuating salinities, that pose a challenge to life. Different geological processes create sandbars and coastal lagoons. These may contain sea water, diluted by terrestrial run-off or concentrated by strong evaporation. The hyperhaline lagoons have often been converted by man into salting pans, but brackish marshes have long resisted agriculture and colonization. Today they are among the few places left as potential nature reserves and havens for numerous itinerant water birds. In some places brackish environments have been used to confine fish, or as places to catch diadromous species as they migrate between sea and rivers. Currently, the suitability of such environments for a progressive aquaculture is being considered.

The development of brackish and hypersaline environments depends upon the geological nature of the coast. The Gulf of Mexico, for example, is almost continuously bordered by a series of lagoons. The Mediterranean shores, occupied by saline lagoons and marshes is estimated to be at least 6,500 km^2, probably half of this in the Western Mediterranean. The nature and development of brackish coastal environments depends on the basic geological structure of the coast, on recent vertical movements of the sea level relative to the land and on the extent of tidal coverage. Such environments are usually transitional and unstable. This instability is probably the reason for the rather limited number of species found in them and for their resilience to environmental changes. A broad distinction can be made between coasts of the Atlantic and Pacific type. Pacific coasts rise against zones of subduction, the shelf is narrow and close to deep water. Pacific rivers excavate canyons and do not form large deltas. Littoral lagoons, although not totally excluded, are not important. The Atlantic type of coast, on the other hand, is receding, with huge accumulation of sediment (sometimes with oil, as in the Gulf of Mexico) and river deltas. The Mediterranean has a very complex structure, but in the western basin, receding coasts of Atlantic type with large deltas (Rhône, Ebro) are relatively important. Relatively recent changes in the sea level and in the pattern of coastal currents, mobilize and transport sand to form deltas and lagoons. A recent fall in the sea level, a few millennia after the postglacial rise, has created complicated patterns, very much related to local configurations of the coast. Finally, tides are very important. If the difference between high water and low water is substantial, important horizontal movements of a great mass of water build a complicated pattern of evacuation channels over the marsh. With tides as small as those in the Mediterranean, the affected areas remain smaller and ecological segregation is enhanced. In large tidal marshes, algae and grasses can develop simultaneously in the same places, periodically covered by the sea. In our modest Mediterranean environments, algae grow in the hollows and herbaceous plants on the convexities of the saline soils.

Brackish coastal environments in the Mediterranean may evolve behind sandbars and have often been influenced by man. The Albufera de Valencia, in eastern Spain, has been de-salinated in historical times, as a result of irrigation. Further south, the secluded Mar Menor has evolved towards a mildly hyperhaline environment. Many lagoons are associated with deltas, which carry sediments and deposit them as walls of ramified channels; the areas between remain as lagoons, more-or-less rapidly silting, often with formations of peat. Springs of artesian water, which become salty when passing through layers of sediments may add to the complexity. A number of such environments are found along the Spanish coast. More extensive ones are found on the French coast, and, to a lesser extent, in Italy, where they are common, especially between Naples and Orbetello. Several lagoons are found also along the North African coast, the most important being Bahiret el Bibane in Tunis, which covers about 300 km^2. All these environments are very shallow, and the bottom could be invaded by benthic vegetation, if not prevented by other reasons (anoxy, or lack of phosphate, or too rapid changes of salinity).

Deltas and lagoons are fascinating ecological boundaries between terrestrial and marine environments. The eventual stress on land ecosystems is transferred, via the rivers, to the coasts, which can be much changed or can act as buffers. More than 1 per cent of the organic production of land goes to the sea, but the coastal environments are active sites of precipitation and denitrification. Coastal marshes and lagoons are sites of intense biochemical activity. Organic matter is in a large proportion absorbed on suspended clay, and the salinity differences (through change of the electrical charge of particles) precipitates the material. The extensive fan-shaped deposits close to the rivers contain much organic matter. Valiela and Teal have described the loss of nitrogen in tidal marshes. Less important is the marine flotsam accumulating in a driftline that has specific scavengers.

1.6.2. Salinity, a key factor

At river mouths and along the estuaries there is a rapid dispersal of the river water in the sea, and this process pulls in sea-water into the estuary at a lower level. Many nutrients carried by the river are retained near the coast and the fertilizing effect of the river is more important in raising the level of subsuperficial water around the mouth and mixing it with the top layers.

The pattern may be retained in coastal lagoons which occupy old, and dead, river beds. In these more static situations the top layer of freshwater and the lower layer of salt water do not mix readily, and remain quite distinct at both sides of a sharp density gradient. The deep, saline layer may be heated by solar radiation (solar lakes) and become anoxic if appreciable organic matter is decomposing in it. In such conditions, a plate of photosynthetic bacteria develops at the boundary.

Most lagoons are well mixed and their salinity depends upon the inflow of freshwater. Salinity fluctuates widely, depending on the level of the river and, ultimately on storms that force marine water into the lagoons, through the inlet channels or 'graus'. The state and permeability of the continuously changing sandbars is an important factor as is also human manipulation (i.e. if the lagoons receive water from rice culture and other agricultural activities). Calcium is more soluble in water of high salinity and thus becomes available to form shells and stromatolites; its solubility is helped by high CO_2 tension; later calcium carbonate can precipitate in the beaches, as beach rock or calcarenites, or in the matrix of organisms. If evaporation is intense, hypersaline water is formed. This eventually flows into the sea, over the bottom, in the opposite direction to the inflow of surface sea-water. Such lagoons are considered as 'negative' estuaries. In fact, the whole Mediterranean, in which high salinity water is also formed, has been compared to a negative estuary. The increased salinity leads to changes in the atomic relations of solutes, since insoluble minerals are formed and separate, beginning with gypsum or calcium sulphate. In consequence, hyperhaline waters are, as well, athalasohaline, that is their ionic proportions depart from that of standard sea-water.

Saline coastal waters are populated by organisms coming either from sea-water or from freshwater. Several marine species can withstand a lowering in the salinity. Still more species of freshwater origin have become so independent of the external medium and able to regulate its internal fluids, that could easily evolve to produce forms able to support extremely high salinities. It is only natural that athalasohaline waters have more colonizers from epicontinental than from marine origin. In a restricted sense, brackish water is marine diluted water, and there are rather a large number of species adapted to just these sorts of environments, although no class of organisms is exclusively confined to brackish waters. This may be a consequence of their transitory nature, which has made it difficult for them to be the cradle for the evolution of any important groups.

Many papers have been published about the distribution of species along gradients of salinity. In the case of diatoms this has led to a classification of the species into a number of groups (halophilous, olihalobes, mesohalobes, and polyhalobes) that do not need further explanation. The primary purposes of these studies was to use diatoms as indicators of environmental conditions. This aim was fulfilled in the analysis of diatoms contained in the different layers of sediments deposited in the Baltic Sea, and in coastal lagoons close to the Mediterranean, an excellent record of fluctuations in the salinities in which the diatoms developed. The method has limitations, however, because the silicified membranes of the diatoms can dissolve to varying degrees.

It has been suggested that definite values of salinity can be considered as boundaries in the distribution of the species (by Remane and other authors), but probably, if such limits are justified, they should have only a regional signification They may, in fact, be related, perhaps, more to the relative frequency of the different environments from which the present faunas and floras were selected.

Although differences in salinity have many physiological implications, the osmotic effect is usually considered to be the most important. In environments of changing salinity survival is dependent on the capacity to conserve constant osmolality of the body fluids in the face of large fluctuations in external salinity. The physiological regulation requires energy, and this means an increase in the metabolism, that is, in oxygen consumption. With pairs of closely related species (one of them marine, and the other brackish) both could, theoretically, coexist in sea-water. But usually one has the advantage in competition: the marine species which, in a narrow range of the salinity spectrum, can dominate because of its lower metabolic requirements. This seems to be the situation in *Sphaeroma* and *Coregonus*, but in *Gammarus* is more doubtful. Different populations of the prawn *Palaemonetes* differ in the relation between salinity and oxygen consumption, and this consumption is always minimal at the salinity under which the population was naturally living.

1.6.3. Economical use and importance

In geographical areas where evaporation exceeds precipitation (as in some places in the Caribbean and the Mediterranean) coastal brackish environments function as negative estuaries. These have been exploited by industry.

Many of the saline lagoons around the Mediterranean have traditionally been used for salt extraction. The most simple procedure is to evaporate the already concentrated water, with a partial separation of gypsum and other salts, and then to evaporate the brine in extensive shallow ponds. More sophisticated techniques were used at Aigues Mortes (Rhône), in the 18th and 19th centuries. The water was led, through horse-driven rotary pumps, to elevated pans where 3 to 4 mm of water evaporated each day. Every 20 days the layers of salt were broken and extracted, and the process started anew. In the 18th century these salines produced 17 to 19,000 tonnes annually, an amount comparable to the yield of the best inland salines (for instance, Wielicza in Poland). Other important and old salines were those of Alacant (Mata), Eivissa (Ibiza) in Spain, Peccais, Hyères, Giraud and others in France, Trapani in Sicily, etc. Natural salt, besides halite, contains many 'impurities', which from the point of view of biology have been a source of oligoelements. Real or fantastic properties attributed to the salt of particular places could be related to the extraction procedures and, sometimes, to local biological factors.

Coastal lagoons and other brackish water systems connected to the lower course of the rivers have been important fishing places. Several models of trapnets and barriers, some of them ingeniously complicated, have been used to catch diadromous fishes, such as eels, in the Mediterranean. Coastal marine fishes enter lagoons to feed, as do the bass (*Dicentrarchus*) and grey mullet (*Mugil*), and traditionally have fallen prey to local fishermen. The particular conditions of such exploitation have, through history, spawned an extremely complicated system of privileges and rules to ensure equitable and acceptable distributions of the bounty.

Only a few species of fishes spend all of their life cycle in the coastal brackish water of the Mediterranean. These fishes are of small size and without economical utility (*Aphanius fasciatus* in Italy and Corsica; *Gasterosteus aculeatus*). The marine fishes that temporarily enter the lagoons are, however, exploited. These include *Mugil* (and the closely-related genera *Liza* and *Crenimugil*), *Dicentrarchus labrax*, *Diplodus*, *Sarpa salpa*, *Sparus auratus*, *Lithognatus mormyrus*, *Solea vulgaris*, *Pleuronectes flesus*, *Atherina boyeri*, *Belone belone*. As these species come to the lagoons to feed and to grow, the use of brackish water in aquaculture has been considered as a feasible commercial development.

A critical requirement is the capacity of brackish coastal environments to support fish life. Primary productivity (based on circumstantial evidence from lagoons at Venice, in southern France and close to the Ebro, in Spain) can be twice that of the average marine production. Moreover, there is an influx of organic matter from land, which provides heterotrophic conditions. The exploitability of brackish environments by alien species, whether marine fishes or birds (ducks, geese, herons, aigrettes, flamingos, etc.), is a consequence of the instability, long fluctuations and lack of variety of the fauna in the high trophic levels, as present in such environments. Long-term stabilization would probably lead to more complex local ecosystems which might be more difficult for aliens to exploit. It must be stressed that if marshland and brackish waters are to be preserved, conservation must maintain random fluctuations and not use stable water supplies.

The greatest success in raising fish in brackish water conditions is obtained when the natural mechanisms are preserved and enhanced. In the Italian Valli, close to the Adriatic (i.e. not in the Western Mediterranean), the basins are filled at high tide, local life develops, and introduced fishes grow rapidly. A similar procedure has been followed in the enterprise of growing shrimps in Ecuador. The shrimps are placed in basins that have been drained, fertilized and then filled, with a rapid development of planktonic life. The crucial task for the developers of aquaculture in brackish basins, is to maintain fluctuating and exploitable environments. These, if possible, should be combined with the preservation of a variety of existing habitats. The addition of food pellets is an ecological heresy which plays havoc with any attempt to relate production to the available area. Production of fish in confined coastal lagoons may amount to $100 - 130$ kg/ha year (in more conventional units $1 - 1.3$ g C/m^2 year). This is more than 20 times higher than the extraction in the Mediterranean proper.

The complex estuaries, deltas, marshes and brackwater lagoons integrate to form a landscape of exceptional beauty. Until recently direct human pressure has not been intense, for the lands were considered poor and unhealthy (chiefly because of malaria). It is a robust ecosystem and is strongly dynamic, so it can repair and adapt easily. These environments are prime candidates for conservation, primarily because of the itinerant populations of migrating birds that use such places as relays. The region of the Camargue, in the delta of the Rhône, has the status of a natural park, and including the area of the Vaccares lagoon (salinity between 6 and 10 per thousand), is a reserve of 120 km^2. There is also some pressure to create a similar reserve in the Ebro delta. Thus the concern for birds, has focused international attention on the conservation and research of brackish and saline water environments.

REFERENCES

Ambio (1977) The Mediterranean. Special issue of *Ambio*, 6, 6. Spanish translation (1979) *El Mediterraneo, un microcosmos amenazado*. Blume Ecologia, Barcelona, 281 pp.
Cialdi, A. (1866) *Sul moto ondoso del mare e su le correnti di esso*. 2nd ed. Tipografia delle Belle Arti, Roma, 693 pp.
Heuss, T. (1940) *Anton Dohrn*. Tubingen. Italian abridged translation: (1959) *"L'Acquario" di Napoli e il suo fondatore, Anton Dohrn*. Edizioni Cassini, Roma, 280 pp.
Report on the Danish Oceanographical Expeditions, 1908 – 1910, to the Mediterranean and adjacent seas. Copenhagen, 1912 – 1939.
Report on the Carlsberg Foundation's Oceanographical Expedition round the World, 1928 – 1930, and previous Dana expeditions. Copenhagen.
Resultats des Campagnes scientifiques accomplies sur son yacht par Albert I, Prince Souverain de Monaco.
Rodriguez, J. (1982) *Oceanografia del Mar Mediterraneo*. Ediciones Piramide, Madrid, 1174 pp.

CHAPTER 2

Evolution of the Mediterranean Basins and a Detailed Reconstruction of the Cenozoic Paleoceanography

ANDRÉS MALDONADO

Instituto 'Jaime Almera', C.S.I.C., Zona Universitaria de Pedralbes, Apto. 30.102, 08028 Barcelona, Spain

CONTENTS

2.1. Introduction	17
2.2. Early evolution of the Tethys and the Mediterranean	18
2.2.1. The birth of the Tethys	18
2.2.2. Evolution of the Atlantic Ocean	19
2.2.3. The alpine belts and the genesis of the Mediterranean basins	21
2.3. Evolution of the Western Mediterranean basins	22
2.3.1. Balearic basin	23
2.3.2. The Alboran Sea	26
2.3.3. Tyrrhenian Sea	28
2.3.4. The stratigraphic record of the Western Mediterranean	30
2.4. The Messinian salinity crisis	32
2.4.1. Global events leading to the onset of the salinity crisis	33
2.4.2. Geological evidences of the salinity crisis	35
2.4.3. A dynamic model of the salinity crisis	37
2.4.4. World response to the salinity crisis	40
2.5. Regional depositional patterns during the Pliocene and Quaternary	40
2.5.1. Primary factors influencing sedimentation	41
2.5.2. Selected sedimentation models in the Western Mediterranean	45

2.1. INTRODUCTION

The Mediterranean Sea is a classic example of the relatively small and deep marine basins surrounded by continents that have been called 'mediterranean' by Shepard (1973). This sea lies in a trapped basin within a collision zone between the African and Eurasian continents, one of the most geologically complex areas of the world. The collision involved several stages of continental convergence and sea-floor spreading that continuously rearranged the relative position of continents and oceans in the ancient Tethys, an east — west trending ocean between two main continental blocks of the supercontinent called Pangea (Fig. 2.1). About 200 million years ago, sea-floor spreading subdivided the supercontinent Pangea into the present continents. Later, the opening of the Atlantic Ocean changed the original opening trend of the Tethys to one of convergence. This evolution resulted in the alpine orogeny during the Cretaceous and Early Tertiary which destroyed most of the marine basins of the Tethys. Following the climax of this orogeny, the basic configuration of the Mediterranean Sea was created about 40 million years ago.

The initial stages in the evolution of the Mediterranean basins were characterized by limited sea-floor spreading, creation of oceanic crust, and vertical tectonic movements of the continental margins. In a few million years during the Tertiary, continuous subsidence due to faulting and thermal cooling of the oceanic crust developed the morphology of the Western Mediterranean basins.

Many theories have been proposed to explain the mechanisms and evolution of these basins, but a consensus is still far from being reached. At present, however, most authors agree that the basic Mediterranean configuration was structured prior to the Late Miocene, when the temporal closing of the existing connections of the Mediterranean with the world oceans caused one of the most spectacular geologic events on the entire Cenozoic. This event known as the Messinian 'salinity crisis' is characterized by the deposit of a thick and extensive sequence of evaporites in the deep isolated basins. Vertical tectonics and a glacio-eustatic fall in sea level may have been the triggering mechanism for this catastrophic event. The reopening of the connection at the Strait of Gibraltar at the beginning of the Pliocene led to the flooding of the Mediterranean and the re-establishment of normal marine circulation. Since that time, the principal factors controlling the evolution of this sea are a new tectonic style and the changes in the oceanographic parameters that influence sedimentation.

In this chapter a synthesis of the evolution of the Mediterranean Sea, from the ancestor Tethys to the present day configuration is presented. It is impossible, however, to fully discuss the many theories and often conflicting dynamic models proposed for this sea by marine and land geologists. Only those models that are more generally accepted or have a greater significance for the paleoceanographic reconstruction will be considered. Alternative models are also presented when in the author's view, the increasing amount of new data of the Mediterranean Sea cannot be satisfactorily explained with the available reconstructions.

2.2. EARLY EVOLUTION OF THE TETHYS AND THE MEDITERRANEAN

The opening histories of the Mesozoic oceans provide some of the basic clues for the reconstructions of the Alpine belts and the origin of the Mediterranean Sea. Following the model for the opening of the Atlantic Ocean by Pitman and Talwani (1972), several attempts at a kinematic reconstruction for the Tethys have been presented (Dewey *et al.*, 1973; Biju-Duval *et al.*, 1977; Laubscher and Bernouilli, 1977).

2.2.1. The birth of the Tethys

The distribution patterns of continents and oceans since the Jurassic is well established by the elimination of successive magnetic lineations, each representing a specific time span of sea-floor accretion. For earlier geological times, when the present oceans were not in existence, the

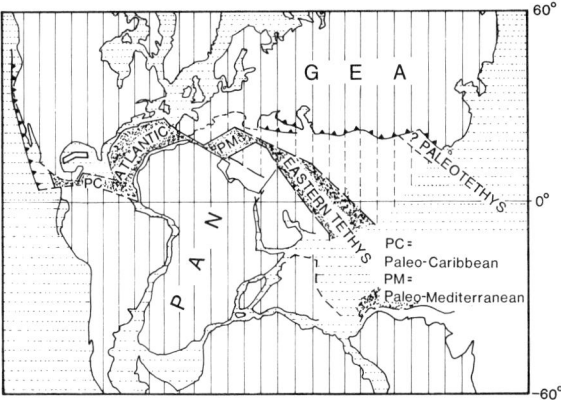

Fig. 2.1. Tentative reassembly of the principal continental masses in the Late Jurassic. The possible remnant of the Triassic Paleotethys and the Mesozoic basins of the Tethys created by the breakup of Pangea are outlined (Modified from Bernoulli and Lemoine, 1980).

reconstruction is based on other evidences such as the shape of the continents and indirect geological data. These reconstructions are increasingly less precise with time since many of the geological evidences may be concealed in orogenic belts affected by several tectonic cycles.

Although these constraints make the reconstruction of the oceans during the Triassic difficult, several models have been proposed (Hsü and Bernouilli, 1978). The best reconstructions for this period showed that an equatorial ocean must have existed between the northern and southern continents of Pangea (Carey, 1958; Bullard et al., 1965; Smith, 1971). This wedge-shaped ocean was open to the east and is referred to as the Paleotethys to distinguish it from its Mesozoic and Cenozoic successors (Fig. 2.1). The Paleotethys was eliminated during the early Mesozoic by subduction under the Eurasian continent (Laubscher and Bernouilli, 1977).

The Tethys proper was created later during the Early Jurassic as an east – west trending equatorial ocean composed of several segments individualized by different styles in the tectonic evolution (Bernouilli and Lemoine, 1980). The Tethys basins oriented in a NE – SW, such as the central Atlantic basin, and in a NW – SE direction, such as the eastern Tethys, were expanding segments flanked by subsiding passive margins. In contrast, the equatorial oriented basins, such as the Caribbean and Western Mediterranean basins, underwent a prevailing left-lateral movement and alternating periods of distension and compression. This evolution was related to the sinistral movement of Africa relative to Europe that was predominant between the Early Jurassic and Late Cretaceous (Fig. 2.2). The movement became dextral during the Tertiary and the closing of the gap between Europe and Africa led to the elimination of much of the original Tethys. Several fragments of the Tethys may have been relatively independent in this evolution, but the main trends were superimposed by the different stages in the opening of the Atlantic Ocean.

2.2.2. Evolution of the Atlantic Ocean

The opening of the Atlantic Ocean from the end of the Jurassic resulted in a second stage in the evolution of the Tethys (Aubouin et al., 1980). The Central North Atlantic began to open about 165 m.y.a. (Middle Jurassic) with the formation of oceanic crust between North America and Africa (Sclater and Tapscott, 1979; Sclater et al., 1977). At this time the Central North Atlantic basin was closed to the north by the juncture of the Guinea nose and the Bahama Platform, but it was connected with the Mediterranean and Caribbean Tethys through a narrow passage (Fig. 2.2). By 120 m.y.a. the North Atlantic had developed an active mid-ocean ridge and had reached depths of up to 4000 m, causing a deep connection with the Tethys.

The opening of the South Atlantic began about 110 – 125 m.y.a. but it was separated from the North Atlantic at the bulge of Africa. During the Middle and Late Cretaceous (95 – 65 m.y.a.), Europe began to separate from North America and deep-water connections were established between the North, South Atlantic and the Tethys. Thus, until the Late Cretaceous, the Tethys and Central North Atlantic formed an equatorial ocean composed of several deep basins isolated by shallow sills. This ocean has an early connection with the world oceans to the east and later in the Cretaceous time with the North and South Atlantic.

The end of the Cretaceous is marked by a major biotic crisis characterized by a degradation of the pelagic communities and lowered diversity (Fischer and Arthur, 1977). Deep water sedimentation was interrupted by intensified current systems and organic-rich, black shales were replaced by red clays. This fluctuation in the marine conditions also coincided with a sharp increase in the opening rate of the Atlantic Ocean north of the Azores — Gibraltar fracture zone. The prevailing left-lateral movement of Africa in relation to Europe became dextral since the early Tertiary (about 65 m.y.a.) and the gap

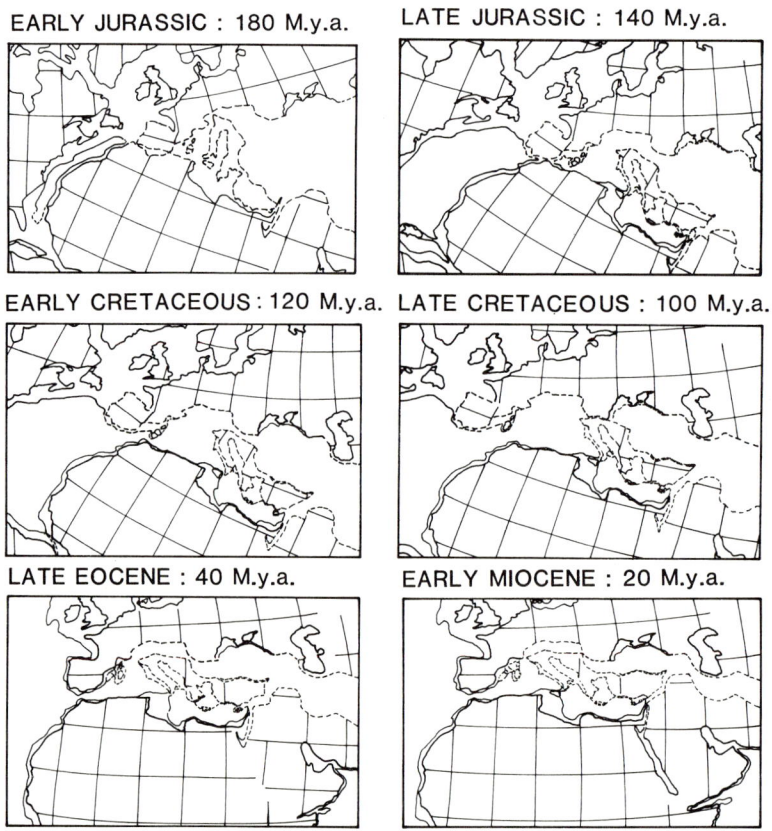

Fig. 2.2. Relative positions of stable Africa and stable Europe at their inferred paleolatitudes throughout Mesozoic and Cenozoic time based on computer-drawn maps of available ocean-floor and paleomagnetic data (From Smith and Woodcock, 1982).

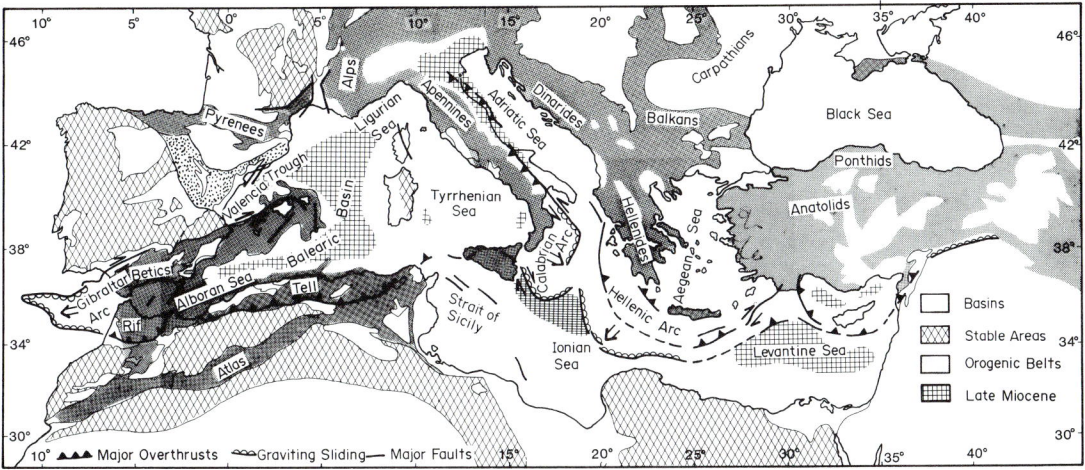

Fig. 2.3. Structural map of the circum-Mediterranean region showing the geotectonic setting of the Mediterranean basins (Modified from Biju-Duval et al., 1978).

between these two continental blocks was progressively consumed. As a result, the Alpine orogeny developed and the Mediterranean basins came into existence (Fig. 2.2).

In this evolution most of the major topographic features of the North Atlantic had formed by 40 m.y.a. (Sibuet, Ryan, et al., 1979). However, Spain and Africa were still significantly separated by a deep channel allowing the connection between the eastern Tethys and the Atlantic. During the Early to Middle Miocene (20 – 15 m.y.a.), overflow water from the Norwegian Sea and Arctic Ocean began to flow in significant quantities into the North Atlantic creating the basic circulation patterns of the modern Atlantic Ocean.

The Atlantic Ocean is in consequence composed of three main provinces. The South and North Atlantic were created by a single opening. In contrast, the Central North Atlantic resulted from two superimposed trends: (a) the equatorial Tethys opening from Triassic to Late Cretaceous, and (b) the longitudinal Atlantic Ocean opening from Middle Cretaceous. From the tectonic point of view the opening of the Atlantic resulted in a Late Jurassic – Early Cretaceous crisis that drastically altered the previous evolution of the Tethys (Aubouin et al., 1980).

Thus the interplay between the Atlantic and Tethys oceans can be summarized in two main stages. First, South America and Africa drifted towards the northern continents following the birth of the South Atlantic. Second, after the Late Cretaceous, a new degree of complexity was introduced by accelerated motion between Europe and Greenland as a consequence of the active spreading in the North Atlantic. The result of this tectonic evolution was the alpine orogeny which accounts for the basic stages in development of the Alpine Belts and the creation of the basic morphologic features of the present Mediterranean basins.

2.2.3. The alpine belts and the genesis of the Mediterranean basins

The Mediterranean Sea reflects the structural relationships between the tectonic belts of North Africa and those of southern Europe (Fig. 2.3). According to Suess (1901) this sea was a relict of the Tethys, while for Argand (1924) the Tethys was completely consumed by the alpine orogeny. Recent geophysical studies and reinterpretation based on plate tectonics, however, have demonstrated that the Mediterranean Sea is composed of both ancient and recent features. Dynamic reconstruction models of this sea show a changing mosaic of small and large plates which produced ridges, trenches, backarc basins and island arcs (Dewey et al., 1973; Alvarez, 1976; Biju-Duval et al., 1977; Cohen, 1980).

The Eastern Mediterranean basins have long been in existence and influenced by compressional forces (Figs. 2.2, 2.3). The active subduction in the Hellenic Trench system illustrates the final fragments of the Tethys Ocean that may be in the latest stages of subduction below the European plate (Le Pichon and Angelier, 1979; Neev et al., 1982). Old oceanic crust is considered to exist in these basins, although the African continental crust is reaching the Hellenic Trench.

In contrast, the Western Mediterranean basins were recently created by extensional rifting during the Cenozoic with the alpine orogenic belts as backarc basins. New oceanic crust developed in the Western Mediterranean following the climax of the alpine orogeny. The oldest Cenozoic basin of the Mediterranean Sea is the Balearic Sea in the Western Mediterranean, dating back to latest Oligocene or earliest Miocene. The youngest basin is the Aegean Sea in the Eastern Mediterranean which mostly developed during the Pliocene and Quaternary.

2.3. EVOLUTION OF THE WESTERN MEDITERRANEAN BASINS

The Western Mediterranean is composed of several basins defined by their structural framework in relation to the alpine orogeny, the nature and age of the crust and their recent sedimentological development (Fig. 2.4). Most of these basins were developed by the process of continental crust stretching and magmatic intrusion of new oceanic crust. However, some basic differences may have existed in the mechanisms involved in the formation of each basin. The models proposed for the

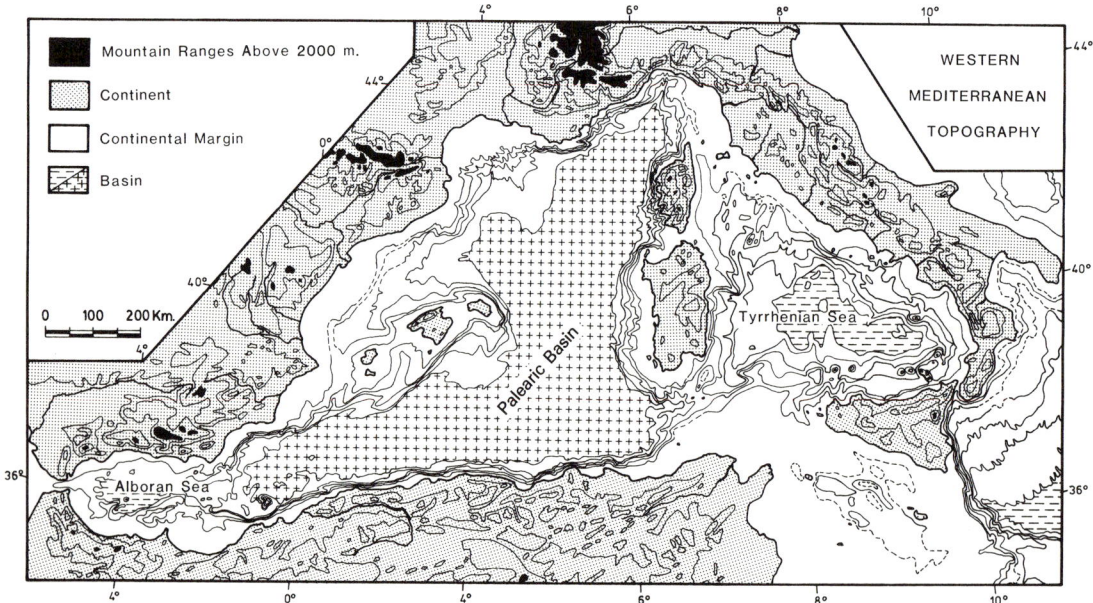

Fig. 2.4. Bathymetric chart and topography of the Western Mediterranean region (From Pannekoek, 1969).

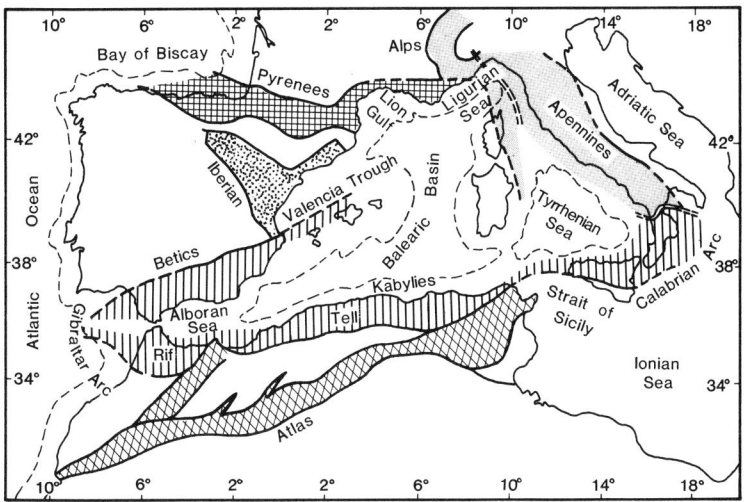

Fig. 2.5. Structural map of the alpine belts and geotectonic setting of the Western Mediterranean (Modified from Durad-Delga and Fontboté, 1980).

development of these basins show several analogies between the Alboran and Tyrrhenian seas, which can be interpreted as backarc basins related to the development of the Gibraltar and Calabrian arcs (Fig. 2.5). The Balearic basin in contrast is a more complex feature with several sectors that may have a different tectonic origin (Biju-Duval et al., 1978).

Another major difference among the Western Mediterranean basins is the age of development. The Balearic basin is the oldest and was basically structured following the climax of the alpine orogeny. The Alboran Sea developed next during the Early Miocene, followed by the youngest basin of the Tyrrhenian Sea which developed from the Late Miocene.

In general, two major phases can be differentiated in the evolution of these basins. The first phase involves the breaking of the continental crust and the structural build-up of the continental margins, followed by the sea-floor spreading and widening of the basin. The first phase clearly differentiated the basins of the Western Mediterranean. The second phase may partially coexist with the previous one and is represented by the subsidence of the basin and the deposition of thick sedimentary sequences, both on the continental margins and in the deeper parts of the basins. Seismic reflection profiles (Biju-Duval et al., 1978), deep sea drilling (Ryan, Hsü et al., 1973; Hsü, Montadert et al., 1978), and conventional deep sea sampling (Mauffret, 1979; Mauffret et al., 1982) show major analogies throughout the area for this second phase. In consequence, the initial phase of development is independently discussed for the three main Western Mediterranean basins: Balearic, Alboran and Tyrrhenian seas. The second phase of this evolution is considered for the totality of the Western Mediterranean and only the major differences will be emphasized.

2.3.1. Balearic basin

The Balearic basin is the largest of the Western Mediterranean basins and contains several sectors which can be clearly differentiated by the nature of the crust and the main tectonic lineations. The best studied sector is perhaps the north Balearic–Provençal basin, including the Valencia Trough and the Ligurian Sea (Biju-Duval et al., 1978; Mauffret, 1979; Rehault, 1981). This basin is characterized by the absence of a well-developed tectonic arc in contrast to the Alboran and Tyrrhenian seas, located within a divergent orogeny (Fig. 2.5). Moreover, this sector may be considered as a type of rifted basin, bounded by passive margins that were created by the partial rotation of Corsica and Sardinia relative to Europe. The southern sectors of the Balearic basin show characteristics that are similar either to the northern Balearic sector or to the Alboran and Tyrrhenian seas. These sectors will not be individually discussed.

The northern Balearic basin developed in a pre-existing craton of folded Paleozoic formations and Mesozoic platform deposits. The alpine belts only extended offshore locally (Durand-Delga and Fontboté, 1980). The passive margins in this area are in a juvenile phase and the evolution of the basin can be established by comparison with this type of margin from other areas. In a comparison with other passive margins, however, we must consider that the Mediterranean margins are more narrow (several tens of kilometres) in breadth than the classical examples of Atlantic passive margins (up to 200 kms or more).

The analysis of the world oceans shows that passive margins are located within a lithospheric plate and, within that plate, straddle the boundary between oceanic and continental crust (Continental Margin Panel, 1979). These margins document the history of the transition between the continent and ocean basin and preserves the information about the genesis of the ocean. Several models for the evolution of passive margins and the creation of a new oceanic crust have been proposed (Bott, 1976). A consensus exists however that favours the following sequence of evolutionary phases: (a) rifting, which represents the breaking and stretching of the pre-existing crust; (b) onset of drifting, that leads to the separation of continental crust as oceanic crust accretes, and (c) post-rift evolution, dominated by the subsidence and sedimentary history of the margin.

The classical models show that early rifts are associated with graben development and extensive volcanism. During the Oligocene, many grabens were created in the north-western Mediterranean area (Rhine and Rhône valleys, Catalonian graben, Ebro basin) with the development of continental, evaporitic and brackish deposits. These grabens extend offshore in the continental margin as shown by the commercial oil wells (Stoeckinger, 1976). DSDP Site 372 cored the lower-most Burdigalian with massive neritic facies east of Menorca in the continental rise. The underlying series filling the grabens, although not reached because of limited drilling capabilities of the *Glomar Challenger*, are attributed to the Aquitanian and Late Oligocene, perhaps with evaporitic facies (Hsü, Montadert et al., 1978). Volcanic activity is also recorded at this time (about 20 to 24 m.y.a.) on the margin (Valencia Trough, DSDP Site 123; Ryan, Hsü et al., 1973) and on land (Provence and Sardinia). The development of these grabens and associated emplacement of volcanic dikes involved the breaking of normal continental crust where basement was substantially older and initiated the development of the Western Mediterranean about 25 m.y.a.

Separation of the rifted continental fragments began later with the emplacement of oceanic crust. Seismic refraction and gravity data seem to indicate that the north Balearic basin floor is oceanic (Hinz, 1972; Morelli, 1975; Le Douaran et al., 1983). Recent studies show that the age of this oceanic crust is about 20.7 to 18.6 m.y.a. (Edel, 1980; Rehault, 1981), which is confirmed by other evidence, such as the age of sediments on the margin not affected by the graben tectonics. The rift-drift transition and the opening of the basin occurred for a short period during the Early Miocene.

Several models have been proposed for the geometry of the opening on the basis of onland paleomagnetic data, sea aeromagnetic charts, bathymetric contours, and structural considerations. The model originally proposed by Carey (1958) and Nairn and Westphal (1968) suggests a single 40° – 50° counterclockwise rotation of the Corsica – Sardinia block with a pole of rotation close to the Gulf of Genova. The original position of Corsica was closed from Provence and Sardinia to the Gulf of Lions. This model was later modified by postulating a large fracture zone between Corsica and Sardinia (Auzende et al., 1973; Bayer et al., 1973). Corsica and Sardinia would have been affected by major differential movements along the fracture and little rotation. Another model by Biju-Duval et al. (1978)

considers a restricted (about 100 kms) east-west translation during the Early Miocene with little rotation (Fig. 2.6). In this model a triple junction may have existed at the entrance to the Valencia Trough. In any case the lack of well-defined magnetic anomalies in most of the Balearic basin does not allow a precise reconstruction of the geometry of the opening (Galdeano and Rossignol, 1977).

Two additional problems must be considered in the origin of the northern Balearic basin: (a) the tectonic significance of the basin in relation to the alpine orogenic belts, and (b) the mechanisms of the accretion of oceanic crust. The orientation of the grabens related to the rifting phase cut indiscriminately across the main trends of the western European alpine belts (Rhône Valley, Catalan depression) and the continental margin as in the eastern Pyrenees margin, eastern Balearic platform, and southern Catalan margin (Julivert et al., 1974). The onset of rifting also occurred after the main Eocene orogenic event that developed the principal tectonic trends. On the basis of these data, the northern Balearic basin could be attributed to a type of rifted basin that developed independently from the alpine orogeny and cut across previous tectonic lineations. A modern analog for this evolution can be seen in the Red Sea.

In contrast, the northern Balearic — Valencia Trough basins can be considered as a type of episutural basins developed inside the alpine megasuture (Bally, 1975). They could also represent backarc basins, associated with A-subduction and floored by continental and oceanic crust. Marginal basins or backarc basins floored by oceanic crust have been explained by Toksöz and Bird (1977) as a necessary consequence of the subduction of oceanic lithosphere that results in an induced convective circulation in the wedge above the slab. The development of the Balearic Sea could be related in this view to an Apennine subduction zone to the east, active during the Early Miocene and marked by the calc-alkaline volcanism of Sardinia (Haccard et al., 1972; Biju-Duval et al., 1978).

A similar evolution of the Tyrrhenian Sea as a backarc basin during the Late Miocene, due to the subduction of the Ionian Sea below the Calabrian Arc, seems to be well established by the high heat

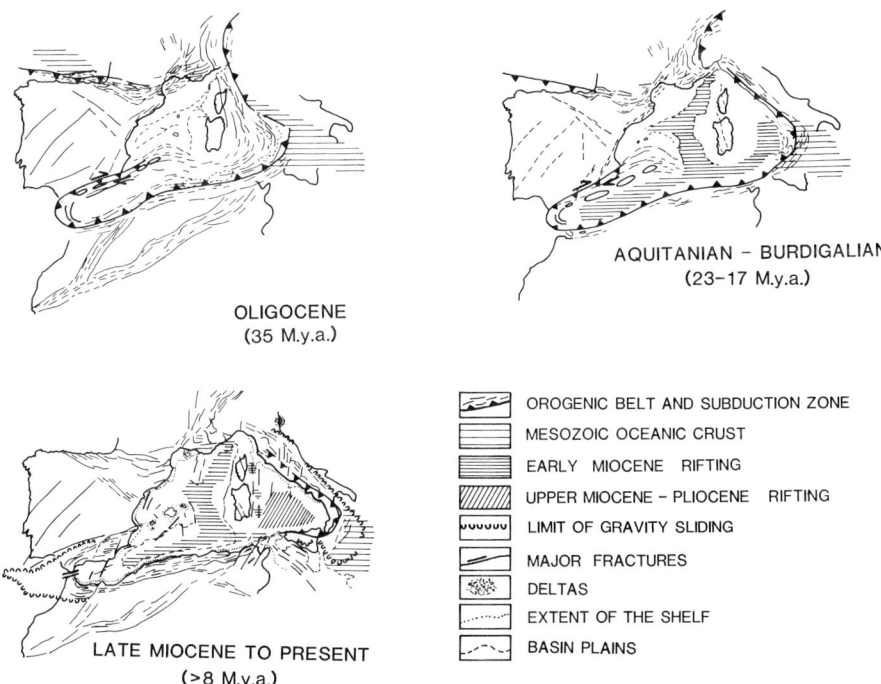

Fig. 2.6. Tentative reconstructions of the Neogene evolution of the Western mediterranean (Modified from Biju-Duval et al., 1978).

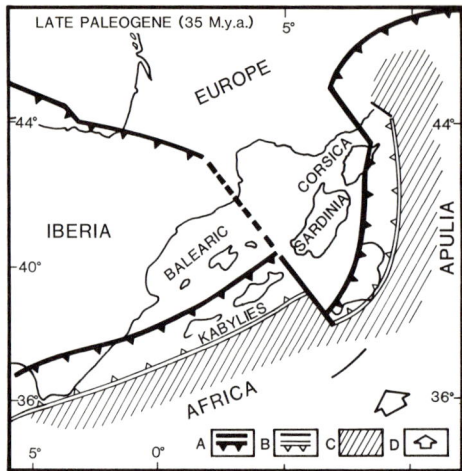

Fig. 2.7. Reconstruction of the relative position of the continental blocks in the Western Mediterranean during the Late Paleogene (about 35 m.y.a.) showing the location of two suture zones. A, Eocene suture; B, Oligocene suture; C, Under-thrusting plate; D, Relative movement of the African and European plates (From Rehault, 1981).

flow distribution (Malinverno et al., 1981). This model may also be of application to the northern Balearic basin, and a mechanism of micro-plate dispersal that implies a jumping of the spreading from the Balearic to the Tyrrhenian during the Miocene has been proposed (Alvarez et al., 1974).

In summary, with the information available at present, two alternative models must be retained for the development of the north-western Mediterranean. In the first model, this basin is a true rifted-drifted basin created by the drifting of the Corsica – Sardinia continental blocks, following the climax of the Alpine orogeny. The second and more fundamental model defines the north-western Mediterranean as a backarc basin. In this view, the kinematic reconstruction implies that most of the Penninic – Ligurian ocean was consumed by subduction during the upper Cretaceous and Paleocene as a result of the African rotation (Fig. 2.7). A second phase during the Upper Eocene and Oligocene led to the closure of the remaining western part of the Mesozoic Penninic – Ligurian ocean and the opening of the present Balearic Sea. This process was accomplished by the eastward swing of the Corsica – Sardinia block in a counterclockwise rotation (Channell et al., 1979; Rehault, 1981). The contemporary calc-alkaline magmatism of Sardinia represent the arc with a subduction zone dipping toward Sardinia.

2.3.2. The Alboran Sea

The initial stages in the development of the Alboran Sea can be dated during the Oligocene when abundant clastic flysch deposits accumulated in the relatively narrow corridors of the Subbetic and Intrarif (Durand-Delga and Fontboté, 1980). However, one of the major debates concerning the origin of this sea is the paleogeographic position of the flysch. Some authors suggest an internal origin and that the flysch was ejected into the external zones by gravity sliding. Others visualize a flysch trough of some hundred kms wide and locally characterized by oceanic crust which separated the African foreland and the external Rif from the internal units (Fig. 2.8). In this view, two flysch types were developed in the same basin as in the present Levantine Sea of the Eastern Mediterranean. To the south the mature

Numidian flysch is derived from the African craton, while in the internal sector of the basin, immature michaceous flysch is generated from the erosion of the metamorphic Rif and Betic massifs.

Independently the paleogeographic position, plate tectonic reconstruction and land geology of the flysch demonstrates that it was consumed by the northward movement of the African – Apulian plate under the Iberian plate (Fig. 2.7). This subduction led to the collision of the internal and external units. The paroxysm of continental collision took place during the Oligocene and Early Miocene when flysch nappes were overthrusted into the external zones. At this time the area presently occupied by the Alboran Sea was elevated, along with most of the internal zone of the Betic belt, which probably was emergent.

A very rapid subsidence of the Alboran Sea seems indicated in order to allow it to accumulate younger Late Miocene deposits. This abrupt reversal in relief in southern Spain, in which formerly positive areas subsided to become basins, has been referred to as the 'Mediterranean revolution' (Pannekoek, 1969) or 'revolution Pontienne' (Andrieux et al., 1971). Post-collisional deformations continued in both margins of the Alboran Sea during the Late Miocene and Pliocene, together with a large amount of volcanism (Araña and Vegas, 1974).

The model suggested for the development of the Alboran Sea is a consequence of the north-westward movement of the African plate related to the opening of the central Atlantic (Fig. 2.7). The compression between the Iberian Peninsula and Africa was resolved by the development of a south-western Mediterranean microplate pushed westwards (Andrieux et al., 1971; Araña and Vegas, 1974; Bourruoilh and Gorsline, 1979). Two phases of compression resulted in a large shortening of the area during the Late Cretaceous – Eocene and Middle Miocene. The movement was taken by subduction along a Benioff plane dipping towards the North in the Rif and by westwards slide along the Iberian block of the internal zone or Alboran plate (Figs. 2.8, 2.9).

The sharp bending and continuity of the Gibraltar arc is explained in terms of further westward movement of the central sector of the microplate overthrusting the thinned Atlantic continental margin. In contrast, the movement was inhibited in the Betic and Rif margins of the plate by continent-continent collision resulting in the ejection of the flysch nappes. The Miocene calc-alkaline volcanism may be an indication of the active subduction during the compressional period. Similar arc development may have occurred in other sectors of the Mediterranean Sea. In this aspect several analogies are observed between the Gibraltar, Calabrian and Hellenic arcs (Fig. 2.10). The age of development of these arcs

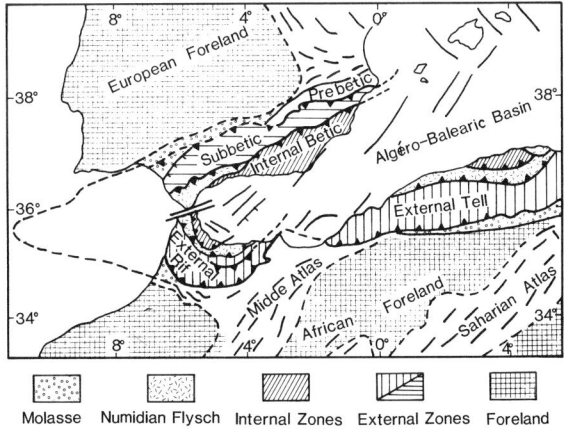

Fig. 2.8. Geotectonic sketch of the Alboran Sea, South Balearic Basin and Gibraltar arc (Modified from Horvath and Berckhemer, 1982).

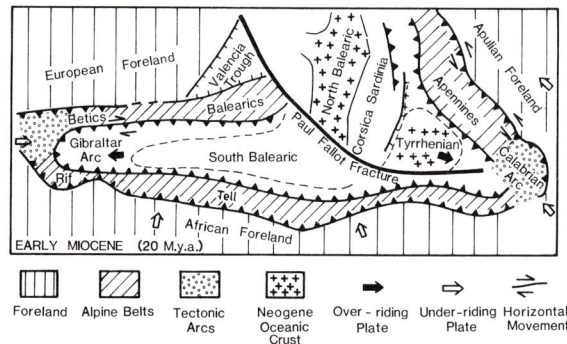

Fig. 2.9. Hypothetical reconstruction of the tectonic trends of the Western Mediterranean during the Early Miocene (Modified from Durand-Delga and Fontboté, 1980).

becomes youngest eastward in accordance with a wedge-shaped closing of the Mediterranean progressively west to east.

Most of the relative movement between the Iberian and African plates was blocked during the Eocene about 48 m.y.a. and between this time and the Early Miocene, the Alboran Sea and Gibraltar arc were basically structured. However, several tectonic phases have affected the region to the present; recent work suggests that a phase of extension occurred from Late Miocene into the Pliocene, and the Alboran Sea may have deepened as a result of this extension (Bousquet and Philip, 1976). Continued extension probably caused the block faulting of the basin and the opening of the Strait of Gibraltar.

A change in the tectonic style from extension to compression is again proposed for about one million years ago (Dillon et al., 1980). This recent faulting probably represents transcurrent shear along some ancient fractures oriented in a N−S or NNE direction. Thus, the Pliocene potassic volcanism of SE-Spain is interpreted as an indication of a senile stage of the subduction, and the deep earthquakes of Granada may be associated with a fossil fragment of the sinking slab.

2.3.3. Tyrrhenian Sea

The evolution of the Tyrrhenian basin has been deduced from marine geological and geophysical studies and information of the surrounding land masses (Selli and Fabbri, 1971; Boccaletti and Manetti, 1978; Görler and Giese, 1978). In contrast to other Western Mediterranean basins, the paleomagnetic

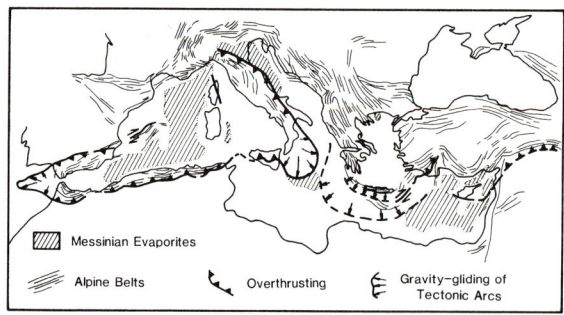

Fig. 2.10. Alpine synsedimentary gravity processes in the Mediterranean basins during the Tertiary (Modified from Biju-Duval, 1974).

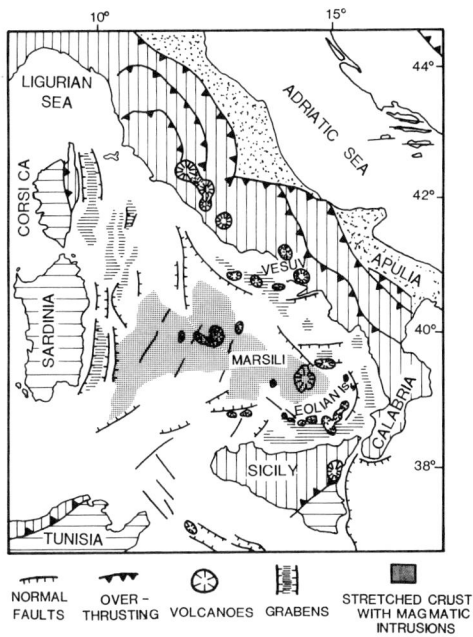

Fig. 2.11. Geotectonic sketch of the Tyrrhenian Sea and surrounding region (Mainly after Boccaletti and Manetti. 1978).

data and the orientation of the tectonic trends play a significant role in the paleogeographic reconstruction (Channell et al., 1980; Fabbri et al., 1980).

The Tyrrhenian Sea is characterized by several tectonic grabens and elongated peripheric basins filled with considerable thickness of Neogene – Quaternary sediments (Finetti and Morelli, 1973). The basins of the eastern margin of Corsica – Sardinia show distensive graben tectonics. In contrast, the depression along the southern Tyrrhenian margin suffered tectonization during the Apennine orogenesis (Fig. 2.11).

The southern Tyrrhenian Sea is the region with the most intense volcanic activity and the densest population of both shallow and deep focus earthquakes in the Western Mediterranean. The volcanoes of the region belong to two main magmatic provinces (Selli et al., 1979; Di Girolamo, 1978). The volcanoes in the central basin plain are of tholeitic nature and genetically related to extension of the crust from 7.5 to 3.5 m.y.a. (Fig. 2.11). The belt of volcanoes in the south-eastern Tyrrhenian from the Eolian islands to the Roman provinces are characteristic of convergent plate boundaries.

The crustal structure of the Tyrrhenian Sea as derived from geophysical data is predominantly composed of sialic material with some magmatic intrusion that may indicate limited sea-floor spreading (Finetti and Morelli, 1973; Barberi et al., 1978). The crust increases in thickness steadily toward the Apennines and Calabria, reflecting the thrusting of the Tyrrhenian crust into the Adria – Apennine crust. Moreover, the seismicity of the Calabrian arc shows a distinct gap between 100 and 200 km depth. This may indicate lithospheric fragments detached from a subducted slab sinking into the mantle rather than a continuous Benioff plane dipping from the Ionian toward the central Tyrrhenian (Cagnetti et al., 1978).

This geological and geophysical information characterizes the structural dynamics of the Tyrrhenian Sea with a detail not attained in other Western Mediterranean basins. However, the interpretation of these data in terms of a genetic evolution has resulted in several subduction-collision models which are discussed in detail by Berckhemer and Hsü (1982). In general this sea is considered a backarc basin largely floored by intermediate and continental crust and locally by oceanic crust. This basin is associated with B-subduction as a result of the eastward displacement of the Sicilian − Calabrian arc above the last fragments of the Mesozoic oceanic crust of the Ionian Sea (Fig. 2.10).

According to these data, the kinematics of the Tyrrhenian Sea can be summarized in the following steps. The initial stage of extensional tectonics as demonstrated by the magmatic activity occurred about 9.5 m.y.a. By the Messinian time the basin was well-developed and extensive evaporatic formations were deposited in the central sectors. Backarc spreading and formation of oceanic crust was active prior to the salinity crisis. However, the main phase of distension in the central and southern Tyrrhenian Sea began in Late Pliocene, about 2.5 m.y.a., with prevailing E−W stretching. About 0.8 m.y.a. the extension shifted to a NE−SW direction. Vertical block movement with sliding characterizes the tectonic style from the late Pliocene, about 2.5 m.y.a., with prevailing E−W stretching. About 0.8 m.y.a. the extension shifted to a NE−SW direction. Vertical block movement with sliding characterizes the tectonic style from the Late Pliocene.

The Tyrrhenian extension was also accompanied by the progressive bending of the Calabrian orocline as shown by paleomagnetic data and the tectonic trends of the Apenninic−Sicilian thrust front (D'Argenio et al., 1980). The recent evolution of this backarc basin is characterized by the foundering of the basin plain and uplifting of the surrounding Calabrian arc.

2.3.4. The stratigraphic record of the Western Mediterranean

Seismic profiles from the Balearic basin plain show a thick sequence of sediments of up to 6 km in thickness above the oceanic crust (Fig. 2.12). This sequence laps over the continental margins where it becomes very reduced or absent (Biju-Duval et al., 1978). The oldest units developed during the rifting phase of the Late Oligocene and Early Miocene fill the grabens in the continental margin. The overlying Lower Burdigalian, (about 22 m.y.a.) sampled at DSDP Site 372, is a thick marine sequence characterized by high terrigenous influx and high sedimentation rates (Hsü, Montadert et al., 1978). New relief on land initiated active erosion and the high subsidence rate of the basin favoured the deposition of a thick sedimentary wedge. During this period graben development ceased and by Middle Burdigalian time (about 20 m.y.a.) the relief was subdued along most of the continental margin and on land. During the Late Burdigalian and Tortonian (from about 18 to 9 m.y.a.), a large transgression occurred over extensive areas of the alpine domain. Most terrigenous input was trapped in the proximal continental margin and deep water sedimentation was characterized by thin hemipelagic units.

At the beginning of the Messinian stage (approximately 6 m.y.a.) the openings to the Atlantic in the northern Betic and southern Rif straits were closed. Excess water lost by evaporation from the isolated Mediterranean Sea led to a basin-wide lowering of sea level and the deposition of thick evaporites during the period of the Messinian 'salinity crisis' (Ryan et al., 1973; Hsü et al., 1978). This spectacular event, one of the most important in the Cenozoic evolution of the Mediterranean and maybe of the world oceans, is discussed in the following section. The end of the salinity crisis coincided with the creation of the permanent connection between the Atlantic and the Mediterranean Sea through the Strait of Gibraltar.

In contrast, the thick sequence of sedimentary units in the Alboran Sea are bounded by unconformities that appear to record a distinct set of tectonic events. The oldest units are attributed to the Miocene and may contain the Late Tortonian (Dillon et al., 1980). The salinity crisis is reflected by an eroded, faulted, angular unconformity on top of the Miocene unit. Salt deposits are thought to occur at great depth, although the Messinian evaporites may be missing in most of the basin (Ryan et al.,

EVOLUTION OF THE MEDITERRANEAN BASINS

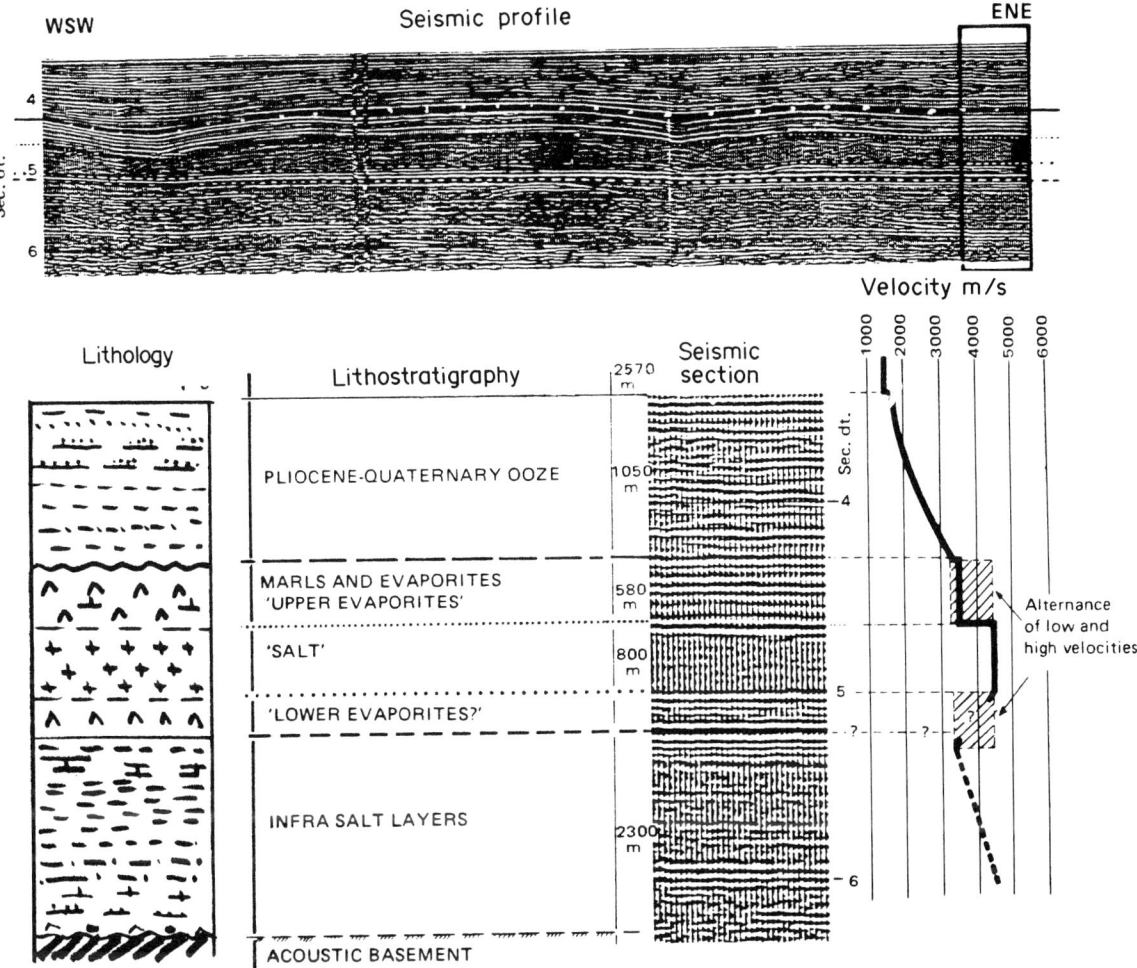

Fig. 2.12. Typical stratigraphic and lithoseismic sequence in the Western Mediterranean (north-western Balearic Basin Plain). (Modified from Montadert et al., 1978).

1973; Auzende et al., 1975). A distinct unconformity and change in sedimentation at the end of the Calabrian individualized the two uppermost units. This unconformity seems to correspond to a rapid transgression that changed the dominantly turbiditic sedimentation into a hemipelagic regime. The transgression might be due to the generally higher sea level of the Late Pleistocene, but the angular nature of the unconformity and associated faults suggest a possible tectonic influence. A recent change from extension to compression for the region has been proposed for the Calabrian, at about 1.8 m.y.a. (Bousquet et al., 1976).

The stratigraphic sequence in the Tyrrhenian Sea is similar to the one in the Alboran Sea. Seismic profiles complemented with the results of DSDP Site 132 define four lithostratigraphic units above the acoustic basement in the central Tyrrhenian basin (Ryan et al., 1973; Selli, 1974). The oldest unit is mostly of Tortonian age, but also comprises the lower Miocene. The evaporitic Messinian sequence above is several hundred metres thick in the bathyal plain and absent in the northern Tyrrhenian Sea.

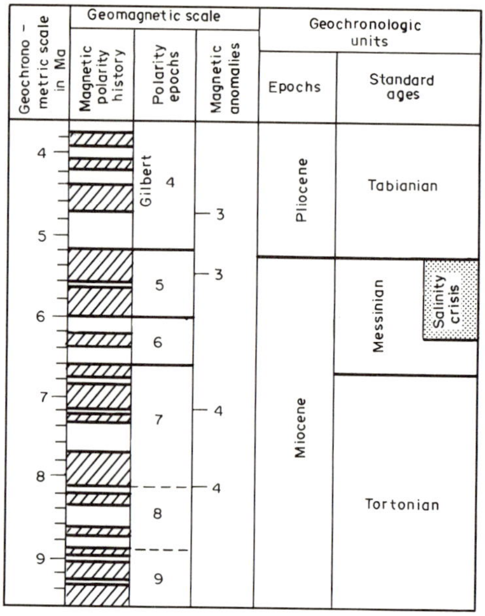

Fig. 2.13. Late Miocene-Early Pliocene stratigraphy in relation to the Mediterranean 'salinity crisis' (Modified from Cita, 1982).

The two youngest units are a lower Pliocene foraminiferal ooze and an upper Pliocene – Quaternary, well-layered muddy unit separated by a small angular unconformity.

Moreover, all the Western Mediterranean basins are characterized by a Pliocene – Quaternary evolution marked by active subsidence (Stanley, 1977). The occurrence of post-Messinian vertical movement is well shown in many sectors of the continental margin (Stanley et al., 1976). Quantitative geophysical calculations of subsidence due to the isostatic loading of sediment, water overburden, and thermal cooling of the oceanic crust show that post-Messinian displacements of several hundred metres may have occurred in the basin plain (Ryan, 1976; Biju-Duval et al., 1978; Rehault, 1981). Subsidence was either in the form of tilting or marked by tilting and faulting. The importance of the subsidence is also related to the nature of the lithosphere. In the continental margins and basins with continental crust, the subsidence was less important than in the deep basins with oceanic crust.

2.4. THE MESSINIAN SALINITY CRISIS

It is now well-established that the entire Mediterranean Sea became isolated from the Atlantic Ocean during the Messinian Stage of the Late Miocene, between about six and five million years ago (Fig. 2.13). A strong regional aridity led to the evaporation of this isolated sea and water level dropped below that of the open ocean. As revealed by the first drilling of the *Glomar Challenger* in 1970, the sea withdrew hundreds of kilometres from the ancient coastline and extensive erosive surfaces developed over most of the continental margin. The Mediterranean was transformed into a series of large lakes in which precipitated a thick and extensive sequence of evaporites including gypsum, halite, and other salts, that exceeds one million cubic kilometres. The net thickness of the salts exceeds 1.5 kilometres in places and their general distribution indicates a preferential accumulation in that part of the sea bottom which is the deepest today.

Sedimentary strata containing the crust and filaments of blue-green algae (Ryan, Hsü et al., 1973) have been found buried below half a kilometre and more of a younger sediment carpet beneath the edges of the present abyssal plain in water depths approaching 3 kilometres. How did these tidal-flat, playa-lake and even terrestrial sediments, which make up an enormous volume, come to be found at such a great depth?

The history of dessication of the isolated Mediterranean basins is complicated and major evaporite deposition occurred in two main phases with distinct paleoenvironmental significance (Hsü et al., 1978). Arguments have centred on whether these evaporites were deposited in a shallow basin during the latest Miocene or a deep basin, as in the present day, and whether sea level was maintained at or far below that of the world ocean. Various theories have been proposed to account for the build-up of these sequences, all of which require a restricted, shallow portal between the Atlantic and Mediterranean.

One model maintains that the deep abyssal plain only came into existence after the deposition of the evaporites when tectonic events led to a collapse of the continental margin and a foundering of the basin centres (Auzende et al., 1971; Stanley et al., 1976; and others). Other hypotheses state that oceanic basins were already in existence at the time of the Mediterranean isolation and that the salt accumulate on the floor of deep depressions dried out from excess evaporation (Hsü et al., 1973, 1977; Ryan, 1976; Cita, 1982). One of the possible alternative models, that of brine reflux in a deep oceanic basin, was refuted because of the discovery of evaporitic facies indicative of tidal environments, and of the areal distribution of the evaporites in the Western Mediterranean in which the most soluble precipitates are restricted to the basin centre (Cita, 1982).

The debate between these contradictory hypotheses is not by any means settled as reported, for example, in an editorial in *Nature* (Volume 242, March 1973), The International Geodynamics Commission Colloquium convened by Drooger (Utrecht, 1973), a book review in *Marine Geology* by Stanley (1975), and a provocative commentary by Sonnenfeld (1974). This controversial subject can be considered one of the most important geological problems to be solved within the next few years. The question of whether the Mediterranean was deep or shallow is not just crucial to the understanding of the genesis of salt deposits in oceanic settings, but indeed the question is fundamental to the mechanism by which small ocean basins come into existence in tectonic settings interior to island arc subduction zones (Boccaletti et al., 1971; Auzende et al., 1973; Boccaletti and Guazzone, 1974). Due to this importance, the several lines of evidence that have been investigated in relation to the origin of the salinity crisis are discussed in the following sections. Although several aspects related to the paleogeography of the Mediterranean basins during the salinity crisis have not been deciphered as yet, a dynamic model is presented for the development of the salinity crisis that seems to have the largest consensus.

2.4.1. Global events leading to the onset of the salinity crisis

Pre-Messinian tectonic evolution

As mentioned previously, much of the equatorial Tethys was eliminated by the alpine orogeny. With the creation of the new Mediterranean basins after the orogeny, the connection between the Atlantic and Indo-Pacific was maintained until the Early Miocene (Fig. 2.2) when it was severed with the joining of Africa and Europe along the Middle East front during the Burdigalian (Hsü, Montadert et al., 1978). During this period the present distribution pattern of land and sea was established. The joining of the two continents permitted the exchange of land faunas and started the differentiation of the Indo-Pacific and Atlantic – Mediterranean marine faunas (Adams and Ager, 1967; Hallam, 1973). The micropaleontological data also suggest that there might have been repeated transgressions into the Mediterranean from the Indo-Pacific until Middle Miocene time (Rögl et al., 1978). The final rupture of the Mediterranean link with the Indian Ocean probably took place at about the Serravallian time (14 – 13 m.y.a.). This was one of the major paleobiogeographical events of the Tertiary.

The joining of the continents also favoured the gradual change in the Mediterranean climate, which tends toward aridity in the low latitude regions on the west side of a continent. This evolution initiated the first steps of the salinity crisis in the Burdigalian time, which became viable in the Serravallian, some 15 million years later (Hsü et al., 1978).

The Paratethys extending to the East was separated from the Mediterranean basins by the newly created mountain chain which stretched from the Alps through the Dinarides and the Hellenides to Anatolia (Fig. 2.14). Several connections have been suggested between these two seas (Hsü et al., 1978). The Burdigalian connection through the peri-alpine depression was eliminated early in Middle Miocene, when the Helvetic Alps rose (Rögl et al., 1978). A last contact was maintained between the Mediterranean and the Paratethys through an opening in northern Italy and north-western Yugoslavia. The complete isolation between the two island seas took place some 14 or 15 million years ago during the early Serravallian.

The last links of the Mediterranean with the world ocean were the Betic and the Rif straits (Biju-Duval et al., 1978). The deep North Betic Strait shoaled after the end of the Middle Miocene. Weak links may have been competent enough, however, to keep the Mediterranean waters normal marine. Sedimentary sequences below the Messinian contains microfossil assemblages indicative of open marine conditions in both the land sections and the deep sea (Cita et al., 1978).

Late Miocene climatic and eustatic oscillations

The change from normal marine to the evaporite formation was a sudden event that does not seem to have a direct correspondence with an abrupt climatic change. A general trend to a drier and cooler climate had been in evidence since the Early Miocene when the Mediterranean became isolated from the Indian Ocean. The climate in the latest Tortonian in the Mediterranean was similar to that of the Messinian as shown by the pollen record of D.S.D.P. Leg 42 (Hsü, Montadert et al., 1978). In consequence, the drastic change to evaporative conditions cannot be directly interpreted as change in climate. Instead, the final closure of the narrow connections with the Atlantic may have been the triggering mechanism. With isolation and the negative hydrographic balance of the mediterranean, the salinity crisis became unavoidable. The Betic and Rif straits were eliminated due to the general tectonic evolution of the alpine system. Vertical upliftings of several hundred metres of the middle Miocene deposits are well known in the Betics (Pannekoek, 1969).

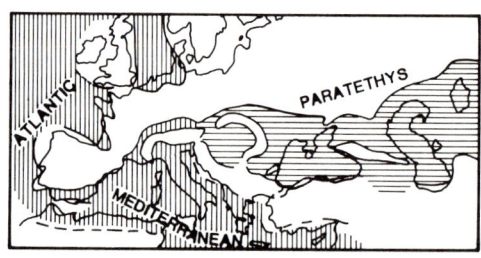

Fig. 2.14. Relationships between the Mediterranean, the Paratethys and the world oceans during the middle Miocene (From Hsü et al., 1978).

A glacio-eustatic fall in sea level has also been invoked as an important factor in closing the Atlantic — Mediterranean connection. Increased glacial activity of Antarctica has been cited as the cause of major lowering of sea level during latest Miocene (Kennett, 1967; Vail and others, 1977). Shackleton and Kennett (1975) calculated on the basis of oxygen and carbon isotopic changes a glacio-eustatic lowering of sea level of about 40 m. A latest Miocene regressive phase is widespread throughout much of the shallow marine areas of the oceans. For example, in the Andalusian stratotype of the Betics, Berggren and Haq (1976) demonstrated a drop in sea level of about 50 m during the latest Miocene.

In summary, the onset of the salinity crisis was an event that drastically modified the depositional conditions throughout the entire Mediterranean. However, the factors leading to that situation, such as the tectonic evolution and the regional climatic conditions were well in existence before the evaporite formation. The salinity crisis of the Mediterranean is only the most spectacular geologic event of the global evolution during the Cenozoic!

2.4.2. Geological evidences of the salinity crisis

Several lines of geological evidences have been investigated to test the validity of the dynamic models proposed for the Messinian event. These include the distribution of erosional surfaces, the biological response to progressive isolation and the nature of the evaporite sequences. The largest amount of evidence has been compiled in connection with the results of the two drilling campaigns of the *Glomar Challenger* in the Mediterranean (Ryan, Hsü et al., 1973; Hsü, Montadert et al., 1978). These results are briefly discussed in the following sections.

Messinian outcrops and erosional surfaces

Information on the paleodepth of the basin can be obtained from the nature and distribution of the Miocene outcrops in the distal continental margin (Hsü, Montadert et al., 1978; Eastward Group, 1981; Mauffret et al., 1982). The Serravallian is composed of dolomitic nanno ooze that is attributed to hemipelagic deposition (Fig. 2.15). Above the Serravallian there is a large unconformity and the Tortonian is generally missing (Cita et al., 1978). The Messinian in the eastern Menorca margin has a large variety of facies, with varved dolomitic marls, iron stained soil horizons and turbiditic sequences. This variety of deposits confirms the evaporite dessication of a deep basin, with evidence of intermittent subaerial exposures and periodic flooding. The underlying Serravallian deposits and the overlying Pliocene hemipelagic oozes show normal deposition in a deep marine basin (Cita, 1973). Unfortunately, facies interpretation alone does not allow a quantitative estimation of water depth and some other evidences must be used in this respect.

The top of the Messinian formation is marked by a characteristic seismic reflector originally noted as Horizon M by Ryan et al. (1970), and also referred to as Reflector H by Alla et al. (1972) and Mauffret et al. (1973). On gently sloping margins such as in the Valencia Trough, or across the Rhône Fan, the interface marking the top of the evaporite formation passes laterally in the landward direction to an erosional discordance where detectable thicknesses of evaporites are absent (Figs. 2.15, 2.16). The erosional surface separates pre-Messinian strata from post-Messinian strata (Ryan, 1973, 1976). It can be traced beneath the continental shelf on the Catalan margin where it has been penetrated at sub-bottom depth in excess of one kilometre by numerous commercial boreholes in the offshore of the Ebro Delta (Maldonado and Riba, 1974; Stoeckinger, 1976). The stratigraphic gap represented by the erosional surface may be considerable and in some oil boreholes and DSDP Site 372 east of Menorca, the hiatus

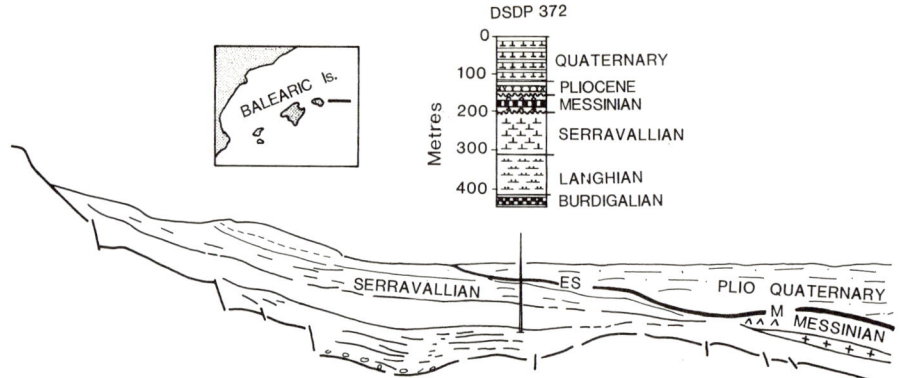

Fig. 2.15. Lithoseismic profile of the distal continental margin to the east of the Balearic Islands showing the stratigraphic section of Site 372 of the DSDP (Modified from Hsü et al., 1978, and Mauffret et al., 1982). M, Messinian M-reflector; ES, Messinian erosional surface.

approaches six million years (Fig. 2.15). This erosional surface is considered the product of subaerial erosion during the evaporative sea level drop of the Mediterranean.

On the passive-type continental margins, like those of the Balearic Sea, the seaward gradients of the erosion surfaces permit the calculation of the paleorelief between the shoreline of the pre-salinity crisis and the basin centre. The calculation corrected for post-Messinian sediment loading, compaction and regional subsidence show a basin plain of more than 2.5 km deep for the Western Mediterranean (Ryan, 1976; Ryan and Cita, 1978). However, quantitative determinations in the Gulf of Lions and Ligurian Sea on the basis of regional geomorphology and thermal cooling of the lithosphere show basin depths significantly shallower. These recent geophysical calculations give depths of about 1.3 km at the end of

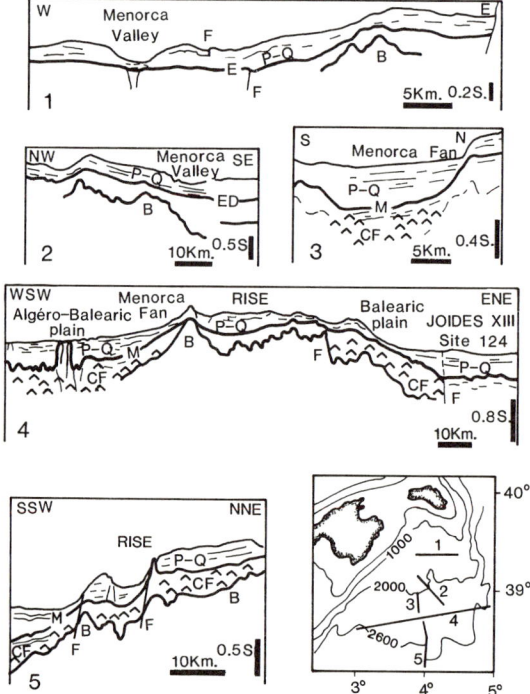

Fig. 2.16. Seismic profiles of the southern Balearic margin. B, Acoustic basement; CF, Messinian evaporites ('couche fluante'); M, Messinian M-reflector; E, Messinian erosional surface; P-Q, Plio-Quaternary; F, fault (Modified from Maldonado and Stanley, 1979).

the Messinian and 2.2 to 2.4 km before the Messinian (Lefebvre and Gennesseaux, 1981; Rehault, 1981). The important aspect of these calculations, in spite of the major differences in depth, is to show that the basin plains were over 2 km deep before the deposit of the Messinian evaporites.

The biological response

The fossil record provides good documentation of a continuous deterioration of the thermohaline circulation and of water mass stratification which recorded the progressive isolation of the Mediterranean. The most significant paleocological indicators of the progressive isolation are diatoms and corals which react in a more sensitive way than foraminifers and ostracods to local environmental changes (Cita, 1982). Diatomaceous sediments ('tripoli') appear directly below the evaporites. The Tripoli Formation was apparently deposited in a gulf which connected westwards with the Atlantic. Periodic severing of this connection caused the cyclic changes between normal marine and hypersaline, stagnant conditions.

In the same line of arguments, Chevalier (1961) showed the progressive decrease in coral species diversity during the Miocene and demonstrated the permanence of coral reefs in the Late Miocene of the Mediterranean. Only red algae and bryozoan reefs with ahermatypic corals were developing at this time on the Atlantic side of France, Spain, Portugal and Morocco. The preservation of these reefs in the Mediterranean is a consequence of the isolation of this basin from the colder Atlantic waters. Recent studies by Esteban (1980) have shown that the distribution in space and time of the different coral reef types is a result of retrogressive ecologic succession since the Middle Miocene in a partially isolated Mediterranean. Increasingly aberrant communities are considered to reflect the progressive deterioration of the marine conditions in the Mediterranean. This author calculated that the cyclic reef complexes of the Messinian were developed as result of sealevel oscillations of several hundred metres.

The evaporative formation

Seismic records suggest a twofold division of the mediterranean evaporites (Montadert et al., 1978): (1) an upper evaporite sequence, with numerous reflectors, up to several hundred metres thick, and consisting of dolomite, gypsum, anhydrite, and some salts, and (2) the main salt sequence, seismically homogenous, up to a thousand metres thick or more, together with a lower sequence of evaporites with several reflectors (Fig. 2.12). The two units are separated by a large angular or erosional unconformity. The main salt unit is restricted to the central sectors of the Mediterranean basins, while the upper evaporite unit may extend into the continental margin.

Seismic profiles display a geometry of the evaporitic bodies characterized by great thicknesses under the basin floors, thinning towards the margins of the basin and pinching out at the base of the slope (Mauffret, 1979). This evidence shows that the basin was in existence when the evaporites were laid down and that the basic morphology predates the salinity crisis. Moreover, the geometry of the M-reflector on top of the evaporites simulates the relief of the present sea floor, indicating that the Messinian configuration was not much different from that of today.

2.4.3. A dynamic model of the salinity crisis

Middle Miocene plate movements between Africa and Europe cut off the Mediterranean openings to the Indo-Pacific. The perialpine depression north of the Alps was also raised and the communication between the new Mediterranean basins and the Paratethys was severed (Fig. 2.14). A considerable fresh-

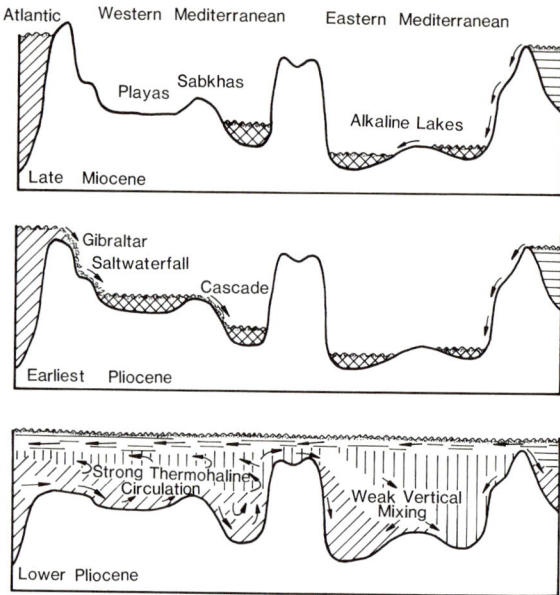

Fig. 2.17. Late Miocene and Pliocene phases of dessication and flooding in the Mediterranean Sea related to the Messinian 'salinity crisis' (After Cita and Ryan, 1973).

water supply from Central Europe was then diverted to the Parathys, and the reorganization of drainages placed considerable strain on the water budget of the Mediterranean.

There is no general consensus about what actually created the final isolation of the Mediterranean basins in the latest Miocene. As mentioned previously the final connection with the Atlantic is believed to have been the alpine straits, entering into the Mediterranean from southern Spain and northern Morocco. Although the closing of these connections is related to large-scale plate movements, its final severing may have resulted from the lowering of sea level in the latest Miocene (Adams et al., 1977). The synchronism between the beginning of the salinity crisis and the global cooling and sea level fall as dated in the Southern Hemisphere strongly suggests that the isolation could have been triggered by a glacio-eustatic change (Loutit and Kennett, 1979).

Soon after the beginning of the Messinian stage, the last openings to the Atlantic through the Betic and Rif straits were closed. It seems that these passageways became progressively constricted during the Middle and Late Miocene, and virtually closed by the Messinian (Benson, 1976), with the southern seaway lasting somewhat longer. Evaporative drawdown of the isolated Mediterranean sea level, leading to dessication, thus became unavoidable since evaporative loss considerably exceeded precipitation and influx from rivers. The modern Mediterranean Sea has a water volume of 3.7×10^6 cubic kilometres, and the excess evaporation is 3,300 cubic kilometres per year. If the Strait of Gibraltar were closed today, the present Mediterranean would dry in about 1000 years!

Facies analysis of the pre-Messinian deposits in the Mediterranean basins and the distribution of the lithoseismic units have demonstrated that the configuration of the Mediterranean basins was well developed before the salinity crisis (Mauffret et al., 1982). Quantitative calculations on the basis of thermal cooling, geomorphological analysis of erosional surfaces and biological paleobathymetric indicators shows that the Balearic basin may have been deeper than 2 km before the Messinian.

The initial desiccation was followed by the deposition of the main salt unit of halite and potash salts. A thick body of salt was built by the evaporation of sea water spilled over from the Atlantic into

Mediterranean brine lakes (Fig. 2.17). The salt section contains several units representing deposition from shallow salina environments to subaqueous environments several hundred metres deep (Schreiber and Decima, 1976). An intra-Messinian desiccation probably due to complete isolation from the Atlantic, led to widespread erosion and recycling of primary halites. As result, a widespread unconformity developed between the two evaporites. Deposition of the upper evaporite followed a marine inundation which may have filled the Mediterranean again to the brim. Periodic influx and evaporite drawdown of the Mediterranean sea level and the gradual concentration of the brines led to formation of evaporites characterized by the 'bull's eye' pattern. Dolomite was the first mineral formed, followed by sulfate deposition. Halite was laid down on the bottom of playas and the potash salts were concentrated in the deepest and most central parts of the Mediterranean basins. Shallow-water diatoms and algal stromatolites prove that the basins were covered by shallow waters during the flooding stage, and soil horizons indicate subaerial exposure during the desiccation stages (Hsü et al., 1978; Eastward Group, 1981).

The evaporite formation is up to 2 or 3 km thick in places; one basin of Mediterranean sea water could not have supplied enough salt for such thick deposits. If 17 m of salt is deposited each time the basin was desiccated, it follows that the Mediterranean must have been desiccated about 40 times during the latest Miocene (Ryan, 1973), withdrawing a total of about 6% of the salt from the world ocean (Fig. 2.17). Hsü et al. (1978) suggested that 11 interludes of infill would have been sufficient to account for the transport of ions from the open sea and thus for the Messinian salt deposition in the Mediterranean.

The youngest Messinian sediments cored in DSDP and outcropping in many land sections of the Eastern Mediterranean are marls deposited in a brackish standing body of water (Hsü et al., 1978). A number of lakes of reduced salinity, referred to collectively as the 'Lago Mare', came into existence. The Eastern Mediterranean was inundated by brackish water probably from a Paratethys origin. At the same time the Western Mediterranean may have remained desiccated until the Atlantic water began to spill over the Strait of Gibraltar. The final flooding and refilling of the Mediterranean took place at the beginning of the Pliocene, as a result of east-west normal faulting perpendicular to the Strait of Gibraltar arc that created the narrow passage. The infill began with the deposition of marls with restricted marine, dwarf faunas, followed by normal marine deposition that has prevailed since that time. Paleobathymetric analysis show that the earliest Pliocene deposits above the Messinian invariably were deposited in water depths in excess of 1000 or 1500 metres (Hsü et al., 1978). This indicates that deep Mediterranean depressions, pre-dating the Messinian salinity crisis were still in existence at the end of the evaporite deposition.

Seismic evidence, however, also demonstrates the occurrence of Pliocene-Quaternary subsidence (Morelli, 1975; Stanley et al., 1976; Biju-Duval et al., 1978). The subsidence was at least in part induced by the isostatic load of the water flooding the basins (Hsü et al., 1973; Ryan, 1976). Additional subsidence is related to thermal cooling of the mantle under the basins and may be also due to continued tectonic activity deepening back arc-basins (Montadert et al., 1978; Mauffret et al., 1982; Rehault, 1981). Considerable Pliocene — Quaternary epirogenic displacements are also recorded in the circum-Mediterranean regions (Stanley et al., 1976; Stanley, 1977). The Messinian basins of Sicily, Calabria, the Apennines, the Betics, etc., were uplifted and emerged, while continental margins as the Rhône, Ebro and Balearics locally underwent subsidence. Such evolution has largely modified the Late Messinian paleogeography.

In addition, recent interpretations of the Western Mediterranean tectonics seem to indicate a change from a distensive to a compressive regime during the Pliocene (Mauffret et al., 1982). This evolution may explain a partial decoupling between the basins and the continents along the distal continental margin. Ancient normal faults may have been transformed into thrust faults due to the post-Messinian compression, which resulted in the foundering of the basins and rising of the continents (Rehault, 1981). Overall, however, the Mediterranean basins owe their origin to earlier Neogene movements, even though their floors may have been uplifted or subsided during the last five million years (Hsü et al., 1978).

In the author's view, the present 'state of the art' does not allow an unequivocal reconstruction of the basin depths in relation to the salinity crisis within a few hundred metres. Most researchers of the Mediterranean paleogeography will agree, however, upon the following points: (a) the Mediterranean Sea was close to the present morphology before the salinity crisis; (b) deep basins were in existence prior, during and after the salinity crisis; (c) the basin plain depth below present sea level in most basins during the Messinian was at least several hundred metres below 1000 m, and (d) there has been a significant foundering of the basins since the Messinian to the present.

2.4.4. World response to the salinity crisis

The deposition of the Mediterranean evaporite, calculated in a volume of more than 10^6 km^3, lowered the salinity of the world ocean by 6%, or a mean reduction in oceanic salinity of 2‰. This salinity decrease in such a brief time may have had important paleoclimatic repercussions by raising the freezing point of sea-water and allowing sea ice to form at slightly higher temperatures (Ryan et al., 1974). In turn, this would have increased the albedo of the earth, impeding the transfer of heat into the atmosphere and favoured the Antarctic ice growth during latest Miocene. The hypothesis of Ryan and others (1974) also suggests that the salinity crisis was the cause, not the effect, of the global cooling and glacial expansion of Antarctica during latest Miocene. However, it may also be interpreted that the Antarctic ice sheets expanded sufficiently to lower sea level enough to isolate the Mediterranean basin. The isolation of the Mediterranean led to a paleoenvironmental feedback which at least temporarily encouraged global cooling, as the well-documented cooling phase of the Southern Hemisphere.

In any case the correlation between the onset of the salinity crisis and the global glacio-eustatic curve is not unequivocal. Isotopic analysis of upper Miocene deposits of the eastern North Atlantic continental margin shows a warm trend at the time of the main salt development (Shackleton and Cita, 1979). More studies will be necessary in order to clearly establish the relationships between global climatic events, glacio-eustatic sea level oscillations and the onset of the salinity crisis.

The salinity crisis is also considered by Thierstein and Berger (1979) as an example of an 'injection event'. The opening of the connection with the world ocean of the temporarily isolated basin should have resulted in the injection of hypersaline waters favouring abyssal stratification and stagnation. However, according to Cita (1982), the injection model seems to be inappropriate to the Messinian event because no evidence exists of any substantial contribution of Mediterranean waters to the ocean during the salinity crisis.

In summary, the global implications of the Messinian event have only begun to be explored, although many hypotheses have been formulated on the response of the oceans to the salinity crisis (see Hsü et al., 1973, 1978; Cita, 1982). A complete account of the consequences of the Mediterranean desiccation is, however, beyond the scope of this chapter and only some of the most significant aspects from the geological point of view have been summarized.

2.5. REGIONAL DEPOSITIONAL PATTERNS DURING THE PLIOCENE AND QUATERNARY

The salinity crisis was interrupted about 5 m.y.a. by the opening of the Strait of Gibraltar and the subsequent filling of the Mediterranean with Atlantic waters at the beginning of the Pliocene. Normal marine conditions were restored and deep-sea hemipelagic sediments started to develop again. Since the Messinian, the Mediterranean did not become isolated even during the sea level drops of more than 100 metres due to the Quaternary eustatic and climatic oscillations. However, these oscillations affected the hydrographic balance of the sea, altering the circulation patterns and the distribution of water masses (Stanley et al., 1975). One of the most spectacular results of this evolution was the deposition of organic-rich, dark sapropel sediments developed under anoxic conditions in the Eastern Mediterranean (Ryan, 1972; Maldonado and Stanley, 1976; Williams et al., 1978). The Quaternary sapropels were restricted to areas east of the Strait of Sicily sill, but dark, organic-rich sediments may be present in equivalent stratigraphic position in the Western Mediterranean (Maldonado, 1978; Canals et al., 1982).

Several other factors have controlled the depositional patterns during the Plio-Quaternary in the Western Mediterranean, including the structural setting and physiography of the depositional province, the sediment input and the physical oceanography (Maldonado and Stanley, 1979). The importance of each of these factors differs for each depositional province. For example, in the coastal zone and continental shelf, the physical oceanography and wave climate may be one of the most important factors controlling sedimentation, while at the base of the continental slope, deposition is largely modified by the structural framework of the basin (Aloïsi et al., 1981; Bellaiche et al., 1981).

Overall, each depositional province is characterized by a suite of depositional processes and controlling factors that for the Western Mediterranean can be summarized in the following schemes:

(1) Terrigenous shelves predominate off major deltas and areas of large sediment supply, although there is a relative abundance of calcareous-rich areas.

(2) Continental slopes are covered by a large variety of deposits due to the interplay between mass-gravity flows and hemipelagic settling. The distribution of these deposits is influenced by short and long term events.

(3) The prevailing sediment facies are fine-grained deposits. There are a large variety of mud facies that show very distinct transitional trends from the shallow to the deep marine environments. Hemipelagic muds predominate on structural highs and areas protected from large sediment supply, while in the basin plains the most abundant deposits are turbiditic muds.

(4) The Mediterranean continental margins are characterized by rapid sediment entrapment off major rivers and sediment bypass to more distal sectors in areas of low sediment supply.

(5) Many different types of base-of-slope depositional models may be controlled by the structural framework and the nature and amount of sediment input. Some of these models can be described in terms of the well-known examples of submarine fans. In contrast, many other Mediterranean fan types show variations that may be more applicable to the interpretation of the geological record in the alpine-type, orogenic belts.

(6) Sedimentation in the distal sectors of the continental margin and in basin plains is largely influenced by salt tectonics of the underlying Messinian evaporites. The predominant sedimentary process in these sectors is ponding by distal turbidity flows.

These patterns serve to distinguish the major depositional provinces of the Western Mediterranean. However, important variations are also observed in several sectors with similar depositional environments due to differences in the factors controlling sedimentation. The Mediterranean Sea, a multi-silled and almost totally enclosed system, serves as a natural sedimentation laboratory which allows new depositional schemes to be formulated and basic principles to be tested (Stanley, 1972). Some of these depositional schemes are discussed in the following sections.

2.5.1. Primary factors influencing sedimentation

Structural and physiographic setting

High resolution seismic profiles reveal three types of continental margins in the Western Mediterranean (Stanley et al., 1976; Serra et al., 1979): progressive, intermediate and abrupt (Figs. 2.18, 2.19). The progressive type is characterized by a thick wedge of sediments with lateral and seaward continuity of gently dipping Pliocene and Quaternary deposits. The deep structures have little or no topographic expression and the sea floor is characterized by a smooth, convex-up physiography. This type of margin corresponds to a mature example of passive margin or continental 'trailing edge' in the geotectonic classification of Inman and Nordstrom (1971).

The intermediate margin presents a marked step-like physiography that is the surface expression of the downthrown blocks of the underlying basement formed by oldest rocks (Fig. 2.19). In this type of margin, the sedimentary cover is very reduced on highs and the deep basement structure controls the surface physiography. The abrupt type of margin shows the steepest, greatest relief as the result of a reduced sedimentary cover that may be totally absent.

Depositional patterns in each type of continental margin is largely controlled by the structural and physiographic setting (Stanley, 1977; Maldonado and Stanley, 1979). Large wedges of seaward prograding sediments in the progressive margin result from continuous subsidence due to the isostatic adjustment to sediment overburden off areas of large sediment input, such as the Rhône and Ebro deltas (Fig. 2.18). In contrast, the deep structure in the intermediate and abrupt margins control the steep sea floor physiography, which in turn had influenced the sediment dispersal and the distribution of the deposits not allowing the development of thick sedimentary wedges..

Margins affected by continuous subsidence allow the preservation in the stratigraphic record of sedimentary units developed during successive sedimentary cycles. This is particularly true for the continental shelves, affected by periodic sea level rising and lowering where the preservation of thick sedimentary wedges is only possible with active subsidence. On non-subsiding margins, every eustatic cycle reworks and eliminates most of the deposits of the previous cycles (Swift, 1976). The preservation of the stratigraphic record in the continental slope is also possible only in areas of rapid entrapment of sediments, such as the progressive continental margins. In other margin types, the deposits rapidly accumulated on the slope during low sea level stands are triggered by earthquakes or large storms and displaced by mass-gravity processes to the distal margin.

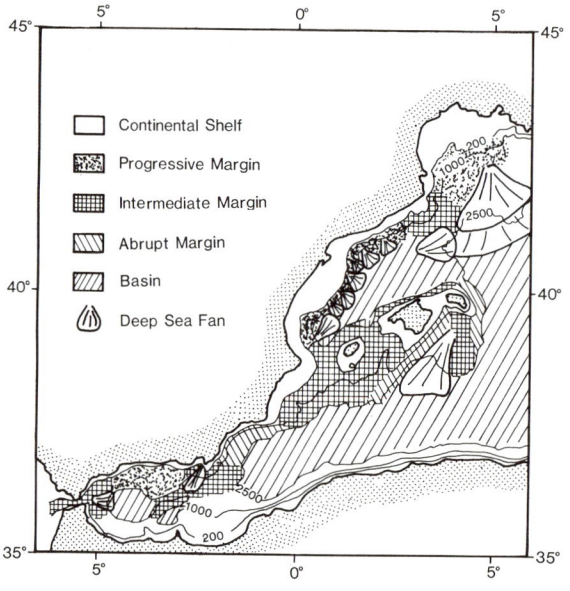

Fig. 2.18. Physiographic sketch showing depositional provinces in the Spanish margin of the Mediterranean.

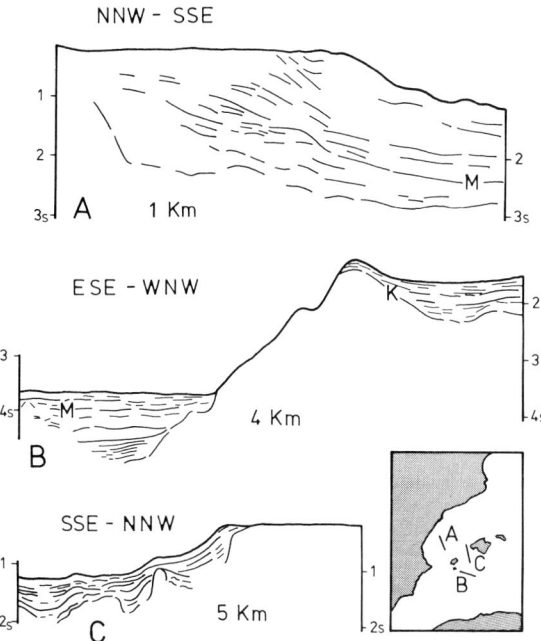

Fig. 2.19. Characteristic profiles of the Western Mediterranean continental margins showing the basic types defined in this sector: A, progressive type; B, abrupt type; and C, intermediate type. Discussion in the text (Modified from Serra et al., 1979).

Sediment input and climatic-eustatic oscillations

Studies of deep-sea cores from the Mediterranean Sea have shown that the Quaternary sedimentary section is characterized by repetitive sequences of sediment types that are organized in a cyclic fashion (Cita et al., 1977; Maldonado, 1978; Stanley and Maldonado, 1979; and others). These lithofacies record repetitive changes induced by major climatic and eustatic oscillations, which are particularly well shown in the sediment facies of the Mediterranean because this is an almost totally enclosed sea. The Quaternary oscillations have affected sedimentation in two principal ways: (a), shifting the location of the depocentres across the continental margin, and (b), altering the headwater extension and regime of the major rivers in relation to the pluvial conditions.

Continental shelf sedimentation is most significantly affected by these climatic oscillations. Rapid sea level drops of more than 100 metres are well-documented in the Mediterranean Sea during the Late Quaternary when most of the continental shelf was subaerially exposed and underwent erosion (Aloïsi et al., 1978). At the same time, the locus of deposition was displaced to the distal continental margin and deep water sedimentation rates sharply increased (Stanley and Maldonado, 1979). During high sea level stands, most sediment input to the marine environments is trapped in deltas and nearshore environments while little materials escape from the continental shelf. These phases are characterized by reduced sedimentation rates in the distal provinces of the margin (Monaco et al., 1982).

Oceanographic parameters and climatic-eustatic oscillations

The oceanographic factors in the Mediterranean Sea, including the distribution of water masses, have fluctuated considerably during the Quaternary as in other ocean basins (see Normark and Piper, 1969; Nelson and Kulm, 1973; Nelson and Nilsen, 1974; Curray and Moore, 1974). At present, the well-

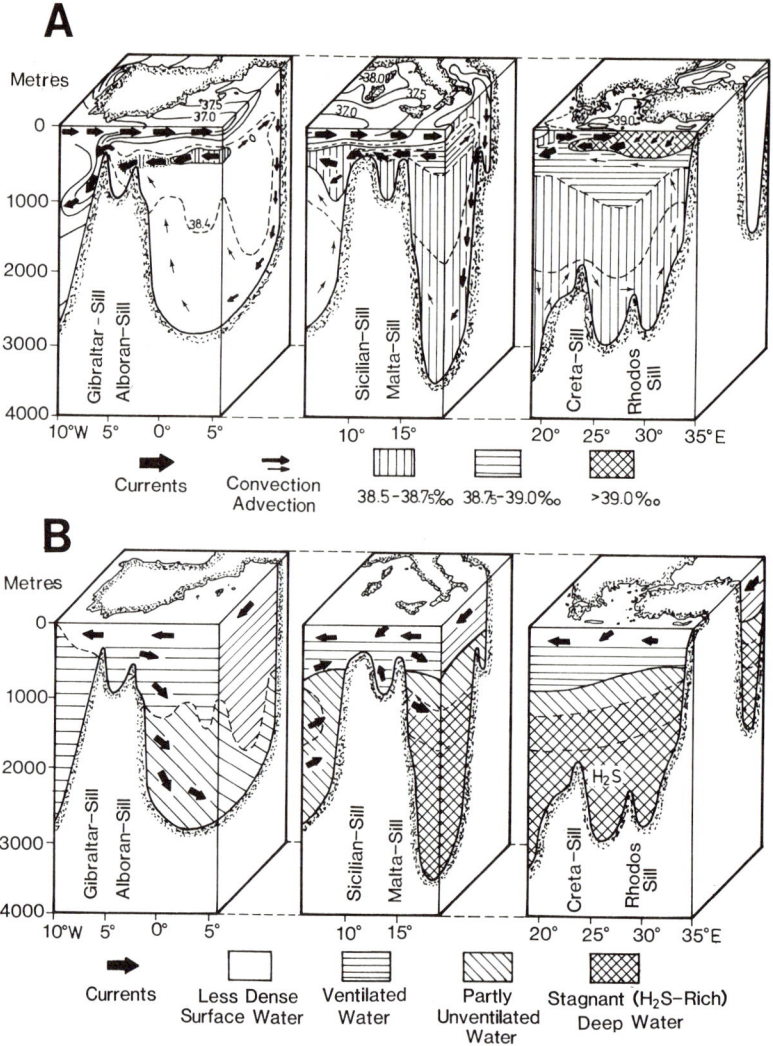

Fig. 2.20. Schematic block diagram showing presently prevailing flow patterns and water masses distribution (A) and possible short-term, early Holocene (B) flow changes and stratification of water masses (Modified from Maldonado and Stanley, 1979).

known Mediterranean circulation pattern is one of less dense water of Atlantic origin flowing eastward above the generally westward flowing intermediate, or levantine water mass (Fig. 2.20A). An early Holocene model depicting a contrasting trend has been proposed. This model implies a short-term current reversal, whereby less dense surface water flowed to the west in response to decreased evaporation rates relative to precipitation and fluvial input in this quasiclosed system (Fig. 2.20B, *see also* Huang and Stanley, 1972; Stanley *et al.*, 1975). This phase of regional flow modification was accompanied by well-developed water mass stratification and anaerobic, H_2S-rich deep water conditions that fostered the accumulation of grey to black, organic-rich sapropel layers in the Eastern Mediterranean.

During the same period, the Western Mediterranean water masses were only partially stratified and the sea floor remained partially oxygenated and locally influenced by bottom currents (Maldonado,

1978; Kelling *et al.*, 1979). Although upper Quaternary sapropels are not recorded west of the Strait of Sicily, analysis of Balearic Basin and Alboran Sea cores reveals dark layers, pyritized test of organisms and changes in the type of bioturbation in equivalent stratigraphic position (*see* Huang and Stanley, 1972; Rupke and Stanley, 1974; Canals *et al.*, 1983). Moreover, evidences of periodic intensification of bottom current activity on the Balearic Rise during the Late Quaternary are also provided by the presence of thin, irregular wavy laminae of silt and mud within hemipelagic muds. These bottom currents resulted in deposition of asymmetric sediment wedges and moderate to large-scale erosion in the vicinity of topographic features such as pinnacles which served as obstacles to regional flow of bottom water (Kelling *et al.*, 1979).

2.5.2. Selected sedimentation models in the Western Mediterranean

Terrigenous, mixed and carbonate shelf sedimentation

Although the Western Mediterranean continental shelves are latitudinally zoned, with predominant carbonate-rich areas in the south and terrigenous deposits in the north, many other factors than climatic belts alone are governing the deposition of biogenic sediments. These shelves are covered by waters that for the most part are considerably colder than 20°C, but extensive sectors, perhaps more than 60% of the shelves, are floored by modern and relict sediments dominated by coarse bioclastic deposits of highly variable composition. Terrigenous sedimentation occurs off major rivers, but the thick prodeltaic wedges are normally restricted to the inner and middle continental shelf. Mixed terrigenous-bioclastic sediment is the predominant type of sedimentation away from areas of prodeltaic influence.

Continental shelf sedimentation off major rivers in the Mediterranean is well known and recent studies have shown that many similar features can be observed in the different depositional systems (Maldonado, 1975; Aloïsi *et al.*, 1975, 1978). Modern deltas such as the Rhône and Ebro deltas were largely developed in a few thousand years during the last high sea level stand by the seaward migration of a depositional wedge. This wedge was built by river-borne sediments deposited near the fresh-salt water interface as a result of a delicate balance between climatic-eustatic oscillations, sediment input from the river and the oceanographic regime.

The terrigenous material is essentially supplied to the prodelta by a bottom nepheloid layer which is initiated at the river mouth in the fresh-salt water contact (Aloïsi *et al.*, 1979). Settling from the bottom layer results in a seaward decrease in grain size and silt content and develops a depositional wedge that thins out from the river mouth. The distributional patterns of facies types and heavy metals demonstrates also the importance of electro-chemical floculation in the development of the prodeltaic muddy deposits (Fig. 2.21). The proximal prodelta is characterized by the rapid settling of the coarsest suspensions while the finest suspensions, defined by heavy metal and organic matter enrichments, are carried offshore by the nepheloid layer to the middle continental shelf (Alonso, 1981).

The facies distribution on continental shelves not influenced by major deltas is, in contrast, controlled by a series of processes which are largely related to the climatic and eustatic oscillations and to the supply of fine-grained materials (Maldonado *et al.*, 1983). Continental shelves in mid-Mediterranean latitudes, such as the Valencia continental shelf, are characterized by an outer continental shelf mud blanket that overlies coarse basal deposits (Fig. 2.22). The basal sequences are inferred to be the product of erosional shoreface retreat as the result of rising sea level and the landward translation of the shoreface (*see* Swift, 1976). The sand generated by this process is swept seaward to accumulate over a lag gravel as a transgressive inner shelf sand. As the water column deepened over the basal deposits due to the progressive sea level rising, fine-grained materials swept from the retreating shoreface came to rest upon

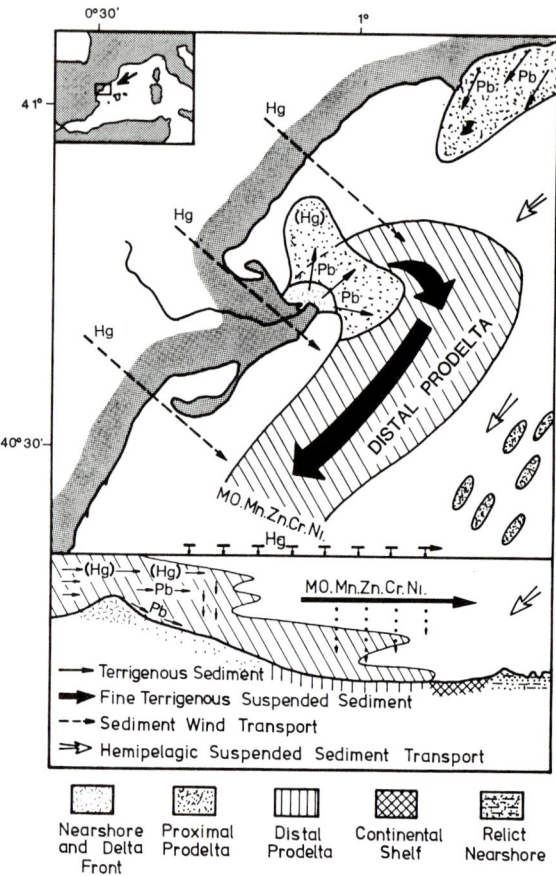

Fig. 2.21. Schematic model of predominant sedimentary processes controlling the distribution of recent deposits in the Continental shelf off the Ebro Delta (Modified from Alonso, 1981).

it. The upward-fining sequences observed over the basal unit are the consequence of the progression of more distal shelf lithotopes into the depositional site as the transgression continued (Maldonado et al., 1983).

The extent and position of mud deposits in this type of Mediterranean shelf are determined by the relative values of fluid power expenditure on the sea floor and the near bottom suspended sediment concentration as in many other continental shelves around the world (McCave, 1972). In this scheme, mud deposits may be localized to a relatively narrow depth zone, or may cover nearly the entire shelf even to the high tide line when the rate of mud input is important relative to the intensity of the wave climate. The Valencia shelf fits into this spectrum near the continuous mud sheet end, although the wave and storm-current climate is rigorous enough to maintain a sandy shoreface facies (Fig. 2.22). Sedimentation in this type of continental shelf is subdivided in two main depositional provinces. The shoreface and inner shelf are affected by jet-like storm flows that are inducing erosional shoreface retreat and generate the basal Holocene sand and gravel. Mud sedimentation prevails on the rest of the shelf. Suspended sediment is being advected down-shelf to the south and is moving offshore by both advection and diffusion processes. On the outer continental shelf the contribution of fine, biogenic particles

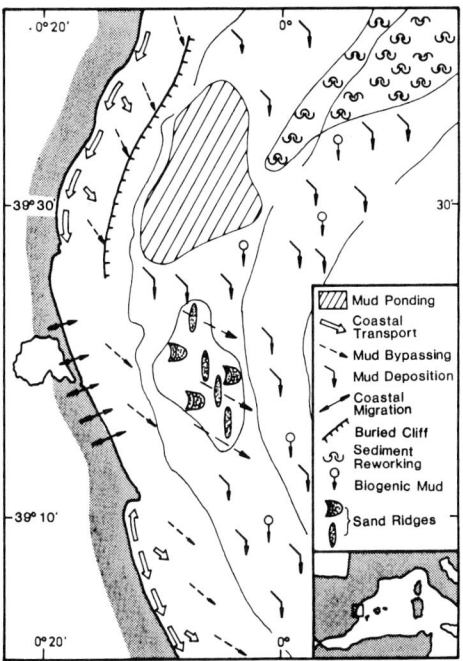

Fig. 2.22. Schematic model of Holocene sedimentation on the Valencia continental shelf showing sedimentary processes (Modified from Maldonado et al., 1983).

becomes more important and accounts for a large proportion of the sedimentary facies developed in that sector.

For these terrigenous and mixed terrigenous-calcareous shelves, dynamic factors exert significant control over the sedimentary processes. In contrast, other environmental parameters may be more influential for the facies distribution on the calcareous shelves. Studies of the Algerian shelf have demonstrated that mud supply may be one of the most important factors (Caulet, 1972). Depth, salinity and biotic factors are also important but an obvious relationship between these factors and the rate of development of specific calcareous facies is not well-determined. In general, calcareous shelves are characterized by a large variety of deposits composed of relict and palimpsest sediment types. Recent studies of the southern Spanish continental shelf have revealed a mosaic distribution pattern of the calcareous deposits. Although the factors affecting the distribution of these deposits are not well known at present, the relict nature of the most bioclastic components seems to indicate that these facies were developed under very specific environmental parameters (Zamarreño et al., 1982; Baena et al., 1982).

Slope facies and the bypassing model

Sedimentary facies of the continental slopes are composed of fine-grained muds considered to be largely homogenous. These deposits have been less investigated than the more diverse deposits of other provinces of the continental margin, but studies in the Mediterranean have demonstrated that the homogeneity of the slope muds is only apparent (Feldhausen et al., 1981; Stanley and Maldonado, 1981; Maldonado and Canals, 1982). Sedimentological analysis of the southern Balearic margin show a large

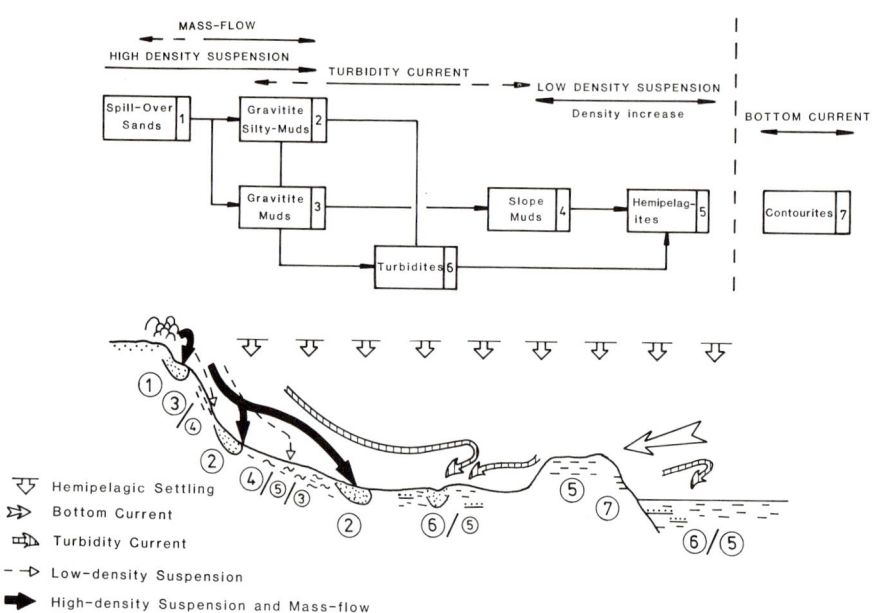

Fig. 2.23. Diagram illustrating the type of sediments and sedimentary processes on the Balearic margin during the Holocene (Modified from Maldonado and Canals, 1982).

variety of genetically-related sediment facies in the different provinces of the slope (Fig. 2.23). Every depositional province is characterized by the predominance of some type of sedimentary process that defines the specific type of deposits. These sediments form a continuous spectrum from deposits largely developed by differential settling of suspensions of diverse densities to mass-gravity flows (*see* Middleton and Hampton, 1973). In addition, bottom currents flowing parallel to the topographic contours modify the previous deposits and generate new facies types.

Three main groups of sedimentary facies are recognized as results of the interplay between these sedimentary processes: gravitites, suspensites and contourites. The gravitite sediments are developed by several types of density currents, including high and low density turbidity currents. These flows can be generated either by sporadic events, such as large mass-failure due to earthquakes, or by periodic events such as seasonal internal waves impinging over the shelf break. This facies group is the most complex and includes a host of sediment types ranging from the muddy-sands of the upper slope, developed by spill-over the shelf break, to the gravitite silty-muds of the middle slope, and to the gravitite muds of the base-of-slope (Fig. 2.23). To this group also belongs the turbidites predominantly located in the distal provinces of the continental margin. The suspensite group is composed of the hemipelagic deposits resulting from differential settling of terrigenous and biogenic particles. This group predominates on highs and areas of the margin protected from large sediment supply. Depending on the proportions of terrigenous and biogenic components, the hemipelagites are transitional to the slope muds, and to the gravitite muds with increased terrigenous influence. The third sediment group includes several sediment types developed under the influence of bottom currents. The most classical sediment types of this group are the contourites, formed by fine laminae of mud and silt deposited from deep turbid currents that flow parallel to the distal continental margin.

A good correlation also exists between sedimentation rates and the sedimentary processes developing the different sediment types. Gravitite sediments have the highest sedimentation rates ranging from more than 80 cm/10^3 yr for the gravitite silty muds to less than 20 cm/10^3 yr for the slope muds. The hemipelagites display the lowest sedimentation rates of all sediment types with ranges of 3 to 5 cm/10³ yr. In general hemipelagic sedimentation rates show a sharp decrease for the last 10,000 yr paralleling the post-glacial eustatic rising of sea level (*see* Stanley and Maldonado, 1979; Maldonado and Canals, 1982).

Additional information on the different types of processes controlling deposition on the continental slope is obtained from the analysis of high resolution seismic profiles. These profiles demonstrate the lack of correlation between the observed stratigraphic sequence and the sedimentary thickness that can be calculated by extrapolation from the sedimentation rates. Except for sectors of rapid sediment entrapment, such as off major deltas, the thickness of the sedimentary units on the slope is largely reduced in comparison to other depositional provinces of the margin and to sedimentation rates. A model illustrating the bypassing of fine-grained sediment by largely gravitative processes toward deeper sectors has been exemplified for the north-western Mediterranean sea by Stanley *et al.* (1980). Although this model emphasizes that mud accumulates at higher rates in more distal regions, the observations of the Balearic margin demonstrate higher sedimentation rates in the proximal sectors (Fig. 2.23 and 2.24).

These contradictions must be explained in terms of two different scales of factors controlling the final geometry of the depositional units in the proximal margin. Short-term sedimentary processes, which are basically related to the environmental conditions of the depositional province such as the flow regime and sediment input, regulate the sedimentary facies including the type of deposits and sedimentation rates. In contrast, the final geometry of the depositional bodies is largely controlled by long-term factors such as the structural framework, or by short-term processes that may occur very sparsely but have great environmental impact, such as large mass-sliding triggered by earthquakes. The interplay between these processes regulates the final characteristics of each type of continental margin.

Model variations for deep sea fans in the north-western Mediterranean

Studies of deep sea sedimentation have generally concentrated during the last decades on understanding the growth patterns of submarine fans, a thick sedimentary wedge often developed off areas of major sediment input at the foot of the continental slope. The classical example of submarine fan

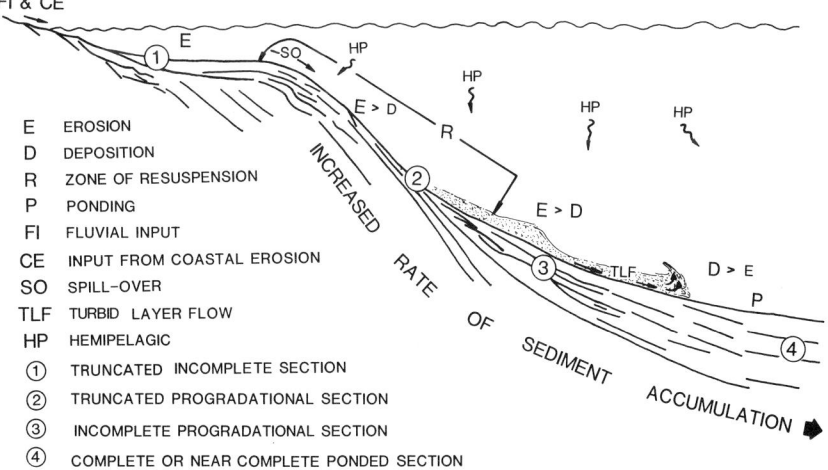

Fig. 2.24. Schematic diagram showing the bypassing model of fine-grained sediments by gravitite processes to the distal provinces of the continental margin (Modified from Stanley *et al.*, 1980).

model represents an idealized spatial and temporal distribution of deposits with clearly defined proximal to distal and central to marginal changes in morphology and depositional facies (Nelson and Kulm, 1973; Mutti, 1974; Normark, 1978; Maldonado and Stanley, 1979). A vertical transition in facies often also occurs, corresponding to either a lateral or a basinward migration of the depositional environments (Pickering, 1983).

The analysis of the four major fans in the north-western Mediterranean allow a better understanding of these depositional bodies and to question the applicability of a single fan model for the wholesale interpretation of the deep water clastic systems. Three of these examples, the Rhône, Ebro and Menorca fans, are developing by overbank flows and interchannel ponding or by fan lobes migrating in the space (Kelling et al., 1979; Aloïsi et al., 1981; Bellaiche et al., 1981). These fans grow perpendicular to the main slope trend as most examples described to date (Fig. 2.25). The Valencia Fan in contrast, is located at the northern end of a small, narrow basin between two land masses with strong morphological restrictions on its growth. This system is developed by different processes than those observed at the base of the continental slope in open basins with few morphological restrictions. The comparative analysis of these depositional systems accumulating in very contrasting settings allows a better definition of the processes and factors governing fan development. Two of these systems, the Ebro and the Valencia Fan, will be analysed in some detail.

The Ebro fan system is located on the continental rise between the lobate Ebro Delta and the Valencia Trough (Fig. 2.25). It is an intermediate-sized fan system of about 50 km length perpendicular down a steep continental margin. Studies of the seismic stratigraphy of this fan show that a series of isolated depositional lobes of different age extend the entire length of the rise from individual slope valleys (Aloïsi et al., 1981; Nelson et al., 1984). The steep continental rise restricts development of distributary channels and mid-fan lobe deposits. This is in contrast to normal fan development that exhibits a main upper fan valley evolving into distributaries and mid-fan or lower fan depositional lobe development (Normark, 1970, 1978; Nelson and Kulm, 1973).

The Ebro channelized bodies are identified by generally transparent seismic facies. After being abandoned, the lobes are blanketed by sediment characterized by numerous continuous acoustic reflectors that represent deposition cycles of overbank thin-bedded turbidite sand, turbidite mud and hemipelagic mud (Fig. 2.26A). Channel-floor and levee sectors include a high proportion of sand layers that decrease laterally across the levee flanks, except at the end of valley systems. Between the coalescing channelized bodies, the interlobe areas fill with ponded sediment bodies; these have transparent seismic character and variable core stratigraphy of mud or sand in different locations (Nelson et al., 1984). Much of the sandy sediment moves through the region of channelized sediment bodies into the deep-sea valley of the Valencia Trough and is displaced to its terminus (Fig. 2.26B). The resultant Ebro Fan geometry is a complex of active, channelized sediment body deposition at the surface, superimposed over older channelized bodies covered by overbank deposition from younger bodies, in between which may be ponded sedimentary wedges developed in lower-lying interchannel regions.

In contrast to the Ebro Fan, the Valencia Fan is not an example of a base-of-slope deposit. This fan develops at the end of the Valencia Valley and is largely detached from and independent of the continental slope (Fig. 2.25). The principal physiographic orientation of this fan is related to the morphology of the distal sector of the Valencia Trough and the junction with the Balearic Plain (Maldonado et al., 1985). The sedimentary processes and the resulting sedimentary facies, clearly identify an upper, middle and outer fan (Fig. 2.27A). The upper fan is predominantly developed by channelized sediment flows, with material gradually infilling the channel from the axis of the fan. Channel migration and overbank flows seem to play a very limited role, because of the morphological constraints which confine the upper fan to a narrow area. The middle fan is built by both channelized sediment flows and overbank flows. This is the only province of the fan where channelized sediment bodies with prominent levees are observed as in most classical modern fans. The processes active in this

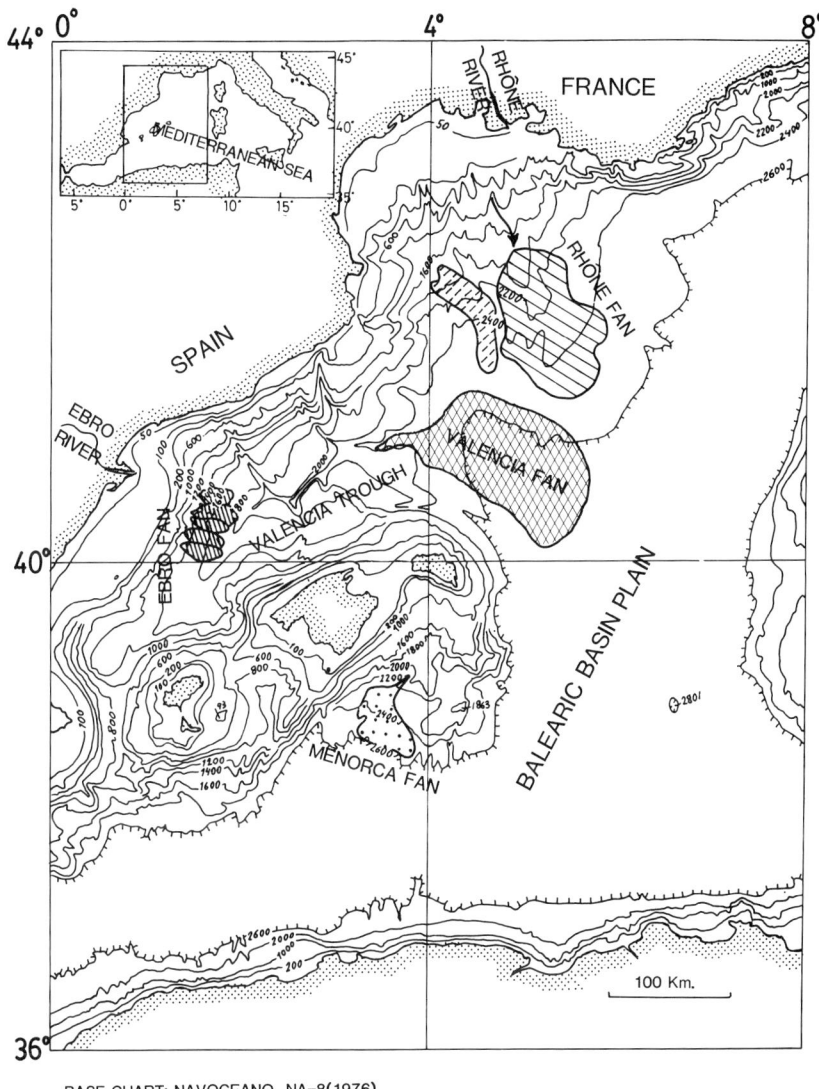

Fig. 2.25. Chart showing the location of the four most important submarine fans in the Western Mediterranean. The Ebro and the Valencia fans are discussed in the text.

province are channelized sediment flows, overbank flows and lateral channel migration. The distal fan is incised by wide, shallow, non-leveed channels with flat bottom. In this area the processes are sheet sediment flows and some low density channelized flows, but the latter are not contributing significantly to the fan growth. The sources for the development of this sector of the Valencia Fan are distal flows from the Valencia Valley, the Rhône Fan and the small fans from the north-western Spanish margin. Reworking and accumulation by bottom current activity may also influence the final distribution of distal fan sediments.

The vertical sequence of seismic units in the Valencia Fan demonstrates a general upward gradation from highly-channelized reflectors at the bottom, to poorly channelized and continuous reflectors at the

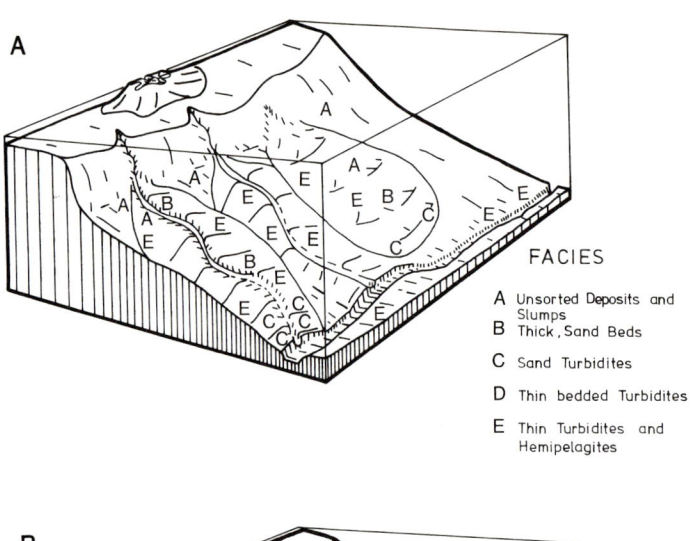

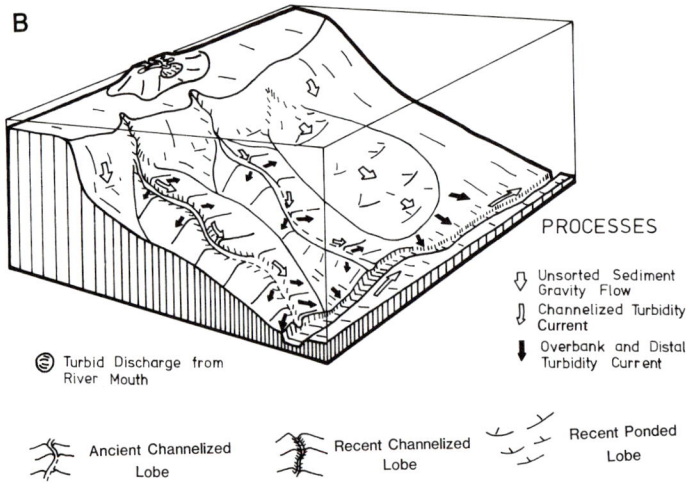

Fig. 2.26. Schematic block diagram of the Ebro deep-sea fan system showing predominant sedimentation patterns and sedimentary processes controlling the development of this system during the Late Quaternary. The different types of depositional lobes are also depicted (From Nelson et al., 1984).

top. This sequence represents a progressive shifting into the depositional site of more distal environments as the result of upslope migration of the fan depocentres. In this evolution the Valencia Valley is filled from the lower end, as the fan migrates upslope (Fig. 2.27B). This evolution is different from the other deep sea depositional systems of the north-western Mediterranean Sea. A general seaward migration of the depocenters occurs in the Rhône Fan, while the Ebro and Menorca fans are characterized by lateral migration of depositional lobes that also have slowly migrated seaward (Kelling et al., 1979; Monaco et al., 1982; Nelson et al., 1984).

Previous comparative studies of Mediterranean fans have demonstrated the importance of the interplay between several factors controlling fan development (Maldonado and Stanley, 1979). These factors include, in order of importance, the structural setting and the associated physiography of the depositional basin, the climatic-eustatic oscillations influencing sediment input and the physical

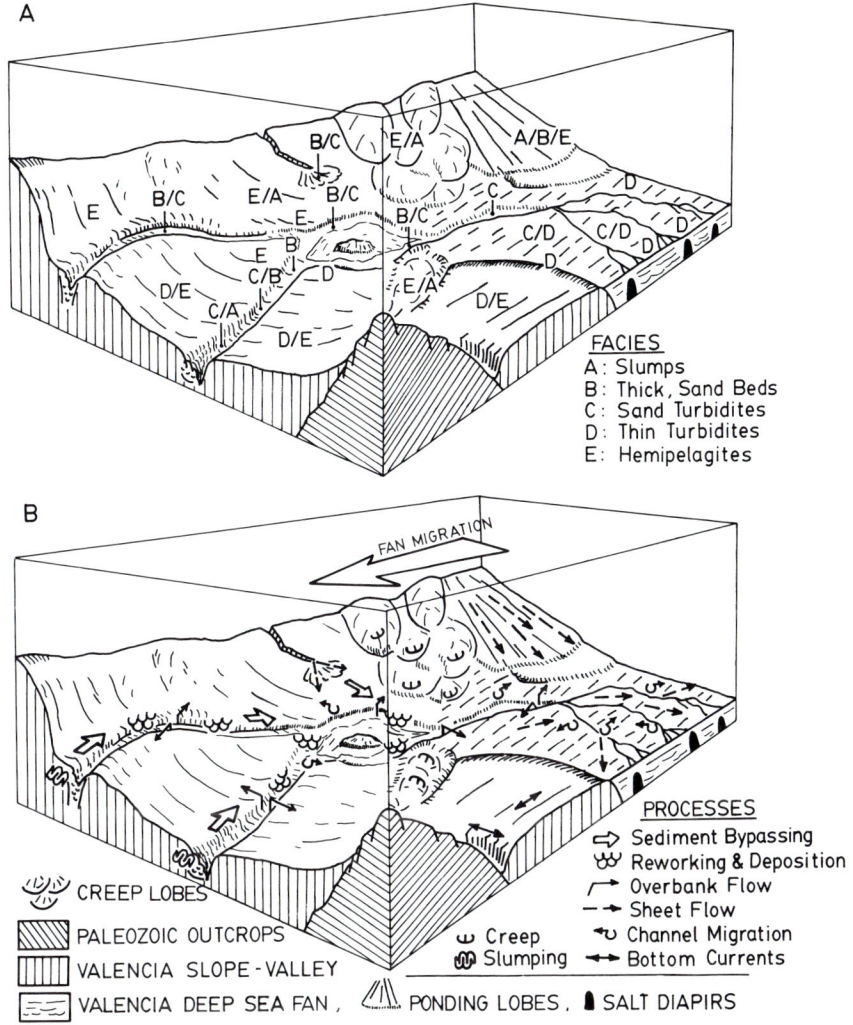

Fig. 2.27. Schematic diagrams showing sedimentary facies and processes in the Valencia Fan. Each diagram depicts the basic morphologic provinces in the area (From Maldonado et al., 1985).

oceanographic parameters (Fig. 2.28). As in other Mediterranean examples, the most significant factors for the overall evolution of submarine fans in this sector are the structural setting and the associated physiography.

CONCLUSIONS

It can be finally concluded that the evolution of the Mediterranean Sea was controlled by the interplay between a host of geological processes and environmental parameters. This quasi-enclosed system allows the analysis of the multiple relationships between the controlling factors and their products which are represented by each depositional example. The sedimentary and structural schemes deduced from this sea may be applicable to more open systems in the world oceans and may serve for the interpretation of ancient marine series in the circum-Mediterranean regions.

Much work remains, however, to be done on such important questions as: (1) the tectonic significance of the Mediterranean basins in relation to the alpine orogeny; (2) the subsiding history of the continental margins and basin plains in relation to the thermal cooling and structural evolution; (3) the nature of the crust and mechanisms of development; (4) the origin and consequences of the Messinian 'salinity crisis'; (5) the evolution of marine faunas and floras in a semi-enclosed basin isolated from parent stocks in the Tethys, Parathethys and ancient Atlantic; and (6) differences and similarities in sedimentation processes of the Western and Eastern Mediterranean basins as a function of the climatic, tectonic and oceanographic controls.

ACKNOWLEDGEMENTS

Many of the ideas summarized in this paper are the result of a sharing of experiences at sea and in the laboratory during several years of continuous work in the Mediterranean Sea. Much of this recent work was conducted in cooperation with Drs. H. Got, A. Mauffret, A. Monaco, H. Nelson, S. O'Connell, W.B.F. Ryan, D.J. Stanley and I. Zamarreño, to which I am indebted. I also thank Drs. M. Julivert and D.F. Williams for the review and improvement of the manuscript and Mrs. M.T. Solans for technical help. This contribution largely summarized the results of a grant of the Joint Spanish — USA Committee for Scientific Cooperation (Ref. 0394/17).

REFERENCES

Adams, C. G. and Ager, D. V. (1967) *Aspects of Tethyam Biogeography*. London, The Systematic Association.
Adams, C. G., Benson, R. H., Ryan, W. B. F., Kidd, R. B. and Wright, R. C. (1977) The Messinian Salinity crisis and evidence of Late Miocene eustatic changes in the world ocean. *Nature* 269 (5627), 383 – 6.

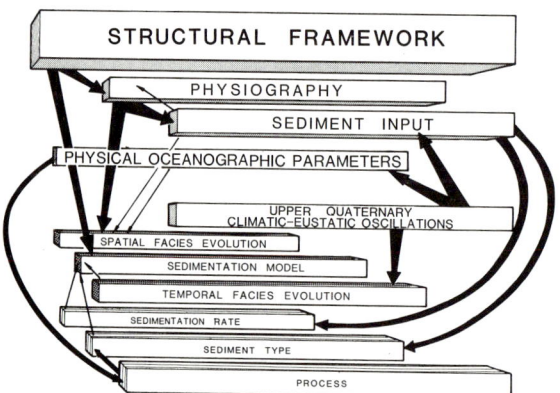

Fig. 2.28. Schematic relationship between factors controlling the development of deep-sea sedimentation models and their relative importance. The applicability of these relationships to other depositional settings in the distal continental margins of the Mediterranean Sea is proposed (Modified from Maldonado and Stanley, 1979).

Aloïsi, J. C., Monaco, A., Thommeret, J. and Thommeret, Y. (1975) Evolution paléogéographique du plateau continental languedocien dans le cadre du Golfe du Lion. Analyse comparée des données sismiques, sédimentologiques et radiométriques concernant le Quaternaire récent. *Rev. Geogr. Phys. Géol. Dyn.* V(XVII), 13–22.

Aloïsi, J. C., Monaco, A., Planchais, S., Thommeret, J. and Thommeret, Y. (1978) The Holocène transgression in the Golfe du Lion, southwestern France: paleogeographic and paleobotanical evolution. *Geogr. Phys. Quat.* XXXII(2), 145–62.

Aloïsi, J. C., Millot, C., Monaco, A. and Pauc, H. (1979) Dynamique des suspensions et mécanismes sédimentogénétiques sur le plateau continental du Golfe du Lion. *C.R. Acad. Sc. Paris* 289, 879–82.

Aloïsi, J. C., Bellaiche, G., Bouye, C., Droz, L., Got, H., Maldonado, A., Mirabile, L. and Monaco, A. (1981) L'eventail sous-marin profond du Rhône et les depôts de pente de l'Ebre: essai de comparaison morphologique et structurale. *Sedimentary basins of Mediterranean Margins.* Ed. F. Z. Wezel, pp.227–38. Tecnoprint, Bologna.

Alla, G., Byramjee, R., Didier, J. *et al.* (1972) Structure géologique de la marge continentale du Golfe du Lion. *22th C.I.E.S.M., Inter. Congress. Symp. Géodyn. Rég. Méditerr.* 22(2a, 38–40.

Alonso, B. (1981) *Microcontaminantes Inorgánicos y Procesos Sedimentarios en la Plataforma Continental de Tarragona–Castellón.* Barcelona, Univ. Barcelona, Tesis de Licenciature, 154 pages.

Alvarez, W. (1976) A former continuation of the Alps. *Geol. Soc. of America Bull.* 87, 891–6.

Alvarez, W., Cocozza, T. and Wezel, F. C. (1974) Fragmentation of the Alpine orogenic belt by microplate dispersal. *Nature* 248, 309–14.

Andrieux, J., Fontboté, J. M. and Mattauer, M. (1971) Sur un modéle explicatif de l'arc de Gibraltar. *Earth Planet. Sc. Lett.* 12, 191–8.

Araña, V. and Vegas, R. (1974) Plate tectonics and volcanism in the Gibraltar Arc. *Tectonophysics* 24, 197–212.

Argand, E. (1924) La tectonique de l'Asie. *13th Int. Geol. Congress Brussels,* 171–372.

Aubouin, H. J., Debelmas, J. and Latreille, M. (1980) Les chaînes alpines issues de la Téthys: Introduction générale. *Mémoire du B.R.G.M.* 115, 7–12.

Auzende, J. M., Bonnin, J., Olivet, J.L., Pautot, F. and Mauffret, A. (1971) Upper Miocene Salt Layer in the Western Mediterranean Basin. *Nature Phys. Sc.* 230, 82–5.

Auzende, J. M., Bonnin, J. and Olivet, J. L. (1973) The origin of Western Mediterranean basin. *J. Geol. Soc. London* 129, 607–20.

Auzende, J. M., Rehault, J. P., Pastouret, L., Szep, B. and Olivet, J. L. (1975) Les bassins sédimentaires de la mer d'Alboran. *Bull. Soc. Géol. Fr.* 17, 98–107.

Baena Perez, J., Garcia-Rodriguez, J., Maldonado, A., Zamarreño, I. *et al.* (1982) Mapa geológico de la plataforma continental española y zonas adyacentes. Almeria-Garrucha Chella-Los Genoveses. A escala 1:200.000. *Instituto Geologico y Minero de España.* Madrid, Ministerio de Industria. 105 pages.

Bally, A. W. (1975) A geodinamic scenario for hydrocarbon occurrences. *Ninth World Petrol. Congress,* Proceedings 2, 33–4.

Barberi, F., Bizouard, H., Capaldi, G., Ferrara, G., Gasparini, P., Innocenti, F., Joron, J. L., Lambert, B., Treuil, M. and Allegre, C. (1978) Age and nature of basalts from the Tyrrhenian Abyssal Plain. *Initial Reports of the Deep Sea Drilling Project* 41(1). Eds K. J. Hsü, L. Montadert *et al.,* pp. 509–14. U.S. Govt. Printing, Office, Washington D.C.

Bayer, R., Le Mouël, J. L. and Le Pichon, X (1973) Magnetic anomaly pattern in the Western Mediterranean. *Earth and Planet. Sc. Lett.* 19, 168–76.

Bellaiche, G., Droz, L., Aloïsi, J. C., Bouye, Ch., Got, H., Monaco, A., Maldonado, A. *et al.* (1981) The Ebro and the Rhône Deep-Sea Fans: First comparative study. *Marine Geology* 43, 75–85.

Benson, R. H. (1976) Miocene deep sea ostracods of the Iberian Portal and the Balearic Basin. *Marine Micropal.* 1, 249–62.

Berckhemer, H. and Hsü, K. J. (1982) Editor's Introduction Alpine Mediterranean Geodynamics Past, Present and Future. *Alpine Mediterranean Geodynamics.* Eds H. Berckhemer and K. J. Hsü, pp. 7–14. American Geophysical Union, Washington D.C.

Berggren, W. A. and Haq, B. (1976) The Andalusian Stage (Late Miocene): Biostratigraphy, Biochronology and Paleoecology. *Paleogeogr. Paleoclimatol. Paleoecol.* 26, 67–129.

Bernoulli, D. and Lemoine, M. (1980) Birth and early evolution of the Tethys: the overall situation. *Mémoires du B.R.G.M.* 115, 168–79.

Biju-Duval, B. (1974) Carte géologique et structurale des bassins tertiaires du domaine Méditerranéen: Commentaires. *Rev. Inst. Fr. Pétr.* 29, 607–39.

Biju-Duval, B., Dercourt, J. and Le Pichon, X. (1977) From the Tethys Ocean to the Mediterranean Sea: A plate tectonic model of the evolution of western Alpine system. *Structural History of the Mediterranean Basins.* Eds B. Biju-Duval and L. Montadert, pp. 143–64. Technip. Paris.

Biju-Duval, B., Letouzey, J. and Montadert, L. (1978) Structure and evolution of the Mediterranean basins. *Initial Reports of the Deep Sea Drilling Project* 42(1). Eds K. J. Hsü, L. Montadert *et al.,* pp. 951–84. U.S. Govt. Printing Office, Washington D.C.

Boccaletti, M. and Guazzone, G. (1974) Remnant arcs and marginal basins in the Cainozoic development of the Mediterranean. *Nature* 252, 18–21.

Boccaletti, M. and Manetti, M. (1978) The Tyrrhenian Sea and adjoining regions. *The Ocean Basins and Margins* 4B, pp. 149–200. Plenum Press, New York.

Boccaletti, M., Elter, P. and Guazzone, G. (1971) Plate tectonic models for the development of the Western Alps and Northern Apennines. *Nature Phys. Sci.* 234(6), 108–111.

Bott, M. H. P. (1976) Sedimentary basins of continental margins and cratons. *Symposium on Sedimentary Basins of Continental Margins and Cratons. Technophysics* 36, 313.

Bourrouilh, R. and Gorsline, D. S. (1979) Pre-Triassic fit and Alpine tectonics of continental blocks in the western Mediterranean. *Geol. Soc. Am. Bull.* 90, 1074 – 83.

Bousquet, J. C. and Philip, M. (1976) Observations microtectoniques sur la compression nord-sud quaternaire des Cordillères Bétiques orientales (Espagne méridionale, Arc de Gibraltar). *Bull. Soc. Géol. France* XXVIII(7), 711 – 24.

Bousquet, J. C., Montenat, C. and Philip, H. (1976) La néotectonique des cordillères bétiques orientales. *Reunión sobra la Geodinámica de la Cordillera Bética y Mar de Alborán*, pp. 59 – 78. Universidad de Granada, Granada.

Bullard, E. C., Everett, J. E. and Smith, A. G. (1965) The fit of the continents around the Atlantic. *Phil. Trans. R. Soc. Lond. A* 258, 41 – 51.

Cagnetti, V., Pasquale, V. and Polinari, S. (1978) Faultplane solutions and stress regime in Italy and adjacent regions. *Tectonophysics* 46, 239 – 50.

Canals, M., Maldonado, A., Mangini, A. and Douglas, F. (1983) Paleoenvironmental correlation between Eastern and Western Mediterranean Basins during Late Quaternary. *XXVIIIth Congress Assem.Plén. C.I.E.S.M. Repp.Comm.int.Mer Médit., 28(4)*, 39 – 40 Monaco.

Carey, S. W. (1958) A tectonic approach to continental drift. *Symposium Continental Drift.* Ed. S. W. Carey, pp. 177 – 355. Tasmania University, Hobart.

Caulet, J. P. (1972) Recent biogenic calcareous sedimentation on the Algerian continental shelf. *The Mediterranean Sea. A Natural Sedimentation Laboratory.* Ed. D. J. Stanley, pp. 261 – 92. Dowden, Hutchinson and Ross, Stroudsburg, Pennsylvania.

Channell, J. E. T., D'Argenio, B. and Horvath, F. (1979) Adria, the African promontory, in Mesozoic Mediterranean paleogeography. *Earth-Science Rev.* 15(3), 215 – 92.

Channell, J. E. T., Catalano, R. and D'Argenio, B. (1980) Paleomagnetism and deformation of the Mesozoic continental margin in Sicily. *Tectonophysics* 61, 391 – 407.

Chevalier, G. (1961) Recherches sur les Madréporaires et les formations récifales miocènes de la Mediterranée occidentale. *Me. Soc. Géol. France* 93, 1 – 562.

Cita, M. B. (1973) Mediterranean evaporite: paleontological arguments for a deep-basin dessication model. *Messinian Events in the Mediterranean.* Ed. C. W. Drooger, pp. 206 – 28. North-Holland, Amsterdam.

Cita, M. B. (1982) The Messinian salinity crisis in the Mediterranean: a review. *Alpine-Mediterranean Geodynamics, Geodynamic Series 7.* Eds H. Berckhemer, K. J. Hsü, pp. 113 – 40. American Geophysical Union, Washington D.C.

Cita, M. B. and Ryan, W. B. F. (1973) Time scale and general Synthesis. *Initial Reports of the Deep Sea Drilling Project.* 13. Edited by W. B. F. Ryan, K. J. Hsü *et al.*, pages 1405 – 1416. U.S. Boevr. Printing Office, Washington D.C.

Cita, M. B., Vergnaud-Grazzini, C., Robert, Ch., Chamley, H., Ciaranfi, N. and D'Onofrio, S. (1977) Paleoclimatic record of long deep sea Core from the Eastern Mediterranean. *Quaternary Research* 8, 205 – 35.

Cita, M. B., Wright, R. C., Ryan, W. B. F., Longinelli, A. (1978) Messinian Paleoenvironments. *Initial Reports of the Deep Sea Drilling Project* 42(1). Eds K. J. Hsü, L. Montadert *et al.*, pp. 1003 – 35. U.S. Govt. Printing Office, Washington D.C.

Cohen, C. R. (1980) Plate tectonic model for the Oligo-Miocene evolution of the western Mediterranean. *Tectonophysics* 68, 283 – 311.

Continental Margin Panel (1979) Continental Margins: Geological and Geophysical Research Needs and Problems. *Nat. Acad. of Sci.*, Washington D.C. 302 pages.

Curray, J. R. and Moore, D. G. (1974) Sedimentary and Tectonic Processes in the Bengal Deep-Sea fan and Geosyncline. *The Geology of Continental Margins.* Eds C. A. Burke and C. L. Drake, pp. 617 – 27. Springer Verlag, New York.

D'Argenio, B., Horvath, F. and Channell, J. E. T. (1980) Paleotectonic evolution of Adria, African promontory. *Mémoire du B.R.G.M.* 115, 331 – 51.

Dewey, J. F., Pitman, W. C., Ryan, W. B. F. and Bonnin, J. (1973) Plate tectonics and the evolution of the Alpine system. *Bull. Geol. Soc. Am.* 84, 3137 – 80.

Di Girolamo, P. (1978) Geotectonic settings of Miocene-Quaternary volcanism in and around the Eastern Tyrrhenian sea border Italy, as deduced from major element geochemistry. *Bull. Volcanology* 41(3), 1 – 22.

Dillon, W. P., Robb, J. M., Greene, H. G. and Lucena, J. C. (1980) Evolution of the continental margin of southern Spain and the Alboran Sea. *Marine Geology* 36, 205 – 26.

Drooger, C. W. (1973) *Messinian Events in the Mediterranean*, North Holland, Amsterdam, 272 pages.

Durand-Delga, M. and Fontboté, J. M. (1980) Le cadre structurale de la Méditerranée occidentale. *Mémoire du B.R.G.M.* 115, 67 – 85.

Eastward Group (1981) Infra Messinian outcroping on the continental margin of Balearic Island. *C.I.E.S.M. Rapports et proces-Verbaux des Reunions* 27(8), pp. 47. Géologie et Géophysique Marines, Monaco.

Edel, J. B. (1980) *Etude paléomagnétique en Sardaigne. Conséquences pour la géodynamique de la Méditerranée occidentale*, Strasbourg. Thése d'Etat, 310 pages.

Esteban, M. (1980) Significance of the Upper Miocene Coral Reefs of the Western Mediterranean. *Palaeogeography. Palaeoclimatology, Palaeocology* 29, 169 – 88.

Fabbri, A., Gallignani, P., Gennesseaux, M. and Rehault, J. P. (1980) Structure superficielle et profonde de la zone de la faille centrale (mer Tyrrhénienne). *XXVIIth C.I.E.S.M. Inter. Congress, Cagliari*, 27, 147 – 8.

Feldhausen, P. H., Stanley, D. J., Knight, R. J. and Maldonado, A. (1981) Homogenization of gravity-emplaced muds and unifites: models from the Helenic Trench. *Sedimentary basins of Mediterranean Margins*, Ed. F. C. Wezel, pp. 203–26. Tecnoprint, Bologna.

Finetti, I. and Morelli, C. (1973) Geophysical exploration of the Mediterranean Sea. *Boll. Geofis. Teor. Appl.* 15(60), 263–342.

Fischer, A. G. and Arthur, M. A. (1977) Secular variations in the pelagic realm. *SEPM Special Publication. Deep-Water Carbonate Environments* 25, pp. 19–50. Soc. Economic Paleontologists and Mineralogists, Tulsa, Oklahoma.

Galdeano, A. and Rossignol, J.-C. (1977) Assemblage a attitude constante des cartes d'anomalies magnetiques couvrant l'ensemble du basin occidental de la Mediterranean. *Bull. Soc. Geol. Fr.* XIX(3), 461–8.

Gorler, K. and Giese, P. (1978) Aspects of the evolution of the Calabrian arc in Alps, Apennines, Hellenides. *Schweizerbardt.* Eds H. Closs, D. Roeder, K. Schmidt, pp. 374–88. Stuttgart.

Haccard, D., Lorenz, C. and Grandjacquet, C. (1972) Essai sur l'evolution tectogénétique de la liaison Alpes-Apennins (de la Ligurie à la Calabre). *Mém. Soc. Geol. Ital.* 2, 309–41.

Hallam, A. (1973) *The Atlas of Paleobiogeography*. Elsevier, Amsterdam.

Hinz, K. (1972) Results of seismic refraction investigations (Project Anna) in the western Mediterranean Sea, South and North of the Island of Mallorca. *Bull. Centre Rech. Pau SNPA* 6. Eds O. Leenhardt et al., pp. 405–26.

Horvath, F. and Berckhemer, H. (1982) Mediterranean backarc basins. *Alpine Mediterranean Geodynamics* 7. Eds H. Berckhemer, K. J. Hsü, pp. 141–73. American Geophysical Union, Washington D.C.

Hsü, K. J. and Bernoulli, D. (1978) Genesis of the Tethys and the Mediterranean. *Initial Reports of the Deep Sea Drilling Project* 42(1). Eds K. J. Hsü, L. Montadert et al., pp. 943–9. U.S. Govt. Printing Office, Washington D.C.

Hsü, K. J., Cita, M. B. and Ryan, W. B. F. (1973) Origin of the Mediterranean Evaporites. *Initial Reports of the Deep Sea Drilling Project* 42(1). Eds K. J. Hsü, L. Montadert et al., pp. 1203–31. U.S. Govt. Printing Office, Washington D.C.

Hsü, K. J., Montadert, L., Bernoulli, D., et al. (1977) History of the Mediterranean salinity crisis. *Nature* 267, 399–403.

Hsü, K. J., Montadert, L. et al. (1978) *Initial Reports of the Deep Sea Drilling Project* 42(1). 1248 pages. U.S. Govt. Printing Office, Washington D.C.

Hsü, K. J., Montadert, L., Bernoulli, D., Cita, M. B., Arickson, A., Carrison, R. E., Kidd, R. B., Melieres, F., Muller, C. and Wright, R. (1978) History of the Mediterranean salinity crisis. *Initial Reports of the Deep Sea Drilling Project* 42(1). Eds K. J. Hsü, L. Montadert et al., pp. 1053–78. U.S. Govt. Printing Office, Washington D.C.

Huang, T. C. and Stanley, D. J. (1972) Western Alboran Sea: Sediment dispersal ponding and reversal of currents. *The Mediterranean Sea: A Natural Sedimentation Laboratory*. Ed. D. J. Stanley, pp. 521–59. Dowden, Hutchinson and Ross Inc., Stroudsburg, Pennsylvania.

Inman, D. L. and Nordstrom, C. E. (1971) On the tectonic and morphologic classification of Coasts. *J. Geology* 74, 1–21.

Julivert, M., Fontboté, J. M., Ribero, A. and Conde, L. (1974) Mapa Tectónico de la Peninsula Ibérica y Baleares. Escala 1:100:000. *Inst. Geol. Min. España*. Ministerio de Industria, Madrid. 113 pages.

Kelling, G., Maldonado, A. and Stanley, D. J. (1979) Salt tectonics and basement fractures: Key controls of recent sediment distribution on the Balearic Rise, western Mediterranean Santhsonian. *Smithsonian Contribution Marine Sciences* 3, 1–52.

Kennett, J. P. (1967) Recognition and Correlation of the Kapitean Stage (Upper Miocene), New Zealand. *New Zealand Journ. Geol. Geophys.* 70(4), 1051–1603.

Laubscher, H. P. and Bernoulli, D. (1977) Mediterranean and Tethys. *The Ocean Basins and Margins,* 4 Eds Nairn, Kanes and Stehli, pp. 1–28. Plenum Publ. Corp., New York.

Le Douaran, S., Burrus, J. and Avedik, F. (1983) Le structure profonde d'une partie du bassin de Méditerranée nord-occidentale: Les resultats de la campagne de sismique a 2 Navires Croc 2. *Congress Assem.Plén. C.I.E.S.M. Repp.Comm.int. Mer Méd.,* 28(4) 71–72 Monaco.

Lefebvre, D. and Gennesseaux, M. (1981) Tectonics and Sedimentation in the Gulf of Lions (Northwestern Mediterranean). *Marine Geology*.

Le Pichon, X. and Angelier, J. (1979) The Hellenic Arc and Trench System: a key to the neotectonic evolution of the Eastern Mediterranean area. *Tectonophysics.* 60, 1–42.

Loutit, T. S. and Kennett, J. P. (1979) Application of carbon isotope stratigraphy to Late Miocene shallow marine sediments, New Zealand. *Science* 204, 1196–9.

Maldonado, A. (1975) Sedimentation, stratigraphy and development of the Ebro Delta, Spain. *Delta Models for Exploration*. ed. M. L. Broussard, pp. 311-338. Houston Geological Society, Houston, Texas.

Maldonado, A. (1978) El estancamientos de las aguas del Mar Mediterraneo. *Investigación y Ciencia.* 23, 32–44.

Maldonado, A. and Canals, M. (1982) El margen continental Sur-Balear: un modelo deposicional reciente sobre un margen de tipo pasivo. *Acta Geol. Hispanica* 17, 241–254.

Maldonado, A. and Riba, O. (1974) Les rapports sédimentaires du Néogène et du Quaternaire dans le plateau continental aux environs du delta de l'Ebre. (Espagne). *Mem. Inst. Geol. Bassin Aquitaine.* 7, 321–9.

Maldonado, A. and Stanley, D. J. (1976) The Nile Cone: submarine fan development by cyclic sedimentation, *Marine Geology* 20, 27–40.

Maldonado, A. and Stanley, D. J. (1979) Depositional patterns and Late Quaternary Evolution of two Mediterranean Submarine fans: a comparison. *Marine Geology* 31, 215–50.

Maldonado, A., Swift, D. J. P., Young, R. A., Han, G., Nittrouer, C., Demaster, D., Rey, J., Palomo, C., Acosta, J., Ballester, A. and Castellvi, J. (1983) Sedimentation on the Valencia Continental shelf: preliminary results. *Continental Shelf Research*, 2, 195–211.

Maldonado, A., Got, H. Monaco, A., O'Connell, S. and Mirabile, L. (1985) The Valencia fan (Northwestern Mediterranean): a variation on a model for distal deposition. *Marine Geology* 62, 295–319.

Malinverno, A., Cafiero, M., Ryan, W. B. F. and Cita, M. B. (1981) Distribution of messinian sediments and erosional surfaces beneath the Tyrrhenian Sea: geodinamic implications. *Oceanologia Acta.* 4(4), 489 – 96.

Mauffret, A. (1979) Etude Géodynamique de la marge del Iles Baléares. *Mém. Soc. Geol. Fr.* LVI(132), 1 – 96.

Mauffret, A., Fail, J. P., Montadert, L., Sancho, J. and Winnock, E. (1973) Northwestern Mediterranean sedimentary basin from seismic reflection profile. *Bull. Am. Assoc. Petrol. Geol.* 57(11), 2245 – 62.

Mauffret, A., Labarbarie, M. and Montadert, L. (1982) Les affleurements de series sedimentaires pré-pliocènes dans le bassin Mediterranéen Nord-Occidental. *Marine Geology* 45, 159 – 75.

McCave, I. N. (1972) Transport and escape of fine-grained sediment from shelf areas. *Shelf Sediment Transport and Patterns*, Eds D. J. P. Swift, P. B. Duame, and O. H. Pilkey, pp. 225 – 78. Dowden, Hutchinson and Ross, Stroudsburg, Pennsylvania.

Middleton, G. V. and Hampton, M. A. (1973) Sediment gravity flows: mechanics of flow and deposition. *Turbidities and Deep-Water Sedimentation.* Eds G. V. Middleton and A. H. Bouma, pp. 1 – 38. Pacific Soc. Econ. paleont. Miner. Section Short Course, Anaheim, California.

Monaco, A., Aloïsi, J. C., Bouye, C., Got, H., Mear, Y., Bellaiche, G., Droz, L., Mirabile, L., Mattielo, L., Maldonado, A., Le Calvez, Y., Chassefiere, B. and Nelson, H. (1982) Essai de reconstitution des mecanismes d'alimentation des eventails sedimentaires profonds de l'Ebre et du Rhône (Méditerranée Occidentale). *Bull. Inst. Géol. Bassin Aquitaine.* (in press).

Montadert, L., Letouzey, J. and Mauffret, A. (1978) Messinian event: seismic evidence. *Initial Reports of the Deep Sea Drilling Project.* 42. Eds K. J. Hsü, L. Montadert *et al.*, pp. 1037 – 50. U.S. Govt. Printing Office, Washington D.C.

Morelli, C. (1975) Geophysics of the Mediterranean. *New Sl. Coop. Invest. Med., Spec. Issue* 7, 27 – 111.

Mutti, E. (1974) Examples of ancient deep-sea fan deposits from circum-Mediterranean geosynclines. *Modern and Ancient Geosynclinal Sedimentation* 19. Eds R. H. Dott and R. H. Shaver, pp. 92 – 105. Soc. Econ. Paleontol. Mineral, Tulsa, Oklahoma.

Nairn, A. E. and Westphal, M. (1968) Possible implications of the paleomagnetic study of Late Paleozoic igneous rocks of Northwestern Corsica. *Paleogegr. Paleoclimat., Paleoecol.* 5, 179 – 204.

Neev, D., Greenfield, L. and Hall, J. K. (1982) Slice Tectonics in the Levantine Basin. *Marine Geology and Geomathematics MGG/2/82*, pp. 1 – 29. Ministry of Energy and Infrastructure. Geological Survey of Israel, Israel.

Nelson, H. and Kulm, L. D. (1973) Submarine fans and channels. *Turbidites and Deep Water Sedimentation*, Eds G. V. Middleton and A. H. Bouma, pp. 37 – 78. Soc. Econ. Paleontol. Mineral. Pacific Sections Short Course, Anaheim, California.

Nelson, H. and Nilsen, T. H. (1974) Depositional trends of modern and ancient deep-sea fans. *Modern and Ancient Geosynclinal Sedimentation* 19, Eds N. Dott and R. H. Shaves, pp. 69 – 91.

Nelson, C. H., Maldonado, A., Coumes, F., Got, H. and Monaco, A. (1984) The Ebro deep-sea fan system. *Geo-Mar. Letters*, 3, 125 – 131.

Normark, W. R. (1970) Growth patterns of deep-sea fans. *Bull. Am. Assoc. Pet. Geol.* 54, 2170 – 95.

Normark, W. R. (1978) Fan valleys, channels, and depositional lobes of modern submarine fans: Characters for recognition of sandy turbidite environments. *American Assoc. Petr. Geol. Bull.* 62, 912 – 31.

Normark, W. R. and Piper, D. J. W. (1969) Deep-sea fan-valleys, past and present. *Geol. Soc. Am. Bull*, 80, 1859 – 66.

Pannekoek, A. J. (1969) Uplift and subsidence in and around the western Mediterranean since the Oligocene: a review. *Verhand. Kin. Ned. Geol. Mijnb. Gen.* XXVI, 53 pages.

Pickering, K. T. (1983) The shape of deep water siliciclastic systems. A discussion. *Geo-Marine Letters* 2, 41 – 46.

Pitman, W. C. and Talwani, M. (1972) Sea-floor spreading in the North Atlantic. *Geol. Soc. Am. Bull.* 83, 619 – 46.

Rehault, J. P. (1981) *Evolution Tectonique et Sedimentaire di Bassin Ligure (Méditerranée Occidentale).* Paris. Université Pierre et Marie Curie. Thèse de Doctoral d'Etat. 132 pages.

Rogl, F., Steininger, F. F. and Muller, C. (1978) Middle Miocene salinity crisis and paleogeography of the Paratethys (Middle and Eastern Europe). *Initial Reports of the Deep Sea Drilling Project.* 42(1). Eds K. J. Hsü, L. Montadert *et al.*, pp. 985 – 99. U.S. Govt. Printing Office, Washington D.C.

Rupke, N. A. and Stanley, D. J. (1974) Distinctive properties of Turbiditic and Hemipelagic Mud Layers in the Algéro-Balearic Basin, Western Mediterranean sea. *Smithsonian Contributions to the Earth Sciences* 13, 40 pages.

Ryan, W. B. F. (1972) Stratigraphy of Late Quaternary sediments in the Eastern Mediterranean. *The Mediterranean Sea: A Natural Sedimentation Laboratory.* Ed. D. J. Stanley, pp. 149 – 69. Dowden, Hutchinson and Ross Inc., Stroudsburg, Pennsylvania.

Ryan, W. B. F. (1973) Geodynamic implications of the Messinian crisis of salinity. *Messinian Events in the Mediterranean.* Ed. C.W. Drooger, pp. 26 – 38. North-Holland, Amsterdam.

Ryan, W. B. F. (1976) Quantitative evaluation of the depth of the western Mediterranean before, during and after the Late Miocene salinity crisis. *Sedimentology* 23, 791 – 813.

Ryan, W. B. F. and Cita, M. B. (1978) The nature and distribution of Messinian erosional surfaces. Indicators of a several kilometre deep Mediterranean in the Miocene. *Marine Geology* 27, 193 – 230.

Ryan, W. B. F., Hsü, K. J. *et al.* (1973) *Initial Reports of the Deep Sea Drilling Project* 13(2), U.S. Govt. Printing Office, Washington D.C. 1447.

Ryan, W. B. F., Stanley, D. J. *et al.* (1970). The tectonics and geology of the Mediterranean Sea. *The Sea.* Edited by A. E. Maxwell, V. 4(2), pp.387 – 492, Wiley-Interscience, New York.

Ryan, W. B. F., Cita, M. B., Dreyfus Rawson, M., Burckle, L. H. and Saito, T. (1974) A paleomagnetic assignment of Neogene stage boundaries and the development of isochronous datus planes between the Mediterranean, the Pacific and

the Indian Oceans in order to investigate the response of the World Oceans to the Mediterranean "salinity crisis". *Riv. Ital. Paleontol. Stratigr.* 80, 631–88.
Schreiber, Ch. and Decima, A. (1976) Sedimentary facies produced under evaporitic environments a review. *Mem. Soc. Geol. It.* 16, 111–26.
Sclater, J. G. and Tapscott, C. (1979) The History of the Atlantic. *Scientific America* 240, 156–74.
Sclater, J. G., Hellinger, S. and Tapscott, C. (1977) The Paleobathymetry of the Atlantic Ocean from the Jurassic to the Present. *J. of Geol.* 85, 509–52.
Selli, R. (1974) Apunti sulla geologia del Mar Tirreno. *Rend. Semi. Fac. Sci. Univ. Cagliari* 43, 327–51.
Selli, R. and Fabri, M. (1971) Tyrrhenian, a Pliocene deep sea. *R.C. Accad. Lincei.* 8, 104–16.
Selli, R., Lucchini, F., Rossi, P. L., Savelli, C. and Del Monte, M. (1979) Geology and petrochemistry of the central Tyrrhenian volcanoes. *C.I.E.S.M. Rapports et procès-verbaux des Reunions, 25/26,* pp. 61–2.
Serra, J., Maldonado, A. and Riba, O. (1979) Caracterización del margen continental de Cataluña y Baleares. *Acta Geol. Hispanica.* 14, 494–504.
Shackleton, N. J. and Kennett, J. P. (1975) Paleotemperature history of the Cenozoic and the initiation of Antarctic glaciation: Oxygen and Carbon isotope analysis in DSDP Sites 277, 279 and 281. *Initial Reports of the Deep Sea Drilling Project. 29.* Eds J. P. Kennett, R. E. Houtz et al., 743 pages. U.S. Govt. Printing Office, Washington D.C.
Shackleton, N. J. and Cita, M. B. (1979) Oxygen and Carbon Isotope stratigraphy of benthonic foraminifera in Site 397: fine-structure of climatic change in the Late Neogene. *Initial Reports of the Deep Sea Drilling Project. 47A.* Eds W.B.F. Ryan, U. von Rad et al., pp. 433–46. U.S. Govt. Printing Office, Washington D.C.
Shepard, F. P. (1973) *Submarine Geology.* 517 pages. Harper and Row, New York.
Sibuet, J. C., Ryan, W. B. F., et al. (1979) *Initial Reports of the Deep Sea Drilling Project.* 47 (2). U.S. Govt. Printing Office, Washington D.C. 787.
Smith, A. G. (1971) Alpine deformation and the oceanic areas of the Tethys. Mediterranean and Atlantic. *Geol. Soc. Amer. Bull.* 82, 2039–70.
Smith, A. G. and Woodcock, N. H. (1982) Tectonic syntheses of the Alpine-Mediterranean region: A review. *Alpine-Mediterranean Geodynamics.* Eds H. Berckhemer and K. J. Hsü, pp. 15–38. American Geophysical Union, Washington D.C.
Sonnenfeld, P. (1974) The Upper Miocene evaporite basins in the Mediterranean Region: A study in paleoceanography. *Geol. Rdsch.* 63, 1133–72.
Stanley, D. J. (1972) *The Mediterranean Sea: A Natural Sedimentation Laboratory.* 765 pages. Dowden, Hutchinson and Ross, Stroudsburg, Pennsylvania.
Stanley, D. J. (1975) Messinian Events in the Mediterranean. *Marine Geology* 18, 339–42.
Stanley, D. J. (1977) Post-miocene depositional patterns and structural displacement in the Mediterranean. *The Ocean Basin and Margins.* Eds A. E. M. Nairn, W. H. Kanes and F. G. Stehli, pp. 77–130. The Eastern Mediterranean V. 4A. Plenum Press, New York.
Stanley, D. J. and Maldonado, A. (1979) Levantine Sea-Nile Cone lithostratigraphic evolution: quantitative analysis and correlation with paleoclimatic and eustatic oscillations in the Late Quaternary. *Sedimentary Geology* 23, 37–65.
Stanley, D. J. and Maldonado, A. (1981) Depositional models for fine-grained sediment in the western Hellenic Trench. Eastern Mediterranean. *Sedimentology* 28, 273–90.
Stanley, D. J., Maldonado, A. and Stuckenrath, R. (1975) Strait of Sicily depositional rates and patterns, and possible reversal of currents in the Late Quaternary. *Palaeogeography, Palaeoclimatology, Palaeoecology* 18, 279–91.
Stanley, D. J., Got, H., Kenyon, N. H., Monaco, A. and Weiler, Y. (1976) Catalonian Eastern Betic and Balearic margins: structural types and geologically recent foundering of the Western Mediterranean basin. *Smithsonian Contribution to the Earth Sciences,* 20, 1–67.
Stanley, D. J., Rehault, J. P. and Stuckenrath, R. (1980) Turbid-layer bypassing model: the Corsican Trough, Northwestern Mediterranean. *Marine Geology* 37, 19–40.
Stoeckinger, W. T. (1976) Valencian Gulf offers dead line nears. *Oil Gas Journal,* 197–204.
Suess, A. (1901) *Das antlitz der Erde.* Ed. F. Temsky. Vienna, 508.
Swift, D. J. P. (1976) Continental shelf Sedimentation. *Marine Sediment Transport and Environmental Management.* Eds D. J. Stanley and D. J. P. Swift, pp. 311–50. John Wiley and Sons, Inc., New York.
Thierstein, H. R. and Berger, W. H. (1979) Injection events in ocean history. *Nature* 276, 461–6.
Toksoz, M. N. and Bird, F. (1977) Formation and evolution of marginal basins and continental plateaus. *Island arcs, deep sea trenches and backarc basins.* Eds M. Talwani and W. C. Pitman III, pp. 379–93. Am. Geophys. Union, M. Erwing.
Vail, P. R., Mitchum. R. M. Jr. and Thompson, S. III (1977) Seismic stratigraphy and global changes of sea level. Part 4: Global cycles of relative changes of sea level. *Applications to Hydrocarbon Exploration.* Ed. C. E. Payton, pp. 83–97. *Am. Assoc. Petr. Geol. Mem.*
Williams, D. F., Thunell, R. C. and Kennett, J. P. (1978) Periodic freshwater and stagnation of the Eastern mediterranean Sea during the Late Quaternary. *Science* 201, 252–4.
Zamarreño, I., Maldonado, A., Canals, M., Diaz, I., Farran, M. and Vazquez, A. (1982) Temperate carbonate sedimentation on the continental shelf of Southern Spain (Western Mediterranean Sea). *XIth Inter. Congress on Sedimentology,* p. 95. McMaster University, Hamilton, Ontario, Canada.

CHAPTER 3

The Driving Machine

JORDI FLOS

Department of Ecology, Faculty of Biology, University of Barcelona, Spain

CONTENTS

3.1. Water, a fluid of unique properties	60
3.1.1. Density of liquid water	61
3.1.2. Compressibility	62
3.1.3. Transport processes in water	64
3.2. The energy from the sun	66
3.2.1. The flux of solar radiation	66
3.2.2. The attenuation of solar radiation in the sea	67
3.3. Thermal change and thermal structure	71
3.3.1. Introduction	71
3.3.2. Turbulent diffusion and models	74
3.3.3. Balance of radiant energy and heat storage	78
3.3.4. Fluxes and the mixed-layer model	79
3.3.5. Geostrophic and wind-driven currents	81
3.3.6. Seasonal vertical mixing in the Western Mediterranean	85
3.4. The weather in the Western Mediterranean	87
3.4.1. Topographical features	88
3.4.2. Large-scale weather systems	89
3.4.3. Western Mediterranean weather regimes	90
3.4.4. Weather and primary production	95

3.1. WATER, A FLUID OF UNIQUE PROPERTIES

As a first approximation, water is a simple molecule, especially when seen 'drawn' or 'written' on a sheet of paper. By analogy with other similar molecules (CH_4, NH_3, HCl, HF) we could infer some of its macroscopic, physical and chemical properties. Without a sophisticated model most of our conclusions on the properties of a set of water molecules would be wrong. This is the reason why it has often been said that the physical and chemical properties of water are surprising and life, which is strongly associated with them, has been considered as an improbable event dependent on a molecular peculiarity. Nowadays, life is more often seen as an unavoidable result of the evolution of matter in our Universe, and as a phenomenon strongly linked to water in all space and time scales — from the molecular and subcellular levels to the ecological and evolutionary ones.

The unusual properties of pure water result from the fact that its molecule has two free couples of electrons with two negative charges at these sites and two positive charges centred in the protons of the hydrogen nuclei. This structure allows the formation of hydrogen bonds and gives a dipole character to the molecule.

Water is able to build polymers of different sizes. Although water polymers $(H_2O)_n$ are very short-lived ($10^{-10} - 10^{-21}$s), the statistical distribution of the number of polymerization is related to temperature and pressure. The considerable surface tension, viscosity, specific heat, latent heat of vaporization and its melting and boiling points, as compared to related chemical compounds, are explained by the great associative power and polymerization of pure water (the boiling point of NH_3, a similar asymmetric compound with one free couple of electrons is $-33°C$).

The fact that ice is less dense than liquid water is one of the most widely known properties of water and is also related to its high associative power. At 4°C, two opposing tendencies compensate each other: thermal expansion by increasing vibratory motion with increasing temperatures and organization (and expansion) to crystallize when thermic agitation is weak enough (with decreasing temperature).

The presence of dissolved salts and other substances interferes with all these properties, as those are related to the possibility of building hydrogen bonds and to the dipole structure of the water molecule.

3.1.1. Density of liquid water

One of the most variable properties is density. It depends on temperature and salinity as well as pressure (water being slightly compressible). The different behaviour of sea-water density compared with pure or fresh water is responsible for the different behaviour of sea, compared with lakes, especially in the vertical thermal structure and processes of stratification and overturning. With increasing salinity, the temperature of the density maximum of sea-water decreases; at salinities of over 17 ppt (parts per thousand) it drops below the freezing point of pure water. Sea-water also shows a depression of its freezing point, as compared with pure water. Yet because the temperature of the freezing point decreases at a slower rate than the temperature of the density maximum with increasing salinity, at a salinity of S = 24.7 ppt both temperatures are equal at $-1.33°C$, the density of this water being ϱ = 1.01985. This is important for seas at high latitudes with a more or less important inflow of fresh water. In fresh water lakes, or in seas with a salinity less than 24.7 ppt (Baltic Sea), cooling of the whole body of water proceeds in winter until the temperature of maximum density is reached (4°C for fresh water, less for saline water). Cooling of surface layers occur, but these do not sink as they become lighter than subsurface water. In this way, an ice sheet is formed at the top of a body of 'warmer' water. In lakes at high latitudes or altitudes, a clear stratification with a negative vertical gradient of temperature can develop in winter (temperature increases from top, 0°C, to bottom, 4°C or less). In waters of salinities higher than 24.7 ppt, vertical convection continues with cooling until the entire body of water has reached the temperature of the freezing point. Under the ice, vertical convection continues as a consequence of the density increase with increasing salinity (the freezing of sea-water segregates salt, the ice being less saline than the surrounding water from which it has formed).

Although for a number of problems the density of fresh water as well as that of sea-water can be taken as being 1 g cm^{-3}, the slight differences in density found in natural water bodies are responsible for a great number of important processes taking place at sea as well as in lakes.

In oceanography $\sigma_t = (\varrho_t - 1) \times 1,000$ is often used instead of ϱ. The reason for using σ is that the differences in density relevant to oceanography affect usually the third and higher order decimals in ϱ. In the Mediterranean, during the period of stratification in summer, when the density differences between surface and deeper levels are greater, this difference is of the order of $2-3 \times 10^{-3}$ g cm^{-3}. For example: with T = 22.56°C and S = 37.85 ppt at the surface, we get ϱ = 1.02624 or σ_t = 26.24; at 100 m depth in the same vertical T = 13.27°C, S = 38.23 ppt, ϱ = 1.02886 and σ_t = 28.86, equivalent to a difference in density of 0.0026 g cm^{-3}. During winter, density differences are obviously much lower. These small differences are responsible for, or are associated with, the vertical and horizontal motions that are of paramount importance for the working of the whole ecological system.

Physical oceanographers need to know to high accuracy the density of sea-water (which is a function of T, S and pressure). In fact, they are more interested in differences than in absolute values and they calculate relative densities (relative to pure water) that can be given with an accuracy of 3 in 10^6, while absolute density is only accurate to about 10 in 10^6. Empirical expressions forming several tables were worked out a time ago by Knudsen (1901), in order to calculate sea-water density (or related quantities) correcting for T, S and p. Usually, the reciprocal of ϱ or specific volume ($\alpha = 1/\varrho$) is computed. To avoid writing a large number of decimals and for practical reasons in dynamic oceanography, the specific volume 'in situ' is divided into two parts by writing $\alpha_{S,T,p} = \alpha_{35,0,p} + \delta$, where $\alpha_{35,0,p}$ gives the specific volume of a 'standard ocean' of constant salinity S = 35 ppt and constant temperature T = 0°C at different pressures (depths). The value δ represents the departures from the standard ocean and it is called the anomaly of the specific volume and we are able to determine it through the appropriate tables or mathematical expressions.

3.1.2. Compressibility

Water is only slightly compressible (the volume reduces by around 0.5% every 100 atm or 1,000 m depth). However, its compressibility influences such important phenomena as sound propagation, adiabatic temperature changes and stability. The relative change of volume, $\Delta V/V$, of sea-water is proportional to the change in pressure. The coefficient of compressibility ($K = -1/\alpha \, d\alpha/dp$) decreases with increasing pressure or depth, increasing temperature and increasing salinity. It is a fundamental quantity for the computation of sound velocity in sea-water and the natural distribution of T and S with depth is responsible for some interesting phenomena in sound propagation at sea.

Compressibility of sea-water is responsible for adiabatic temperature changes with pressure changes. The 'in situ' temperature at a given level is higher than that which would be measured in the same body of water if it was raised to the surface without exchanging heat with the surrounding waters. Work is then performed by the water at the expense of its heat content, thus resulting in a decrease in its temperature. Conversely, if water is carried adiabatically from the surface into a deep layer, work is performed on it by the external pressure and the result is a rise in temperature. As the specific heat capacity and the coefficient of thermal expansion of sea water depend on T, S and p, the adiabatic temperature gradient ($\Delta T/\Delta z$) is also dependent on these variables. At depth, where the adiabatic rate is relatively more important than at the surface, it increases from about 0.14°C/1,000 m at 5,000 m, to 0.19°C/1,000 m at 9,000 m depth.

The temperature of a water sample expressed as the temperature attained when raised adiabatically to the surface is called potential temperature (θ). As the exchange in temperature affects the density, a σ_θ is defined in the same way. It is useful to give potential temperatures or σ_θ profiles instead of 'in situ' values when dealing with water bodies with low vertical stability as it is the case in the north-western Mediterranean during deep sea-water formation events.

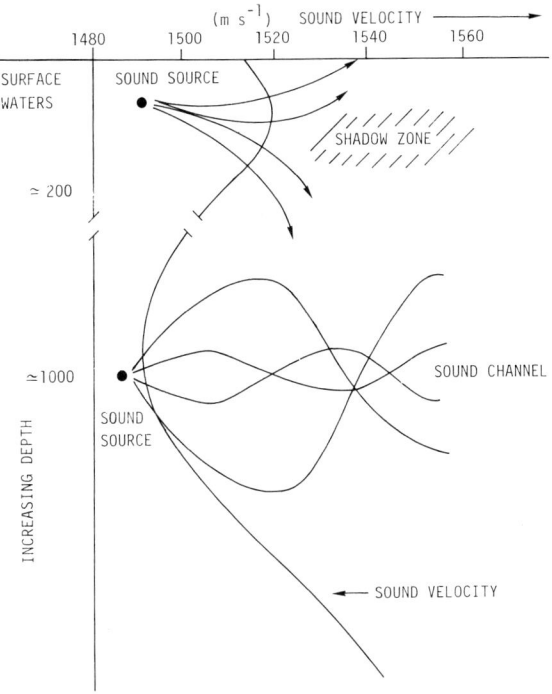

Fig. 3.1. Sound propagation in the sea. Sound rays bend towards the low-sound velocity levels.

Compressibility is also responsible for the ability of sound waves to propagate in water. Sound is a longitudinal pressure wave and its velocity is given by $V = (1/\varrho K)^{1/2}$, where K is the compressibility coefficient and ϱ is density. Sound velocity in water is greater than in air because of the much smaller compressibility of water. The density and the coefficient of compressibility depend on T, S and p, and sound velocity increases with the increasing of every one of the three parameters. The speed of sound at a salinity of 34.85 ppt (deep-water average) and 0°C is 1,445 m s^{-1}. It increases by approximately 4 m s^{-1} per °C rise of temperature, by 1.5 m s^{-1} per 1 ppt increase in salinity and by 18 m s^{-1} per 1,000 m increase in depth. The wave equation $V = \lambda n$ connects speed, wave length (λ) and frequency (n). So, the wavelength for a frequency of 10 kilohertz is about 14 cm and for a frequency of 100 kHz is about 1.4 cm.

The distributions of T and S with depth in the sea make it anisotropic with respect to the absorption and refraction of sound (the direction of travel of the sound is altered by the changes or inhomogeneities in the water properties). Absorption results in a decreasing of the intensity of sound energy with increasing of the distance from the sound source. Low frequency waves are able to travel for longer distances but high frequency sound is better suited for echosounding in near distances with higher resolution. Actual total absorption depends on other properties of the medium too, as air bubble content and organisms, which may cause scattering.

Sound refraction occurs in the stratified ocean as a result of the variation of sound velocity following the changes in T, S and p (Fig. 3.1). The sound energy emitted horizontally from a source is refracted and bends to the lower velocity layers. This is the reason for the appearance of a sound shadow effect. Waves emerging from a sound source placed at a level where a minimum of velocity of sound is found, are trapped around this level as they are refracted and bent towards this minimum level. This situation constitutes a sound channel and sound can travel for very long distances in it (thousands of miles). The development of underwater technology (underwater measuring devices) and the use of powerful computers have provided the basic tools in order to use sound as a means of 'seeing' through the sea. Ocean acoustic tomography, for example, is now possible. This technique is able to detect and measure mesoscale phenomena or structures in the ocean. The information gathered by a few transmitters and a few receivers is enough to reconstruct a three-dimensional picture of the velocity of sound for an area a few hundred miles wide (Munk and Wunsch, 1979; Spindel, 1982).

Marine organisms of all sizes have evolved and dwelt in the ocean for millions of years. Sound is often as important for them in the sea as is light for us in the air. Sound scattering and sound production by marine organisms have been topics for study since the beginning of echosounding at sea (*see* Hersey and Backus, 1962; Schevill, Backus and Hersey, 1962).

3.1.3. Transport processes in water

The high specific heat of water is the basis of the marked differences in the behaviour of land and water masses under the solar flux of energy. Water surfaces are heated slowly and only moderately by the sun, while land temperature rises higher and more rapidly than water temperature. Water is also able to diffuse heat through convection of different scales (turbulent diffusion) while land cannot, because of its rigid structure. This differential property of water against land is responsible for a number of meteorological and climatological phenomena of different size scales.

Not only specific heat of water is high, but its latent heat of vaporization is also high (about 585 g-cal g^{-1} at 20°C, the amount of heat accumulated or released by a unit mass of water for the change of its state from liquid to vapour or from vapour to liquid). Estimated values for evaporation in the Western Mediterranean range from 100 to 175 cm yr^{-1}, with a mean value around 1,200 mm yr^{-1}. This evaporation represents an outward heat flux of the order of 70 Kcal yr^{-1} cm^{-2}. The mean value for downward radiation reaching the sea surface in the Western Mediterranean is of the order of 140 Kcal cm^{-2} yr^{-1}. The heat of evaporation alone thus accounts for half of the total incoming solar radiation. Evaporation is a function of wind velocity and is usually higher in winter than in summer.

The most important fluxes (of momentum, latent heat and sensible heat, for example) occur across the sea surface, through which it exchanges energy and matter with the surrounding systems (atmosphere, shores, bottom). They represent the driving forces, and their assessment is an important although difficult task (Bush, 1977). It is important because the working of the sea as a mechanical and thermodynamical machine depends upon these fluxes, and it is a difficult matter because so many processes occur simultaneously. Moreover, in another time scale, the surface, and the sea as a whole, act as a physical basis for memory in the form of undulatory movements and chemical heterogeneities, not to speak of living organisms. New information always impinges upon the old and some information is lost in the interactions. The result is a composite picture which is difficult to interpret, for the effect of the driving force at the surface, depends on past history, in other words, on the information already recorded at the interfaces.

The molecular viscosity of water is responsible for the damping of motions (internal friction), the exchange of momentum between different neighbour layers being the result of molecular motion. According to Newton, $\tau = \mu \, du/dx$, where the frictional stress τ is a force per unit area directed parallel to the surface (perpendicular to x) and du/dx the velocity gradient in the x direction. μ is the proportionality coefficient called dynamic viscosity (molecular), and its value for sea water (35 ppt and 10°C) is of the order of 1.4×10^{-3} kg m^{-1} s^{-1}. Molecular viscosity of water depends on the temperature and only slightly on salinity and pressure. However, another striking property of water is that, unlike most other liquids, its viscosity decreases with increasing pressure (below 50°C). Pure water at room temperature must be compressed under more than 1,000 atm (equivalent to a depth of 10,000 m) in order to make its viscosity to increase with increasing pressure.

Other properties, as heat or salt, diffuse through water and the most familiar transport laws for them are the Fick's diffusion law (for molecular substances)

$$F = -D \, dC/dx$$

where F is the flux density (g m^{-2}s^{-1}) of the diffusing substance, D is the molecular diffusivity (m^{2}s^{-1}) and C is the concentration (g m^{-3}) and dC/dx is the concentration gradient in the x dimension; and the Fourier's heat-transfer law

$$H = -K \, dT/dx$$

where H is the heat flux density (Wm^{-2}), K the thermal conductivity and dT/dx the temperature gradient.

The transport of momentum can also be given in a diffusion form and the Newton's law of viscosity can be changed by

$$\tau = \nu \, d(\varrho u)/dx$$

where ϱ is the fluid density, ϱu is the concentration of momentum and $\nu = \mu/\varrho$ is a momentum diffusivity called kinematic viscosity and its value is of the order of 1.4×10^{-6} m^2 s^{-1}.

Heat transport can also be given as a thermal diffusivity:

$$\varrho H = D_H \, d(\varrho c_p T)/dx$$

where ϱc_p is a volumetric heat capacity and the thermal diffusivity $D_H = K/\varrho c_p$; molecular diffusivity for heat in sea-water is around 10^{-7} m^2 s^{-1} and for salt is 10^{-9} m^2 s^{-1}. This difference is obvious if we think that molecules (salt) can go back and forth, in neat diminishing progression, while heat progresses in an irreversible manner. In general, diffusion of momentum and heat is higher than that of salt or gases. Momentum is finally dampened and dissipated, and the patterns observed in the spatial distribution of temperature and salinity (or other substances) reflect patterns of present or past motions.

At the surface, molecular viscosity and surface tension between water and air are responsible for the generation of initial small waves (capillary waves). These waves (with a maximum length of 4 – 5 cm) may grow under the wind stress to give the big gravity waves of a well-developed sea. The generation of waves on the sea surface has been much studied (*see*, for example, Phillips, 1969). The problem with waves is not the simple mechanisms involved in their generation and growth, but the construction of a satisfactory model embracing all the characteristics of waves. Fluctuating motions within the turbulent atmospheric boundary layer must be described statistically. Modern techniques, using Doppler radar for example, can help to describe and to understand the motions at the sea surface. Surface waves affect the fluxes across the upper boundary of the sea in different ways and by different amounts. Some processes relating wind, waves, surface currents, turbulent mixing and energy transportation and dissipation are very important but not yet fully understood (Phillips, 1977).

Another speciality of water is its surface tension, which is the highest of all liquids at normal temperatures (72 – 76 dina cm^{-1} at 20°C). The arrangement of water molecules at the interface is related to the reinforcement of their dipole character. This is important for the development of capillary waves and also for those organisms living 'standing on' or 'hanging from' the sea-air interface. Naturally occurring tensoactive substances (released by organisms themselves) modify the surface tension of water (even by 30% in the sea). In Langmuir circulations and related surface flow structures (with convergence and divergence spots or strips) these substances are concentrated in the convergences and are detected by their effect on the behaviour of water surface under light wind action. Man dumps large quantities of tensoactive substances into the sea: directly and through rivers. These substances which accumulate at the interface interfere with the processes taking place there (i.e. fluxes across the boundary layer and life forms adapted to dwell there) and can also adhere to particles (or organisms). The formation of foam and bubbles (in the sea) and droplets (in the air) is affected by the substances accumulated at the top of the sea surface which form a thin film that inhibits capillary waves, enhances the attenuation of short waves and retards evaporation.

Possibly because humans are relatively large organisms, we are often more interested in flow properties than in the properties of water itself (at the microscopic or molecular levels). Some of these properties are recognized as fundamental only because they are at the basis of the macroscopic properties of water masses. However, a more objective way of looking at ecological phenomena, or processes affecting marine organisms should consider all dimensional and time scales appropriate to particular organisms. This is especially true for small and microscopic organisms for whom molecular viscosity, surface tension and the types of flow where these two properties are dominant, give a peculiar structure to their environment and determine their subjective 'vision' of the physical world.

3.2. THE ENERGY FROM THE SUN

3.2.1. The flux of solar radiation

The energy that moves our fluid environments comes from the Sun, in an almost constant flux, as electromagnetic radiation. The solar constant (the radiation flux at the fringe of the atmosphere across a surface normal to the Sun's beam) is approximately 2 g-cal cm^{-2} min^{-1} (or around 136 mW cm^{-2}, with fluctuations of 1.5% due to fluctuations in solar activity and changes of 3.5% through the year due to changes of Earth-to-Sun distance). However, as the solar beam is intercepted by a circle of area πr^2, but distributed daily over a surface of area $4\pi r^2$, the mean flux normal to the Earth's surface is around 34 mW cm^{-2}.

The way in which the radiative energy of the Sun is distributed over the continuous spectrum of wavelengths (Fig. 3.2) corresponds to the spectrum characteristic of a black-body radiating at a temperature of about 5,700–5,800°K, with a corresponding wavelength of maximum energy emittance of 0.48 μm. However, the spectrum observed at ground level is much more varied, showing the effects of the absorption due to the atmosphere's oxygen, ozone, water vapour and CO_2 contents (Fig. 3.3).

The reflection coefficient of the Earth is about 0.36, so that only 22.5 mW cm^{-2} enter the atmosphere. 1 mW cm^{-2} corresponding to wavelengths shorter than 0.29 m and larger than 0.7 μm are

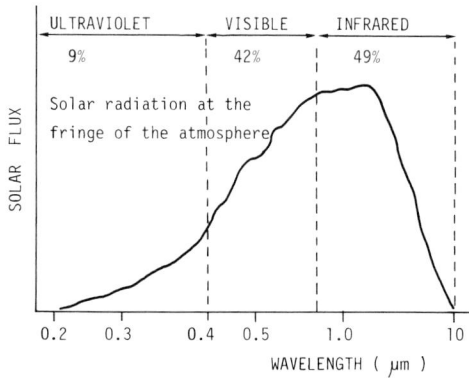

Fig. 3.2. Spectral distribution of the solar constant. The percentages of the total flux of energy are shown for ultraviolet, visible and infrared bands.

THE DRIVING MACHINE

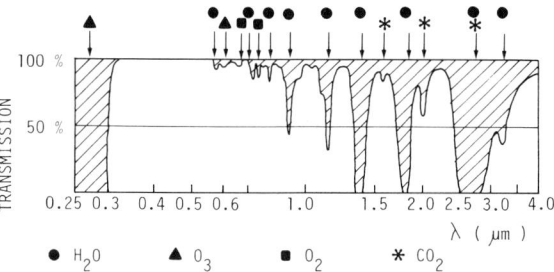

Fig. 3.3. Spectral transmission of the atmosphere in the visible and the near-infrared as a function of wavelength. The absorption bands of water vapour, carbon dioxide, oxygen and ozone are shown.

absorbed in the high atmosphere and 4 mW cm^{-2} in the troposphere. So, only 16.5 mW cm^{-2} are available at the bottom of the atmosphere to be absorbed by the Earth's surface.

The atmosphere receives the direct solar radiation which is in part reflected, absorbed and scattered, but it also receives the back-scattered radiation from the land and underlying water masses. The differential global distribution of radiative energy fluxes through the atmosphere as the basic driving force for its circulation and although a general balance is accepted for the Earth (it does not cool or warm) the local differences are associated to weather and climatic regimes.

Apart from some obvious differences in absorption capacity, the sea can be considered in the same way as the atmosphere: the two media being fluid masses in which radiation is absorbed, scattered by molecules and particles and reflected over the surfaces present. The differences in absorption are obvious: the atmosphere is mainly heated from below (at the bottom) the sea is heated from above (at its surface).

The magnitude of the energy reaching sea level is a function of the thickness of atmosphere penetrated (Sun's elevation, which depends on the latitude, the month of the year and time of the day) and of cloudiness. Assessing the amount and the form of radiative energy reaching the sea surface is important to understand and to model the interactions between the atmosphere and ocean. Clouds not only decrease the radiation reaching the sea surface and change its spectral composition, but also alter the proportion of direct to diffuse radiation. The ratio of skylight to solar radiation for a clear sky is a function of wavelength and solar elevation. Cloudiness and the cloud-type affect this ratio and a number of empirical expressions (giving the total resulting downward irradiance — flux per unit area — as a function of the irradiance under clear sky, cloudiness and cloud type) have been published (Ivanoff, 1977). It is possible to obtain empirically adjusted models to estimate the total solar irradiation, as a sum of the direct beam solar radiation, scattered solar radiation from clear sky and reflected and transmitted solar radiation from clouds (Satterlund and Means, 1978). These formulae may be useful, when properly calibrated in a given locality, to use as the input of solar radiation for marine physical or ecological models, related to the absorption or attenuation of radiation in water.

3.2.2. The attenuation of solar radiation in the sea

Some of the radiation striking the surface is reflected back into the atmosphere while the rest enters the sea, where it can be absorbed (transformed into heat), by molecules in selected spectral bands, scattered by particles (photons are not absorbed but diverted) or merely transmitted without interacting with the medium. The sea then shows an upward radiation flux which is the sum of back-scattered radiation and the radiation emitted by molecules. The absorbed and scattered parts constitute the attenuated part of the radiation energy.

The albedo is defined as the ratio

$$A = (I_r + I_u)/I_d$$

where I_d is the incident downward solar flux (radiation per unit surface area), I_r is the reflected flux at the sea surface and I_u is the back-scattered radiation (from below the sea surface). The reflectance, R, is the ratio between the reflected radiation and the input (downward radiation). The flux absorbed by the sea is thus the difference between that which is not reflected and the back-scattered radiation. The reflectance can be calculated, for an optically flat sea, from Fresnel's formula and it is a function of solar elevation and it is independent of the absorption and scattering properties of water. For a uniform sky (as a perfect overcast sky would be) an integration of reflectance can be made, obtaining values around 6.6%. When the sky radiation is not uniform, total reflectance can be lower or much higher (as is the case for blue sky with the Sun low on the horizon, total reflectance being then of the order of 30%).

The situation in rough sea is more complicated, for the angle of incidence of direct solar rays varies with the slope of the waves, the case being more difficult to deal with when the Sun is low in the horizon and shadow zones come into play. For solar elevations greater than about 25° the surface roughness leads to a slight increase in reflectance but for lower angles it produces a strong reduction, although for blue skies it can still reach 20% for a solar elevation of about 10°.

To calculate the albedo, the absorbing and scattering of radiation in the sea must be taken into account. However, the maximum ratio I_u/I_d to be added to the reflectance, in order to calculate the albedo, is only about 2% (water very clear or strongly scattering) but in general is less than 0.5%. In fact, experimental values of the albedo are very close to those calculated for the reflectance. For overcast weather it has an average value of 6%; for clear skies Laevastu (1960) proposed a simple formula

$$A(\%) = 300/\zeta°$$

$\zeta°$ being the solar elevation. Ivanoff (1977) comments that $250/\zeta°$ would be closer to the truth for solar elevations less than 50°.

For the Mediterranean, Calathas (1970) gave a number of albedo measurements for clear sky with different sea states and solar elevations (*see* Ivanoff, 1977).

From the surface to the depths, solar radiation suffers an attenuation by absorption and scattering. Absorption consists of a conversion from radiant energy into heat, while scattering does not imply a direct energy loss of photons but it increases their path between the sea surface and the depth, the process leading to an increased absorption and additional energy loss.

The attenuation of solar radiation in sea-water can be expressed by the equation

$$I_z = I_o e^{-kz}$$

where I_o is the solar irradiance at the surface (in the water), k is the attenuation coefficient and z is depth. The attenuation coefficient depends on the wavelengths considered and may be decomposed into a sum of an attenuation coefficient due to absorption and one due to scattering by the different compounds found in natural waters (i.e. water itself, dissolved salts, organic substances in solution and suspended matter). Absorption and scattering coefficients for pure or natural waters have been measured by several authors (see Tyler (ed.), 1977). In Figures 3.4 and 3.5 the absorption coefficients for pure water for different wavelengths are given. In sea-water, substracting the absorption due to pure water, one notices the absorption bands of the chlorophyll pigments around 0.44 and 0.675 μm and the increasing absorption towards the short wavelengths due to organic products of decay often called 'yellow substances'.

Molecular scattering follows a law close to $\lambda^{-4.3}$ while that of scattering by particles is around λ^{-1} and is directly dependent on the particle size (λ is the wavelength). The total scattering at sea is much more selective when its particle content is low (molecular scattering is the most selective) and particle sizes are below 1 μm. In fact, particle scattering in surface waters is produced chiefly by large particles (>2 μm) and thus is almost independent of the wavelength (Jerlov, 1968). Because absorption strongly increases towards long wavelengths, scattering is much less important than absorption at wavelengths greater

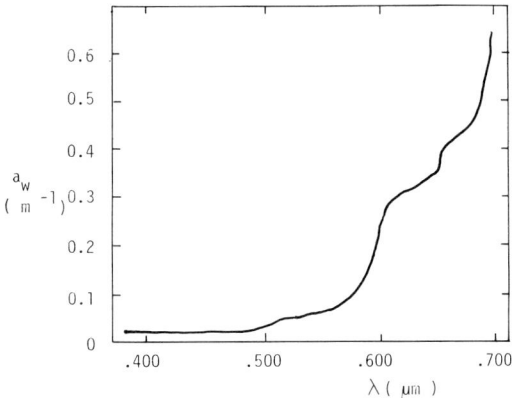

Fig. 3.4. Absorption coefficient of pure water in the visible band (after Morel and Prieur, 1975).

than 0.7-0.8 μm. In clear oceanic water, scattering by water molecules produces only about 7% of total scattering. Its prominent blue colour is due almost entirely to selective absorption by water molecules but, usually, the selective scattering by water molecules is so heavily masked by the non-selective scattering by large particles that total scattering in the sea is virtually independent of wavelength. This is the reason why sea-water colour depends very much on the nature and distribution with depth of suspended particulate matter, although dissolved substances can play an important role, specially in coastal waters.

The radiant energy of about 0.48 μm wavelength is the most penetrating in the sea and at a more or less important distance from the surface, underwater light becomes bluish or greenish. Another point is that direct beams reaching the sea surface change their direction due to refraction. The refractive index of water in front of air is about 1.44. This means that direct beams entering sea-water (flat surface) show a maximum angle (measured from the vertical or zenital point) of 48.6° (daylight is confined to a cone with an aperture angle of about 97°). So, seen from below, the sky dome is compressed into a bright circle. Outside the circle light comes from scattering. When the sea surface is disturbed by waves, dark patches appear in the bright circle and bright patches are seen outside it. The contrast between direct

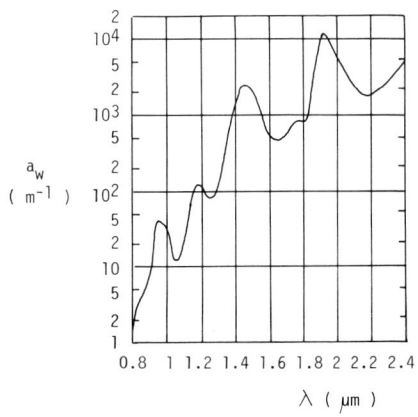

Fig. 3.5. Absorption coefficient of pure water in the near infrared (redrawn from Ivanoff (1977), after several authors).

skylight and diffused light diminishes with increasing depth (Smith, 1974). Divers are warned about the 'blue-wall' phenomenon which occurs when, at a particular depth, the diver is no longer able to distinguish where the light comes from. So, undisturbed direct beams are only reaching the objects immersed in the first few metres. In clear Mediterranean waters, the effect of waves on the bright zenital circle is only detected at depths less than about 7 m.

For a long time, biologists, limnologists and oceanographers, have used the Secchi disk (see Cialdi and Secchi, 1968) as a simple and inexpensive tool to measure light attenuation in water. The 'standard' Secchi disk is a white circle of 30 cm in diameter which is lowered into the water until it is out of sight. The reading of the instrument is the depth of 'Secchi disk visibility'. Some attempts have been made to relate this depth to an averaged attenuation coefficient, adjusting the exponential function

$$I_z = I_o e^{-kz}$$

assuming that $k = q/D_{sd}$, where q is a constant and D_{sd} is the Secchi disk visibility (in metres). Several values for q have been published in the literature (1.7, Poole and Atkins, 1929; 1.41, Gall, 1949; 1.39, for the Mediterranean, Weinberg, 1975). However, q changes very much with the type of water, its vertical structure and distribution of suspended and dissolved matter in layers. Weinberg (1976) gives an empirical hyperbolic function adjusting the data of Poole and Atkins, of Gall and his own:

$$k = 2.6(D_{sd} + 2.5) - 0.048$$

to be used for D_{sd} values between 5 and 35 m. The Mediterranean gives Secchi disk readings of even more than 40 m.

From a purely physical point of view, the absorption of radiant energy in the sea is particularly important as it represents an input of heat. The annual radiation balance of the ocean surface is everywhere positive. That is, the radiant energy input from the Sun, clouds and atmosphere is greater than the output radiant flux from the sea (emitted at a wavelength appropriate to its temperature). This balance of radiant energy on an area of the ocean surface is distributed mainly between ocean-atmosphere transfers of sensible and latent heat, storage of heat in the column of water and an horizontal divergence of heat by sea currents.

From a biological point of view, the attenuation of radiation in the sea is not only important because of its physical consequences (thermal and density structure, generated motions) but also directly because a certain, although small, proportion of total incident radiation is absorbed by green plants and provides biochemical energy to the whole ecological system. In fact, only about 0.1% of the radiant energy is spent in this way (i.e. it is absorbed by the sea as the result of photosynthesis).

Data obtained by Béthoux in September 1969 in the Western Mediterranean and cited by Ivanoff (1977) show that the top metre absorbs 55% of the solar energy which is about the proportion of infrared radiation present (Fig. 3.6). This value corresponds to an attenuation coefficient of 0.8 m^{-1}. At depth, this coefficient rapidly reaches an almost constant value of 0.075 m^{-1} (after 7 m). At depth and in very clear water, the attenuation coefficient is about 0.03 m^{-1} and for turbid coastal waters, during a period of plankton bloom, it is about 0.3 m^{-1}. Making some reasonable assumptions, Ivanoff (1977) concludes that the first 5 m of water absorb 69% of the solar energy, in the case of the clearest water, and 89% in turbid water (73% and 98% respectively for 10 m of water). This result is conceptually important because it shows that the first metres act as a cold filter (they allow the passage of 'cold light' and retain a large proportion of the total radiant energy reaching the sea surface). In other words, most of the energy is transferred as heat to water molecules at the very top of the column and is thus retained there, while radiant energy with wavelengths around 0.48 µm penetrate much more deep in the water. The photosynthetic active radiation (PAR, the spectral band between 0.350 and 0.700 µm after the recommendations of the SCOR; SCOR Working Group 15, 1974) attenuation curve lies between the curve for total irradiance and that for 0.48 µm.

The first centimetre of water absorbs around 15% of the solar energy the first 50 cm almost 50%. These numbers mean that, with a surface flux of 0.08 W cm^{-2} (equivalent to midday during summer in

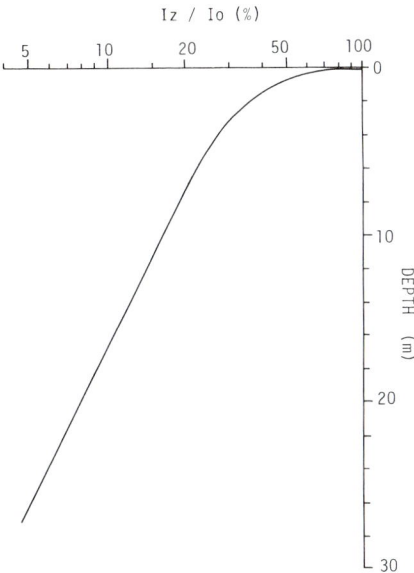

Fig. 3.6. Attenuation of irradiance with depth (downwelling irradiance is given as a fraction of the surface value after taking account of the albedo). Note that upper scale is logarithmic but data do not follow a straight line. (Redrawn from Ivanoff (1977) on original measurements by Béthoux in the Mediterranean, at 42°14'N, 5°35'E on September 23, 1969).

the Western Mediterranean) the first centimetre absorbs 0.012 W cm^{-2} or 0.17 g-cal cm^{-3} min^{-1}, equivalent to a heating of 0.17°C min^{-1} or 10.2°C hr^{-1}. The first metre would, by the same reasoning, increase its temperature by 0.37°C hr^{-1} (Ivanoff, 1977).

In fact, there is always enough mixing in the uppermost layers of the sea to prevent the formation of strong vertical gradients of temperature. Heat is not transported downward by radiation, it must be transferred, molecule-to-molecule or by water movement. The appearance of a gradient of temperature at deeper layers (not just at the surface) is much more a consequence of the mixing processes than of the '*in situ*' attenuation of radiation. This is an important point in properly modelling the upper ocean structure.

3.3. THERMAL CHANGE AND THERMAL STRUCTURE

3.3.1. Introduction

Vertical structure and static stability

In a water mass every part is subjected to a buoyant force equal to the weight of the surrounding displaced fluid (Archimedes' principle). So, in a column of water at rest, densities must increase towards the bottom so that a volume displaced a short distance downward will tend to return to its original position. Because of its inertia, this moving water will tend to overshoot its original position and then to oscillate about it. This effect is related to the occurrence of internal waves. If the column of water is neutrally stable, a displaced volume will tend to remain in its new position (the density of the parcel at its new position equals that of the surrounding fluid). Negative stability occurs when the displaced volume tends to continue its displacement (causing water overturning).

The vertical static stability (E) of a water column may be measured as the ratio of the restoring acceleration of the displaced volume to the acceleration due to gravity. Because density is a function of temperature and salinity and that the restoring force is a function of the differences in density between that of the displaced parcel and that of the surrounding water,

$$E = -1/\varrho \left[\partial\varrho/\partial S \cdot \frac{\partial S}{\partial z} + \partial\varrho/\partial T (\partial T/\partial z + \Gamma) \right] m^{-1}$$

where $\partial\varrho/\partial S$ and $\partial\varrho/\partial T$ are the rates of change of sea-water density due to changes of salinity and temperature that can be found in tables (Neumann and Pierson, 1966), while $\partial S/\partial z$ and $\partial T/\partial z$ are the 'in situ', actually measured, rates of change of salinity and temperature in the actual body of water for which we want to calculate E at a given level, and Γ is the adiabatic rate of change of temperature related to the change in depth or pressure (see Pond and Pickard, 1978).

So, all the processes affecting the temperature and salinity of water, may affect its buoyancy and its static stability, enhancing or decreasing it.

The equation of motion

Simply stated, in a fixed point of a body of water the rate of change of the velocity vector (speed and direction) must be the result of external and internal forces acting at this point. Gravitation, wind stress and atmospheric pressure are primary forces. Gravitation is a body force because it acts on the total mass of water while the other forces act at the boundaries although they may affect the interior of the water body. Secondary forces appear because of motion. They are the Coriolis force (an apparent force due to the interaction of local forces with rotation of the Earth observed relative to the rotating Earth) and friction, which acts at the boundaries and within the fluid, tending to oppose its motion, making the motion more uniform and converting kinetic energy into heat. Coriolis force is a body force while friction is mainly a surface acting force although its effects penetrate into the body of the fluid.

For most general purposes in dynamic oceanography, water may be considered as being uncompressible and the equation of motion (the Navier-Stokes equation) may be expressed as:

$$d\vec{v}/dt = \vec{F} - 1/\varrho \, \vec{\nabla}p + \nu\Delta\vec{v}$$

where $\vec{F}$ are external forces (per unit mass), ϱ is density, $\vec{\nabla}p$ the pressure gradient (the pressure spatial rate of change in the x, y and z directions), ν is the kinematic coefficient of viscosity and $\Delta\vec{v}$ are the second spatial derivatives of velocity. For example, for the component of the velocity in the χ direction (u), the equation becomes:

$$(1) \quad \frac{du}{dt} = -1/\varrho \frac{\partial p}{\partial x} + 2\Omega \sin\varphi \, v - 2\Omega \cos\varphi w + \nu \left(\frac{\partial^2 u}{\partial x^2} + \frac{\partial^2 u}{\partial y^2} + \frac{\partial^2 u}{\partial z^2} \right) + F_x$$

where Ω is the angular speed of rotation of the earth about its axis ($= 7.29 \times 10^{-5} \, s^{-1}$), φ is the latitude, v and w are the components in the y and z direction (z is directed upwards), and F_x are some external forces. The second and third terms on the right are the Coriolis terms.

On the other hand, if we take velocities at fixed points as function of time (the so-called Eulerian form), the total rate of change of u with time is the sum of its local rate of change due to time variation and the advective rates of change due to motion. Then the equation 1 becomes:

$$du/dt = \frac{\partial u}{\partial t} + u\frac{\partial u}{\partial x} + v\frac{\partial u}{\partial y} + w\frac{\partial u}{\partial z} = \text{(right terms of equation 1)}$$

The thermal vertical profile

The thermal structure of the sea in its upper layers depends very much on the fluxes that take place at the sea surface. For greater depths the thermal structure is more dependent on the circulation of the seawater. Water masses found at different levels result from the mixing of waters which got their characteristic salinities and temperatures near or at the sea surface, and which are, consequently, the final products of air-sea interactions.

Let us examine how the surface layers heat or cool and how the temperature distribution with depth is shaped and changes with time. The problem is not easy to deal with, and it is enough to look through the book *Modelling and prediction of the upper layers of the ocean* edited by E. B. Kraus (1977) for example, to realize that no definitive models exist which can explain all the possible structures given in the upper layers of the sea, nor the importance achieved by the different processes that can act at every spot.

In the Western Mediterranean, as almost everywhere in temperate latitudes, the thermal structure of the upper layers of the sea is homogeneous in winter. The temperature in the Western Mediterranean is surprisingly constant at this season, around 13°C from top to bottom, while the salinity increases gently from the surface to the lower layers.

When a body of water is heated from above, a thermal gradient and, consequently, a stabilizing density gradient, is formed. In the sea, most of the solar radiation is absorbed in the first few metres of water and, in the absence of wind, the gradient would be very steep. From spring to summer, the solar action and the mixing effect of wind, shape the temperature profile, while precipitation, river run off and evaporation affect the salt content in the upper layers. In general, 'turbulence' is responsible for the mixing of the upper layer and the lowering of the steep temperature gradient (thermocline) while leaving a uniform layer above it. However, the deeper the thermocline, the weaker is the effect of wind on it, unless the mixed layer is very homogeneous, in which case some turbulence can flow from the surface to the bottom of the mixed layer, where it can do some work (deepening the thermocline and enhancing the entrainment of cold water into the mixed layer).

With constant winds and insolation a single steep thermocline at some depth is favoured, while variable winds and changes in the insolation result in a more complicated and not so steep stair-like profile (Fig. 3.7).

When the heat flux across the water surface is strongly inverted, the process is highly destabilizing and can produce an important vertical mixing and eroding of the thermocline. Water cools at the surface and becomes denser. It then tends to sink and to be replaced by deeper water, giving rise to a convective motion and mixing over a large range. The heat flux is inverted not only with the change of the season but also, in a smaller scale, during the night or in overcast days. Wind is also important in this context. In the Western Mediterranean the rate of evaporation is rather constant (when averaged monthly) but it shows strong daily fluctuations. The maximum computed values usually correspond to very cold days with strong winds, a situation which is more frequent in winter. So, its cooling effect adds to the outgoing flux of sensible heat when the sea temperature is higher than the temperature of the air. Moreover, the salinity increase due to evaporation also helps the density of the surface water to increase, thus enhancing vertical mixing.

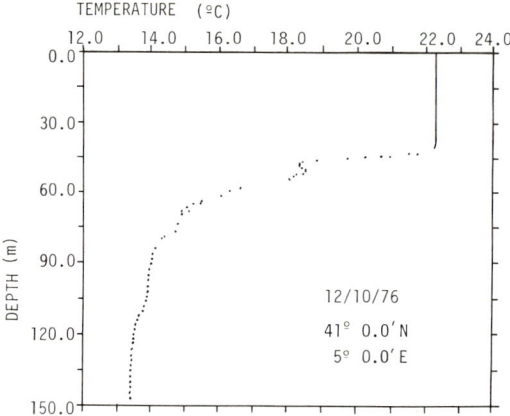

Fig. 3.7. (a) Temperature profiles showing a single and two upper mixed layers. Data from 'Mediterraneo I' cruise (October 1976), of the Instituto de Investigaciones Pesqueras of Barcelona.

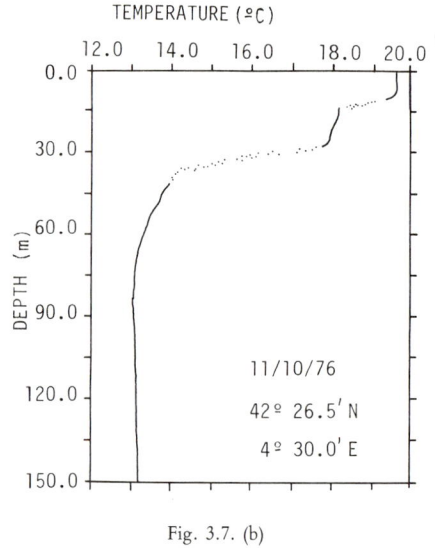

Fig. 3.7. (b)

At the end of the summer the thermocline has deepened and the upper layer becomes well-mixed and loses buoyancy. During autumn the thermocline breaks down and further cooling and evaporation give the final homogeneous winter column.

3.3.2. Turbulent diffusion and models

Molecular diffusion cannot explain quantitatively the transfer of either momentum, or heat or salt between neighbouring water bodies or across the air-sea boundary. In fact, a turbulent or eddy diffusivity occurs in the sea, its value being of the order of 10^5 m^2 s^{-1} for the horizontal plane and 10^{-1} m^2 s^{-1} for the vertical direction. It is important to note that the eddy diffusion is a property of the flow while molecular diffusion is a property of the water compounds. However, were it not for molecular viscosity, the turbulent eddy motions in fluid flows could not be dissipated into heat, nor turbulent stirring and mixing of different kinds of water could produce a relatively homogeneous end product. Eddy viscosity is usually introduced in textbooks of oceanography as an extension, by analogy, of the concept of molecular viscosity. As we shall see later, the coefficient appears in the parameterization of the Reynolds stresses (*see* Pond and Pickard, 1978).

In the Navier and Stokes equations, the friction term appears, as was hypothesized by Newton (*see* 3.1.3 and 3.3.1) as a viscosity coefficient multiplying the spatial derivatives of velocity. Internal friction and friction at the boundaries are responsible for the damping of motion, transforming kinetic energy into heat. Expressing the frictional effects in terms of the velocity, a closed system of equations can be obtained and, at least in principle, the set of equations can be solved.

In the basic equations (based on the Newton's laws of motion) some non-linear terms appear (with squares of the velocities or products between different velocity components and their derivatives) (*see* 3.3.1). These 'non-linear' terms are responsible for the growth of small perturbations into large fluctuations. The dynamics of most natural systems can be seen, in general, as a composition of several processes that often act at very different rates. By the process of scaling or ordering the different terms in the equations, some of them may be neglected (thus simplifying the analysis) without any practical loss in precision. However, the neglected terms may become the driving terms when the conditions change. The transition from one regime to another can be rather abrupt.

Turbulence can be regarded as a disorderly sequence of eddies of different sizes, the smaller ones riding or travelling on the larger ones. When the non-linear terms of the classic equations of motion become 'more important' than the friction terms the flow becomes unstable and can become turbulent. The Reynolds number (R_e) is a measure of the ratio of the non-linear terms to frictional ones. When R_e = 1,000 the non-linear terms become dominant and the flow can become turbulent. The transition to turbulence is abrupt and it is very likely to occur with R_e 10^5-10^6, although some stabilizing influence, such as density stratification may prevent the transition (this is commented on later).

The problem presented by turbulence is classically approached by focussing the interest in the mean values of velocity (or density, pressure...). In other words, if we are interested in the hourly or daily displacement of a drifting boat, for example, it is wise to use a formal description of this movement where ups and downs, and generalized movements to and fro associated with turbulence occur only in a very simplified form. The equations for the average motion are derived by substituting in the simple equations every variable by its mean value plus a fluctuating part (e.g. $u = \bar{u} + u'$). After some manipulation, the final equations are similar to the simple ones, with mean values in place of the instantaneous values, except for the terms where mean products of fluctuating parts (variances) appear (e.g. $\overline{u'u'}$, $\overline{u'v'}$, $\overline{u'w'}$...). Note that the mean of a fluctuating part is zero ($\overline{u'} = 0$) but the mean products of fluctuating parts are not necessarily zero. The later mean products of fluctuating parts are the Reynolds stresses. Molecular viscosity (the fluid property that makes the transport of momentum in a direction normal to the direction of the flow possible) is classically viewed as produced by molecules bouncing back and forth and exchanging momentum. It is, then, easy to think of volumes of fluid moving back and forth and exchanging momentum with the surrounding fluid. Similarly, molecular viscosity is the proportionality coefficient between a stress (force per unit area in the direction of the surface) and the gradient of velocity. An eddy viscosity is defined by equalling each Reynold stress to the product of an eddy viscosity and the gradient of the mean velocity (e.g. $\overline{u'u'} = A_x \partial \bar{u}/\partial x$; $-\overline{u'w'} = A_z \partial \bar{u}/\partial z$). In this way the fluctuating parts disappear from the equations, and with some further assumptions the new equations for the mean flow are analogous to the old ones except that the coefficient of molecular diffusion has become of eddy diffusion. In this way, the complicated motion due to turbulence and its effect on the average displacement (of the drifting boat, for example) is reduced to a new and simple parameter.

This parameterization has been widely used. One of the problems is that eddy coefficients have different values for the horizontal and vertical directions, and that they are not constant for they represent properties of the flow, not of the fluid. The parameterization process for viscosity can be followed for other transport coefficients (obtaining pseudo-Fickian transport coefficients for heat, salt,...). Since all these coefficients change with the characteristics of the flow, in the models they are often introduced in terms of the velocity and density distributions which are described deterministically by them. The problem is then to deduce empirically the relationships between the transport coefficients and the flow characteristics.

Turbulence is practically assimilated to the variance of the individual motions (or in general, punctual measures of variables). In laminar flow, each molecule follows the same motion of the whole set of molecules; in turbulent flow, the velocities of the different volumes of fluid (direction and speed) are variable. The distance between any pair of water volumes moving in a laminar flow changes little with time while the distance between volumes of fluid in a turbulent flow changes much more with increasing time, initially larger distances changing more than initially small ones. The variance of the distances is a function of the distance itself.

Taking the reciprocal of the distance as a frequency, the variance for a given frequency may be a measure of the turbulent energy involved in the motions corresponding to that frequency. In general, the turbulent energy is cascading from low frequency (large eddies) to high frequency (small eddies) motions, to finally be dissipated by molecular viscosity into heat. The following relationship between frequency (F) and the variance or turbulent energy (E) was proposed by Kolmogorov.

$$E = A F^{-5/3}$$

but in the oceans the energy input has preferent frequences at scales of approximately 1 m, 10 km and 1,000 km (Fig. 3.8), and at these points (scales) the variance is maximum, so that the spectrum of turbulence is not smooth but shows several peaks (Fig. 3.8).

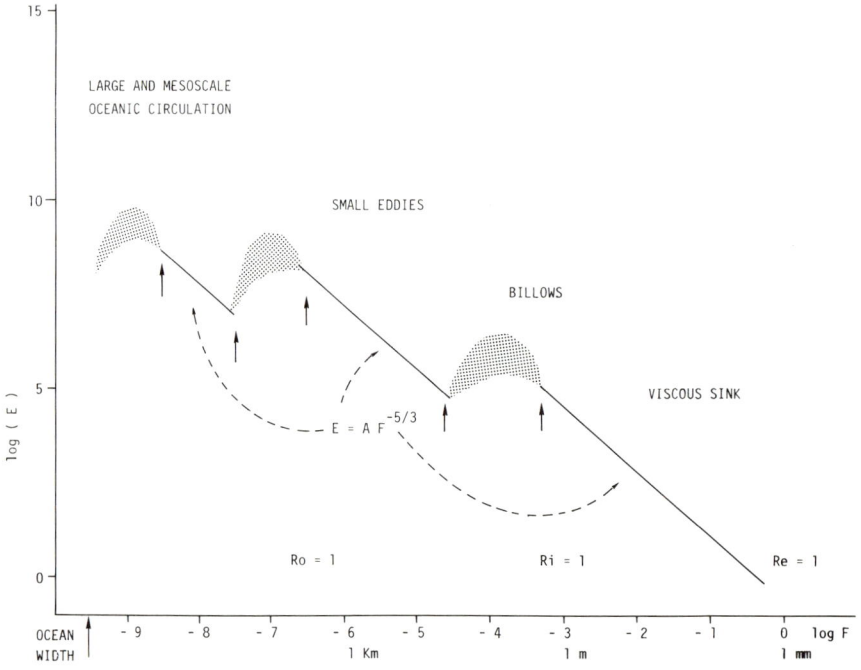

Fig. 3.8. Schematic kinetic-energy spectrum of ocean motions with indications on the principal scales of turbulence eddies. The places where Rossby (R_o), Richardson (R_i) and Reynolds (R_e) Numbers are close to 1 are shown. Zones between arrows satisfy the -5/3 law of Kolmogorov.

The characterization of a flow is usually done by scaling it, that is, by evaluating the relative importance of the different terms involved (as it was done to define the Reynolds Number). For example, in large-scale flows, non-linear terms and friction effects are very small and the Coriolis term (due to the rotation of the Earth) becomes dominant. The classification of flow types is helped by the consideration of ratios between non-linear and Coriolis terms, and between friction and Coriolis terms. These ratios are called the Rossby Number (R_o) and the Ekman Numbers (E_x, E_y, E_z) respectively (there are several Ekman Numbers, the horizontal and the vertical, because eddy viscosity has different values for horizontal and vertical transports of momentum). For the interior of the sea, where usually the Coriolis term is dominant, R_o is approximately 10^{-3} and $E_z = E_h \times 10^{-3}$ (E_h is horizontal, either E_x or E_y). For large scale circulation, values of the order of 1 are an upper limit.

The problem of determining when a flow will become dynamically unstable (so that it may break down into irregular small-scale motions leading to friction effects largely due to the molecular nature of the fluid) was partially solved by the introduction of the Reynolds Number (R_e). The partiality comes from the fact that density variations affect the dynamic stability. A water column showing a particular density stratification and a gradient of velocity will remain stable if several processes which are into play compensate each other. Turbulence leads to a more uniform density distribution in the vertical because light fluid is mixed down and heavy fluid up. As the centre of gravity of the column rises, the gravitational potential energy of the system increases at the expense of the kinetic energy of turbulence. So, turbulence acts against static stability and a source of density variation enhancing static stability (through buoyancy) may compensate its action. Then, if the rate of turbulent energy loss exceeds the rate of gain, the turbulence will die out. If the rate of buoyancy gain is not enough to compensate for the effect of turbulence, the vertical component of velocity fluctuations is enhanced.

The generation of turbulence requires a velocity gradient, that is, for dynamic instability to occur, a spatial change in velocity (shear stress) must be present. Then, the Richardson Number (R_i), defined as the ratio of the energy provided by buoyancy forces to the energy produced by shear stress, can be used as a criterion to evaluate the relative importance of static stability (a function of the density gradient) and the tendency for instability due to the non-linear terms:

$$R_i = g E / (\partial u / \partial z)^2$$

where g is the gravity, E is the static stability (*see* 3.3.1) and $\partial u / \partial z$ is the vertical gradient of horizontal velocity.

The exact value of the 'critical' Richardson Number must be determined experimentally. It seems that for R_i larger than 0.25, turbulence cannot be generated by vertical gradients of velocity (those that can be taken into account in one-dimensional vertical models). So, if horizontal gradients of velocity are present, fluctuations of horizontal velocity may develop even for Richardson Numbers much larger than the critical value for damping vertical component fluctuations.

There is a clear effect of density variations on the turbulence, modifying the coefficients of eddy diffusion (for the transport of momentum, heat or salt). These can be parameterized in several ways through experimentally derived expressions. The Ekman's model for wind-driven currents, for example, predicts absurdly small influence of wind stress on the water movement if the coefficient of molecular viscosity is used. Based on empirical observations, Neumann and Pierson (1966) suggested an expression for total energy dissipation in a fully-developed sea applicable to the Ekman model

$$A_z = 0.1825 \times 10^{-4} \, U^{5/2} \text{ g cm}^{-1} \text{ s}^{-1}$$

U is the wind velocity (cm.s^{-1}), and A_z is the eddy viscosity coefficient. Presently, one of the ways to parameterize the eddy diffusion consists of finding experimental expressions relating the coefficients to the Richardson Number, though it is a difficult task.

Although a set of equations (involving implicitly all the solutions and thereby all the structures appearing at different scales of motion) could be devised, they could not be solved by computer integration. Woods (1977a) comments that a computer model for the upper layer with more than 10^{16}

times greater resolution than those available at present would be needed to follow all the motions deterministically. So, primitive equation models are forced to cover only a narrow spectral window amid the vast band of natural variability. When the largest eddies and waves are resolved, motions with scales smaller than the resolution of the model are neglected. If the model has a spectral window inserted at the middle of the natural band, the intermediate scales of motion are described deterministically while larger scale motions are introduced as boundary conditions and smaller scale motions are incorporated as averaged perturbations into the model equations.

The motions not resolved by the model usually produce significant transport of momentum, heat and dissolved and particulate constituents of seawater. Small scale motions cause changes in those predicted by the model and progressively degrade the accuracy of the predictions, while larger, not predicted motions, extract energy from them. The parameters used take account of these diffusive, dissipative and interactive effects and it is necessary to describe the motions and processes at a given scale.

The lower portion of the atmosphere and the upper layers of the ocean (with the air-sea interface in the middle) are coupled systems which are difficult to model, for they show several different flow regimes. Mixing, for example, is essentially an intermittent process (Woods, 1977) which is not simulated adequately by most of the current parameterization methods. However, modelling the thermal structure of the upper layers of the ocean is important for it must help to specify the sea surface boundary conditions in numerical models of the general oceanic and atmospheric circulations, and, from an ecological point of view, the structure and dynamics of the upper layers constitute the medium where primary producers develop and hence, the fundamental link between marine food webs and all other systems working in the Earth.

3.3.3. Balance of radiant energy and heat storage

Some of the radiant energy entering the Earth reaches the sea surface where it is absorbed or reflected. The sea emits radiation of wavelength appropriate to its temperature and absorbs long wave radiation transmitted downwards from clouds and atmosphere. The balance of radiant energy (R_N) at the sea surface can then be written

$$R_N = (1 - A) R_s - (R_E - R_B)$$

where A is the surface albedo, R_s the total short-wave radiation and R_E and R_B the emitted and absorbed long wave radiation.

The balance of radiant energy on an area of the sea surface is distributed chiefly between ocean-atmosphere transfers of sensible and latent heat (Q_H and Q_E), storage of heat in the water column (Q_s) and horizontal divergence of heat by currents in the sea (Q_{vo})

$$R_N = Q_H + Q_E + Q_s + Q_{vo}$$

The uneven temporal and spatial distribution of solar flux over the Earth is the driving force for air and water currents, and makes the relative contribution of the different terms in the expression of the radiation balance to change from one place to another and with time.

Daily balance of radiant energy can be positive or negative. The upper portion of the water column stores heat during spring and summer and loses heat during autumn and winter. Surface temperature follows a similar trend but it is not in phase with the upper layer depth. For the same surface temperature, total heat content in the water column is higher in autumn than in spring (Fig. 3.9). This is a consequence of the non-constant nature of forcing functions at the surface (heating or cooling and working by the wind). The upper warmer layer is shallower as heating increases and it is expected to reach its smallest value when heating is at its maximum (at the summer solstice or at noon for the diurnal evolution). At the end of the heating season the layer deepens. Later, it loses buoyancy and becomes better mixed while it gets thicker. This evolution causes a smoothening of the weather, as heat stored in the sea during the warm season is given back again in autumn and winter.

THE DRIVING MACHINE

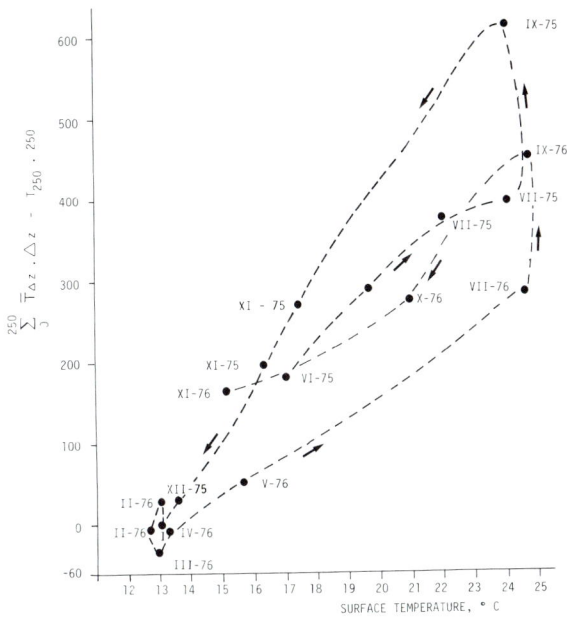

Fig. 3.9. Rough integration in the upper 250 m of water, of the temperature differences between those at z metres and that at 250 metres depth, corresponding to data shown in fig. 3.13.

Calculations made on the thermic cycle data offshore from Barcelona (Fig. 3.9) show well this loop. The hysteresis appears regularly in the cycles for different years but the absolute values change from year to year and from place to place.

3.3.4. Fluxes and the mixed-layer model

For most purposes the upper ocean layers may be regarded as being statistically homogeneous along the horizontal plane. Temperatures or salinities tend to vary more along a vertical distance of a hundred metres than along a horizontal distance of a thousand kilometres. Vertical exchange processes between the air and the sea, as well as vertical mixing within the water column, are expected to affect local conditions more rapidly than horizontal advection or mixing. So, as a first approximation, we can consider that every column of water is like any other one in the neighbourhood. This approximation allows us to simplify the real picture, but at the expense of missing some important phenomena involving horizontal motions and stresses.

One-dimensional models that consider only the vertical axis are well-suited to represent the thermal upper structure of the ocean, and among them, the best-suited to explain the typical thermal structure in the Mediterranean is, probably, the mixed-layer model (*see* Niiler and Kraus, 1977). In the Mediterranean a more-or-less uniform mixed layer, its differential characteristics with respect to the water below and the sharpness of its bottom boundary, depend on the characteristics of fluxes across the sea surface.

The four unknowns in these models are the mixed-layer temperature, the horizontal velocity components and the layer thickness. These variables are functions of time and they are governed by the overall budgets of heat, horizontal momentum and turbulent kinetic energy. The time rate of change of the turbulent kinetic energy is the sum of surface flux from the atmosphere through surface-wave breaking and the rate of production by the shear of the mean flow, minus the rate of potential energy, the rate of dissipation by friction in the interior of the mixed layer and the flux of energy lost by internal gravity waves. The time required for redistribution of turbulent kinetic energy in the mixed layer is small compared to the time over which the energy input from wind changes. It is thus reasonable to consider a steady state situation and to neglect the time rate of change of kinetic energy. On the other hand, the downward flux by internal gravity waves is also often neglected.

In the mixed-layer models the surface fluxes of sensible heat, salinity and buoyancy can be approximated by bulk aerodynamical formulae as functions of the prevailing atmospheric and mixed-layer conditions.

Fluxes of latent and sensible heat are often estimated by similar expressions:

$$Q_E = K_1(U_z)(e_o - e_z) \text{ and } Q_H = K_2(U_z)(T_o - T_z)$$

where the K's are empirical functions of the form $(a + bU)$ that prevent a null flux in calm weather, when there is no wind; e_o is the saturation specific humidity at the temperature T_o (that of surface water) and e_z, T_z are the specific humidity and temperature of the air at z metres above sea level, where U_z is the wind velocity. Budyko (1963) used this kind of expression in his Atlas of Heat Balance of the Earth's Surface and the World Meteorological Organization (Laevastu, 1976) recommended the use of these expressions because of the difficulties existing otherwise in measuring several parameters in extended marine areas.

Once the flux of heat and the change of salinity are described, the rate at which buoyancy is removed from the water column (or available potential energy is supplied) by surface cooling and salinity changes is almost straightforward.

The magnitude of the downward flux of momentum from the surface is the wind stress (τ) and it can be also given as a function of wind velocity (U) and a surface drag coefficient (ζ, empirical). The explicit expression has the form

$$\tau = \zeta \varrho_a U^2$$

where ϱ_a is the density of air. The direction of the stress is equal to the wind direction close over the surface. To give an idea of magnitudes, when the wind blows with a speed of 8 m s^{-1} at a height of 10 m, the friction velocity (that of water at the surface) is around 1.0 cm s^{-1}. For winds not so strong, ζ is independent of U and a function only of the vertical stability of the atmospheric boundary layer and aerodynamic roughness of the surface. Since the latter is determined by wave conditions and these depend upon the wind situation, ζ is indeed a function of wind-speed. However, a relationship describing satisfactorily the variation of ζ with U has not yet been found. It is now accepted that ripples and wavelets are important elements in the creation and maintenance of drift currents, probably accounting for almost all the drag component due to waves. So, contamination of the sea surface (modification of surface tension of water) might cause a significant diminution in strength of oceanic drift currents with consequences in climatic patterns over areas subjected to maritime influences.

In the mixed-layer models a number of proportionality factors are needed but they are not necessarily constant and their numerical values depend on the layer depth and the magnitude of the forcings. Actual values must be established empirically from laboratory experiments or field observations.

Turbulence energy (supplied or generated indirectly by wind near the surface or supplied by a mean shearing motion at the bottom of the mixed layer) is used to work against gravity at the bottom of the layer where it entrains water from below which is as agitated as the mixed-layer water. When there is insufficient turbulence energy available to overcome the stratification at the base of the layer, this becomes decoupled from the ocean interior.

One-dimensional models provide useful and simple approximations of the upper thermal structure and can yield quite realistic simulations of diurnal and seasonal temperature changes (applicable to climate and biological models). However, horizontal variations which affect vertical mixing are not always negligible in the real ocean. Langmuir circulations, for example, cannot be explicitly used in these models. Langmuir circulations consist of alternate left and right-handed helical roll vortices aligned more or less along wind. They form lines of convergence, where, along the wind, surface velocities are strongest, alternating with lines of divergence. Particulate and dissolved matter and oil films collect at the convergences where they modify the surface tension of water, preventing the formation of wavelets under light winds and giving place to easily observable surface bands or slicks. Row spacing is variable (from 2 to 300 m in the sea) and cells of several different scales can exist together. Although Langmuir circulations do not usually occur in winds of less than 3 m s^{-1}, they can nevertheless, exist in calm seas when there are pronounced swells. Then they are less linear and show a more or less reticulated pattern. A number of theories have been proposed for these circulations since Langmuir first described them in the thirties. What seems sure is that Reynolds stresses associated with Langmuir circulations are potentially very efficient at redistributing momentum through the mixing layer (Pollard, 1977).

Other sources of turbulent energy affecting the thermal structure are inertial and internal oscillations which can cause shear instability in the seasonal thermocline, causing intermittent turbulent patches. Internal waves, for example, can be generated locally by turbulence or externally by distant storms. Waves radiate energy away from the area where they are being generated and can cause a local energy loss from the mixed layer under the storm but also, those which propagate into a region, can make additional energy available there (increasing the rate of layer deepening and blurring its bottom boundary) (Kantha, 1977).

The Reynolds Number of the ocean is large enough to ensure that it is always and everywhere turbulent. Eddies of all sizes and turbulence form the structural support for the cascading of turbulent energy through all the scales — from ocean width to the molecular level. They are dissipative structures, sometimes ephemeral, sometimes rather persistent, with lifetimes increasing, on average, with eddy size.

Spatial and temporal orders of magnitude for Gulf Stream meanders and for rings detached from the main flow are 100 − 200 nautical miles and from several months to 1 year (Ring Group, 1981; Fofonoff, 1981), while smaller structures in the range of 10 km usually develop or change in a few days. A general relationship between spatial horizontal variability of oceanic structures (physical and ecological) and life span could be given by an expression where a spatial numerical value in kilometres approximately equated a time numerical value in days (Steele, 1981).

In fact, many more or less transient structures have been identified in the oceans at all scales that need to be solved or conveniently parameterized (Woods, 1977). The real problem is not to find models which correctly predict some field observations (the temperature of the surface for example) but discover the nature of the processes. Woods (1977b) comments that until now the most important error in upper ocean models has been to consider an essentially isotropic or homogeneous turbulence and to introduce buoyancy in terms of the Richardson Number. He emphasizes that there is strong evidence that the distributions of physical, chemical and biological properties of the upper ocean are highly discontinuous on all scales in space and time and that data gathered at a particular location and time must be treated as a sample of a non-stationary phenomenon. The Mediterranean (with a complex and irregular climatology, where land masses and water are so intimately connected) non-stationary phenomena or structures may be the most commonly expected. The Mediterranean is wide and deep enough to reveal the mechanisms which tend to initiate stationary regimes. But it is also small, and weather fluctuations are strong and unpredictable enough to prevent this, thus resulting in a long-term cyclical (stationary?) succession of transient regimes.

3.3.5. Geostrophic and wind-driven currents

On a world-wide scale, the atmosphere and the ocean are coupled so that the static and dynamic properties (distribution of mass and circulation) in each system depend upon those properties which exist in the other. Similar structures can be found in the two systems though their appropriate descriptive time and length scales are different, mainly because of the different densities of the two fluids, air and water. However, the end product of the interaction shows some permanent or quasi-permanent structures that may be described as stationary, independently of the primary causes of the given situation. So we can often talk about permanent oceanic currents in equilibrium with the distribution of mass. Thus it is supposed that the mean velocity is not changing with time, that the mean vertical component of the velocity can be neglected, that the ocean is frictionless and that there are no net external forces acting upon it. The equation of motion then becomes quite simple and the only forces acting are those due to the gradient of pressure and the Coriolis term due to motion relative to the rotating earth. That is, for the x, y and z directions:

$$-1/\varrho \; \partial p/\partial x + 2\Omega \; v \sin \varphi = 0$$
$$-1/\varrho \; \partial p/\partial y - 2\Omega \; u \sin \varphi = 0$$
$$-1/\varrho \; \partial p/\partial z = g$$

These stationary currents in equilibrium with the distribution of mass are called geostrophic. The model was first derived by Bjerknes in 1898 and applied to the atmosphere, and was later applied to the ocean by Sändström and Helland-Hansen in 1903. Note that the equation for the vertical reduces to the hydrostatic equation.

This model, when applicable, allows us to compute the mean horizontal current at a given depth relative to the current at another depth, from field salinity and temperature data distributions (we know the density or specific volume distribution from T, S and p values).

The pressure at a given level is the result of the weight of the column of water above this level. If the pressures at the same level for two stations some miles apart are different, a net force must be present acting from high to low pressures. Then, if the pressure difference is to be maintained, a water flow must exist at that level in the direction of the isobars (with the high pressures on the right and the low pressures on the left in the Northern Hemisphere) so that the Coriolis force balances the force due to the horizontal component of the pressure gradient.

The surface of water at rest is horizontal and all the isobaric surfaces are parallel to each other and parallel to level surfaces. If the isobaric surfaces are parallel to each other but all of them inclined with respect to level surfaces, then the water is in motion and the sea surface, that is an isobaric surface is also tilted. Then the force system associated to the motion is called barotropic (Fig. 3.10). If the isobaric surfaces are inclined with respect to each other, then the system of forces is called baroclinic. Usually it can be expected that both components of the geostrophic flow are present (Fig. 3.11).

Geostrophic calculations only show the baroclinic component of the velocity, and a common problem is to find a level of reference with a known velocity or of no-motion, or a level where a barotropic situation may be expected. If the latter situation is found (from a given level down to the bottom) then the barotropic component may be regarded as zero. Several methods for the election of the level of reference have been proposed. For a discussion of the level of reference in the Western Mediterranean see for example Font and Miralles (1978) who finally decided to choose that of 500 decibars, the same level chosen by Allain (1960).

The proximity of coasts and of bottom, reduces the applicability of the geostrophic model to offshore, deep enough areas. Near the coast, on the platform, or at the surface, where friction is present, another force enters the balance and water tends to move somewhat across the isobars from high to low pressures (the velocity is diminished so that the Coriolis term no longer balances the pressure gradient). It is usually considered that wind action is a factor affecting transient surface currents only, but in a sea with plenty of coasts and with a variable climatology, the general currents, calculated with the geostrophic method show not to be strictly permanent, and show seasonal changes. These seasonal situations may be good enough to be used as boundary conditions for models of the circulation on the

THE DRIVING MACHINE

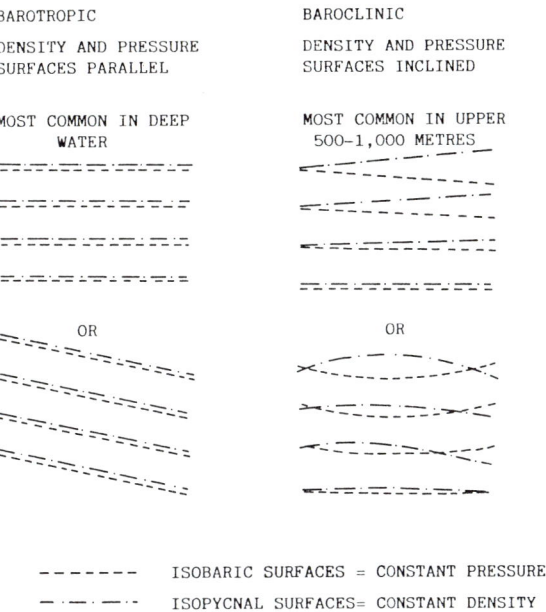

Fig. 3.10. Schematic examples of barotropic and baroclinic fields of mass and pressure (from Pond and Pickard, 1978).

platform (Font, 1983). However, some general features can be given for the Western Mediterranean (*see* Figs. 4.7 and 4.9 in Chapter 4).

At the edge of the platform, the velocities estimated by the geostrophic method are of the order of $10-15$ cm s^{-1} and where they have been compared to direct measurements of moored current metres, the discrepancy at 100 m depth has been found to be less than 30° in direction and 2 cm s^{-1} in speed (Font, 1983).

The Ekman's theory of wind-driven currents attempts to link wind stress with water movement. The model shows that the surface induced current is directed at 45° to the right (or to the left, depending on the hemisphere), and that the net water transport is directed at right angle. The current progressively

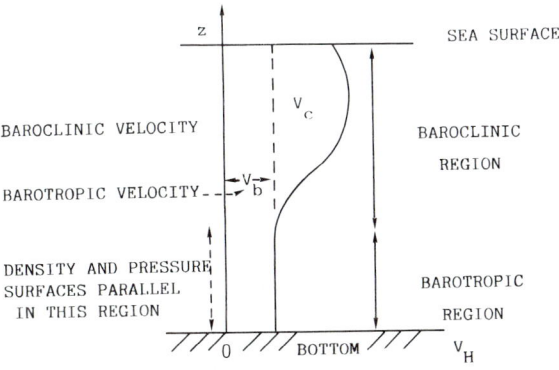

Fig. 3.11. Horizontal speed (V_H) as a combination of barotropic (V_b) and baroclinic (V_c) parts. Geostrophic calculation does not show the barotropic components (from Pond and Pickard, 1978).

deviates to one side and its strength decreases with depth (following the 'Ekman spiral' model). At the depth of 'frictional influence' the current strength is $V_o e^{-\pi}$ (approximately $v_o/23$), and its direction is exactly opposite to that of the surface current (V_o). Although the action of wind is expected to be very important in the upper mixed layer, current and wind measurements on the platform have proved that the relationship is not straightforward. With stratified waters (summer) the upper warm layer follows the fluctuations of wind and its motion is almost decoupled from that of the deeper water. When stratification decreases and gradients smooth, the action of wind reaches deeper levels and the persistence of current velocities decrease at depth.

At the surface, with measured currents of the order of 12 cm^{-1} the net flow is of 7 cm s^{-1}, while at more than 50 m depth, intensities of 10 cm s^{-1} can give net flows of only 1 cm s^{-1}, because almost all the measured current has an inertial origin. That is, the water at a certain depth, almost decoupled from external forcings, shows an oscillatory movement (like one of a rigid horizontal plane where all its points show a circular motion) with a period which is half the 'pendulum day'. One pendulum day is the time required for the plane of vibration of a Foucault pendulum to rotate through 2π radians and it amounts 1 sidereal day/sin φ (where φ is the latitude). The inertial force in the motion of water relative to the rotating Earth is offset by inertial force in the transport motion (the rotation of the planet). For the oscillations shown in Figure 3.12, the radius of the horizontal orbital circulations, calculated by equating the centrifugal and Coriolis forces, is of the order of 1 km.

At the surface, currents do not always follow the wind and water, often flow in the opposite direction to that which would be expected from the transport effects of wind action. The sea can apparently take a few hours and even days to react, and the effect of coasts is not negligible. The open areas are often not very wide in the Mediterranean and the barotropic effect of wind stress, the returning

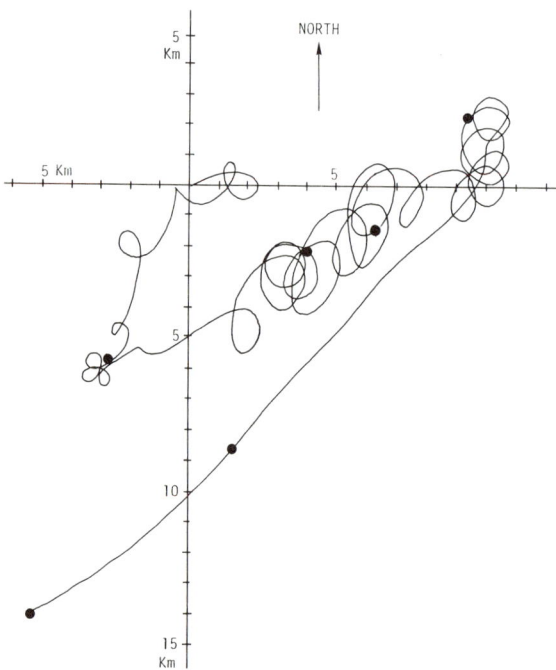

Fig. 3.12. Progressive vector for the current measured by a moored current metre (50°23'N, 0°42,5'E) at 50 m depth, 8 m above sea bottom, near the Ebro mouth. Black dots represent 5 days intervals (after Font, 1983).

flows due to the need for continuity of mass and the Ekman transport are, all of them, important. They act on a given (say seasonal) distribution of mass in presumed equilibrium with a geostrophic flow, but their action is important and, with variable winds and the shore proximity, one can expect the different mechanisms to be successively dominant. So, at least in the upper layer, mesoscale phenomena (tidal and inertial fluctuations) and small scale phenomena (surface and internal waves and turbulence) must be very important in the Mediterranean because their periods (of less than a few days) and length scales (less than 100 km) are the only ones that can fully develop. Synoptic scale phenomena (the geostrophic flow) is probably maintained because of the forcing by river run-off, precipitation, evaporation and fluxes through the straits. The coupling of sea and atmosphere through prevailing winds for example, seems not to be dominant because the weather is not steady enough nor the sea wide enough. However, the geographical distribution of shores and batimetry represents a physical constraint that must help to shape the general circulation forcing it to follow preferent paths.

3.3.6. Seasonal vertical mixing in the Western Mediterranean

We will now consider a typical thermic cycle in the Western Mediterranean, and the parallel distribution of nutrients and chlorophyll. The latter, being an estimate of phytoplankton biomass, reflects the primary production.

Figure 3.13 represents an annual cycle offshore from Barcelona, where there are no special climatological conditions or river discharges. Cycles at other places in the Western Mediterranean would look essentially similar, but would differ in some absolute values and other factors due to specific weather conditions and currents in the area (*see* in other figures the distribution of winds and gales for the Western Mediterranean).

Beginning the cycle in summer, with a well-established thermocline, the water temperature difference between surface and 60 m depth is about 10°C (0.17°C m^{-1}) at the end of July. The depth of the thermocline is shallower and most intense in July. Its mean depth changes from year to year (between 20 and 40 m) and can show oscillations of more than 15 m of amplitude during the summer in coastal areas. At least one of these is often detected in mid-August, before the thermocline begins to deepen and to get weaker. By the end of August, the weather becomes more windy and produces breakdown of the thermocline. The decrease in insolation and the autumnal storms finally disrupt and disperse the thermocline between the end of October and the end of November. During winter the water temperature remains around 13°C and never below 12.7°C (except locally and occasionally in some very cold days, near the coast). During spring, insolation increases and wind blows more frequently. The weather during this season actually shapes the future thermocline during June and July (*see* Fig. 3.13 for the temperature regime during the summers of 1975 and 1976).

Seasonal vertical mixing introduces a rhythm in phytoplankton production. Usually, the principal chlorophyll maximum occurs at the end of January, or during February, when the concentration reaches, at most, 2 µg Chlorophyll l^{-1} in open waters. In winter or early spring, phytoplankton can develop quite well in the upper 60 – 70 m as the newly-mixed water is nutrient rich and clear. In spring, suspended matter accumulates and nutrients are almost depleted, so chlorophyll maximum values are found only in the upper 30 m. As the establishment of the thermocline proceeds, the weak, but effective flux of nutrients to the photic zone ceases and productivity decreases. During summer, some maximum values are found deeper in open waters (at 70 m or more). These deep chlorophyll maxima present some interesting scientific problems and studies to elucidate the functional structure of the stratified Mediterranean in summer are underway. These deep chlorophyll maxima attain concentrations of more than 2 µg Chlor. l^{-1}, but the thickness of the phytoplankton rich layer is only about 10 m and does not extent landwards to the shore.

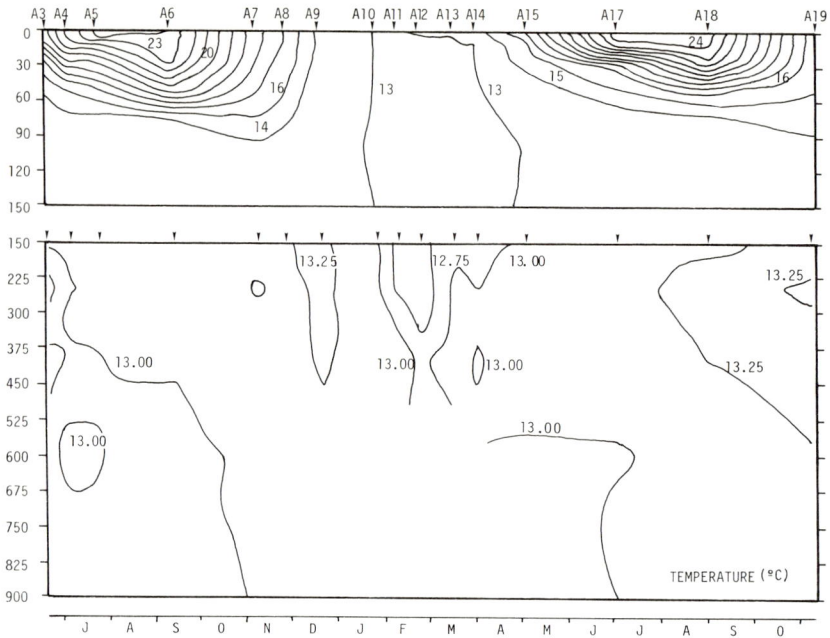

Fig. 3.13. (a) Annual evolution of temperature, salinity and nitrate at a station offshore of Barcelona (41°6.0'N, 2°23.0'E). The maxima of chlorophyll (of less than 1 g l^{-1}) are shown on the figure of nitrate. Data from the Instituto de Investigaciones Pesqueras of Barcelona.

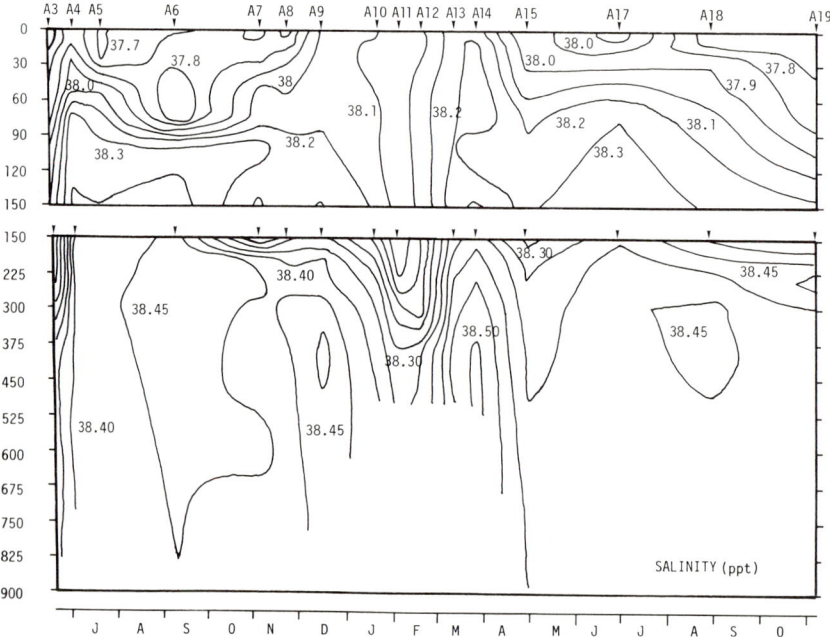

Fig. 3.13 (b)

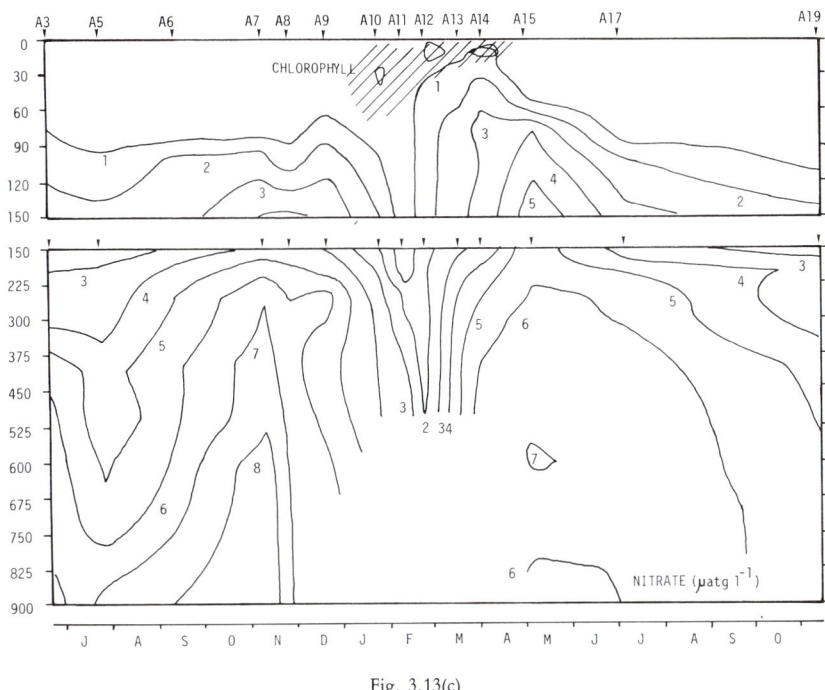

Fig. 3.13(c)

Another secondary maximum is sometimes observed when the thermocline weakens and vertical mixing begins in autumn. However, when mixing is at its highest, long range turbulence prevents the accumulation and permanence of phytoplankton in the photic zone and production decreases again.

Offshore from Barcelona, the first maximum of nitrates in the photic layer (winter) gives rise to a well-developed bloom of phytoplankton. A second nitrate maximum observed in spring is followed only by a much lower increase of plankton. The first, but not the second, nitrate maximum is associated with an increase in phosphate. The available dissolved phosphate is used during the first maximum and becomes particulated. Since then, most of it is rapidly recycled in the upper layers and is no more apparent as dissolved phosphate. During winter, phosphate decreases more than nitrate does in the upper 400 m or so. Nitrate recovers during spring. After stratification, phosphate accumulates slowly in the reservoir of deep water.

One of the possible reasons for this nutrient cycle offshore from Barcelona is the proximity of the Gulf of Lions, an area of violent winter mixing, from where superficial winter water (of high salinity and density) has been enriched with nutrients but has had its phosphates used before spreading to the nearby areas, where productivity is then phosphate limited. Primary production, induced by seasonal vertical mixing, is as irregularly distributed as are external or auxiliary energy inputs.

3.4. THE WEATHER IN THE WESTERN MEDITERRANEAN

The Aerospace Science Division (1968) published a *Catalog of European large-scale weather types* which gives characteristic local weather conditions in 29 different areas in the Western Mediterranean basin (Reiter, 1975).

Everybody, especially the weather forecasters, agree that the Mediterranean is a difficult area and recent multi-national efforts have been undertaken to clarify the mechanisms involved in atmospheric evolution and their relative importance. One example is the Alpine Experiment (ALPEX) carried out in 1982 in the north-western Mediterranean, during late winter and early spring (under the Global Atmospheric Research Program).

The different weather situations in the Western Mediterranean basin as well as their complexity, arise from two facts of a static nature: the orography of the surrounding land masses and its geographical situation in the Globe.

Local topography have intricate and complex modifying effects on the region's synoptic scale weather systems. This topography is responsible for the prevailing wind systems in the area.

The geographical latitude is the same of important topographic features as the Rocky Mountains and the Himalayan highs, which affect the large-scale flows on a world-wide basis.

Finally, orographic features also exercise their influence on the atmosphere on smaller scales, controlling flow phenomena with dimensions of the order of 10 to 100 km.

3.4.1. Topographical features

The complex topography behind the shores of the Western Mediterranean, constitute barriers against air currents and openings or gaps between mountains which impose prevailing wind directions. From the northern edge, going clockwise, there is the Massif Central and the Alps, with the Rhône valley between them, with a meridional orientation and opening into the Gulf of Lions. To the east, the Appenine Mounts separating the western basin (Tyrrhenian Sea) from the Adriatic (in the Eastern Mediterranean). To the south-east, between Europe and Africa, the Strait of Sicily is the natural limit for the Western Mediterranean basin and the connection between western and Levantine waters. In the southern edge, from Tunisia to Morocco, the large Atlas system forms an important barrier which leaves only the Biskra gap in the Department of Aurès (Algeria). The Strait of Gibraltar which connects the Atlantic to the Mediterranean forms also a gap between mountain ranges. To the north the Sierra Nevada reaches elevations of more than 3000 m, and the centre of the Peninsula is an extended plateau between 500 and 700 m high, with the mountain systems higher than 1000 m. Between Gibraltar and the Pyrenees are the Jucar and Ebro valleys, separated by the Sierra de Gudar (the second gap being the most important because it directly connects the Spanish mountain ranges of the Bay of Biscay to the Mediterranean, with the NW – SE orientation). The Pyrenees, with elevations of more than 3000 m, and a W – E orientation, constitutes an important barrier which is separated from the Massif Central by the Carcassonne gap, that connects the SW French low lands (on the Atlantic side) to the NW Mediterranean (Gulf of Lions).

The importance of mountains, and gaps between them is that the air flow is strongly modified from that which would be predicted from the geostrophic model. In the friction or Ekman layer (usually about 1000 m thick above terrain level) frictional drag causes an inflow across the isobars towards lower pressure. This departure from geostrophic wind is usually less than 45° (about 20 or 30°) over flat and smooth terrain, but in narrow mountain gaps, the angle may reach 90°, that is, the air may fall down the fall-line, almost like water would do. So, an ageostrophic flow can pass freely and at a considerable strength into the Mediterranean Basin through those gaps below some 1000 m elevation. In fact, the gaps indicated before are associated with wind systems well known and named by fishermen and people in general many centuries ago. However, if the air flow along the pressure gradients in the lower and middle troposphere is strong enough to force the flow over the mountains rather than around them, strong and gusty winds on the lee side will result (Föhn winds).

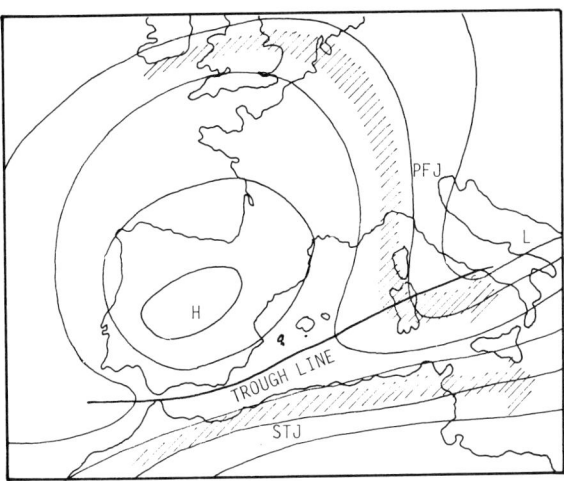

Fig. 3.14. Typical 'blocking' high which brings about the interaction of the Polar and Sub-tropical Jet Streams on the Western Mediterranean (from Reiter, 1975).

3.4.2. Large-scale weather systems

During winter and spring, the Mediterranean regime is characterized by extensive cyclonic activity, and the rainy season consequently extends into winter. One of the large-scale factors which bring about this low-pressure activity is the interaction between the Polar Front Jet Stream and the Sub-tropical Jet Stream. During late winter and spring, blocking anticyclones are frequently observed over the eastern Atlantic. They can remain for several days or even weeks, and are associated with a semi-permanent, low-pressure trough in the mid-troposphere, extending along the axis of the Mediterranean basin. This trough leads to cyclogenesis when vorticity maxima move into its region of influence. The latter are associated with wave disturbances which travel along the Polar Front Jet Stream, extending around the blocking high and taking a north-westerly, or northerly, direction over the Rhône gap and lead to cyclogenesis in Genoa (Fig. 3.14). However, these vorticity maxima may also be associated with the Sub-tropical Jet Stream which crosses the high-pressure bridge between the blocking high and the sub-tropical high-pressure belt.

There are certain regions over which the Polar Front and the Sub-tropical Jet Streams interact quite frequently, giving rise to very strong, combined jet maxima and rapid cyclogenesis and bad weather. The Sub-tropical Jet Stream belt shows a planetary, triple wave pattern, flowing at higher latitudes over the east of the United States, Japan and the east-southern Mediterranean area. In fact, jet maxima in the Sub-tropical Jet Stream are located slightly downstream from regions of major cyclogenetic activity (Fig. 3.15). The generation of cyclones in the lee of the Rocky Mountains is a well-known phenomenon. Downstream from the east China Sea, frequent and strong cyclogenesis occurs along the polar front that separates the cold air sweeping around the barrier of the Himalayas and the Plateau of Tibet and the warm air originating over south-east Asia. In the Mediterranean region and in winter, the Gulf of Genoa is one of the most cyclogenetically active centres of the World (Fig. 3.16). The Sub-tropical Jet Stream does not trigger frontal cyclogenesis along the polar front, but it interacts with the Polar Front Jet Streams as soon as pronounced cyclogenesis in the region occurs. In fact, cyclogenetic activity in the Gulf of Genoa area is mainly due to the orographic effect of the Alps and also to the concave shape of mountains and coastline in the Gulf of Genoa.

The Sub-tropical Jet Stream wind maxima near the Mediterranean may be explained as a resonant planetary wave between the two waves produced by the Rocky Mountains and the Himalayas (in an appropriate position relative to the Mediterranean region of preferred cyclogenesis). As is well-known the streamlines turn cyclonically on the lee side of a N – S oriented mountain range receiving a westerly flow (or in the lee side of an E – W oriented chain under a northerly flow). This may be the reason why rather persistent troughs are found in the westerly flow regime (especially during winter) in the lee of the Himalayas and the Rocky Mountains. The Pyrenees and the Alps offer insufficient resistance to the mid-latitude westerlies to cause the generation of semi-permanent planetary-scale trough, and the European trough situations are less stationary and less stable than the two orographically influenced planetary troughs. However, an enhanced cyclogenetic activity is to be expected over northern Italy, in the lee of the Alps when northerly Jet Stream impinges upon them. The blocking of cold fronts along the northern rim of the Alps also has a dynamic effect upon cyclogenesis in northern Italy. The orographic barrier holds back the pressure rise pattern associated with the inflow of cold air behind the cold front and allows surface pressures to continue falling over northern Italy, thus helping cyclogenetic activity.

3.4.3. Western Mediterranean weather regimes

Three categories of flow can be used to briefly classify the weather types in the Mediterranean. Each of these is related to the position of the sub-tropical high (Fig. 3.17). With a general zonal flow, the sub-tropical high is in its normal position near 35°N; with a mixed flow, it is displaced towards the north or north-east to approximately 50°N; when a blocking high lies between approximately 50 – 70°N there is a meridional flow. In the latter case, the stationary high which 'blocks' the progress of migratory cyclones and anticyclones may vary its position as it does not find any effective orographic 'anchoring'. These changes in longitude cause drastically different weather patterns over Europe and the

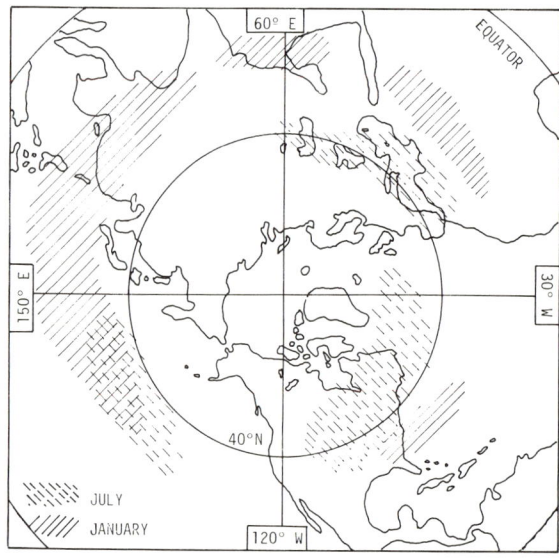

Fig. 3.15. Mean location of the Sub-tropical Jet Stream maxima (dashed areas) in January (40 m s^{-1}) and July (18 m s^{-1}). An approximate three-lobed pattern may be recognized, one of them on the Mediterranean area.

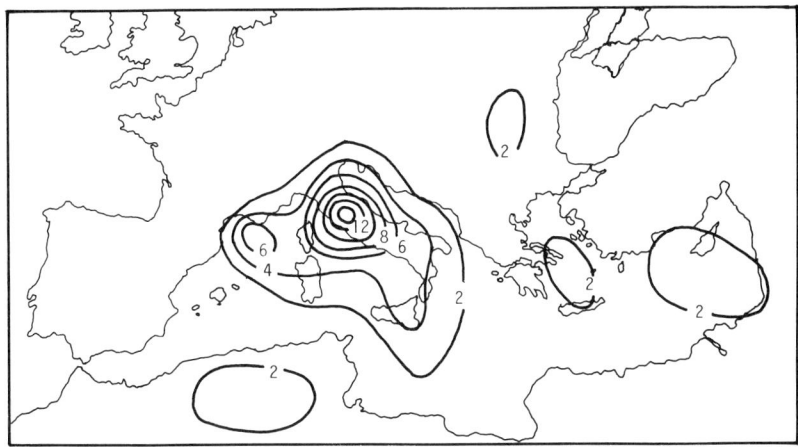

Fig. 3.16. Frequency distribution of cyclogenesis in January – February (after Frank and Eliott, 1953).

Mediterranean region. In fact, more than half of the typical weather patterns occurring in Europe are characterized by a blocking high somewhere in the region. Usually, most blocking anticyclones develop west of Greenwich between January and May, giving place to cold winds in the Mediterranean. This kind of situation, which may last for two weeks or even more, prevents the flow of humid and warm Atlantic air into Europe and helps the intrusion of cold air along the east side of the block into the Mediterranean, where it maintains the weather unsettled over most of the region.

The Genoa depression

They are low-pressure systems which characteristically form in the Gulf of Genoa and in the Ligurian Sea (Fig. 3.18). These are often included with those from the upper Po valley. Several factors seem to play an important role in the development of these depressions.
1. The thermal contrast between land and sea, which affects the pattern and development of surface pressure. During the cold season, the waters of the Mediterranean are significantly warmer than the continental land surfaces and this temperature contrast tends to produce an effect of 'heat low' over the Mediterranean basin. One could say, in general, that a monsoon effect modulates the wind systems in the Western Mediterranean, so that in summer they are deflected to the coasts (when the land has been strongly heated by Sun) and in winter they are reinforced in the seaward direction (when the sea is warmer than land).
2. The interaction between the Polar Front and the Sub-tropical Jet Streams.
3. The enhancing effect of the Alps on cyclogenetic activity along the southern slopes, under a northerly flow.
4. The effect of concave terrain features on cyclone formation.
5. The blocking of cold fronts along the northern rim of the Alps.

Winds and gales

A frequency distribution of winds and gales are shown in Figs. 3.19 and 3.20. The strongest winds in the north-western Mediterranean are those with a northern component; Mistral, Tramontana and Gregal.

Mistral. This has a variety of names, depending on the region. It blows from different directions according to local orography. In southern France it is called Mistral, Mangofango, Sécaire, Maistreau, Magistral; in northern Catalonia it becomes a kind of tramuntana, while in the rest of Catalonia it is called Mestral and it is strongly influenced by the Ebro valley where it is called Cierzo. It is a NW wind which develops usually with a high on the western side of the Iberian Peninsula and a low over Scandinavia or the Northern Sea. This low tends to progress over Europe to Italy and Greece. In the Mediterranean this wind is strong, with forces around 4–6 Beaufort scale, reaching at times forces 8 and even 10 (the latter with the entry of the cold front). In the Gulf of Lions it makes the sea rise (waves of more than 4 m, shorter than those formed in wider seas or oceans). The strong swell extends to the Balearic Islands and swell, with intervals of strong swell, south.

Tramontana. Tramontana (northern wind) generally occurs in NW and NE situations. A low-pressure centre goes to the south, while the Atlantic high extends to the north. The northern wind commences after the passage of the cold front.

Gregal. This is a NE wind, which usually follows the north situation, when the tramontana has been strong. The low is then in the Mediterranean, in Italy, in the Gulf of Genoa or between Sicily and

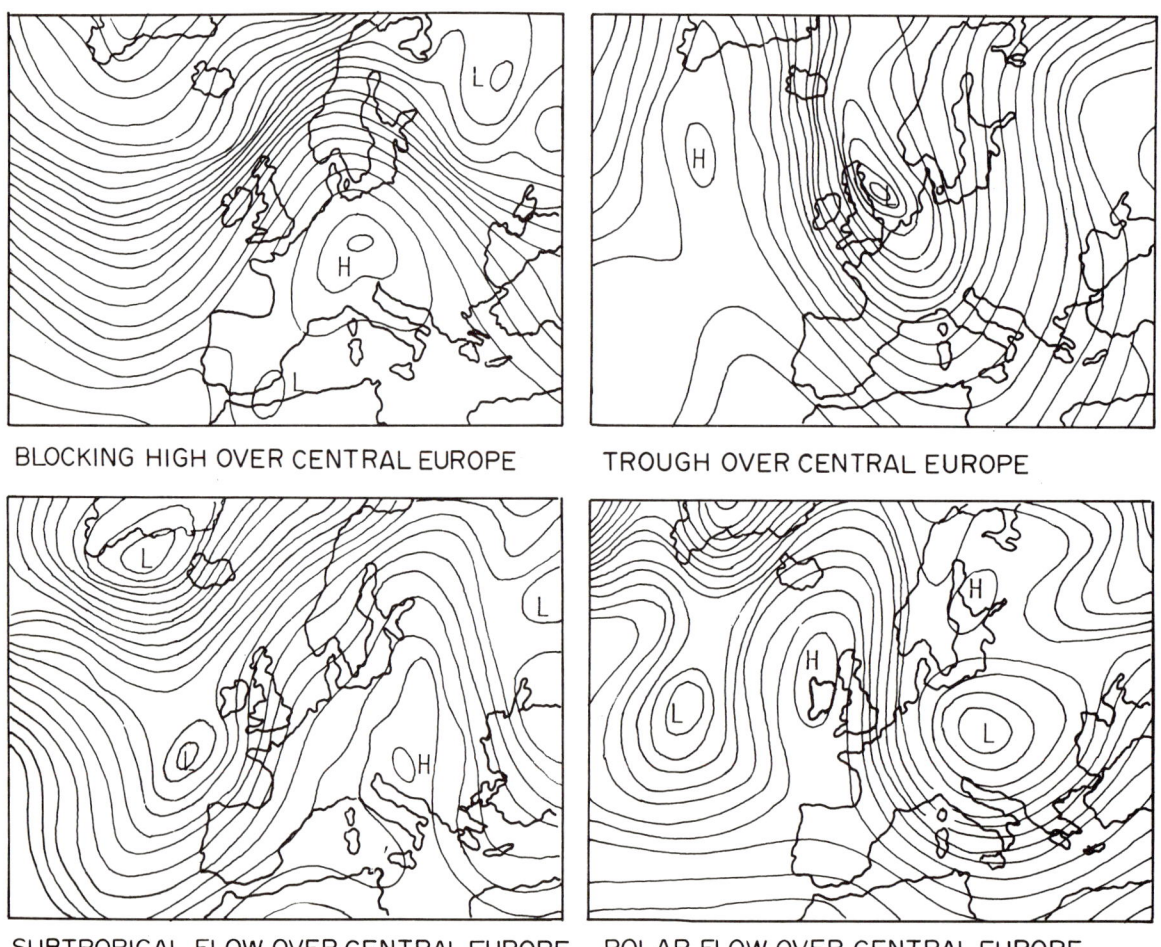

Fig. 3.17. (a) Typical synoptic weather situations in the Western Mediterranean (heights at 500 mb) (from Aerospace Science Division, 1968).

THE DRIVING MACHINE

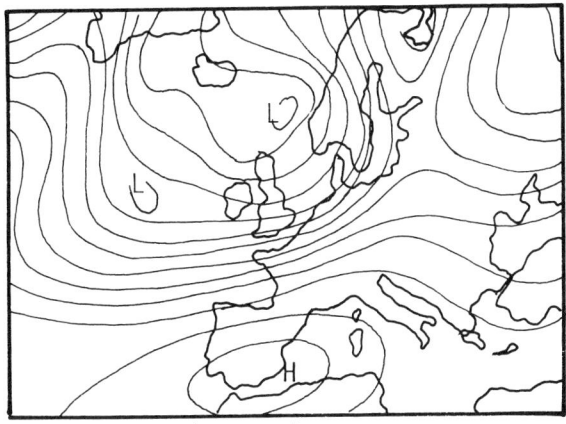

ZONAL FLOW (SUMMER)

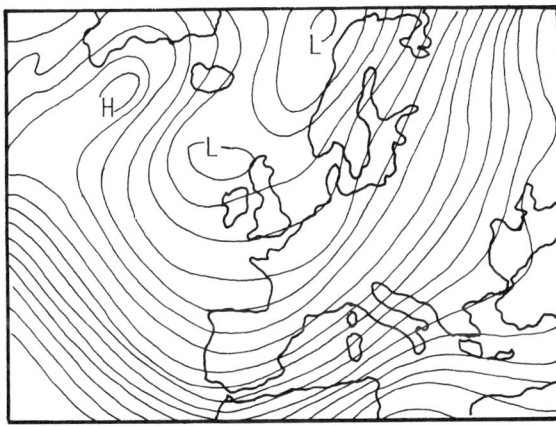

ZONAL FLOW (WINTER)

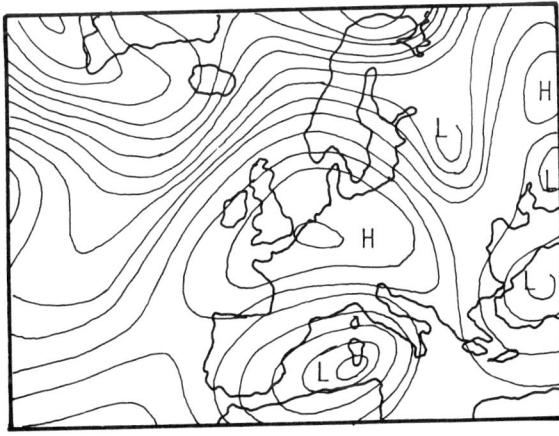

MIXED FLOW

Fig. 3.17. (b)

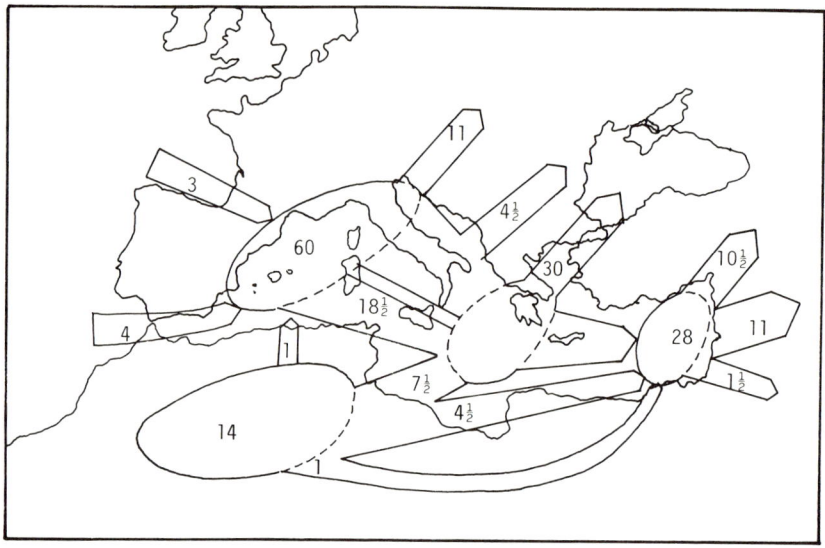

Fig. 3.18. Principal tracks of the Mediterranean depressions. Average annual frequencies are shown (from Air Ministry Meteorological Office, 1962).

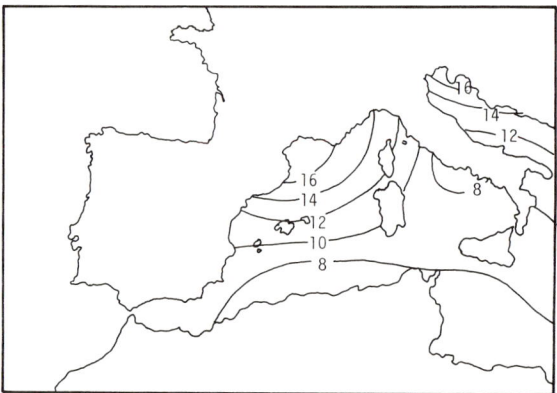

Fig. 3.19. Mean annual percentage of frequency of strong winds (Beaufort force ≥6) (from Air Ministry Meteorological Office, 1962).

Sardinia. The air is very cold and may bring snow and frosts to the coastal areas. Arriving quite dry at the sea, it takes some moisture from the relatively warm sea. Special situations are when there is a cold drop in higher levels in the atmosphere, which creates unstable conditions which lead to heavy snowfall or rain (depending on the season).

Other winds. Other important winds include the Levante (or Llevant), which is an east wind, between 80° and 100°. It is moderately frequent but forces of 6 or higher show a frequency of 1%. In general, these winds appear with a high over Europe, with an east-west orientation, and a low over the north of Africa. Two extreme situations can be described. One of them gives the Bora wind (ENE) over Yugoslavia and Italy, when the high is centred in France (around 45°N) and the low is well in Africa (around 30°N). The other situation gives southern wind (Sirocco) in Sicily, with the high around 50°N and the pressure over northern Africa (affecting Spain). The east wind can produce strong swell which affects the Spanish coasts of the Mediterranean. Usually, the air travels for long enough over the sea to give cloudy weather and rain over the coast of Spain. The worst situations appear when the low in North Africa is reinforced for a time by the rest of a low over southern Italy. Then the winds can reach forces from 7 to 9 between 42° and 37°N.

The SE (Sirocco, Siroco or Xaloc) and South (Migjorn in the Balearic Islands and Garbí in Catalonia) winds blow when there is a low in the southern-western edges of the Iberian Peninsula (between 5 – 10°W and 35 – 40°N) moving eastward. These winds reach forces 3 – 4 Beaufort and only in 0.3% of the times S winds reach force 6 or more. Being originally warm, and dry, the S winds pick up moisture and may reach the islands and the northern coasts with fog and rain.

The SW wind, called Abrego in the centre of Spain, Lebeche or Llebeig in the Mediterranean coasts (Leveccio in Italy), often comes with a frontal depression passing from the Atlantic to the Mediterranean with its centre on 45°N. Sometimes the low crosses the north-western corner of Spain on its way to France.

When there are several lows passing one after the other, the SW becomes vendaval. It is strong and gives place to very high sea in the Western Mediterranean, and strong gale in the Strait of Gibraltar and Alboran sea. The SW is able to produce some upwelling in the Spanish coasts.

All the situations described hitherto are fairly frequent during autumn, winter and spring. In summer, the pressure over the Western Mediterranean is kept relatively high, with poor gradients, to give pleasant sunny weather and a regime of breezes (land and sea breezes) with winds veering with the movement of the sun during the day.

3.4.4. Weather and primary production

The cyclogenetic path in the NW Mediterranean

One of the ways in which the upper layers are enriched with nutrients is the suction effect of atmospheric depressions. In a low-pressure system, the horizontal component of the geostrophic or thermal wind follows the isobaric curves. Near the sea surface, the wind tends to be slowed down by

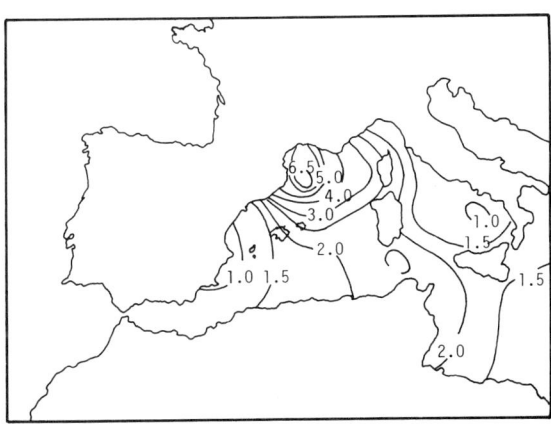

Fig. 3.20. Mean annual percentage frequencies of observations of gales (from Air Ministry Meteorological Office, 1962).

friction. Under these conditions, the force due to the pressure gradient is no longer balanced by the Coriolis force, which is proportional to wind velocity, and the direction of the air flow is slightly deviated to the centre of the cyclonic gyre. However, and according to the Ekman model, the water mass tends to go out of the gyre in a direction of about 90° to the right of surface wind direction (Northern Hemisphere). So, the resultant net water transport in the upper layer of the ocean, under an atmospheric low pressure is to the higher ones, out of the cyclonic gyre.

The outflowing surface water must be replaced by water from below and the isopycnals bend upwards. If water is highly stratified, with a strong thermocline, for example, a rise in the thermocline and a lowering of the sea surface level can be detected at the centre of the system (the latter through geostrophic calculations). If the cyclone is strong enough, nutrient rich water can reach the photic layer even when there is some stratification. When this is weak, as during autumn and winter in the Mediterranean, or if several successive low-pressure systems have weakened it, then the upwelling of rich water may be really important.

The north-western Mediterranean is a well-known cyclonic spreading centre where Atlantic lower pressures combine with polar fronts entering the Mediterranean. This is the reason why the NW Mediterranean is richer than the rest of the basin.

Deep-water formation and global productivity

The Gulf of Lions and Ligurian Sea are often swept by strong northern winds. The most frequent cause of these winds is the cyclogenesis in the Gulf of Genoa, where an interaction between a cold front (linked to a general depression) and the Alps occurs. This mechanism generates the Mistral-Tramontana wind system. These strong cold and often dry winds have a dramatic effect on the water vertical stability. The sea cools in its surface and salinity increases because of intense evaporation. The density rises in some areas to a σ_t greater than 29.1 at the surface, disappearing the vertical stability and mixing the column of water in 2.000 m. The mixing of surface and intermediate waters give deep water. The events responsible for deep-water formation usually take place during February and March but their specific characteristics differ slightly from year to year (Lacombe *et al.*, 1981). Once the water in the column has reached the characteristics of deep water, it sinks and spreads at the corresponding density level. The whole process is better explained by Hopkins in this same publication.

Deep-water formation in the north-western Mediterranean is a process which rapidly transforms surface waters into deep ones. A 'pump effect' can be thought of as the limit between the upper and lower compartments and has to raise its level accordingly. This rise makes a proportional quantity of nutrients available to the upper layer which mixes seasonally. It seems that cold winters are associated with higher catches of pelagic fishes, although an analysis of time series of data would be necessary to confirm this. One of the causes is the general autumnal and winter vertical mixing which can reach greater depths in colder winters. Also, colder winters may be associated with greater volumes of newly formed deep water and hence, greater quantities of nutrients in the upper layer.

According to Béthoux (1981), some 13.52×10^{12} m^3 of surface water mix each year with intermediate waters to form deep water. This volume of surface water, spread under the whole western basin (8.6×10^{11} m^2), can cause the upper limit of the lower compartment to rise by about 15 m. As this water spreads only under the north-western Province (3.0×10^{11} m^2), the rise might even reach 45 m. The actual pump effect can be placed safely between the two estimates with a mean annual rise of the nutricline of 30 m, due to deep-water formation. At a rate of 0.25 mg at P m^{-3}, this represents a production of about 10 g C m^{-2} y^{-1}. So, may be 10 to 15% of the observed production could be explained by this effect and the variation in the total amount of deep-water formation could be reflected in interannual variations of global productivity in the Western Mediterranean basin.

Coastal upwelling in the Western Mediterranean

Several kinds of rich water upwelling to the photic zone have been described for the World Oceans. In a broad sense, 'upwelling' is the phenomenon by which a significant mass of nutrient-rich and relatively cold water (from below the level of the seasonal thermocline, or equivalent depths) rises in mass, by advection, to the surface. In general, a vertical upward directed flux of water occurs at some depth and a horizontal flux of water must exist at the surface for the sake of continuity of mass or volume.

In the Western Mediterranean the general circulation is neither strong enough nor sufficiently well-oriented to favour upwellings such as the major coastal ones, except perhaps in the Alboran Sea, near the Strait of Gibraltar. There, an anticyclonic gyre forced by the Atlantic inflow and moduled by tides exists between the Iberian and African coasts.

However, coastal upwelling events have been described in the South of France, associated with strong northerly winds (Millot, 1979). There are also suspicions of coastal upwelling in the Mediterranean coast of Spain, at least, off the coast of Castellón (Margalef and Herrera, 1963), with winds from the third quadrant. In these cases, the driving force is a strong wind directed parallel to the coast or from land to sea. The Ekman transport, directed to the right of the wind direction in the Northern Hemisphere, can cause upwelling along the shore when the wind blows parallel to it with land on its left. This mechanism can be simplified in two-dimensional models where the entrained surface water to the offshore region enables the underlying waters to reach the surface. However, more complicated mechanisms and three-dimensional models, can explain other cases of coastal upwellings where wind blows from land to sea (O'Brien et al., 1977). The latter case could be frequent in the Western Mediterranean in all those coastal areas where strong winds channelized by mountain gaps, reach the sea. However, all upwellings in the Western Mediterranean are dispersed and sporadic. The extension of their enriching effect is not large, being limited to nearshore areas, and in the conditions when some nutrient-rich water is present at appropriate levels.

A somewhat similar enriching effect can be produced by river discharge (Ebro, Rhône), where the less saline plume entrains some subsurface water and forces an undercurrent to the coast. River discharge has an enriching effect due to the direct contribution of nutrients of terrestrial origin but it also helps some nutrient-rich marine water to reach the upper layer through mixing by shear stress (Herrera and Margalef, 1963). This kind of phenomenon occurs in the Strait of Gibraltar too, where the Atlantic inflow entrains some Mediterranean underlying water so that the entering water is enriched along its path through the Alboran sea to the East. This phenomenon is also important because it means that some recycling of nutrients occurs before the Mediterranean water leaves the western basin so that the impoverishing effects of a negative-estuary are somewhat diminished in the Mediterranean.

In the east coast of Spain, it becomes difficult to separate the three enriching mechanisms hitherto presented: seasonal mixing, the 'pump effect' and upwelling. The latter is often not important because enriched water does not reach the surface in mass. However, salty-rich waters can rise and are able to approach the coast and to extend into the platform and shallower waters, where vertical mixing is more vigorous, especially in very cold winters, and where interactions with bottom or river discharges may occur.

REFERENCES

Aerospace Science Div. (1968) *Catalogue of European Large-scale weather types. Revised by H.Q. 2nd Weather Wing. Aerospace Sci. Div.*, 68 pp.

Air Ministry. Meteorological Office (1962) *Weather in the Mediterranean Vol. I, General Meteorology*, H.M.S.O. 372 pp., London.

Allain, C. (1960) Topographie dynamique et courants généraux dans le bassin occidental de la Méditerranée. *R. Trav. Inst. P. Mar.* 24, 121–145.
Béthoux, J. P. (1981) Le phosphore et l'azote en mer Méditerranée, bilans et fertilité potentielle. *Mar. Chem.* 10, 141–158.
Budyko, M. I. (ed.) (1963) *Atlas teplovogo balansa zemnogo shara*. Akad. Nauk, SSSR, Moscow.
Bush, N. E. (1977) Fluxes in the Surface Boundary Layer Over the Sea. In *Modelling and Prediction of the Upper Layers of the Ocean*, Ed. E. B. Kraus, pp. 72–91, Pergamon Press.
Calathas, J. (1970) *Contribution à l'étude de la réflexion de la lumière du jour à la surface de la mer*. Thèse 3ème cycle. Université de Paris VI.
Cialdi, A. and P. A. Secchi (1968) On the transparency of the sea. Translation by Professor Albert Collier. *Limnol. Oceanogr.* 13, 391–394.
Fofonoff, N. P. (1981) The Gulf Stream. In *Evolution of Physical Oceanography*, Eds. B. A. Warren and C. Wunsch, pp. 112–139, The Massachusetts Institute of Technology Press, Cambridge, Massachusetts and London.
Font, J. (1983) Corrientes permanentes en el borde de la plataforma continental frente al Delta del Ebro. *Estudio oceanográfico de la plataforma continental. Seminario Científico. Cádiz 15–18 marzo, 1983*, pp. 230–248.
Frank, S. R. and Elliott, R. D. (1953) *Operational Weather of the Mediterranean Area, Area 1*. AROWA 09-1053-097.
Gall, M. H. W. (1949) Measurements to determine extinction coefficients and temperature gradients in the North Sea and English Channel. *J. mar. biol. Ass. U.K.* 28, 757–780.
Hersey, S. B. and Backus, R. H. (1962) Sound scattering by marine organisms. In *The Sea. Vol. I. Physical Oceanography*, Ed. M. N. Hill, pp. 540–566, Wiley & Sons.
Herrera, J. and Margalef, R. (1963) Hidrografía y fitoplancton de la costa comprendida entre Castellón y la desembocadura del Ebro, de Julio 1960 a Junio 1961. *Inv. Pesq.* 24, 33–112.
Ivanoff, A. (1977) Oceanic absorption of solar energy. In *Modelling and Prediction of the Upper Layers of the Ocean*, Ed. E. B. Kraus, pp. 47–71, Pergamon Press.
Jerlov, N. G. (1968) *Optical Oceanography*. Elsevier Publishing Company, Amsterdam, London, New York.
Kantha, L. H. (1977) Note on the role of internal waves in the thermocline erosion. In *Modelling and Prediction of the Upper Layers of the Ocean*, Ed. E. B. Kraus, pp. 173–177, Pergamon Press.
Kraus, E. B. (ed.) (1977) *Modelling and Prediction of the Upper Layers of the Ocean*, Pergamon Press.
Lacombe, H., Gascard, S. C., Garella, J. and Béthoux, J. P. (1981) Response of the Mediterranean to the water and energy fluxes across its surface, on seasonal and interannual scales. *Oceanol. Acta* 4, 247–255.
Laevastu, T. (1960) Factors affecting the temperature of the surface layer of the sea. *Societas Scientiarum Fennica. Commentatisnes Physico-Mathematicae*, Vol. 25, no.1, Helsinki.
Laevastu, T. (1976) *Oceanic Water Balance*. V.M.O. no. 442.
Margalef, R. and Herrera, J. (1963) Hidrografía y fitoplancton de las costas de Castellón, de Julio 1959 a Junio 1960. *Inv. Pesq.* 22, 49–110.
Millot, C. (1979) Wind induced upwellings in the Gulf of Lions. *Oceanol. Acta* 2, 261–274.
Munk, W. and Wunsch, C. (1979) Ocean acoustic tomography: a scheme for large-scale monitoring. *Deep Sea Res.* 26A, 123–61.
Neumann, G. and Pierson, W. J. (1966) *Principles of Physical Oceanography*. Prentice Hall, New Jersey.
Niiler, P. P. and Kraus, E. B. (1977) One dimensional models of the Upper Ocean. In *Modelling and Prediction of the Upper Layers on the Ocean*, Ed. E. B. Kraus, pp. 143–172, Pergamon Press.
O'Brien, J. J., Clancy, R. M., Clarke, A. J., Crepon, M., Elsberry, R., Gammelsrod, T., MacVean, M., Röed, L. P. and Thompson, D. (1977) Upwelling in the Ocean: two and three dimensional models of Upper Ocean dynamics and variability. In *Modelling and Prediction of the Upper Layers of the Ocean*, Ed. E. B. Kraus, pp. 178-228, Pergamon Press.
Phillips, O. M. (1969) *The Dynamics of the Upper Ocean*. The University Press, Cambridge.
Phillips, O. M. (1977) Entrainment. In *Modelling and Prediction of the Upper Layers on the Ocean*, Ed. E. B. Kraus, pp. 92–101, Pergamon Press.
Pollard, R. T. (1977) Observations and models of the structure of the Upper Ocean. In *Modelling and Prediction of the Upper Layers on the Ocean*, Ed. E. B. Kraus, pp. 102–117, Pergamon Press.
Poole, H. H. and Atkins, R. G. (1929) Photo-electric measurement of submarine illumination throughout the year. *J. Mar. Biol. Ass. U.K. (N.S.)* 16, 297–324.
Pond, S. and Pickard, G. L. (1978) *Introductory Dynamic Oceanography*. Pergamon Press.
Reiter, E. R. (1975) *Handbook for Forecasters in the Mediterranean: Weather Phenomena of the Mediterranean Basin*, Environmental Prediction Research Facility, Naval Postgraduate School, Monterrey, California.
Ring Group (1981) Gulf Stream cold rings: their physics, chemistry and biology. *Science* 212, 1091-1100.
Satterlund, D. R. and Means, J. E. (1978) Estimating solar radiation under variable cloud conditions. *Forest Sci.* 24, 363–373.
Schevill, W. E., Backus, R. H. and Hersey, J. B. (1962) Sound production by marine animals. In *The Sea. Vol. I. Physical Oceanography*, Ed. M.N. Hill, pp. 540–566, John Wiley & Sons.
SCOR Working Group 15 (1974) Photosynthetic radiant energy. Recommendations. *Sci. Com. Oceanic Res. Proc.* 10, 37–42.
Smith, R. C. (1974) Structure of solar radiation in the upper layers of the sea. In *Optical Aspects of Oceanography*, Eds N. G. Jerlov and E. Steeman Nielsen, pp. 95–119, Academic Press, London, New York.
Spindel, R. C. (1982) Ocean acoustic tomography: a new measuring tool. *Oceanus* 25(2), 12–21.
Steele, J. H. (1981) Some varieties of biological oceanography. In *Evolution of Physical Oceanography*, Eds B. A. Warren and C. Wunsch, pp. 376–383, The MIT Press, Cambridge, Massachusetts, London.
Tyler, J. E. (ed.) (1977) *Light in the Sea*. Dowden, Hutchinson & Ross, Inc., Stroudsburg, Pennsylvania.

Weinberg, S. (1975) Ecologie des octocoralliaires communs du substrat dur dans la région de Banyuls-sur-Mer. Essaie d'une méthode. *Bijdr. Dierk* 45, 50 – 70.
Weinberg, S. (1976) Submarine daylight and ecology. *Marine Biology* 37, 292 – 304.
Woods, J. D. (1977a) Parameterization of unresolved motions. In *Modelling and Prediction of the Upper Layers on the Ocean*, Ed. E. B. Kraus, pp. 118 – 142, Pergamon Press.
Woods, J. D. (1977b) Information theory related to experiments in the Upper Ocean. In *Modelling and Prediction of the Upper Layers on the Ocean*, Ed. E. B. Kraus, pp. 263 – 287, Pergamon Press.

CHAPTER 4

Physics of the Sea

TOM SAWYER HOPKINS

Oceanography Sciences Division, Brookhaven National Laboratory, U.S.A.

CONTENTS

4.1. Introduction	100
4.2. Thermohaline characteristics	101
4.2.1. Balances	101
4.2.2. Classification	103
4.3. Budget calculations	103
4.3.1. Salinity	104
4.3.2. Dilution	105
4.3.3. Volume flow	106
4.3.4. Budget example	107
4.4. Circulation	109
4.4.1. Surface water	110
4.4.2. Intermediate water	112
4.4.3. Deep water	115
4.5. Deep-water formation	116
4.6. Tides and high frequency motions	122

4.1. INTRODUCTION

The Mediterranean is a small ocean, almost completely severed from the waters of the world ocean. As such, it offers a fascinating scale model for oceanographers. Yet it is surprising to realize that this well-defined and hospitable sea has received little study as a unit. Our knowledge of its physical behaviour is deduced and extrapolated from observations taken opportunistically, or with only regional focus. The reasons for this are largely political and monetary, and fortunately various efforts to overcome these obstacles are now being made by international scientific organizations. The Mediterranean presents an opportunity to study oceanic processes on manageable time and space scales.

The fact that the Mediterranean is unique does not limit its relevance. The physics of the oceans are universal. Within each environment the set of dominant physical phenomena varies; and a diversity of environments is healthy for the advancement of our understanding. Many of the physical anomalies encountered in the Mediterranean present what one might call controlled environments, that is, unique physical situations in which certain dynamics are exaggerated by comparison with the norm. Some of these anomalies are worth noting at the onset of this chapter: The Mediterranean is a zonal ocean, not meridional like the Atlantic, Pacific, or Indian oceans, and its thermohaline circulation is driven by evaporation, not by polar cooling. The meteorology that forces it is much more continental in origin than marine, in contrast to the larger oceans. It is remarkably homogeneous throughout, even in the vertical, if the effect of seasonal surface heating is disregarded. Despite this homogeneity, the slight distinctiveness that persists between its water masses is vital to its circulation. These water masses themselves are circumstantial to the unique Mediterranean bathymetry and meteorology. A total loss of vertical stratification during certain winter situations leads to one of the most extensive vertical circulations known to occur anywhere in the world's oceans. The horizontal circulation is now slow, despite the fact that the Mediterranean is cut off from any oceanic forcing. The tidal currents are slight, with some exceptions, but the coastal environments are surprisingly energetic, primarily because of the anomalously narrow continental shelves and the existence of wind energy at the mesoscale frequencies (half to five days) to which the shelf waters are responsive.

4.2. THERMOHALINE CHARACTERISTICS

The Mediterranean Sea can be characterized oceanographically as a 'negative' or a 'concentration' basin, because its fresh water balance is defined as negative. A more fundamental definition would specify that such basins are bathymetrically distinct subdivisions of larger bodies of water within which the resident deep water is consistently denser than that found at comparable depths outside. The Mediterranean is a subdivision of the Atlantic, and within the Mediterranean system itself there are further subdivisions, for example, the Western Mediterranean (WMED) and Eastern Mediterranean (EMED), and within these the Tyrrhenian, Adriatic, and Aegean seas. The sill is a critical bathymetric restriction preventing free access between waters of the same depth inside and outside the basin, and therefore being causal to the separate evolution of water types inside and outside the basin. When a suite of meteorologically related circumstances promotes the formation of a denser water behind the sill, or within the basin, it is then classified oceanographically as having a negative thermohaline circulation, i.e. $\varrho_{out} - \varrho_{in} < 0$, where ϱ is density. The WMED is one of the largest and most important of such basins.

4.2.1. Balances

It may be useful to outline the basic tenets of basin thermohaline circulation. The two simplest extremes are depicted in Figs. 4.1(A) and 4.1(B), where the positive and negative circulations are associated with positive and negative water exchange through the surface, respectively. In Fig. 4.1, this is denoted as (D), for dilution, and as such accounts only for haline forcing. We note that differential cooling inside the basin may affect its deep-water density relative to that outside and constitute a thermal forcing. In fact, alternating dominance by thermal or haline forcing can lead to seasonal thermohaline circulation reversals or the vertical coexistence of both circulations. The tendency for such thermohaline complexity is assisted by a deep sill and/or by large river runoff, as for example, the Adriatic. In such cases simple definitions are not so easily applied. The schematic of Fig. 4.1(C) indicates that the WMED has an important complication in its thermohaline circulation relative to that of Fig. 4.1(B); namely, its interior has a three-layer flow with the Levantine intermediate water interceding between the surface and deep waters. The deep water (DW) generally moves towards the basin mouth, since it is formed at the opposite end. Very significantly, this is not true in the EMED where the DW is formed at the same end of the basin as the sill.

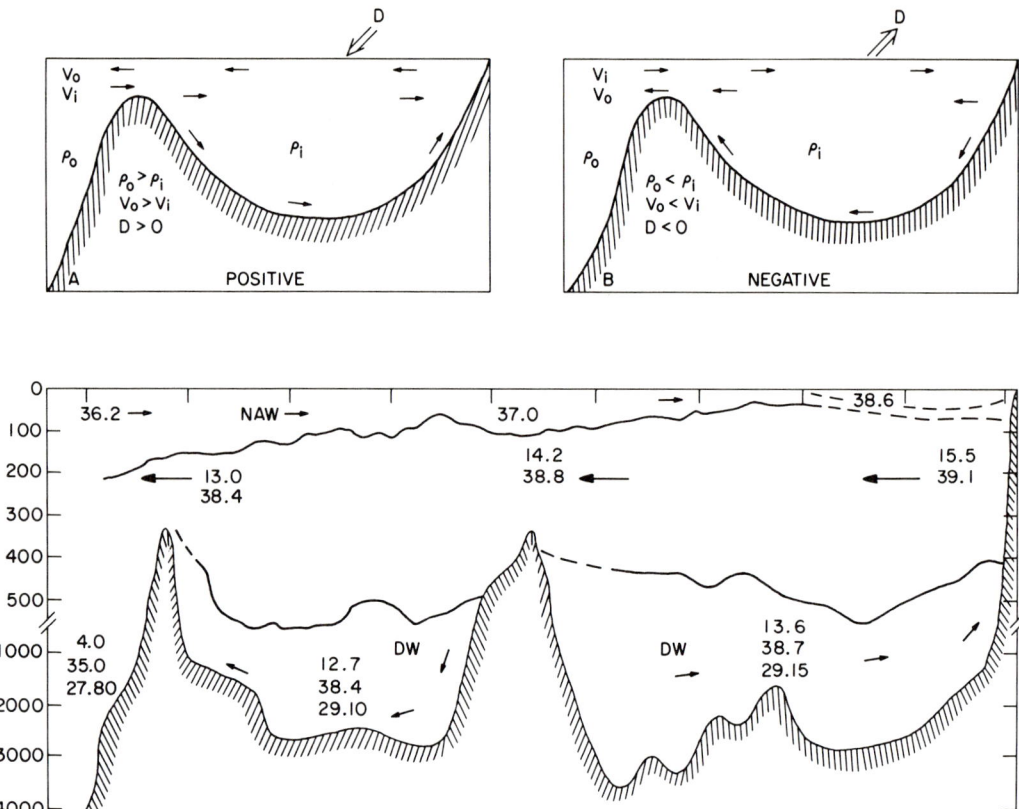

Fig. 4.1. Fundamental thermohaline circulations: (A) for a positive basin (or estuary), (B) for a negative one, and (C) for the Mediterranean. The symbols are defined in the text.

The consequence of a negative dilution is that the inflow (V_i) exceeds the outflow (V_o). The net rate of inflow (km^2/yr) may be expressed as

$$V_i - V_o = -D, \quad (1)$$

where $D = [(P+R-E]\cdot BA$, P is precipitation, R is run-off from land, E is evaporation, and BA is the basin's surface area. This expression is derived from the simple requirement that the basin's water volume be conserved, and it holds for time scales greater than the barotropic response of several hours.

For large negative basins, such as WMED, D is only a small percentage of V_i or V_o, as we shall see below, whose magnitudes are determined by the pressure gradients across the sill, but are difficult to measure. To facilitate determination of these flows, oceanographers have made a less rigorous assumption, that the amount of salt in the basin remains constant over at least an annual cycle, (i.e. the amount of salt entering equals that leaving), therefore,

$$\varrho_i S_i V_i = \varrho_o S_o V_o, \quad (2)$$

where S is the concentration of salt in ppt and ϱ is the density. Equations (1) and (2) are combined to give an expression for the inflow,

$$V_i = D \frac{\varrho_o S_o}{\varrho_o S_o - \varrho_i S_i}. \quad (3)$$

We see that the vigour of the sill exchange is doubly dependent on the evaporation: directly through D, and indirectly through S_o. The value of S_i, we assume, is independent of the internal dynamics of the basin. For large basins, the value of S_o is buffered against annual variations in evaporation, because the amount of salt in each year's production of deep water is only a small portion of that in the deep-water reservoir. However, annual variations in D are directly reflected in the sill exchange. Nevertheless, these simple relationships provide an important guide to the basin's gross thermohaline circulation.

4.2.2. Classification

Ultimately, the questions always asked are how isolated is a basin from its parent waters outside, and how long will it take to significantly flush the basin's deep waters? Clearly, the negative basins have the advantage in terms of turning over the bottom waters because their downward vertical exchange is controlled by convection rather than diffusion as in the positive basins.

The two most critical characteristics in this regard are the volume of dense water formed and the areal extent of the sill opening. The former depends directly on the dilution and indirectly on its internal circulation. The latter is important primarily with respect to the depth of the sill relative to the depth of the basin, that is, the degree of vertical restriction; it is also important in terms of its width, because the wider the sill opening the less is the constraint for unidirectional flow at any depth.

It is useful to compare the WMED with other major negative basins. Hopkins (1978) has classified these basins on the basis of characteristic parameters. The rate at which water enters the basin is characterized by the dilution, D, and the restriction to exchange by the cross-sectional sill area, SA. The basin volume, BV, is another essential characteristic. The ratio of H/D, from BV/(D.BA), defines an atmospheric replacement time where H is mean depth, and BA the basin area. This time approximates the length of time required for the basin to dry up. Another ratio, BV/SA, defines the length of water required to pass through the sill opening in order to replenish the basin. Furthermore, the ratio of these two parameters defines a replacement sill speed, which can be interpreted as the rate at which water must enter to maintain the BV. The logarithms of the two ratios of H/D and BV/SA for a number of negative basins are plotted against each other in Fig. 4.2. Contours of the winter density difference between the incoming surface water and the outgoing water over the sill are drawn on to the plot. These contours roughly parallel the lines of equal replacement sill speed which implies that those basins having greater stratification require greater net inflow and consequently their sills will be more active.

The insert in Fig. 4.2 demonstrates qualitatively the behaviour of a basin with respect to these parameters. The MED is the least sensitive to variations in its deep water and its vertical stratification is second largest after the Red Sea. The WMED, in particular, and the EMED both have very constant deep waters, because D is such a small proportion of H. The position of WMED is more ambiguous because its characteristics cannot easily be uncoupled from the EMED. At the opposite extreme are Elefsis Bay (Greece), the northern Gulf of California, and the Gulf of Suez, all of which are more susceptible to interannual variability in their deep water. The Tyrrhenian and Aegean seas have relatively little vertical structure and less net sill flow, while the Red Sea differs greatly in these respects.

4.3. BUDGET CALCULATIONS

Simple though the above conservation equations may seem, there are a number of observational difficulties that make any determination of the five variables, V_o, V_i, S_o, S_i, and D, uncertain. The trick is to use the three for which observational techniques are most effective, and to use equations (1) and (3) to estimate the other two. The values for the WMED are more sensitive to error since they are inextricably linked to the other internal Mediterranean basins.

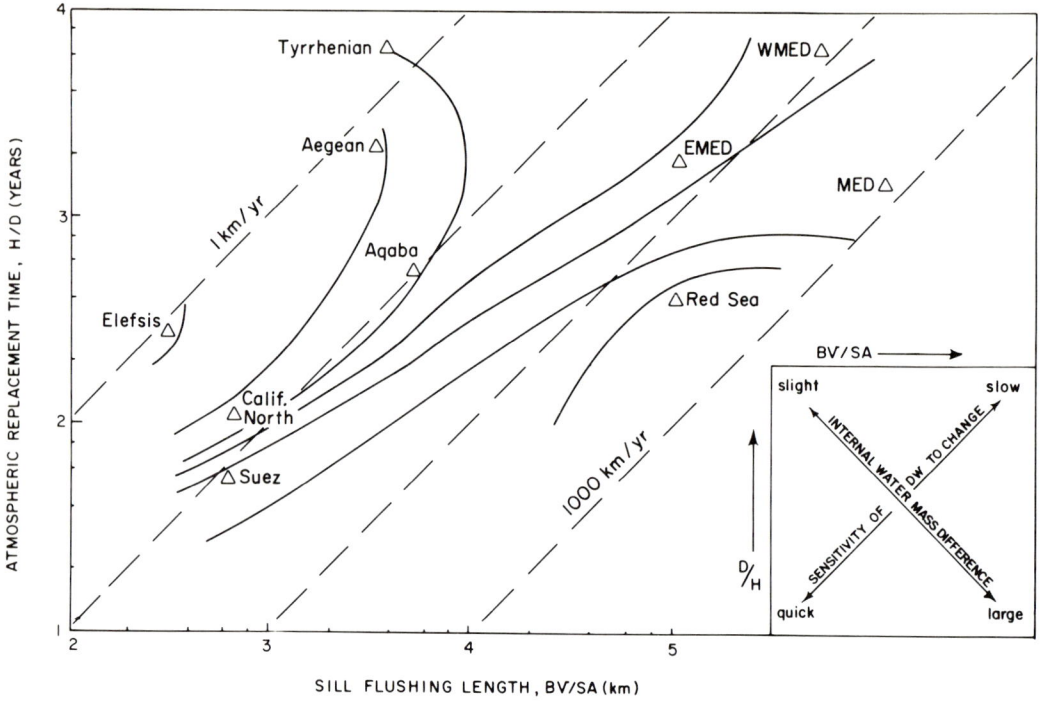

Fig. 4.2. The atmospheric replacement time (H/D) versus the sill flushing length (BV/SA) for a number of negative basins, plotted on a log-log scale with the numbers on the ordinate and abscissa representing powers of ten. The winter density difference over the sill is contoured, and lines representing equal sill speeds are dashed.

4.3.1. Salinity

Generally, S_o and S_i are deduced from observations in and about the straits. These values should represent the salinity associated with the annual mean flux of salt out (or in), which is not quite the same as the annual mean salinity of the entering and exiting waters. But even so, the mean salinities are difficult to estimate because the interface between entering Atlantic and departing Mediterranean waters has large vertical tidal excursions and large vertical salinity gradients which make it difficult to calculate spatial and temporal averages. In a recent review of the Strait of Gibraltar, Lacombe and Richez (1982) have taken values of S_i = 36.15 ppt and S_o = 37.90 ppt; and in a detailed budget of the Mediterranean, Béthoux (1979) uses more or less the same values, S_i = 36.18 ppt and S_o = 37.90 ppt. These averages were taken during the summer, outside the sill, at Station A4 in Fig. 4.3, and they do not represent an annual mean. Winter time values are higher; for example, Bryden and Stommel (1982) cite values for S_i and S_o of 36.50 and 38.45 ppt, respectively. However, the flux of salt is less in winter and greater in summer, for reasons stated below, which must be reflected in any estimate of an annual mean.

Another problem in estimating these salinities peculiar to Gibraltar is the existence of a subsurface salinity minimum ($\approx$ 35.9 ppt) in the adjacent Atlantic caused by an intrusion of the North Atlantic central water mass between the Atlantic surface water and the Mediterranean outflow. Although it seldom intrudes into the Mediterranean, it acts to separate the incoming and outgoing waters, particularly in the more exterior portions of the strait. It tends to contaminate estimates of the outgoing S_o more than those of the incoming S_i, since any vertical mixing that occurs is proportional to the difference in salinity which in this case is much greater below with S_o than above with S_i. This situation

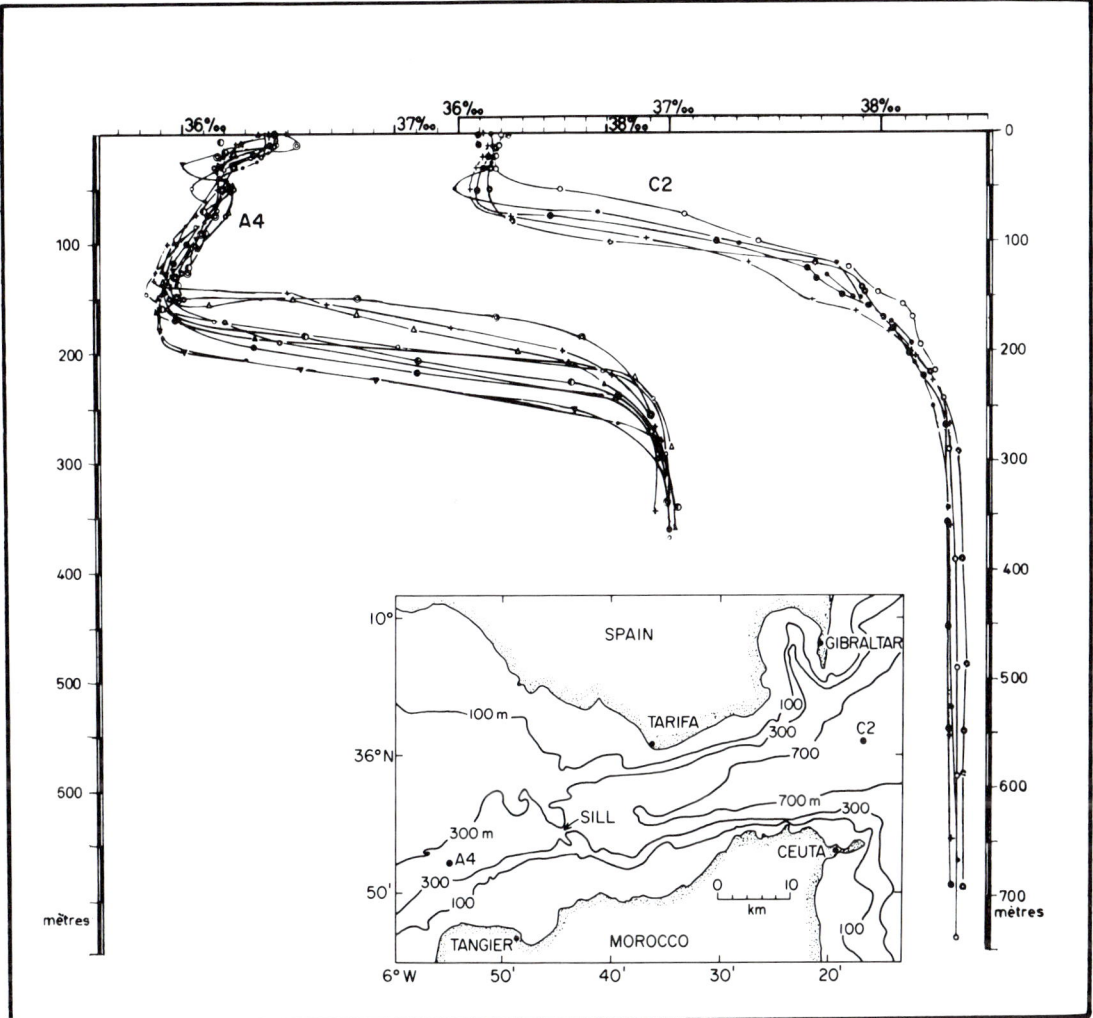

Fig. 4.3. Vertical profiles of salinity from western (A4) and eastern (C2) extremes of the Gibraltar Strait (from Lacombe and Richez, 1982).

is illustrated in Fig. 4.3 which shows a series of salinity profiles taken from just inside (Station C2) and just outside (Station A4) the Strait during the summer of 1960.

Assessment of S_o and S_i at Sicily appears more straightforward, perhaps because fewer observations have been made there than at Gibraltar. It is wider and bathymetrically more complicated in the cross section which makes spatial averages more difficult to obtain, but the tidal action is less and thus temporal averaging is easier. Morel (1971) has estimated that the outflowing surface water changes from 37.05 to 37.30 ppt in transit and the incoming subsurface water from 38.80 to 38.74 ppt.

4.3.2. Dilution

If the salinities are given, the choice remains of estimating either D or one of the Gibraltar flows in order to solve equation (3). Both approaches have been taken, e.g. the water budget by Béthoux (1979), and the Gibraltar volume outflow by Lacombe et al. (1964).

An evaluation of the dilution term requires estimates of its three components: evaporation, precipitation, and run-off. A recent review by Cruzado (1979) summarizes the various estimates of these climatological factors (Fig. 4.4). The subtotals are as follows:

TABLE 4.1. Water balance values expressed in km^3/yr

	WMED	EMED	Total
R	150	479*	629
P	285	498	783
E	1044	2804	3848
D	609	1827	2436

*Black Sea contribution 189 km^3/yr.

Regardless of the methods used to derive these values,* there are several trends that have important consequences to the thermohaline structure and circulation.

Particularly since the great reduction in the Nile discharge, the Mediterranean runoff is strongly concentrated on the northern shores, and while the precipitation is more evenly distributed, the net effect is a relative decrease in surface salinities in the northern regions. However, as we will see later, this is masked in terms of the surface salinity contours (Fig. 4.6) by the residence of the low-salinity North Atlantic water (NAW) along the African coast.

The variability in evaporation is not great, as shown in Fig. 4.4. However, there are important causal and seasonal differences which are not evident in the annual means. The main contribution along the North African coast is associated with Sirocco wind events, which are most prevalent during the cool season, and in the northern sections the maximum evaporation is linked to the strong outbreaks of continental air (the Mistral, the Bora, the Tramontane) occurring during winter. In the Aegean, a summer surface divergence is created as a consequence of mainland Greek and Turkish sea breeze systems, which result in descending air and the strong, dry Meltemi winds. A similar effect occurs in the northern Ionian, and perhaps the Tyrrhenian. The importance of winter evaporation in terms of deep-water formation tends to obscure the fact that the net effect over the entire Mediterranean is for a summer evaporative maximum. Although a seasonal breakdown is not available, the seasonal variation in the WMED is apparently less than in the EMED where, because of its area and its summer evaporative bias, it predominates. The response to evaporative events is barotropic, which in the deep Mediterranean basins is only a matter of hours, and results in an increase in the basin inflow. That is, the time dependence in D [equation (3)] is reflected in V_i, at least for periodicities of a half day or longer. Such a high-frequency causal relationship has not been confirmed by observation, but observations of the amount of Atlantic water entering Gibraltar do show a definite summer maximum (Ovchinnikov, 1974), at least substantiating a summer maximum for the water balance deficit.

4.3.3. Volume flow

Estimates of V_o or V_i are certainly no easier to come by than those of D, but for very different reasons. The entering NAW at Gibraltar may attain speeds in excess of 100 cm/sec, of which the predominantly semidiurnal tide may reach ±60 cm/sec (Defant, 1961). Discounting the tide, the

---

* A more detailed discussion is given in Chapter 5.

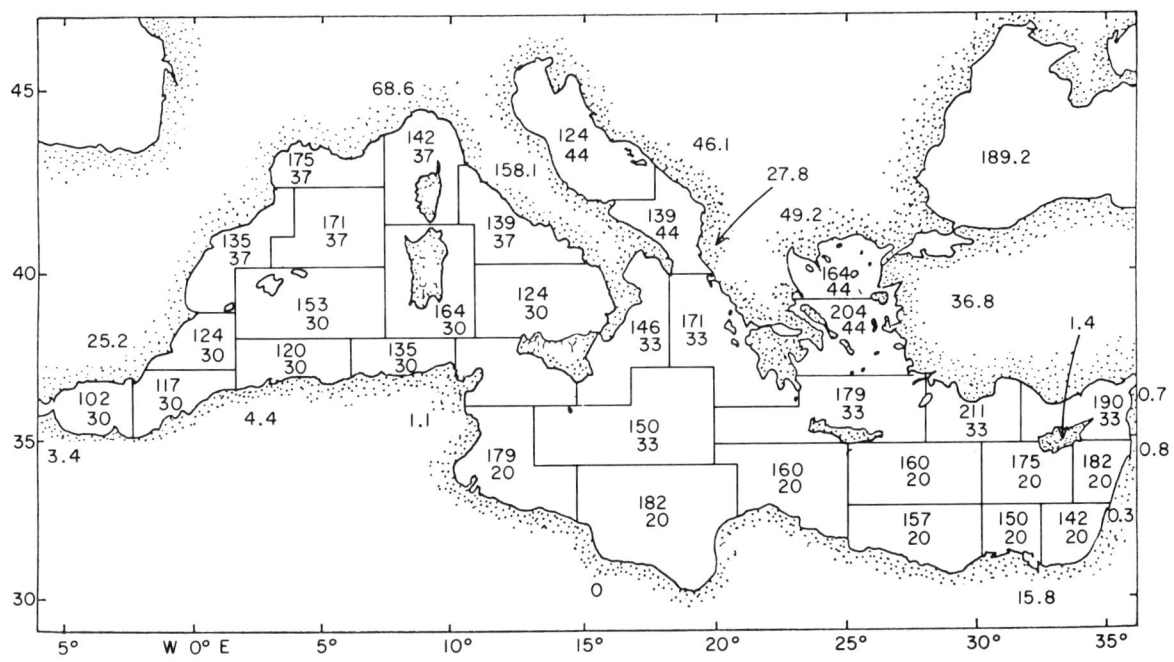

Fig. 4.4. The estimated values for evaporation, precipitation, and run-off of the entire mediterranean broken into subregions. The values given on land indicate the river runoff and agricultural discharges, while those over the sea represent evaporation (upper) and precipitation (lower). The value given for the Black Sea is a net run-off. The runoff is in km³/yr and the other two are in cm/yr cm⁻² (from Cruzado, 1979, and from Béthoux, 1977).

residual flow is not steady as it responds sensitively to differential sea level changes between the Mediterranean and the Atlantic, e.g. variations in D. The passage of atmospheric pressure systems over the Mediterranean is another important cause. Of the nontidal fluctuations, only the seasonal variability has been quantified. Ovchinnikov's (1974) observations indicated that the excess inflow season extends from March to October, with a total of 10,400 km³ and a late summer maximum. The excess outflow season spans from November to February and totals 8,800 km³. His figures give a net inflow of 1600 km³/yr, a bit on the low side compared to other estimates.

The high variability implies that any direct measurements should be made over at least one cycle of each periodicity with a significant amplitude, which in this case is at least seasonal. So far, this has not been possible. However, Lacombe (1971), working with a set of current observations at a point A4 (see Fig. 4.3), averaged over approximately two weeks in May and June, 1961, obtained a mean value of V_o of $\approx$ 36,300 km³/yr. The variation per tidal cycle was approximately 40% of the mean value, and no estimate of the lateral variability was available. Even so, his calculated value for D is 1,740 km³/yr, slightly less than, but reasonably close to, most of the values obtained by estimating D directly.

4.3.4. Budget example

At this point it would be instructive to discuss the qualitative thermohaline behaviour of WMED dynamics through a specific budget example. The interested reader might refer to a more detailed work presented by Béthoux (1980) in which he has subdivided the Mediterranean in order to obtain better spatial resolution of the budget parameters. Let us begin with estimated values for S_i, S_o, and D. For the Gibraltar salinities, we choose values at the sill intermediate between summer and winter (but favouring summer); for Sicily, we use those of Garzoli and Maillard (1979); and we use the values of D given in Table 4.1. These initial values are indicated in Fig. 4.5. From these and equation (3), we calculate the values of the Gibraltar flows (V_i^S, V_o^L) and the Sicily flows (V_o^S, V_i^L). Note that all volume transports are expressed as 10^3 km^3/yr; and that for numerical simplicity we have neglected the estimates of density. Our first conclusion is obvious: that the NAW increases salinity in transit; and our second conclusion is less obvious: that more surface water is leaving than entering the WMED, that is, it is divergent. Together these mean that for the surface layer the transport of salt out is greater than the transport in. This is not necessarily nor generally true; for example, if the salinity at Sicily were less (S_o^S = 37.05), the outflow would be less (V_o^S = 41.95), resulting in a convergence and a salt gain.

Invoking a salt and volume balance on the surface layer uniquely determines the upwelling W_u^L and the downwelling ($W_d^S + W_D^S$) transports. In this case, the loss of salt to the EMED is supplied via a net salt upwelling from the LIW. This would appear to deplete the WMED deep-water salt content, except that exporting more surface salt to the EMED eventually results in more salt import from the EMED via the return flow. The point demonstrated is that variations in salt exchange are ultimately self corrective, although over long time scales. The only independent salinity is that entering at Gibraltar, the seasonal variation of which will be reflected in the NAW but not in the deep waters.

Variations in D are also independent and have a more significant effect than those of the Atlantic salinity. Daily or lower frequency changes in D are directly reflected in the volume transports, and

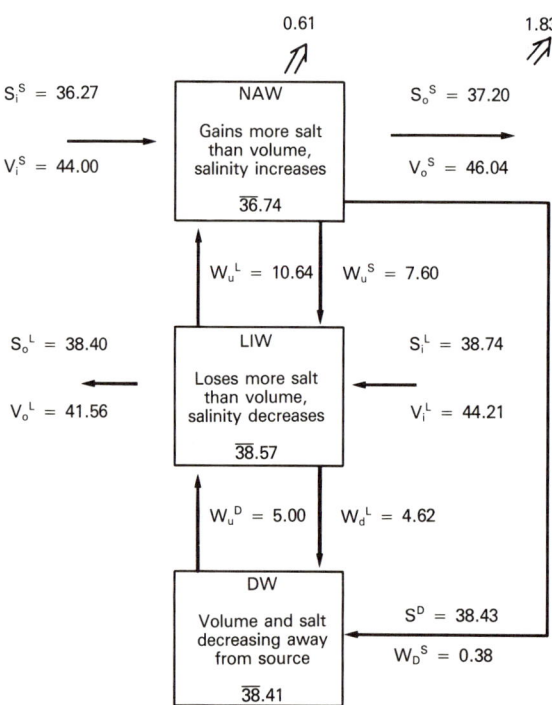

Fig. 4.5. Schematic of WMED conservation of salt and volume calculations. The V represents horizontal transport and W vertical. The salinities are in ppt and the transports in 1000 km^3/yr. The superscripts indicate the donor water mass and the subscript indicates direction of transport, i.e. 0 = outflow, i = inflow, u = up, d = down, S = surface, L = intermediate, and D = deep. The overbars indicate spatial means.

consequently salt transports. A situation of convergence or divergence in the WMED surface layer is directly controlled by D. A decrease in D in the EMED and an increase in the WMED relative to their annual means, as presumably occurs during winter, would result in a convergence, and a downwelling of NAW to depth would be enhanced. For example, in Fig. 4.5 if D is 1.63 in EMED and 0.81 in WMED, W_D^S is reduced and net downwelling is changed. Of course, the salt conservation assumption is less valid on a seasonal scale, but the conservation of volume certainly remains valid. Although qualitative, the point is important: it means that during winter the WMED has a relatively stronger negative thermohaline circulation.

Information on the winter supply of NAW to the EMED, and its distribution there, is not comprehensive enough to document such a seasonability. Hopkins (1978) has commented on the winter intermittency of the NAW in the EMED; perhaps further investigation will show important correlations between the seasonal differences in D in the two basins, the seasonal supply of NAW to the EMED, and the WMED DW production.

Returning to Fig. 4.5, some of the NAW (W_D^S) is mixed directly into the deep water. This is obvious because the DW salinity is less than that of the LIW. We can solve for the total downwelling ($W_D^S + W_d^S$) = 7.98, but cannot separately evaluate these or the DW to LIW exchange (W_u^D and W_d^L) without some independent information. To complete these in Fig. 4.5, we have taken the amount of DW formed to be 5.0 as estimated by Sankey (1973) and the newly formed DW to have S = 38.43 ppt based on MEDOC* 1969 observations. These values give an indication of the degree of vertical exchange between layers: The NAW/LIW interface is about twice as active as the LIW/DW interface. The direct circuit of NAW to DW as represented by W_D^S is critical to the maintenance of a three-layered system. It complexes the thermohaline circulation in the sense that, during the short time that it persists, it imposes a strong negative thermohaline circulation, and during the rest of the year a weak positive circulation (upwelling) prevails.

4.4. CIRCULATION

Those waters that do enter the Strait of Gibraltar to become the North Atlantic water mass (NAW) are destined for a long and complicated journey before finally returning to the Atlantic in the form of a subsurface water mass. In the transit of the Strait, some portion of NAW is mixed down immediately into the exiting waters. However, most of the loss occurs within the basin. The above calculations suggest that about 17% by volume of the NAW that does enter the WMED eventually is mixed into the LIW and subsequently may exit, and about 0.9% is incorporated directly into the DW where it must slowly upwell to the LIW before exiting. The portion of NAW that enters the EMED is lost to its LIW and DW before re-entering the WMED as LIW.

Of course what appeared to be dramatic speeds for the NAW within Gibraltar become much more gentle within the broad Mediterranean. However, the NAW does not spread evenly over the WMED after entering, because locally generated pressure gradient forces and the Coriolis force cause it to flow in somewhat defined streams within the surface layer. This can also be said of the LIW. This is an important three-dimensional fact that is not evident in the two-dimensional schematics of Fig. 4.1. Such lateral concentrations are evidence that the mixing in the NAW and LIW layers occurs more slowly than the advection through these layers and shows up in the identifying isohalines as in Figs. 4.6 and 4.8.

---

* An acronym for a detailed, international study of the wintertime formation of deep water in the northern portion of the Western Mediterranean Sea (MEDOC, 1970).

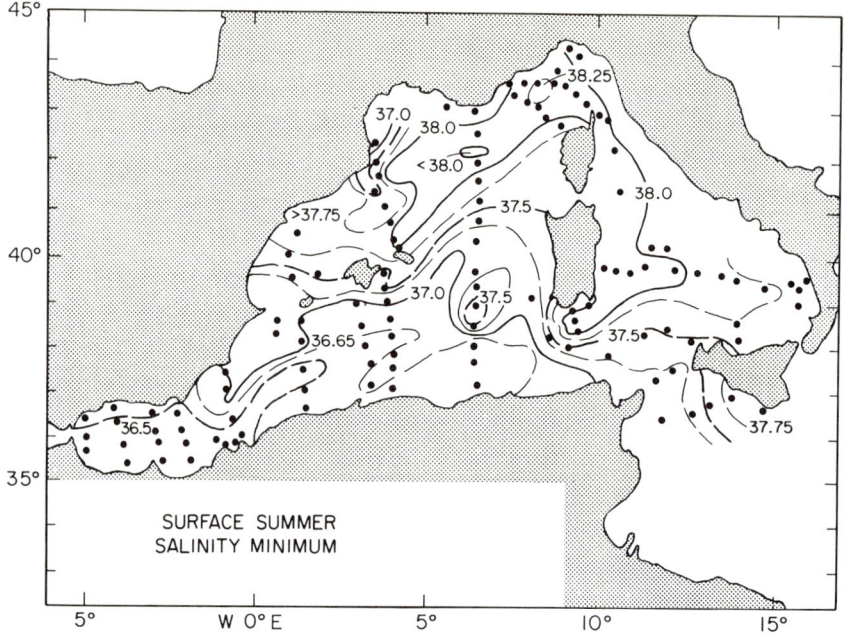

Fig. 4.6. Contours of the surface salinity minimum in the Western Mediterranean from composite data (Lacombe and Tchernia, 1960).

4.4.1 Surface water

The first major circulation feature to which the NAW is exposed is a large anticyclonic gyre occupying the Alboran Sea. This has been well described by Allain (1960) and Lanoix (1974) among others. The flow is more intense on the Spanish side, up to 90 cm/sec eastward, and on the Moroccan side, about 60 cm/sec westward. Typical of anticyclones, the centre portion has a deeper surface layer at ≈ 5 220 m, and the perimeter portions are shallower. In fact, on the Spanish coast, upwelling of subsurface water is often reported.

The mechanisms for this Alboran circulation are not completely understood. One might expect the entering NAW to adhere to the Moroccan shore because of the Coriolis effect. Clearly, this does not occur. It appears that a combination of bathymetric, inertial, and baroclinic effects prevail. The Gibraltar axis is directed slightly north of east; and the Moroccan coastal corner at Ceuta is quite severe having a radius of curvature of about 10 km, which is too sharp for the fast flowing NAW to follow. Thus, the bulk of the entering waters continues as directed by the Strait. The necessity for the subsurface waters to exit Gibraltar imposes another constraint. The dynamics for westward subsurface flow and eastward surface flow require that the interface rise to the north towards Spain which is only compatible with the interfacial structure of an anticyclone.

Bathymetry again plays a role in defining the extent of the anticyclone in that the isobaths on the Spanish side curve southward forming the Alboran basin west of the shoal area of Alboran Island and correspondingly influence the NAW to veer southward towards the North African coast. Hopkins (1978) estimated that about 65% of the NAW is recirculated within the Alboran gyre. The shoal also acts to straighten out that flow which escapes eastward over it because some cyclonic turning is generated on passage to the deep area to the east (Lanoix, 1974). The escaping portion moves eastward towards Cape Aiguille where the Algerian coastline appears to deflect the NAW stream north-eastwards. A circulation scheme given by Allain (1960) shows a branching of the NAW stream subsequent to this deflection, with one part continuing north-eastward towards the Balearic islands, and the other part continuing eastward along the North African coast (Fig. 4.7). This marks the first in a sequence of cyclonic breakaways throughout the entire Mediterranean from the otherwise eastward zonal route of the NAW. Their causes are not well understood. Clearly, those portions of NAW lost to northward routes remain in the WMED and are not exported via Sicily.

The next major turning occurs in the vicinity of 7°E just before the Sardinian Channel. The portion moving north along the south-western Sardinian coast splits again and recirculates cyclonically in the southern Balearic Sea (Fig. 4.7). The other portion continues northward, joined by the NAW of the Balearic islands, then flows past the northern coast of Corsica where it cyclonically circuits the shores of the Ligurian Sea, and finally returns to the south-west along the French and Spanish coasts. For this important branch of the NAW, the low salinity is obscured by riverine inputs along the northern shores of the WMED. That NAW reaching Corsica and entering the Ligurian can be $\geqslant 38$ ppt, but along the Riviera south-eastward it can be fresher at $\leqslant 38$ ppt as a result of local run-off. Béthoux and Prieur (1983) have noted the importance and effect of the drought of the 1981 season on the surface salinities and on that season's DW formation. Thus some of the NAW completes a cyclonic peripheral circuit of the WMED as it merges near the Balearic islands with that north-eastward flowing branch of the NAW that turned northward leaving the Alboran Sea.

The portion of NAW that continues through the Sardinian channel suffers another bifurcation with the larger portion exiting through the Sicily Strait to the EMED and the smaller portion into the Tyrrhenian. Béthoux (1980) estimates this as a two thirds and one third split, respectively. Bathymetry again plays a role in selecting which, and undoubtedly how much, NAW leaves the WMED. The African coast is fairly uniform west to east until the Tunisian coast and the Strait of Sicily, which are oriented about a southeast-northwest axis. Off the north-eastern tip of Tunisia lies the broad, shallow

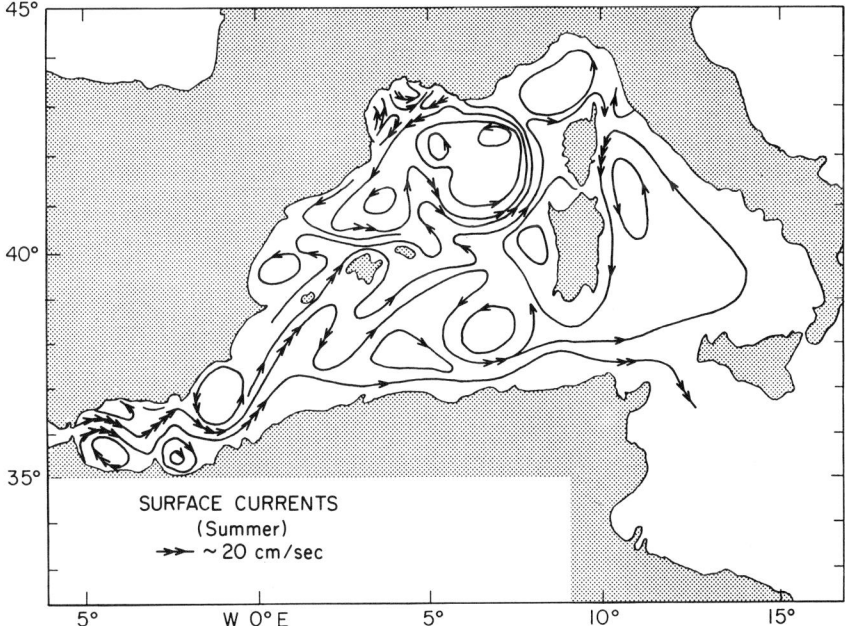

Fig. 4.7. The pattern of surface currents for the Western Mediterranean as estimated by Allain (1960).

Sherki Bank which acts to deflect the deeper, saltier portions of the NAW to the north-east into the Tyrrhenian, while the shallower portion traverses this bank and successfully escapes the WMED. The Tyrrhenian portion favours a cyclonic circuit of that sea. There is no detailed study of the fate of this segment of the NAW. Many of the broad-scale studies (e.g. Lacombe and Tchernia, 1960, or Ovchinnikov, 1966) show a surface convergence south of Sardinia such that the flow southward along the eastern coast of Sardinia is blocked from wrapping around the southern tip of Sardinia and re-entering the Balearic Sea, although Allain (1960) does show this (Fig. 4.7).

This general surface circulation has, in fact, much greater detail. It is well known that the summer surface flow is more complex owing to more diverse and smaller-scale wind regimes. The increased complexity in summer is often manifested in smaller gyre-like flow cells; for example, the Tyrrhenian may have several cyclonic surface features in contrast to one basin-side winter cyclone.

4.4.2. Intermediate water

The WMED has two types of intermediate water: the LIW which is imported from the EMED via the Strait of Sicily and which occupies the $\approx$ 200 to 700 m depth range; and the Western Mediterranean intermediate water (WMIW) which is formed locally in varying quantities and water types throughout the WMED as a result of local buoyancy extraction during the cooling season. The WMIW is not as dense as the LIW and resides between the surface water and the LIW at depths of approximately 50 to 200 m. In some of the more southern regions with less winter cooling exposure, there may be no distinct WMIW, but rather a gradation in water types from NAW to LIW. Unless the winter cooling is sufficient to cool the surface water type lower than the local temperature of the LIW, there will be no temperature minimum (13° to 14°C) expression above the LIW in the water column. In many of the more northern areas, sufficient cooling occurs during winter to cool the surface water below the LIW, but because of its relatively low salinity it remains less dense. In these situations the upper portions of the LIW may be affected. In fact, through such convective exchanges, the LIW loses salt but gains volume as required by the WMED budget. At the depth range of the WMIW there is considerable interregional and even interannual variability, which makes it difficult to trace the movement of this layer on the basis of the distribution of a distinct water type.

The LIW differs greatly in this regard. A number of researchers have employed the 'core method', in which its distinct salinity maximum is contoured, to study its distribution. The shape of the isohalines is interpreted as indicating the prevailing path of the LIW core under the assumption that weak gradients indicate preferred movement and vice versa. The LIW core often is accompanied by a temperature maximum as it is warmer than the underlying deep water and often warmer than the overlying WMIW. The T−S values change in a gradual linear fashion from their entry values of 14°C, 38.74 ppt, at Sicily, to their exit values at Gibraltar of 12.8°C, 38.40 ppt (Wüst, 1961). The amount of exchange is correlated with the length of time the LIW has resided in the WMED. That is, the longer the LIW remains in the WMED the greater its exposure to dilution through mixing with the waters above and below it. In the western Alboran Sea, the LIW salinity maximum is greater than that of the DW by less than 0.1 ppt. The LIW salinities converge to that of the DW at Gibraltar through freshening from above and below.

An example of the salinity maximum distribution is shown in Fig. 4.8. In a gross sense, the LIW appears to move around the western tip of Sicily into the Tyrrhenian Sea and to move westward south of Sardinia where some of it appears to move north on into the Ligurian Sea and some of it to continue westward towards the Alboran Sea. It is also possible to get a broad view of the LIW flow pattern by using dynamic height calculations of composite data. This has been done by Ovchinnikov (1966) who revealed a similar picture (Fig. 4.9) except that no south-westward flow of LIW was calculated between Sardinia and the Balearic Islands.

The supply and characteristics of the LIW in the Ligurian Sea are of great interest since it is this LIW which is critically involved in the WMED DW formation. The core LIW rings the Ligurian at about 500 m in a cyclonic sense. Béthoux and Prieur (1979) have shown that the south-west transport off the Riviera is greater than the north-east flow off Corsica. The difference is made up by a discharge of LIW northward through the Corsican channel out of the Tyrrhenian Sea.

The route of the LIW immediately on reaching the WMED has also received considerable attention, e.g. Frassetto (1964). The Strait of Sicily has two channels, one slightly to the east of the other with sill depths of 430 and 365 m, respectively. The deeper easternmost sill accommodates a greater volume of entering LIW. Its channel is directed north favouring entry to the Tyrrhenian Sea; the other is directed west of north towards Sardinia. The central question is the extent to which entering LIW must be exposed to mixture and dilution in the Tyrrhenian before entering the WMED proper. This question is complicated because of a slight variability in the entering LIW itself. There is a tendency for the denser LIW to enter via the eastern channel so that the Tyrrhenian receives a colder and saltier, or a less mixed, version of LIW. The portion of this that exits the Tyrrhenian south of Sardinia, via some cyclonic internal route, has a lowered $T-S$ characteristic similar to that LIW which perhaps travelled directly from the western Sicilian channel to Sardinia.

The LIW that flows westward along the North African coast appears to be offset northward away from the coast and not directly under the eastward flowing NAW. Its path westward from the Sardinian passage to $\approx 4°E$ is certainly not linear, where the resident LIW appears to be either incorporated into a northerly drift towards Sardinia or a westerly drift paralleling the North African coast. This was suggested by at least one set of observations. From an *Atlantis II* section from Gibraltar to Sicily, Katz (1972) showed a comparatively large gap in the $T-S$ values for LIW to the east and

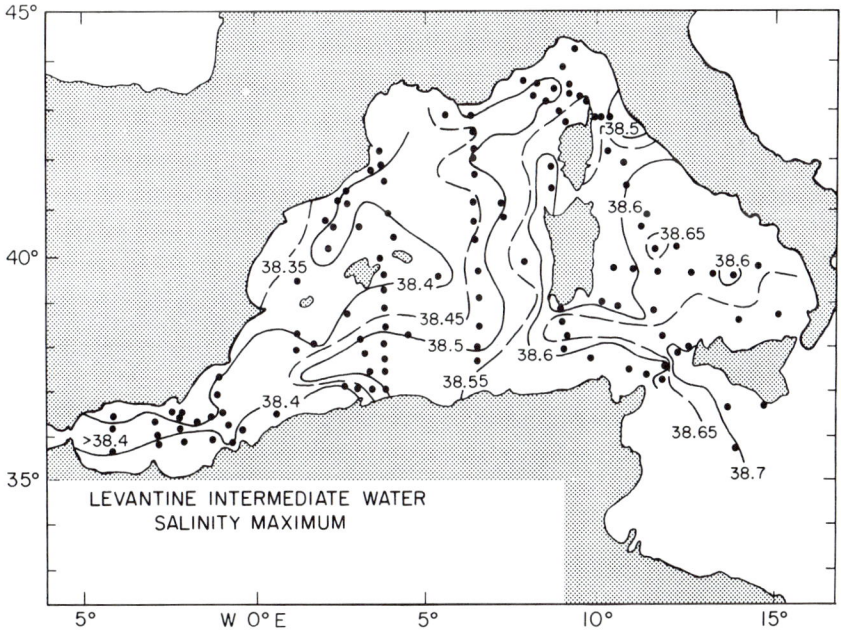

Fig. 4.8. Contours of the salinity maximum, or core LIW, in the Western Mediterranean from composite data (Lacombe and Tchernia, 1960).

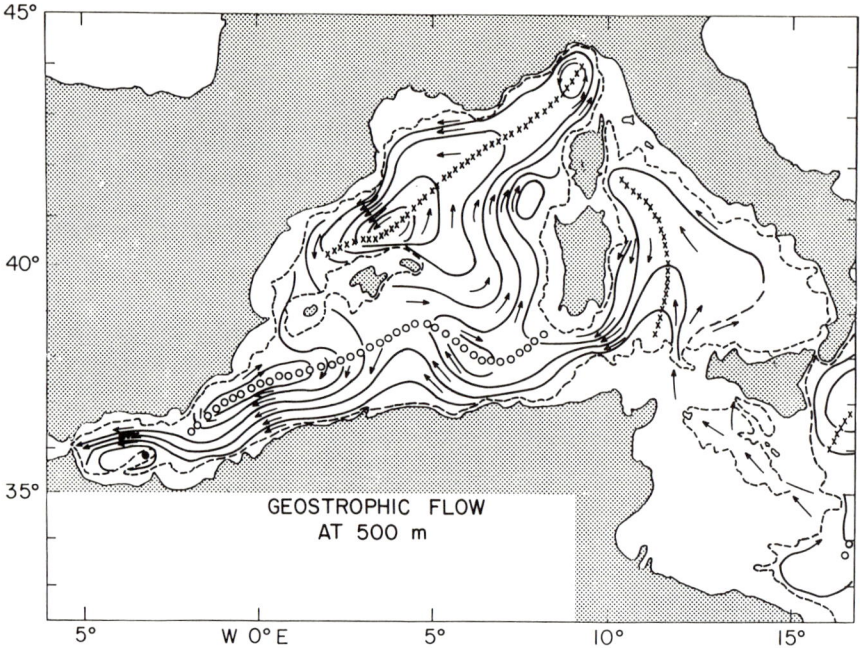

Fig. 4.9. Composite geostrophic flow in winter 500/1000 dbar redrawn from Ovchinnikov (1966). The solid lines are the dynamic height isolines, the arrows show the direction of the geostrophic flow, the open circles and crosses indicate convergences and divergences, respectively.

west of 7°E. This could be explained by a northward movement of LIW to the east of 7°E, and by returning older LIW southward to the west of 7°E. This area is particularly susceptible to quasi-stationary circulation features, possibly precluding any definitive spatial distribution of the LIW flow (Garzoli and Maillard, 1979), but the important aspect of northward flow west of Sardinia LIW seems to be substantiated.

The LIW salinity maximum contours of Fig. 4.8 are taken from Lacombe and Tchernia (1960). They are presented as a schematic distribution of the LIW; the absolute values may be low (≤0.05 ppt) in comparison with other data, (e.g. Miller et al., 1970). Nevertheless, several points are worth noting. The east-west trend of lower salinities is clear. There are relatively small gradients in the Tyrrhenian, which may act as a sort of reservoir for incoming LIW. The tendency for the more undiluted LIW to circulate around the periphery of the WMED leaves the central areas occupied by much weaker LIW signature. In fact, the 38.4-ppt contour of Fig. 4.8, which is probably ≈ 38.43 ppt, extends from Gibraltar to the Gulf of Lions. This should not be interpreted as a region of connected flow, but rather of weakened vertical stability.

Near Gibraltar, both the bathymetry and dynamics favour a more northerly exit for the LIW. The 400-m-depth contour extends toward the Strait north of the centre axis, and the shelf on the Moroccan side is shallow enough to prevent access to the strait of waters from depths 400 m. For westward flow, the concentration of LIW must be against the Spanish slope, and the interface between the LIW and NAW tilts up in that sense. This was observed by Lanoix (1974); however, recent observations show westward flow and upwelled isopycnals along the Moroccan slope (Bryden and Stommel, 1981). These were at about the 500 m depth within the LIW mass layer. They suggest that this water could have been destined for Gibraltar exit; however, this is difficult to reconcile with the fact that the LIW salinity maximum is greater (at 400 to 500 m) toward the centre and north of the Alboran (*see* Bryden *et al.*, 1978, or Lanoix, 1974) and hence must indicate westward flow. Furthermore, the water type found on the sill more closely matches this northern and central LIW type.

4.4.3. Deep water

If the Strait of Gibraltar were closed, the salinity of the entire Mediterranean would increase by approximately 0.03 ppt per year; and if the WMED were isolated, its salinity would rise by ≈ 0.02 ppt per year. The atmospheric replacement time for the WMED is 4,400 years, lowering at a rate of ≈ 75 cm/year. This situation is hypothetical, at least within our geological time frame. If we take estimates of the amount of DW formed each year, we can determine its replacement time by dividing the production rate into the DW volume. Using Sankey's (1973) DW production estimate of 5,000 km^3/yr gives 160 years.

If this entire annual production were formed in the Ligurian-Provençal Basin, then in passing the section between Sardinia and Menorca it would have a net speed of 9 km/yr (0.03 cm/sec). Similarly, it would move through the narrow channel at 3°W (north of Alboran Island) at about 370 km/yr (1.2 cm/sec). This reasoning is valid only if this total amount exits laterally into the Atlantic, which apparently it does not. Very little unmixed DW (< 38.41 ppt) is found in the Alboran Basin, undoubtedly because of this narrow channel with a sill of ≈ 900 m. But some DW does exist, at least as a DW – LIW mixture, and this amount must be expressed as a net south-westerly drift for the DW.

The remainder of the DW that does not reach the Alboran must move vertically instead of laterally. It is as if a volume of new DW were poured into the DW reservoir which then overflows by the same amount, either laterally (Gibraltar) or vertically. However, the surface or interface of this DW reservoir is not level to the degree that the sea surface is level. It has, in fact, large distortions created by the weak difference in weight between the DW and the overlying LIW and NAW. These distortions rise underneath the centres of surface cyclones and under the perimeters of anticyclones, and they lower under the centres of anticyclones and the perimeters of cyclones. A mere upward distortion of the DW interface would not necessarily imply loss of DW unless it is accompanied by mixing across the interface (*see* discussion of Fig. 4.8, p. 114).

The net movement of the DW may be slow, but speeds greater by an order of magnitude are not at all inconceivable. For example, the flow may be oppositely directed on either side of some section, but have only a slightly greater time invariant mean in one direction. Such an increase in the energy environment would enhance the spatial homogeneity of the DW through mixing. Recall from the budget discussion that, unlike the NAW and LIW layers, the DW is flushed much more slowly and hence has more time to become well mixed. We can imagine that without mixing the DW would consist of a sequence of annually produced blobs of DW moving in an orderly fashion from their source to their ultimate sink. Our knowledge of deep-water dynamics and the few observations preclude this, even though the expected differences between annual crops of DW are almost within observational error and may be detected just subsequent to their production.

Some evidence for DW movement can be obtained from the distribution of potential temperature (the temperature corrected for pressure effects) as determined by Wüst (1961). On the assumption that higher potential temperatures imply older DW (more mixing with the warmer LIW), he deduced a not surprising general movement towards the Alboran from the Ligurio-Provincial Basin with ancillary movement into the dead ends of the Catalan and Tyrrhenian seas. This is corroborated also by the DW oxygen values. The concentration of oxygen is not conserved in the DW waters because of its slow utilization due to the oxidation of organic matter raining down from the surface layer. The newly produced DW is always nearly saturated in oxygen because it is formed as a result of surface processes coincident with atmospheric equilibrium concentrations. The Tyrrhenian DW has oxygen values noticeably lower, around 4.4 ppm, compared to the values of ≈ 4.5 ppm found in the Balearic Basin and ≈ 4.6 ppm in the Ligurio-Provincial Basin. The Tyrrhenian DW is also warmer and saltier than water at comparable depths in the Balearic. The Tyrrhenian DW remains nearly vertically unstratified as mixtures of LIW and entering DW form a slightly heavier water type which then mixes downward. This action helps shorten the Tyrrhenian DW residence time and counteracts a build up of isolated denser water behind the Tyrrhenian sill. We do not expect a similar mechanism operative in the Catalan Basin, since its access to the Ligurian-Provençal Basin is deeper than its interior depths, which precludes any retention of a denser bottom water.

The splitting of the southward-moving DW, north of Algeria, into a larger Gibraltar-bound component and a smaller Tyrrhenian-bound component causes a net DW divergence which, it is important to note, coincides with the LIW convergence observed by Katz (1972). The subtleties of DW movement await more precise observations; until then the route and age of the DW remain an important uncertainty.

4.5. DEEP-WATER FORMATION

When buoyancy is extracted through exposure to atmospheric cooling and evaporation, the very surface waters become denser and sink. Their sinking in turn exposes new waters, and so forth. The newly-formed water sinks to a depth where its density matches the ambient density, (i.e. where it is neutrally buoyant). The ambient waters usually do not have the same temperature and salinity and because of the nonlinearity of the sea-water density function, any subsequent mixing can result in an ever slightly dense water. If this entire process continues and if the atmospheric conditions are severe enough, the newly-formed water may be denser than, or as dense as, the densest underlying water, in which case it sinks to the bottom and becomes bottom water.

Certain meteorological conditions are obviously requisite to the process of dense water formation, and, less obviously, certain oceanographic conditions must precede and accompany them. The Mediterranean, as its name implies, is situated within a land mass and is exposed to more continental weather than waters of the Atlantic at the same latitude. Continental air flows from northern and eastern Europe are cool and dry, and those from North Africa are warm and dry, both enhancing the potential for dense water formation through evaporation and cooling (in the north). The meteorological characteristics are described in Chapter 3 of this volume. We will concentrate on summarizing the oceanographic setting for DW formation in the WMED.

The site or sites of dense water formation can be located anywhere within a basin; in the WMED, the site is located at the opposite end from the sill providing the longest possible route for the DW. The favoured site is that location where both meteorological and oceanographic conditions are optimal, and for the WMED this is in the Ligurio-Provincial Basin and more specifically off the Gulf of Lions.

The primary oceanographic condition is a circulation which upwells the densest waters available to the surface and thereby exposes them to the atmosphere. This is often referred to as a preconditioning phase. In a geographical sense, the optimal site depends on the circulation. The waters arriving at the site must have been preconditioned by already experiencing considerable buoyancy loss. For the WMED site, there are two important aspects of the preconditioning phase concerning the two major water masses involved in the dense water formation: the local surface water and the LIW.

The LIW is a water mass formed in the eastern Levantine Sea as a result of atmospheric buoyancy extraction processes there. It is produced in large volumes, but is not as dense as the colder but fresher

EMED DW which is formed in the Northern Adriatic. By the time the LIW crosses into the WMED, and proceeds as described, its water type has changed through mixing and its density has increased slightly. We would refer to the LIW entering the WMED as already strongly preconditioned, in fact, almost optimally so because of its relatively warm, high-salinity water type which makes it most susceptible to buoyancy extraction. On reaching the Ligurian Sea, the LIW has cooled and freshened slightly, again through mixing, yet it still retains a relative salinity maximum.

The second preconditioned water mass is the local surface water. As a result of general autumn winter cooling, the surface waters are homogenized to their winter water type. In the Ligurian cyclonic circulation, the surface waters along the Riviera suffer an increased buoyancy loss since the greatest extraction of buoyancy occurs closest to shore (that is, in a coastal zone before the offshore air equilibrates in heat and water vapour with the sea). Hence, on arriving at the Gulf of Lions area, this coastal boundary flow is seasonally preconditioned and has been subjected to more than its regional mean of buoyancy extraction. This excludes local run-off effects that may counteract any buoyancy loss by decreasing the surface salinity. Over the Gulf of Lions shelf, a different water type is formed during winter, which is of nearly the same density but cooler and fresher. The Riviera boundary current eventually encounters this shelf water in the off-shore vicinity of the Gulf of Lions.

At this point the large-scale cyclonic circulation of the region plays a role in sustaining a low-pressure (high-density) central region. In the terminology of the MEDOC experiment, this is referred to as a dense-water patch. During the winter of 1969, it was surveyed several times. Defined by the isopycnal contour of 29.06 σ_t (Fig. 4.10a) or the isohaline of 38.4 ppt, its centre remained geographically fixed despite several storms, but it began to extend to the north-east by late February – early March (Fig. 4.10b). In these cases, the patch was not so much the central waters of a cyclonic gyre, but rather a distinct water mass about which the coastal surface waters turned cyclonically. Fig. 4.10a gives some impression of this. Thus, the patch is tightly enclosed on three sides by surface preconditioned waters.

Accompanying the horizontal cyclonic flow is a secondary vertical movement such that the waters are circulating clockwise looking downstream in a plane normal to the flow. As a consequence, the LIW is brought to the surface around the edges of the patch. The perimeter flow about the patch retains its vertical structure while the patch itself becomes more and more homogeneous. By late January 1969, the centre patch deviated only 5×10^{-3} σ_t units from the DW potential density, whereas the perimeter waters deviated by ten times this much.

At this point, from say mid-February, the stage is set for what the MEDOC Group called the violent mixing phase. The preconditioning and circulation have consorted to form the patch and expose it to the atmosphere. This deep column of homogeneous dense water is now precariously unstable and with only a slight density loss it will become as dense as the deep water. The extra buoyancy extraction needed to achieve this comes with the onset of the Mistral winds. The vertical extent of the core and the peripheral structure are nicely illustrated by the cross sections in Figs. 4.11a and 11b. The isolines of both temperature and salinity between 1500 to 1000 m are downbowed directly underneath the centre patch, while between 1000 to 500 m the LIW is severed horizontally, and from 500 to the surface the isolines are also nearly vertical.

Because of the neutrally stable conditions in the centre of the patch, large and vigorous vertical motions can be initiated with little forcing. Measurement of vertical motion is not easy in the open waters. However, neutrally buoyant devices that can be tracked acoustically have been used effectively with rather surprising results. These meters, while moving in a generally cyclonic track at about 15 cm/sec, experienced both intensive upward and downward movements of approximately 2 cm/sec (Voorhis and Webb, 1970). There appeared to be some correlation with increased upward motion during an intensification of the wind followed by nearly as strong a downward motion when the Mistral abated. It is clear that at such speeds the entire mixed patch could not simultaneously have the same vertical movement but rather the vertical motions are governed by smaller scales, roughly 10% of the

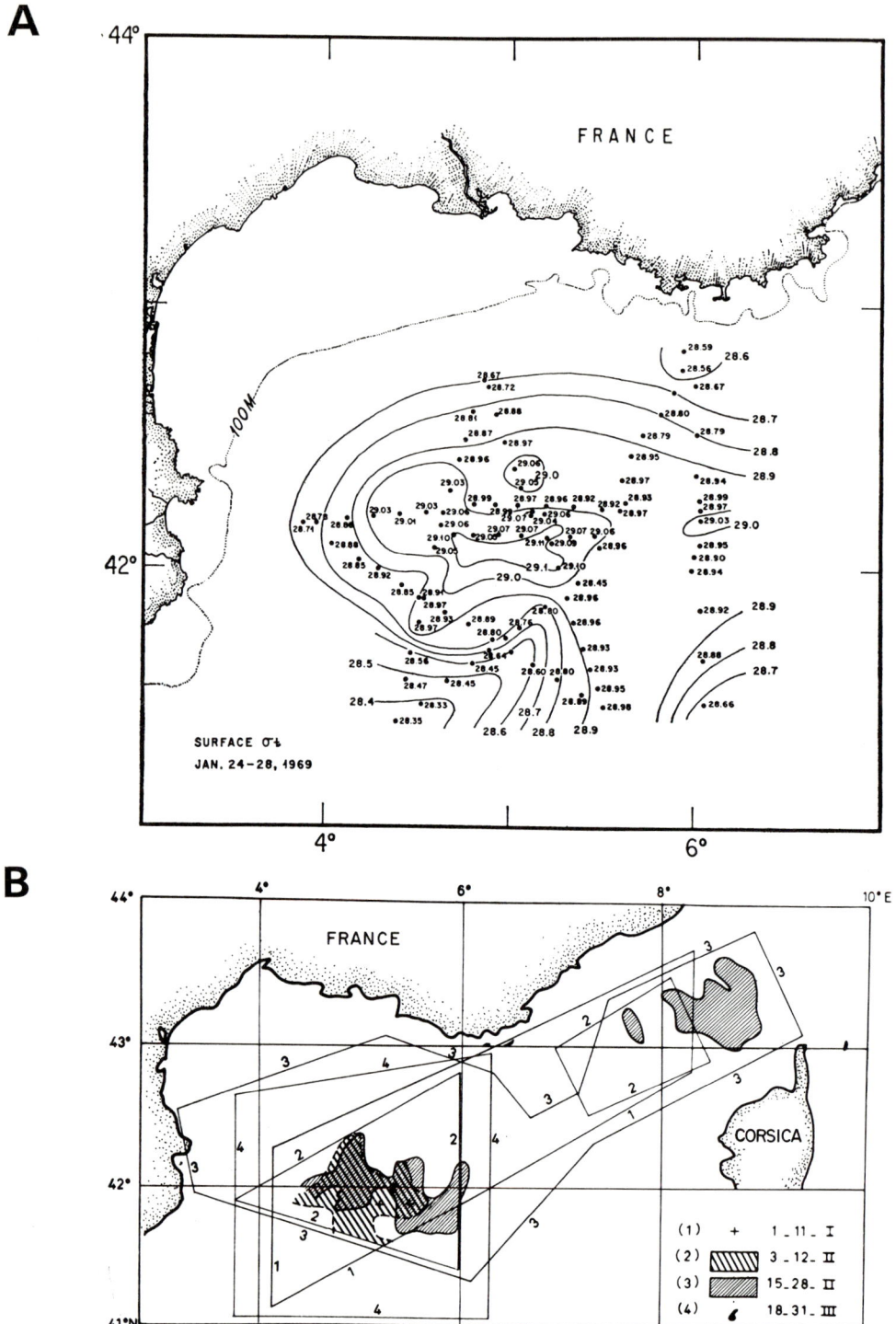

Fig. 4.10. (a) The surface density in the Gulf of Lions before the onset of the Mistral as recorded by *Atlantis II* (after Stommel, 1972). (b) Patch regions enclosed by the 38.42 ppt isohaline during four different surveys. The total survey area by the numbered polygons (from MEDOC Group, 1970).

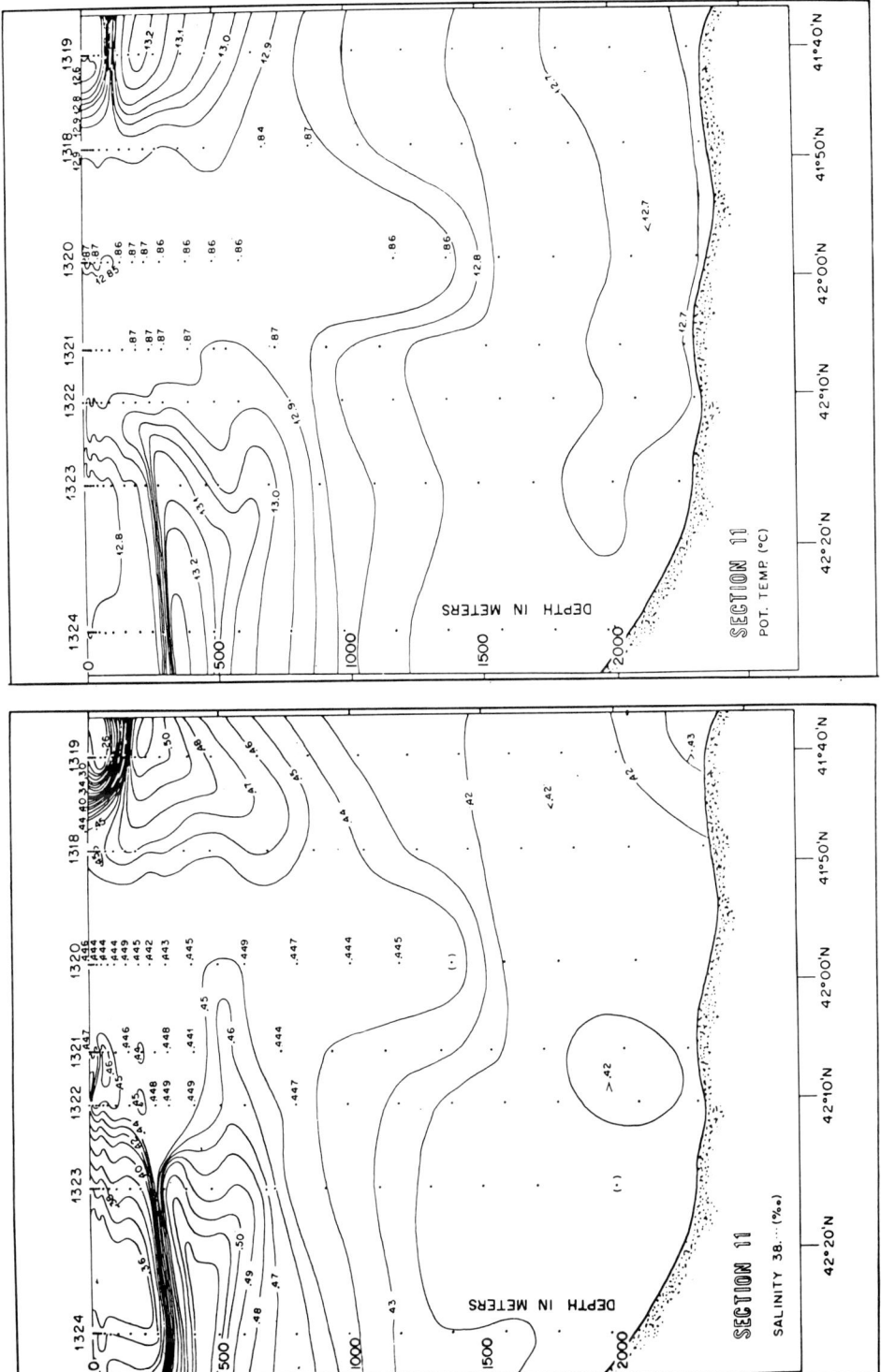

Fig. 4.11. (a, left panel) Salinity and (b, right panel) potential temperature through the *Atlantis II*, 10–11 February, Centre patch section (from Anati and Stommel, 1970).

width of the patch. These have been referred to as 'holes' or chimneys within the centre patch. Conceptually, one might think of an upside-down boiling process with columns of water moving both up and down. Another useful analogy was suggested by Gascard (1978) who described the similitude between the air masses along a cyclonic meander in the polar front and the water masses around the MEDOC vortex. The smaller-scale vertical motions that mix intermediate and surface waters are a manifestation of the baroclinic instability of the vortex and are triggered by heat fluxes and wind stresses.

With such vertical motions and an observed deepening of the central patch over the Mistral season, it is natural to speculate as to whether the growing mixed patch water is comprised of additions of DW. This is a very important point relative to the DW replacement, for if the newly-formed DW is not composed of any appreciable amount of old DW, then the effective renewal of the DW is necessarily greater than if portions of it were involved. This question is not entirely resolved, but there seems to be more evidence that the newly formed DW is independent of the old underlying DW. Stommel (1972) reported a distinct water-type difference between the new and the old DW. We may consider that the new DW varies slightly each year, but is always close enough in water type and small enough in volume that it is soon blended into the existing DW reservoir.

If the centre patch water does not include old DW, it must then be composed of surface water and LIW. Hopkins (1978) suggested that it mainly comes from the surface of the northern landward side which is blown into the patch area and upwelled LIW water. The patch is divergent under Mistral conditions; this means that strong offshore winds tend to move the southern limb of the perimeter flow further to the south, thereby decreasing the curvature and expanding the patch area. Also, upwelling of LIW is enhanced and greater portions of surface waters are driven in from the north. Hence, with each Mistral event the patch grows in size. The admixture of LIW causes the centre-patch water to be slightly warmer and saltier, and the Mistral on encountering the patch water more effectively extracts heat (air-water temperature difference is greater), and with slight increased cooling the saline patch water becomes as dense as the underlying DW. This process is depicted in Fig. 4.12 where it is shown how a 50-50 mixture of northern surface and LIW can form, if exposed to typical Mistral buoyancy extraction, the observed new DW. The reader will recall from the budget calculations that a much lower percentage of surface water was assumed to be incorporated in the DW production. This is because the surface waters resident in the Ligurio-Provincial Basin at the time of production have been pre-conditioned by admixtures of LIW and are not representative of the entire WMED surface layer as required by the budget calculations.

Periods of Mistral may occur in March, but by then the solar insolation is increasing rapidly and the opportunity for severe buoyancy extraction diminishes. In 1969, the centre patch reached its greatest vertical extent of 2150 m over a 2300 m bottom by 18 February, and the violent mixing phase gave way to the sinking and spreading phase. By 20 February, a slight density break began at 900 m, and three days later the surface showed a perceptible warming. Through March, the stratification struggled to re-establish itself, and was set back to a homogeneous state when in late March another Mistral occurred. But this was not sufficient to return the water column to the violent mixing phase. By mid March, the LIW layer began tentatively to span the centre patch area at the 250 m depth. Normal stratification below this, that is, the faint structure normally observed in the LIW to DW transition, took longer to become horizontally uniform, because to do so involved the evacuation or spreading of the volume of new DW now occupying these depths. This volume, which began to settle around the 2000 m depth, was estimated to be $\approx$ 5000 km^3, roughly equivalent to the volume of the February centre patch.

Conclusive evidence for the movement of the new DW has been difficult to acquire without direct current measurements and given the slight T-S differences involved. The difference in dissolved oxygen concentrations is often more striking with the new DW, having values $\geqslant 4.6$ ppm as opposed to the 4.5 ppm ambient values. However, this O_2 tagging was limited, because of coverage and variability within

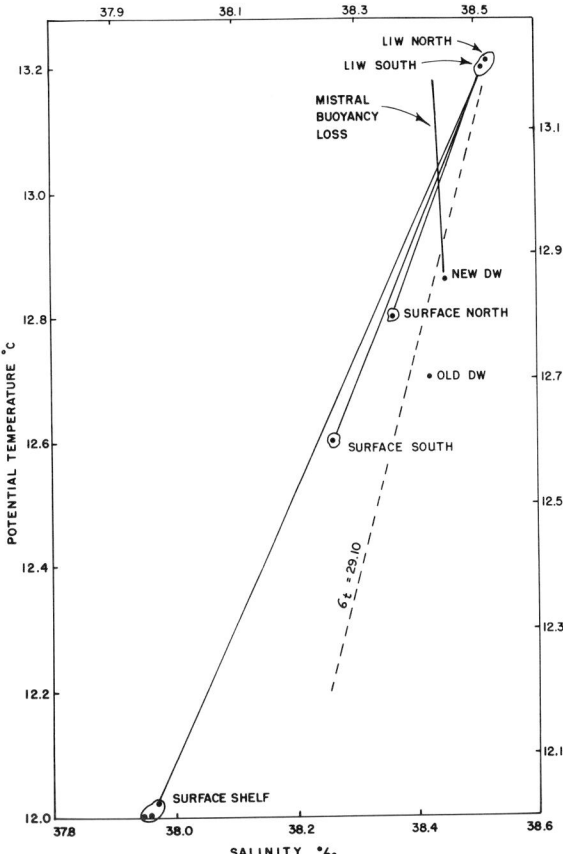

Fig. 4.12. Potential temperature-salinity diagram. Water types are from *Atlantis II*, 11 February, 1969, except for the surface shelf water types which are from CHARCOT stations 74–76, 21 February 1969.

the new DW, so that it was not sufficiently quantitative to monitor changes in the new DW volume. It did show a sequence of patches, four to the south-west and one to the north-east. The speed of the spreading using T−S observations gave ≈ 1.5 cm/sec toward the south-west; use of buoyant floats gave speeds up to 4 cm/sec in the same direction. It is the regional sea level slope that forces translations of the new DW volume from its place of origin, and simple translations do nothing to mix this water. Rather it is the differential velocities caused by the density differences between the new and old DW that ultimately create distortions in the patch to the degree that small-scale diffusive processes can assimilate it into the old DW.

By the sequence of DW formation that we have described, we cannot expect uniformity in the amount or water type of the new DW. The observations of 1963, 1969, and 1973 all indicated a tendency for the new DW to be warmer and more saline. This could be a coincidence, a trend, or a part of a cycle. Our observations are too limited to ascertain. At the production volume of the 1969 year (5000 km^3), a 160-year replacement time is implied for the DW, and it is estimated that about 100 years would be required to make these distinctions. Certainly, the obvious dependence on atmospheric variables suggest that certain of the stronger climatological cycles could be reflected in the DW production. In a long trend, changes in the salinity of the DW should be detectable but would be slighter than the historical accuracy of that measurement; for example, if the new DW were being produced at a rate of 5000 km^3/yr and at a salinity of 38.44 ppt (instead of 38.41 ppt) since the 1908 THOR expedition, the DW would have increased $\approx$0.01 ppt, which is below the accuracy of measurement in 1912 if not today.

4.6. TIDES AND HIGH FREQUENCY MOTIONS

As a large zonal ocean, the Mediterranean has the potential for generating tides independent of the world ocean. This was the original assumption expressed by Darwin; however, it has not been entirely supported by tidal observations within the WMED. In fact, it became apparent (e.g. Sterneck, 1915) that an additional tide existed, which was formed by co-oscillations with the Atlantic through Gibraltar and the EMED through Sicily, and was larger than the independent tide (at a ratio of 5:3 in the Alboran sea). The co-oscillating tide is essentially a standing barotropic wave, the Gibraltar wave reflected from the Italian peninsula and the Sicily and Messina waves reflected from the Iberian peninsula. These co-oscillations and the independent tide contribute to give the observed WMED tide.

Defant (1961) has given a composite value for the tidal harmonic constants in several regions of the WMED (Table 4.2). The semidiurnal components are approximately 180° out of phase (6 hrs) between the Alboran and Tyrrhenian seas, whereas the diurnal components are closer to 90° out of phase. However, for the diurnal tide the independent and co-oscillating portions interfere with each other, producing a fairly weak diurnal tide. The discrepancy between the diurnal and semidiurnal tide varies, since the diurnal tide has an antinode near Alicante, Spain, approximately where the semidiurnal tide has a node.

In any case, the Mediterranean tidal amplitudes are small by world ocean standards. This, together with the existence of narrow continental shelves, results in very little tidal amplification along its coasts. Consequently, the Mediterranean is often considered a tideless sea, which, perhaps, gives a wrong impression; for while the tidal elevations are nearly insignificant, the energy of the tides is not. The tidal movements themselves generate very little net motion and are not considered as contributing to the net circulation. However, in terms of energy the tide and the thermohaline circulation are comparable in magnitude; for example, at Gibraltar the semidiurnal tidal transport is about 70 km^3 (Defant, 1961) whereas that of the mean flow is $\approx$ 60 km^3. The important difference between the two is that the tidal volume changes direction with time while the mean flow changes direction with depth. At Gibraltar the tidal speeds, being equivalent or larger, strongly influence the observed flow; as a result, the surface currents reverse direction with the tide, and the bottom currents also reverse or nearly reverse depending

TABLE 4.2. Harmonic constants for separate oscillating areas in the Mediterranean. H is the amplitude in cm and $\varkappa^1$ is the phase referred to 15°E

		M_2	S_2	N_2	K_1	O_1
Cadiz (Atlantic)	H	92.8	32.3	16.4	6.1	6.1
	$\varkappa^1$	82°	111°	74°	49°	314°
Gibraltar	H	38.3	14.1	7.4	3.0	1.1
	$\varkappa^1$	77°	106°	64°	142°	174°
Alicante	H	1.8	1.0	0.4	3.7	2.2
	$\varkappa^1$	89°	112°	58°	179°	124°
Ligurian Sea	H	7.0	2.8	1.4	3.5	1.8
	$\varkappa^1$	255°	267°	245°	199°	118°
Tyrrhenian Sea	H	10.4	4.1	1.8	3.0	1.2
	$\varkappa^1$	266°	287°	253°	212°	126°

on the local bottom layer cross-sectional area. Lacombe and Richez (1982) describe the results of current observations at a number of stations in the Strait of Gibraltar; for example, at a sill station, the upper layer flowed out at a rms amplitude of $\approx$ 85 cm/sec for $\approx$ 4 hrs and changed to inflow at about 1 hr before local high water at a rms amplitude of 125 cm/sec for $\approx$ 7 hrs. The lower layer is approximately in phase but is skewed towards outflow, 60 cm/sec for eight and a half hours compared to 40 cm/sec for four hours for the inflow.

Some of the tidal energy generates large distortions in the water mass interface. The upper layer has its minimum thickness at the end of the outflow cycle. For example, the 7 May, 1967 *Charcot* time series data show the 37 ppt isohaline fluctuating from a depth of $\approx$ 80 m to $\approx$ 220 m at the sill station (Lacombe and Richez, 1982). Analyses of the tidally generated movements of the water mass interface of the Strait of Messina show a situation that is analogous to, although more extreme than, that found in the Straits of Gibraltar and Sicily (e.g. Hopkins *et al.*, 1983). There, the water mass interface, normally at $\approx$ 150 m, intersects the surface during the stage of the tide when the lower layer (LIW) is flowing from the Ionian into the Tyrrhenian Sea. The potential energy required to lift the interface appears to be derived from the flux of tidal kinetic energy against the bathymetry of the Strait. For the case at Gibraltar, Stommel *et al.* (1973) have suggested that this interfacial uplift is a result of the Bernoulli phenomenon aspirating deep water up over the sill. Regardless of mechanism, the hydrographic significance is to allow deeper access for the exiting waters and thereby effectively deepen the surface layers of the basin.

A portion of the internal energy generated by the interaction of bathymetry and strong tidal currents is radiated away as internal waves, (e.g. Frassetto, 1964b, or Lacombe, 1971). The wave trains emanating from the Strait of Messina have been observed clearly with satellite-borne synthetic aperture radar (Alpers and Salusti, 1982). The eventual dissipation of these internal waves contributes to the interfacial mixing.

The relatively warm temperatures of the intermediate and deeper waters and the relatively low run-off and precipitation cause the WMED surface waters to be less buoyant than for comparable latitudes in the ocean. This means a weaker summer stratification, and often neutral stability during winter, with the result that large vertical motions and baroclinic instabilities are more common. An interesting example of this was recently noted by Crepon *et al.* (1982), who, using satellite imagery, identified undulations in the Ligurian water mass interface, which intersect the surface during winter. They show that these undulations are likely to be baroclinically unstable long waves propagating cyclonically about the Ligurian. An impulse of wind transporting water normal to the front might have been a generating mechanism.

Virtually all of the energy introduced by the wind is dissipated within the basin, and direct wind-generated energy is not imported or exported into the WMED. The only exceptions would be wind transport through the Straits, but this is relatively insignificant. Wind duration and basin length scales impose a low-frequency cutoff in direct wind-forced flow, and wind energy will be either dissipated in smaller-scale flows or converted to barotrophic or baroclinic energy. For example, on a 500-km-length scale, all surface water will have encountered a boundary within a month. This suggests that wind energy at seasonal or lower frequency is ineffective in producing a signal in the velocity spectrum, except to the degree that low-frequency wind stress is converted to geostrophic energy, which, flowing parallel to the bathymetry, is only minimally inhibited by the basin length scales. For example, a seasonally positive wind stress curl over the Tyrrhenian would contribute to a low-frequency cyclonic circulation in that sea.

The higher-frequency and spatially incoherent wind energy is primarily dissipated in the surface tens of metres, in contrast with the tidal energy which is primarily dissipated at the bottom. The amount of energy available from the wind is comparable to that found at similar latitudes in the ocean, whereas the Mediterranean tidal energy is less by approximately an order of magnitude. Consequently, the

Mediterranean subsurface waters are relatively impoverished in the amount of tidal energy available for dissipation. Fortunately, the deep convective nature of the Mediterranean circulation counteracts this. But in the shallower coastal regions, this anomaly becomes ecologically significant, since it suggests, all other things being equal, that considerably less energy is available in the bottom layer for mixing, suspension of particulates, or dispersal of larval forms.

REFERENCES

Allain, C. (1960) Topographie dynamique et courants généraux dans le bassin occidental de la Méditerranée. *Rev. Trav. Inst. Peches Merit.* 24(1).
Anati, D. A. and Stommel, H. (1970) The initial phase of deep water formation in the northwest Mediterranean, during MEDOC '69 on the basis of observations made by *Atlantis II. Cahiers Océanographiques* 22, 343 – 351.
Alpers, W. and Salusti, E. (1983) Scilla and Charybdis observed from space. *J. Geophys. Res.* 88 (C3), 1800 – 1808.
Béthoux, J. P. (1977) Contribution a l'étude thermique de la mer Méditerranée. Rapport No. 20, Laboratoire de Physique et Chimie Marines, Université Pierre et Marie Curie, Paris, 199 pp.
Béthoux, J. P. (1979) Budgets of the Mediterranean Sea. Their dependence on the local climate and on the characteristics of the Atlantic Waters. *Oceanol. Acta* 2(2), 157 – 163.
Béthoux, J. P. (1980) Mean water fluxes across sections in the Mediterranean Sea, evaluated on the basis of water and salt budgets and of observed salinities. *Oceanol. Acta* 3(1), 79 – 88.
Béthoux, J. P. and Prieur, L. (1979) Evaluation des flux d'eaux de la circulation du nort-est du bassin occidental. *Rapp. Comm. int. Mer Médit.* 25/26, 67 – 68.
Béthoux, J. P. and Prieur, L. (1983) Mediterranean sea dynamics, forcing by drought (pers. comm.).
Bryden, H. L., Millard, R. C. and Porter, D. L. (1978) CTD observations in the Western Mediterranean Sea during cruise 118, Leg 2 of R/V CHAIN, February, 1975. WHOI Tech. Rept. 78 – 26, 38 pp.
Bryden, H. L. and Stommel, H. M. (1982) Origin of the Mediterranean Outflow. *J. Mar. Res.* 40 (Suppl.), 55 – 71.
Crepon, M., Wald, L. and Monget, J. M. (1982) Low-frequency waves in the Ligurian Sea during December 1977. *J. Geophys. Res.* 87(C1), 595 – 600.
Cruzado, A. (1979) Climatology and hydrology of the Mediterranean Region. In: *Report on the state of Pollution in the Mediterranean Sea.* UNEP, Chap. 2.
Defant, A. (1961) *Physical Oceanography*, Pergamon Press, Oxford, II, 598 pp.
Frassetto, R. (1964a) A study of the turbulent flow and character of the water masses over the Sicilian Ridge in both summer and winter. *Rapports et Procès-Verbaux, CIESM* 18, 812 – 815.
Frassetto, R. (1964b) Short period vertical displacements of the upper layer in the Strait of Gibraltar. Saclant ASW Res. Center, Tech. Rep. No. 30.
Garzoli, S. and Maillard, C. (1979) Winter circulation in the Sicily and Sardinia straits region. *Deep-Sea Res.* 26(8A), 933 – 954.
Gascard, J. C. (1978) Mediterranean deep water formation, baroclinic instability and oceanic eddies. *Oceanol. Acta* 1(3), 315 – 330.
Hopkins, T. S. (1978) Physical processes in the Mediterranean basins. In *Estuarine Transport Processes*, Ed. B. Kjerfve, pp. 269 – 309, Univ. South Carolina Press, Columbia.
Hopkins, T. S., Salusti, E. and Settimi, D. (1984) Tidal forcing of the water mass interface in the Strait of Messina. *J. Geophys. Res.* 89 (C2), 2013 – 2024.
Katz, E. J. (1972) The Levantine intermediate water between the Strait of Gibraltar. *Deep-Sea Res.* 19, 507 – 520.
Lacombe, H. (1971) Le détroit de Gibraltar: Océanographie physique. *Notes & M. Serv. géol. Maroc,* No. 222 bis: 111 – 146.
Lacombe, H., Gascard, J. C., Gonella, J. and Béthoux, J. P. (1981) Response of the Mediterranean to the water and energy fluxes across its surface, on seasonal and interannual scales. *Oceanol. Acta* 14(2), 247 – 255.
Lacombe, H. and Richez, C. (1982) *The Regime of the Strait of Gibraltar.* Elsevier.
Lacombe, H. and Tchernia, P. (1960) Quelques traits généraux de l'hydrologie Méditerranée. *Cahiers Océanographiques* 12, 527 – 547.
Lacombe, H., Tchernia, P., Richez, C. and Gamberoni, L. (1964) Deuxième contribution à l'étude du régime du détroit de Gibraltar (Travaux de 1960). *Cah. Océanogr.* 16, 238.
Lanoix, F. (1974) Projet Alboran, étude hydrologique et dynamique de la mer d'Alboran. *Rapport Techn. OTAN* 66, 70 pp.
Medoc Group (1970) Observation of formation of deep water in the Mediterranean Sea, 1969. *Nature* 227, 1037 – 1040.
Miller, A. R., Tchernia, P., Charnock, H. and McGill, W. D. (1970) Mediterranean Sea Atlas of temperature, salinity, oxygen profiles and data from cruises of R/V ATLANTIS and R/V CHAIN with distribution of nutrient chemical properties. *WHOI Atlas Series* 3, 90.
Morel, A. (1971) Caractères hydrologiques des eaux échangées entre le bassin oriental et le bassin occidental de la Méditerranée. *Cahiers Océanographiques* 24(4), 329 – 342.
Ovchinnikov, I. M. (1966) Circulation in the surface and intermediate layers of the Mediterranean. *Oceanology* 6, 48 – 59.

Ovchinnikov, I. M. (1974) On the water balance of the Mediterranean Sea. *Oceanology* 14, 198–202 (250–255 Russ.).

Sankey, T. (1973) The formation of deep water in the northwestern Mediterranean. In *Progress in Oceanography*, Ed. B.A. Warren, vol. 6, pp. 159–179, Pergamon Press, Oxford.

Sterneck, R. V. (1915) Hydrodynamische Theorie dar halbtägigen Gezeiten des Mittelmeeres. *S. B. Akad. Wiss. Wien (Abt. IIa)* 124, 905.

Stommel, H. (1972) Deep winter-time convection in the Western Mediterranean Sea. In *Studies in Physical Oceanography, a Tribute to Georg Wüst on his 80th Birthday*, Ed. A. L. Gordon, vol.2, pp.207–218, Gordon and Breach.

Stommel, H., Bryden, H. and Mangelsdorf, P. (1973) Does some of the Mediterranean outflow come from great depth? *Pageoph* 105, 879–889.

Voorhis, A. D. and Webb, D. C. (1970) Large vertical currents observed in a winter sinking region of the northwestern Mediterranean. *Cahiers Océanographiques* 22, 571–580.

Wüst, G. (1961) On the vertical circulation of the Mediterranean Sea. *Journal of Geophysical Research* 66, 3261–3271.

CHAPTER 5

Chemistry of Mediterranean Waters

ANTONIO CRUZADO

Instituto de Investigaciones Pesqueras, Passeig Nacional s/n, Barcelona 08003, Spain

CONTENTS

5.1. Composition of sea-water	126
5.1.1. Major elements	127
5.1.2. Minor elements	128
5.1.3. Trace elements	129
5.2. Processes controlling the chemistry of sea-water	129
5.2.1. Liquid phase processes	130
5.2.2. Interface processes	131
5.2.3. Boundary processes	131
5.2.4. Nitrogen cycle	132
5.2.5. Phosphorus cycle	133
5.2.6. Silicon cycle	134
5.2.7. Carbon and oxygen cycles	135
5.3. The Mediterranean: a concentration basin	137
5.3.1. Water balance of the Mediterranean Sea	138
5.3.2. Precipitation and land run-off	138
5.3.3. Evaporation	140
5.4. The Mediterranean, an oligotrophic ecosystem	144
5.4.1. Eutrophication	145

5.1. COMPOSITION OF SEA-WATER

Despite the massive flux of materials through the oceans, the biological and geological evidence together suggest that the composition of sea-water has remained astonishingly constant for the last 2000 million years. Residence times of the elements in the ocean vary from 100 to 100 million years, while the residence time of the water in the ocean varies from 10,000 to 100,000 years. Residence time is the ratio between the actual volume of a defined compartment, and the amount that yearly moves in and out of the compartment.

The Mediterranean Sea, although of a semi-enclosed nature, does not have a chemistry of its own. The residence time of the water in the Mediterranean is about 100 years. Most of the elements therefore have plenty of time to tour the Mediterranean Sea. However, some differential features do exist in the Mediterranean Sea, when compared to the rest of the ocean. The most outstanding is the relatively high salinity of Mediterranean waters which results from evaporation being greater than precipitation and run-off. Only the Red Sea and some completely enclosed seas like the Dead Sea show higher salinities.

Another well-known characteristic of the Mediterranean Sea is the relatively low concentration, even in the deeper waters, of some biologically important chemical constituents. This is caused by the continuous wash-out through the Strait of Gibraltar which receives poor surface Atlantic water and exports relatively rich deep Mediterranean water. Land run-off and sewers (although creating important enrichments in certain areas) cannot balance the net loss of such chemical constituents as nutrients. Section 5.4 deals with the oligotrophic character of the Mediterranean Sea.

Of particular interest, also, are the high oxygen concentrations of Mediterranean waters — a consequence of the recent character of the deep water.

The chemical composition of sea-water is not easily determined, because of the large differences that exist between the concentration of the various chemical species and to the interferences encountered in many of the analytical procedures. If particulate matter and organisms are excluded, sea-water can be regarded as an aqueous solution of solid and gaseous substances.

The results of the earliest worldwide oceanographic studies showed that the concentrations of the various elements maintained constant proportions as if only pure water was necessary to change the salinity in different parts of the oceans (weight in grammes of the solids dissolved in one kilogramme of water, UNESCO, 1981). These findings led to the formulation of a principle which, for a long time, constituted the basis of the commonly accepted relationships of chlorinity:salinity:density.

The only obvious explanation of this fact is that the chemical constituents of sea-water (almost all known elements) must be in physicochemical equilibrium with the land and air surrounding the oceans. This certainly applies to dissolved atmospheric gases, the $CO_2/HCO_3^-/CO_3^=$ system and other constituents of sea water. However, a large number of elements (especially the most important from the biological standpoint) depart from such a behaviour. In fact, all exchanges of matter between the ocean and the atmosphere, land, sediments and contained organisms would alter the relative concentrations of the elements involved were it not for their long residence time in the oceans and the active circulation of the water. In consequence, the situation approximates to a steady state, when looked at from our time scale.

Chemical equilibria in sea-water do not, by themselves, explain completely the actual concentrations found. A large number of the elements are bound, directly or indirectly, to biological cycles. Therefore, other less obvious and more subtle explanations (such as the steady state concept) must be applied to account for the low equilibrium concentrations shown by some of the minor and trace elements.

It is customary to classify the sea-water elements into three groups, according to their relative concentrations. A number of more or less coherent tables with the chemical composition of sea-water (*see* Table 5.1) are available (Sverdrup *et al.*, 1943; Horne, 1969).

5.1.1. Major elements

The major elements are those which basically determine the physicochemical characteristics of sea-water as a medium. In particular its density and its electrical, thermal and colligative properties.

Hydrogen and oxygen, as constituents of the water molecule, are the main elements followed by the halogens — especially chlorine (with 55% by weight of all the dissolved materials) — and then alkaline

TABLE 5.1. Elements Present in Solution in Sea-Water. Dissolved gases not included. Mainly from Brewer, 1975

The classical textbook figures have been modified in recent years by new methods of analysis and are also still undergoing corrections.

	mg/l		µg/l		ng/l		ng/l
Chlorine	18,800	Zinc	4.9	Xenon	50	Lanthanum	3.0
Sodium	10,770	Argon	4.3	Cobalt	50	Neodymium	3.0
Magnesium	1,290	Arsenic	3.7	Germanium	50	Tantalum	2.0
Sulphur	905	Uranium	3.2	Silver	40	Yttrium	1.3
Calcium	412	Vanadium	2.5	Gallium	30	Cerium	1.0
Potassium	399	Aluminum	2.0	Zirconium	30	Dysprosium	0.9
Bromine	67	Iron	2.0	Bismuth	20	Erbium	0.8
Carbon	28	Nickel	1.7	Niobium	10	Ytterbium	0.8
Strontium	7.9	Titanium	1.0	Thallium	10	Gadolinium	0.7
Boron	4.5	Copper	0.5	Tin	10	Praseodymium	0.6
Silicium	2	Caesium	0.4	Thorium	10	Scandium	0.6
Fluorine	1.3	Chromium	0.3	Helium	7	Holmium	0.2
Lithium	0.18	Antimony	0.2	Hafnium	7	Thulium	0.2
Nitrogen	0.15	Manganese	0.2	Mercury	7	Lutetium	0.2
Rubidium	0.12	Selenium	0.2	Beryllium	6	Indium	0.1
Phosphorus	0.06	Krypton	0.2	Rhenium	4	Terbium	0.1
Iodine	0.06	Cadmium	0.1	Gold	4	Samarium	0.05
Barium	0.02	Tungsten	0.1	Lead	2	Europium	0.01
Molybdenium	0.01	Neon	0.1				

and alkaline-terreous metals (sodium, potassium, magnesium, calcium and strontium). Unlike sodium, potassium and the divalent ions are actively taken up by organisms and their concentrations may thus show important variations. In particular, the Ca/Mg ratio is much lower in the sea than in continental waters, not only because of the precipitation of $CaCO_3$ in estuarine areas, but also because of the differential uptake by shell-forming organisms. Calcium plays an important role in the equilibrium among the different forms of inorganic carbon, $CO_2 + H_2O : H_2CO_3 : HCO_3^- : CO_3^=$.

Carbon, another major element, is crucial for marine life and is also very important as a regulator of the sea-water pH, although careful studies suggest that other elements, such as boron and silicon and the organisms themselves may play a role in pH-regulation.

Sulphur, mainly in the form of sulphate ions, plays an important role, especially in areas of higher oxygen consumption, particularly by heterotrophic bacteria. When oxygen concentrations are low, the sulphate ion substitutes oxygen as an electron acceptor and is transformed into sulphide ion. Anoxic regions (where the sulphide ion is normally present) generally occur in estuarine sediments, in deep waters of stagnant water bodies such as in the Black Sea or under highly productive open-ocean regions.

5.1.2. Minor elements

The minor elements are those which are present at relatively high concentrations but which have only small effects on the physico-chemical characteristics of sea-water. Three groups may be identified: dissolved gases (oxygen, CO_2, nitrogen and argon), nutrients (nitrogen, phosphorus and silicon) and metals from the crustal materials (aluminium, iron, manganese, etc.).

Gases dissolved in sea-water are normally in equilibrium with atmospheric gases and are thus present in saturating concentrations in the surface layers. Nitrogen and argon concentrations exhibit practically no variation by processes other than mixing and can thus be considered conservative except for nitrogen fixation and denitrification. On the other hand oxygen and carbon dioxide are strongly influenced by biochemical and biological processes in the water and sediments (processes are classified as conservative when they maintain a constant ratio with salinity).

Nutrients are minor elements which are, nevertheless, indispensable for the formation of organic materials in photosynthesis. Among these are various forms of nitrogen (nitrate, nitrite, ammonia), phosphorus (as orthophosphate) and silicon (as orthosilicate). Redfield *et al.* (1963) established the principle that the elements involved in the production of organic matter in the oceans (C:O:N:P:Si) are taken up and regenerated by organisms in similar proportions to their actual concentrations in sea-water. The regeneration of the elements in organic matter occurs primarily in strata below the region of active exchange of oxygen with the atmosphere. A closer relation is found, therefore, between the regeneration of inorganic nutrients and the consumption of oxygen than with its production of photosynthetic processes. However, the proportions given for the world ocean do not hold exactly for the Mediterranean Sea.

The third group of minor elements is that represented by metals present in crust materials, which have low solubility in oxygenated sea-water. Some of these are essential for many biochemical processes. However, their concentrations do not seem to be affected in a noticeable manner by these processes and thus maintain relatively constant proportions except in coastal water which is influenced by river discharges.

5.1.3. Trace elements

Most, if not all of the other natural elements present are found in trace concentrations in sea-water. Trace elements are mainly heavy metals which form highly insoluble compounds. In some cases, however, they occur in relatively high concentrations, due to their tendency to form complexes with Cl^- and F^- ions. Many of the components of this group of elements would go unnoticed in the environment, were it not for the potential hazards to marine organisms and to human health. This is especially the case with mercury in the Mediterranean Sea where the concentrations of this element are comparable to those in the open ocean, but are known to be much higher than normal in some organisms, mainly in high-level predator fishes.

5.2. PROCESSES CONTROLLING THE CHEMISTRY OF SEA-WATER

The sea is a dynamic and highly fluctuating system which seldom reaches an equilibrium. The existing concentrations of the components of the system cannot completely define it. The flow of matter through the various compartments, including the dissolved phase, must be evaluated to be able to do this. In particular, the relative importance of the processes taking place in the marine ecosystem can only be assessed after taking into account its general framework, the sources and sinks of nutrients and organic matter and the biological and/or geochemical processes which affect the specific components. Three categories of general processes can be recognised in the marine system:
(1) Liquid phase processes, (2) interface processes (occurring at small-scale discontinuities) and (3) boundary processes (occurring at large-scale discontinuities).

The Liquid Phase Processes (mainly diffusion and advection) act over a wide range of scales from the molecular to the basin-size level. They act mainly within the water masses (i.e. in physical entities which can be considered as a continuum).

The Interface Processes are mainly biological and/or geochemical. The uptake and regeneration of nutrients depend on the presence of particulate matter, either as living phytoplankton or decaying detritus.

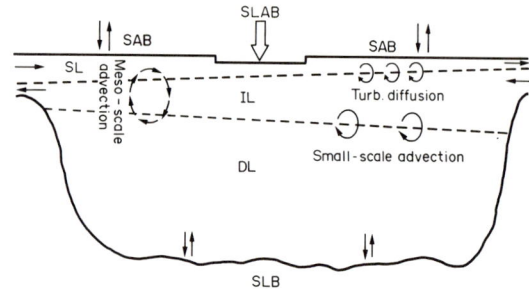

Fig. 5.1. Hydrographic structure of a model Mediterranean basin. SL, surface layer; IL, intermediate layer; DL, deep layer; SAB, sea air boundary; SLAB, sea-land-air boundary; SLB, sea-land (sediment) boundary.

The Boundary Processes control most of the input and output of matter and energy to the system. The boundaries are: the sea-air boundary (SAB) at the free surface of the sea, the sea-land boundary (SLB) at the bottom of the sea, and sea-land-air boundary (SLAB) along the coastline (Fig. 5.1).

5.2.1. Liquid phase processes

Advection and diffusion are the main processes which take place in the continuum phase. They are linked to the hydrographic structure of the Mediterranean (described as being approximatively three-layered) and to the lateral openings (sills) into the adjacent seas (Fig. 5.1). The surface layer (SL) (i.e. down to approximatively 100 m) is of low density and nutrient concentration and contains well-aerated water. Its circulation appears to be primarily formed by cyclonic gyres in all the sub-basins of the Mediterranean. The intermediate layer (IL) contains high density water with medium concentrations of nutrients. In general it shows a westward motion, with some restriction across the sills. The deep layer (DL) (below 500 m) is rather stagnant within each basin and consists of water with a density comparable to that of the intermediate layer, but with slightly higher concentrations of nutrients.

The chemistry of the surface layer is partly controlled by exchanges across the sea-air and sea-air-land boundaries through which the net input of nutrients, organic matter, and pollutants occur. The thin surface film is of especial importance, for its chemical composition and biological populations deviate quite markedly from that of the major water mass. Exchanges between the surface and the intermediate layers are rather limited in the Mediterranean Sea (especially during periods of strong thermal stratification). These exchanges nevertheless control primary production. Similar exchanges also affect the flow of nutrients and of organic matter into the adjacent seas. Direct exchanges between the surface and the deep layers occur only during the peculiar circumstances of deep-water formation. However, it may constitute an important mechanism for transferring nutrients to the surface and pollutants to the depth. Exchanges between the intermediate and the deep layers are easier, because their densities are not so different.

Advection and diffusion act over a wide range of scales (even the distinction between these processes is a matter of scale). Large and meso-scale advection control mainly the horizontal transport of the dissolved substances; as typical scales we can take 10 – 100 km for the meso- and 1000 – 10000 km for the large scale movements. Such transport takes place with the displacement of the entire water mass. Both types of processes act either at the surface or in the intermediate layer and thus do not favour the transport of pollutants into the deep layer or the transport of nutrients into the surface one. The large-scale components may contribute to export pollutions from the basins, while the meso-scale ones may promote the transfer of pollutants from highly polluted zones into areas that would otherwise have a negligible input. Meso-scale circulation also plays an important role in vertical mixing and deep-water formation. Small-scale advection overlaps with turbulent diffusion in controlling the vertical mixing of the various water masses that brings about the transport of the dissolved substances. At the high-frequency end, molecular diffusion overlaps with the interface processes and may constitute a rate-controlling step in many biological processes.

When the sources and sinks of a given substance are in different layers and the vertical spreading of that substance in the marine environment is dependent on turbulent diffusion and small-scale advection, the relative concentrations in the various layers provide an estimate of the diffusion rates. Such estimates may be complicated, among other factors, by processes that act with opposing effects in the same layer. This can be the case with nitrogen fixation, or *in situ* regeneration of nutrients by planktonic organisms, in the surface layer, where nutrient uptake by photosynthesis is the major process expected.

5.2.2. Interface processes

Biological and/or geochemical processes are always associated with particle interface and may therefore be grouped under this heading. They are carried on everywhere in the marine environment, but some of them are more prominent at the boundaries. Synthetic processes take place mostly in the upper layers and are related to the sea-air boundary. For example, photosynthesis requires radiant energy coming through the surface and releases oxygen. If the oxygen becomes oversaturated, part of it may escape into the atmosphere. Regenerative processes, on the other hand, take place mostly in the lower layers and are related to the sea-land boundary. For example, nutrients are released from the organic matter by biochemical reactions. This occurs mainly in the sediments, although in the Mediterranean, not much organic material becomes initially incorporated in the sediment.

The condensed phase of the system, mainly particulate organic matter, tends to sink from the upper layers into the bottom and to disequilibrate the system: oxygen is produced in excess at the surface and depleted in the deep layer. Downward transportation of oxygen from the surface by advection and turbulent diffusion tends to re-equilibrate the system. Upwards transport of nutrients from the deep and intermediate layers promotes photosynthetic production of organic matter and therefore constitutes a negative feedback mechanism that enhances the supply of organic matter to the lower layers. The more active the system, the more intense is the disequilibrium. In oligotrophic ecosystems, downwards transport of oxygen (or oxygen available in water) is sufficient to oxidize all the organic matter produced in the surface layer or introduced through the upper boundaries from the atmosphere and/or land. However, in eutrophic or fertile ecosystems, oxygen may be partly lost to the atmosphere and the oxidation of organic matter accumulated in the bottom sediments may be extremely difficult.

5.2.3. Boundary processes

Evaporation, rainfall, spray and fall-out are the most important mechanisms by which water and airborne particles are exchanged between the sea and the atmosphere, although radiation and gas exchanges also occur at the surface. At the sea-land-air boundary, land drainage unidirectionally introduces into the sea large amounts of water and organic and inorganic materials, both dissolved and suspended. At the sea-land boundary dissolved organic and inorganic substances diffuse into the bulk of the water after being released from the organic matter buried with inorganic sediments.

The most important chemical and biochemical processes controlling the productivity of the marine ecosystems are those involving the cycles of limiting nutrients (nitrogen, phosphorus), of carbon, and of oxygen.

5.2.4. Nitrogen cycle

The main forms of nitrogen (N) occurring in marine waters (excluding gaseous N_2) are nitrate, nitrite, ammonia, and organic N. Total concentration of inorganic N is usually between 12 and 25 times the concentration of total available phosphorus. The concentrations of the various forms of N in the water depend not only on the physical rates of supply but also on biological processes such as the uptake by phytoplanktonic algae and the mineralization or immobilization by bacterial micro-organisms both in the water and in the sediments underlying it. In sediments, as in soils, N is primarily in an organic form arising either from particulate material formed in the waters above, or brought in by terrestrial run-off. Fig. 5.2 shows in a very schematic way the biotic and abiotic paths along which N can circulate in the marine ecosystem. The most important pathways are those determined by plant and bacterial micro-organisms but some animals may contribute in a no negligible way to the mobilization of nitrogen (Dugdale and Goering, 1967). Due to the many complex, competing, biological reactions occurring in a given ecosystem, it is extremely difficult to determine the relative importance of each individual microbial process. However, some of them which are likely to be most important in determining the availability of N are ammonification, nitrification, denitrification and nitrogen fixation.

Ammonification: Comparatively little information is available on either the micro-organisms or the environmental characteristics controlling the formation of ammonia from organic matter in the marine systems. Early studies show that ammonia can be produced by the decomposition of sinking detritus in freshwater lakes, in the waters below the thermocline, as well as being released from the sediments. Recent work has shown that the process is favoured by anaerobic conditions. On the other hand, ammonia N can also be produced by excretion in fish and smaller animals (Whitledge, 1972) and even in sediments where nitrate is reduced, in goes nitrate and out comes ammonia.

Nitrification: This process, resulting in the conversion of ammonia to nitrate, is carried on by obligate aerobic bacteria of the genera Nitrosomonas (ammonia to nitrite) and Nitrobacter (nitrite to nitrate) and has been extensively studied in terrestrial systems and in sewage purification processes. In sea-water, the trophic level may determine whether the process is functional or not and its molecular oxygen

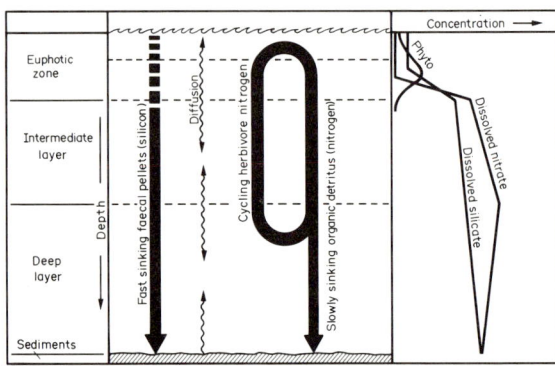

Fig. 5.2. Diagram of the main processes controlling the vertical distribution of nitrate and silicate in Mediterranean waters.

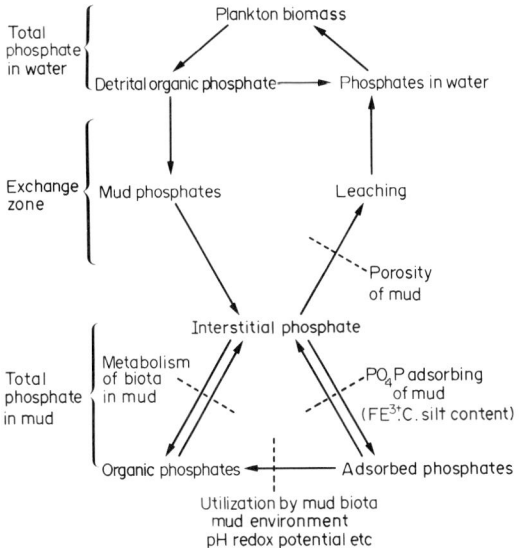

Fig. 5.3. The various forms in which phosporus can be found in the marine ecosystems.

requirements result in an overall increase in BOD (Biochemical Oxygen Demand) of waste-laden waters. The reverse process (nitrate to nitrite) has been described in Mediterranean waters (Blasco, 1972) as being carried out by sedimenting phytoplanktonic algae, when exposed to very dim light and substantially high nitrate concentrations, conditions normally existing in summer at the base of the photic zone, around 100 − 110 m depth.

Denitrification: In this process, nitrate and nitrite are biologically reduced to gaseous nitrogen oxides (N_2O and NO) and to molecular N_2. Nitrogen oxides have attracted attention only recently and are present in very small concentrations in the marine environments. Reduction to gaseous N_2 is commonly observed in sewage sludges and in highly eutrophic systems. It may be difficult to detect if it is not through changes in the ratio N:Ar. As a consequence of denitrification, highly eutrophic systems lose compounds of N and can experience a considerable limitation in the amount of available inorganic compounds of nitrogen.

Nitrogen fixation: This is the reverse process to the latter in the sense that molecular N_2 is converted to organic N. The process appears to be light-dependent and therefore coupled to photosynthesis although certain blue-green algae can also fix nitrogen in the dark. A considerable volume of data is available to show that nitrogen fixation occurs, usually in tropical waters (Goering et al., 1966). Its occurrence in the Mediterranean Sea has been suggested, although evidence has not been reported so far, if it were not for the presence of organisms able to do such operation (*Richelia*, for example, usually associated with diatoms).

5.2.5. Phosphorus cycle

Unlike nitrogen, which is present in several inorganic forms, phosphorus occurs in sea-water only as the orthophosphate. Fig. 5.3 schematically reviews the various forms in which phosphorus can be found in the marine environment. Phosphorus bound to organic matter is released very rapidly. However, even less work has been published on specific bacteria active in the regeneration of phosphate than in the case of ammonia. One would assume that there are many micro-organisms which will bring about the solubilization of phosphate from organic phosphorus compounds present in detrital material. Anyway, the necessary enzymes (phosphatases) have been identified almost everywhere. The average concentration of phosphorus in deep Mediterranean water is only 0.2 μM (6 mg/m^3), in surface layers much less, down to indetectable. In the major oceans the average phosphorus concentration in deep water is around 2.4 μM.

In marine sediments, phosphorus is considered to exist as interstitial phosphate which can be leached to the water above, adsorbed phosphate which can be released by chemical processes, insoluble phosphate bound by such ions as Ca^{+2} and Fe^{+3} or bound to organic matter. A large fraction (up to 20%) of the phosphate dissolved in fresh water entering the sea is lost to the sediments through the formation of insoluble phosphates deposited as silt on the bottom (Perkins, 1974). Insoluble ferric phosphate is strongly bound to the sediments but under low redox potentials, iron is reduced and in the form of ferrous phosphate may leach out to the aquatic phase. Thus, an oxygen deficiency created by bacterial action in the sediment may be expected to facilitate the release of phosphate by the substratum, especially when shallow sediments are stirred by gales.

Phosphorus recycles in water, with an eventual loop in the sediment. Nitrogen cycles equally in water and organisms, but with a final loop in the atmosphere. Such difference in the cycles of the most important limiting nutrients is related to the solubility of most nitrogen compounds and the existence of gaseous nitrogen, against the ease with which phosphorus forms highly insoluble materials. This difference has had an importance in the evolution of the atmosphere of the Earth and, without doubt, has conditioned the evolution of life.

5.2.6. Silicon cycle

Unlike nitrogen and phosphorus, silicon is not so much in demand by living organisms. Only diatoms, silicoflagellates, radiolarians and sponges require the presence of this element for their growth. The monomeric form of orthosilicic acid is thought to be the normal state of dissolved silica in natural waters. Polymeric silica, crystallizing out as quartz, is stable in supersaturated solutions with respect to amorphous silica, but normal concentrations found in sea-water are well below saturation levels. Orthosilicic acid dissociates in solution but, at the pH of sea-water, undissociated orthosilicic acid is the stable form. Electrolytes in sea-water do not affect truly dissolved silica but coagulate colloidal silica which may be depolymerized in a few days, provided total silica concentration is below about 100 ppm (parts per million or mg/l). The ordinary content of Si in sea-water is only of 2 to 4 ppm (60 – 120 μM), going down to 0.05 – 0.5 ppm (1.5 – 15 μM) in surface waters depleted by diatom growth. It is of the same order than the concentration of inorganic compounds of nitrogen.

Particulate silica may be crystalline or amorphous. Crystalline silica may take the form of quartz, chert or aluminosilicates such as kaolinite, montmorillonite, illite and glauconite. Orthosilicic acid is rapidly released by clays to silica-deficient water (Mackenzie et al., 1967). This release is governed by formation of some ill-defined aluminosilicate of increased Al/Si ratio that ends up as kaolinite. Some authors have claimed sea-water silica concentrations to be controlled by alteration of clay minerals or by dissolution of finely divided quartz in the interstitial water of sediments, since solubility of quartz increases considerably when the particle size is below 10 μm and quartz is unknown in the fraction below the 0.1 μm. This is, however, very unlikely since interstitial waters can hardly be considered silica-deficient and amorphous silica and opal, mostly biogenous, are the most soluble species with a saturation concentration of about 120 ppm at 15°C for both, though opal has a much lower dissolution rate than amorphous silica.

The average residence time for silica in the ocean is approximately 4000 years. The sources of particulate silica are river discharges, wind transport and glacier transport and the main sink is sediment burial. As far as dissolved silica is concerned, biological uptake (8×10^{15} g/yr) by far out-balances the input through river outflow (4×10^{14} g/yr) and re-solution of sediments (3×10^{12} g/yr) (Harris, 1966). Since the rate of formation of biogenous silica is about 20 times larger than river discharges and 3000 times larger than leaching from non-biogenic sediments, one may conclude that the cycling of biogenous silica controls the concentrations of this nutrient in the sea-water. Silica uptake takes place mainly at the euphotic zone by diatoms (Fig. 5.2). Re-solution of silica, directly from undigested diatom tests, is presumed to be a strictly inorganic hydrolytic process (Grill and Richards, 1964). However, as soon as diatoms are ingested by grazers, silica goes practically unchanged to the faecal pellets that are part of the detrital component, rapidly settling to the bottom where decay of the organic envelope produces regeneration of dissolved silica. A large part of the detrital silica may be re-solved in the water column during sinking of detritus or at the water-sediment interface by further biological decomposition of the detrital material and does not accumulate in the sediments unless the sedimentation rate is so high, that over-balances re-solution. Biogenous silica materials are converted through diagenesis into opal and even crystallize out as quartz or chert below highly productive areas.

5.2.7. Carbon and oxygen cycles

An amount of oxygen equivalent to the carbon fixed in the photosynthetic process is released to the water and eventually to the atmosphere. Inorganic carbon (as CO_2, HCO_3^- or $CO_3^=$) is normally in excess in sea-water even in the extreme case of large algal blooms. Besides, it has a further reserve in the atmosphere. We should not, therefore, concern ourselves with the details of its pathways in the marine ecosystem, as carbon rarely is limiting, and never in the Mediterranean.

Oxygen, although not itself a growth factor, plays a major role in the process of biological or chemical oxidation of the dead or living organic matter dissolved or dispersed in sea-water, since only in the presence of this element can organic matter be converted by respiration and aerobic bacterial action into CO_2 and simple inorganic salts. Oxidation of organic matter can however proceed beyond the availability of oxygen under anaerobic conditions by using NO_3^-, NO_2^- or even $SO_4^=$ as oxygen donors, although this occurs only in extreme eutrophic conditions and, so far only the bottom of the Black and Baltic Seas and some other smaller and localized areas in the world ocean are known to be affected by anoxy.

Oxygen, and to some extent carbon dioxide dissolved in sea-water, are mainly controlled by the combined action of several processes:
(1) direct exchange between the sea-water and the air above it;
(2) turbulent mixing with adjacent water layers;
(3) photosynthesis carried out by plants, mainly phytoplankton; and
(4) respiration and other biological and chemical processes.

Process 1, acts both ways across the sea surface. Oxygen enters the sea when consumption in respiratory processes lowers its concentration below saturation, and is lost to the atmosphere when production by photosynthetic organisms causes supersaturation. Oil films or detergents dispersed in the surface layer may decrease the rate of exchange of oxygen and CO_2 through the surface of the water by as much as 20% but it may be doubled by strong wave action (Perkins, 1974). Process 3, takes place only in those waters that have a good balance between the rate of supply of nutrients and the intensity of the incoming solar radiation within the adequate spectral range. Owing to the reduced nutrient supply in highly stratified waters, Process 3 is only of some importance between the thermocline and the compensation depth. This depth may be established at or near the 100 m in highly stratified waters due to their generally high transparency, producing a maximum oxygen layer between 50 and 75 m depth, typical of the oligotrophic regions of the oceans. On the other hand, vertical turbulence may strongly depress photosynthetic production by keeping the organisms for too long a time below the compensation depth.

Process 2, complemented by currents, is the only one which actually supplies oxygen to the waters lying below the compensation depth. Process 4, acts everywhere in the sea and is the one responsible for the general decrease in dissolved oxygen concentrations with increasing depth in the world ocean, or with increase in elapsed time since the water mass lost direct communication with the atmosphere, as well as in relation with some specific features found in certain regions such as the oxygen minimum layers or the anoxic basins. The process is highly dependent on the temperature of the water and on the amounts of organic matter, dead as well as living, produced in the euphotic zone or discharged into the sea from nearby land-based sources.

Most of the naturally occurring organic matter in the marine environment has its origin in planktonic detritus, forming relatively large aggregates with bacteria, and in faecal pellets that sink quickly to the bottom. Locally produced organic matter and humic compounds derived from the land are rapidly assimilated by the ecosystem. However, introduction with river run-off or sewage discharges of large amounts of organic matter (detritus) tends to encourage the development of the heterotrophic decomposers component whilst introduction of nutrient salts favours the growth of the producers component. The response of biological systems to increasing concentrations of trophic materials is an increase in the respiratory demand in the waters below the euphotic zone and in the sediments.

Oxygen consumption in sea-water

The chain of animal life in the deep waters feeds upon the rain of organic material, relying on the dissolved oxygen for respiration. Below the euphotic zone there is a net consumption of oxygen by respiration of animals and bacteria that decompose organic matter. These processes are accompanied by the release of CO_2 and nutrients to the water that are often re-utilized by photosynthetic producers. A useful term to express the amount of organic matter originally present in the sea-water is the oxygen consumption or apparent oxygen utilization (AOU) given by the difference between the actual oxygen concentration and the equilibrium saturation value. Apparent oxygen utilization is used to estimate the changes in oxygen concentration which have taken place since the water left the surface layers, where it was in equilibrium with the atmosphere. The rates of oxygen consumption estimated by Riley (1955), Wyrtki (1962) and Packard *et al.* (1977) show an exponential decrease with depth.

$$R_z = R_o e^{-az}$$

where the rate at the surface (R_o) is between 0.01 and 0.03 ml per litre of water and 24 hours, z is depth in metres, and a is between 0.003 and 0.004. R is equivalent to the concept of Biochemical Oxygen Demand (BOD) used in the evaluation of oxygen consumption by water and sewage.

In most of the world ocean, circulatory processes replenish the depths with oxygen-bearing water at rates such that oxidative consumption does not exceed oxygen renewal. Some dissolved oxygen is thus retained.

Oxygen consumption in marine sediments

Most of the organic matter in marine deposits is mixed with fine particles of inorganic sediments which may have been trapped in undisturbed parts of the sea floor. Coarse sandy sediments tend to be poor in organic matter because they accumulate in agitated, well-ventilated waters, where the light organic matter is mostly washed away and the supply of oxygen is sufficient to result in swift decay of organic materials. The organic carbon in coarse sediments is generally less than 1%. In finer sediments there is commonly a higher organic content (in many cases of several per cent) because light organic remains and clay both tend to come to rest in quiet regions of the sea floor, where there is poor ventilation of the bottom water and a low rate of oxygen exchange with the sediment. Thus, below a thin cover of oxidized sediment, anaerobic conditions tend to develop, due to bacterial activity and lack of oxygen supply by diffusion. However, various mud-feeding animals stir the surface stratum and burrow and plough their way through newly-deposited sediment or seek protection by living below the surface. Plant roots (e.g. mangroves) also disturb the lamination of sediments in shallow water. As a result decomposition of organic matter is facilitated by bioturbation of sediments.

Many marine sediments are anoxic. Natural conditions favourable for the formation of sediments rich in organic matter are found in regions of upwelling or near estuaries. In these areas, high primary production results in accumulation of detrital material on the sea floor and in development of anaerobic conditions. Much organic matter can thus be preserved, in spite of ventilation of the overlying waters. Contents of 10% organic carbon in sediments are however rare, always being associated with anoxic conditions in the water (Black Sea). With the reducing conditions (low values of the redox potential, Eh), the thermodynamic drive is diminished and organic materials tend to accumulate in the sediments rather than to be oxidized. In consequence, sediments laid down under anoxic conditions are considerably richer (up to 10-fold) in organic matter than are sediments deposited under oxygen-bearing water. In the water trapped inside the sediment, diffusion is low and approaches molecular values. As a consequence in the space of a few cm, or even mm, a number of biochemical processes can be segregated and layered, according to the gradient of redox potential. For instance, in the top there is reduction of nitrate, then of nitrite, deeper down reduction of sulphate to sulphide occurs, and further down, perhaps, of CO_2 to methane.

When H_2S is produced in the sediments, lethal conditions may be created for the fauna and flora in the overlying water. Sulphide also combines with the oxides of iron and blackens the anaerobic layer. The blackened sulphide-containing sediments are of a widespread distribution, but the sulphides, formed under anaerobic conditions, are oxidized rapidly in the presence of oxygen. The depth at which black sulphides occur indicates the depth to which significant amounts of oxygen penetrate, either by diffusion or by bioturbation.

5.3 THE MEDITERRANEAN: A CONCENTRATION BASIN

If the Mediterranean basin as a whole is regarded as a chemical reactor, it appears that the first unit process taking place is a concentration of all the substances entering it with the 1.2 million m^3/s of surface Atlantic water. After approximately 75 to 100 years, almost all the substances dissolved in this surface water have undergone an increase in concentrations of about 4.7% and then they flow back into the Atlantic Ocean at depth. Their higher concentrations allow the Mediterranean waters to be traced as far as the subequatorial Atlantic or the Bay of Biscay.

The water budget for the entire Mediterranean basin is negative (Table 5.2). The high evaporation rates caused by the strong heating of the surface water and the dryness of the continental winds flowing across the basin from neighbouring continental masses over-run the weak rates of precipitation and run-off from the rivers draining areas of Europe, Asia and Africa that are relatively large, although rather arid.

To balance the loss of water from evaporation (which exceeds that gained by precipitation and run-off) approximately 15 times this amount of Atlantic surface water has to flow in and nearly the same amount of Mediterranean deep water has to flow out to maintain a salt mass balance.

TABLE 5.2. Water budget for the Mediterranean Sea (in $m^3 s^{-1} \times 10^{-3}$, or thousands of cubic metres per second)

Process	McGill 1969	Tixeront 1970	Morel (*) 1971	Lacombe et Tchernia 1974	Béthoux 1977
Evaporation	−92(2)	−95(1)	−70(1)	—	−122
Precipitation	33(2)	28(3)	13(3)	—	25
Run-off (**)	14(2)	16(4)	9(4)	—	16(4)
Net Inflow Through:					
Dardanelles	6(5)	6(5)	6(5)	6(5)	6(5)
Sicilian channel	—	—	42(6)	—	52
Gibraltar	40(7)	45(6)		54.4(8)	75
Total Inflow Through:					
Dardanelles	—	—	—	12.5	—
Sicilian channel	—	—	1000	—	—
Gibraltar	—	—	—	1187.5	—

(*) Only for the eastern basin
(**) MED X estimate for total run-off is 13.250 for the whole Mediterranean and 8.663 for the eastern basin
(1) Estimated from studies made in Tunisia (Berkaloff, 1952)
(2) Based on Carter (1956)
(3) Estimated from pluviometric charts and sailors' accounts
(4) Direct measurements on some rivers and from pluviometric charts and area of all other river basins
(5) Based on Merz and Moeller (1928)
(6) Estimated by balance
(7) Based on Schink (1967)
(8) Estimated by salinity difference between incoming and outgoing waters and mean flow

The incoming Atlantic surface water (with salinity slightly above 36 $°/_{\infty}$) is transformed into outgoing deep Mediterranean water (with a salinity of 38.4 $°/_{\infty}$ or more). This turns out to be one of the most exciting oceanographic phenomena encountered in the world's oceans. Studies carried out in the Mediterranean Sea have helped to understand equivalent processes occurring in other regions where deep ocean waters are formed.

5.3.1. Water balance of the Mediterranean Sea

The exact amount of water flowing through the Strait of Gibraltar and the Dardanelles is difficult to estimate, largely because the actual flow through the Straits is of the order of 1.2 million m^3/s in each direction, and the net flow represents only about 5% of this figure. Merz and Moeller (1928) estimated the net flow of water through the Dardanelles in about 6000 m^3/s and nobody has tried so far to improve this estimation. Lacombe and Tchernia (1974) made an estimate of 54,500 m^3/s (based on the salinity difference between the incoming and outgoing waters and on the mean flow) for the net gain of water through the Strait of Gibraltar. This value is only slightly above that made earlier by McGill (1969) on the basis of data supplied by Schink (1967). Béthoux (1977) estimated the net flow as approximating nearly 75,000 m^3/s (based on the thermal advection). This is consistent with his values for evaporation, precipitation and run-off.

Uncertainties in measuring actual mean flow rates and in attributing salinities to incoming and outgoing water masses can only be eliminated with more active oceanographic research in the field. This, unfortunately, is not taking place.

5.3.2. Precipitation and land run-off

Summer drought, a characteristic of the Mediterranean region, arises primarily from a low rainfall. Precipitation varies considerably from place to place and from year to year. There is a general decrease in yearly rainfall from west to east and from north to south. Rainfall varies from more than 1500 mm/yr

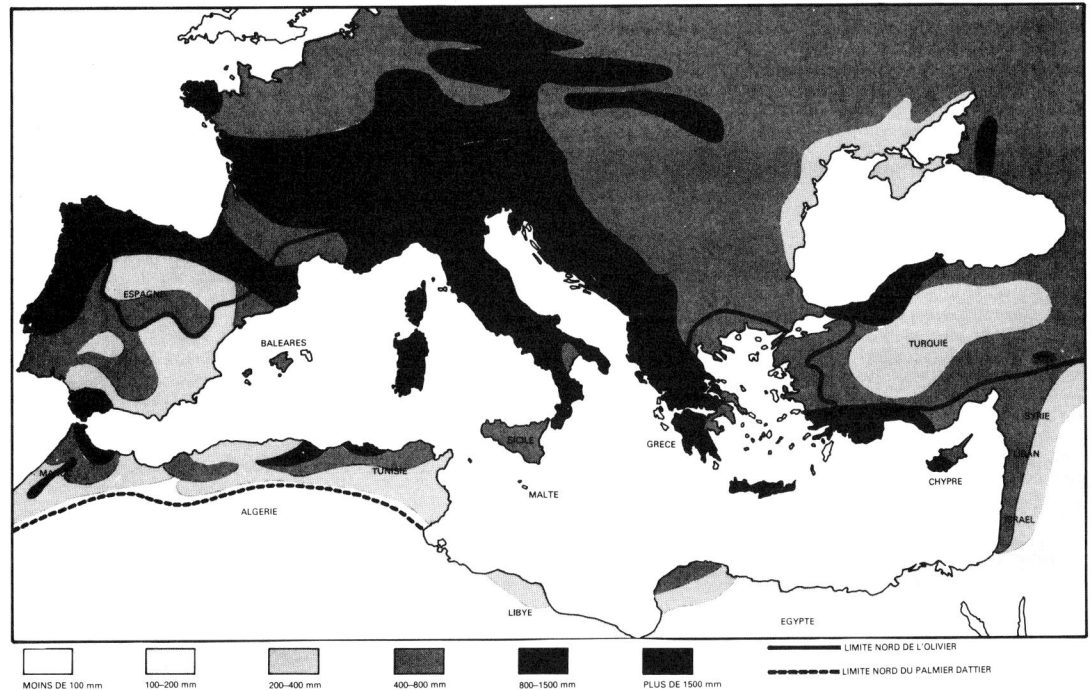

Fig. 5.4. Distribution of rainfall in the mediterranean region (from Henry, 1977).

(along the western part of Yugoslavia and over the mountain ranges of the Alps and the Pyrenees) to less than 100 mm/yr (in the interior highlands of northern Africa and western Asia, as well as over some islands of both the eastern and western basins) (Fig. 5.4).

Carter (1956) computed precipitation from pluviometric charts derived from the rainfall recorded at shore stations in about 33,000 m^3/s for the basin. Tixeront (1970) gave a somewhat lower figure (based on sailors' accounts), but precipitation at sea is still an unsettled question.

Extrapolation of measurements made on the shore to adjacent marine areas does not seem to be feasible. Recent comparisons of rainfall over the sea with real observations made on the BORHA laboratory, moored in the north-west Mediterranean Sea (Béthoux, 1977) gave a much larger deviation than had been accounted for previously. The most likely estimate of precipitation over the Mediterranean Sea amounts to 310 mm/yr with a slightly larger value for the western basin, 335 mm/yr, and a somewhat lower one for the eastern basin, 295 mm/yr. These amounts correspond to about 25,000 m^3/yr over the whole basin, significantly lower than previous estimates.

The overall limit of the natural catchment areas of the rivers draining into the Mediterranean Sea is about 4 million km^2. Rivers with catchment areas partially outside the Mediterranean region have a more regular regime as well as higher flow rates than those entirely running in the Mediterranean region since they are fed by the slow melting snows of the mountain ranges of the northern reaches of the Pyrenees (Ebro) and Alps (Rhône and Po), or by the abundant rains of tropical Africa (Nile). The Nile River however, should be considered in the group of typical Mediterranean rivers.

The continental run-off estimated by Tixeront (1970), from pluviometric charts and some direct measurements, turned out to be rather high when compared to that of Carter (1956). The best current estimate of river flow rates was made in the framework of MED POL X (UNEP, 1977), giving a value for total run-off of about 14,000 m^3/s (Tables 5.3) which is close to Carter's estimate.

TABLE 5.3. Flow of fresh water from river run-off and other sources adapted from UNEP (1977)

	Region	Approx. area 10^3 km²	Total m³ s⁻¹	Domestic %	Industrial %	Agriculture %	Rivers %	Specific discharge mm	Population × 10^3
I	Alboran	68	197	1.9	2.3	34.8	61.0	91	2690
II	North-west	287	3013	0.8	2.5	12.2	84.5	331	8870
III	South-west	264	290	1.2	2.8	57.0	39.0	35	4439
IV	Tyrrhenian	231	1225	1.1	1.8	59.1	38.0	167	8131
V	Adriatic	133	5195	0.2	0.7	32.9	66.2	1228	3625
VI	Ionian	151	1103	0.3	0.9	77.5	21.3	230	1883
VII	Central	583	105	2.8	3.9	93.3	0.0	6	2904
VIII	Aegean	143	1463	0.4	1.5	40.0	58.1	321	4584
IX	North Levantine	177	778	0.1	0.1	15.3	84.5	139	1320
X	South Levantine	500	550	1.1	0.8	8.8	89.3	35	5388
Total western basin		850	4725	1.0	2.3	28.0	68.6	173	24130
Total eastern basin		1687	9194	0.3	0.8	37.1	61.7	170	19704
Total Mediterranean		2537	13919	0.5	1.4	32.9	65.2	173	43834

There is a large imbalance between run-off in the northern shore (Fig. 5.5), which drains 92% of the water that flows into the Mediterranean Sea, and the southern shore, which drains only the remaining 8%. This difference arises mainly from differences in yearly precipitation since the size of the drainage areas are similar. The area that receives a larger input through river run-off from the entire Mediterranean Sea is the Adriatic Sea, followed by the north-western Mediterranean, receiving among them nearly 70% of all the river discharges. These two areas are followed by the Aegean Sea, the Tyrrhenian Sea and the Ionian Sea (20%) and the North African coast, including run-off from the Nile, receives less than 10%.

5.3.3. Evaporation

If the previous estimates are imprecise, assessment of the rate of evaporation (itself the result of various complex interacting mechanisms) is absolutely impossible. Evaporation rates are controlled by the difference in the vapour pressure of water in the sea (sea surface temperature) and in the air above it (relative humidity), as well as by the wind velocity at the boundary layer.

As a result of the restricted water exchange with the Atlantic Ocean, sea surface temperature is controlled basically by radiation balance. Cold water of polar origin is totally absent from the Mediterranean Sea and subsurface and deep Mediterranean waters are, therefore, warmer than those of the Atlantic Ocean at equivalent latitudes.

Taking into account the solar constant and atmospheric transmission coefficient of 0.7 (Perry and Walker, 1977), the distribution of the solar radiation for the latitudes of the Mediterranean region may be drawn in comparison with other areas of the world (Fig. 5.6). According to these computations, the amount of solar radiation in summer is higher for the Mediterranean latitudes than for those of the tropics, due to the relatively high sun elevation and to the increased length of the daytime as compared to those of, for example, the Equator. If the real cloud cover is taken into account for the various locations around the basin, the distribution of solar radiation reaching the Mediterranean region (Fig. 5.7) varies between about 100 and 200 kcal/cm³ yr increasing from north to south and from west to east with an average of 158 kcal/cm² yr for the western basin and 163 kcal/cm² yr for the eastern basin (Béthoux, 1977).

The total solar radiation reaching the Mediterranean region is, in the summer solstice, one of the highest in the world with values above 22 kcal/cm² month on the southern shore of the Levantine basin,

CHEMISTRY OF MEDITERRANEAN WATERS 141

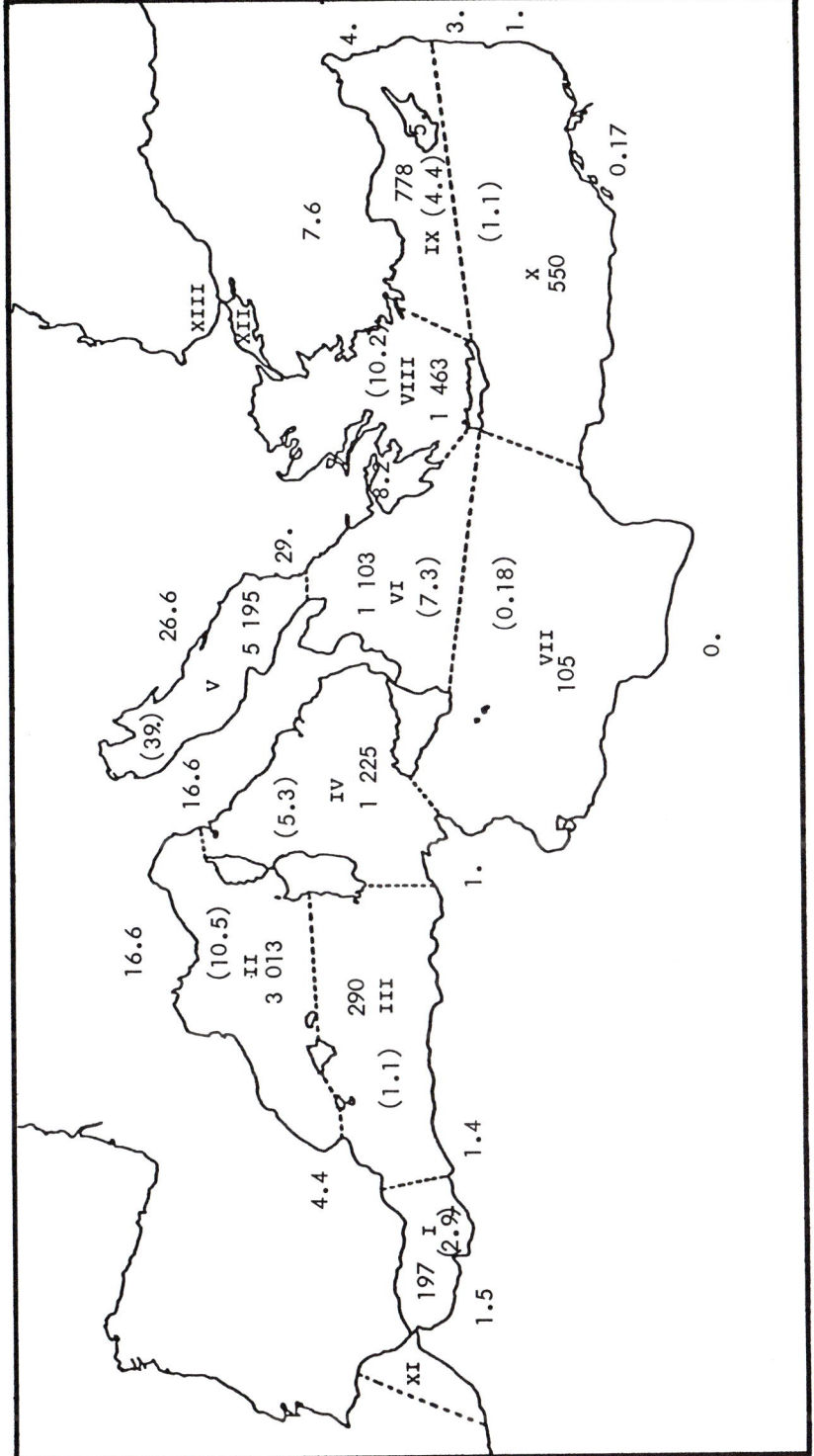

Fig. 5.5. River outflow into the Mediterranean Sea. Figures on the sea represent input of river water to the area (m³/s⁻¹). Figures on land represent specific discharge for the country (m³ s⁻¹ km⁻² × 10³). Figures in brackets represent specific discharges for the area (m³ s⁻¹ km⁻² × 10³) (adapted from UNEP, 1977). See also Fig. 4.4.

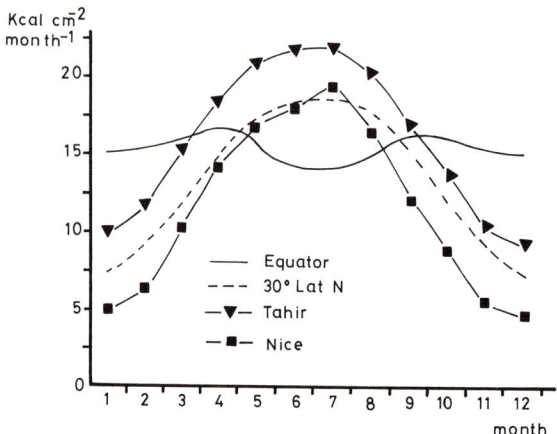

Fig. 5.6. Solar radiation at various latitudes (adapted from Béthoux, 1977).

falling only below 18 kcal/cm² month on the north-western Mediterranean. This, and the fresh water flowing at the surface contribute to the stabilization of a strong seasonal thermocline and thus to the maintenance of the high surface temperatures appearing in the Mediterranean during the summer. Loss of heat to the atmosphere occurs in summer through sensible heat transfer. However, except in the Gulf of Lions and the northern part of the Aegean Sea where gusts of dry wind are frequent even in summer,

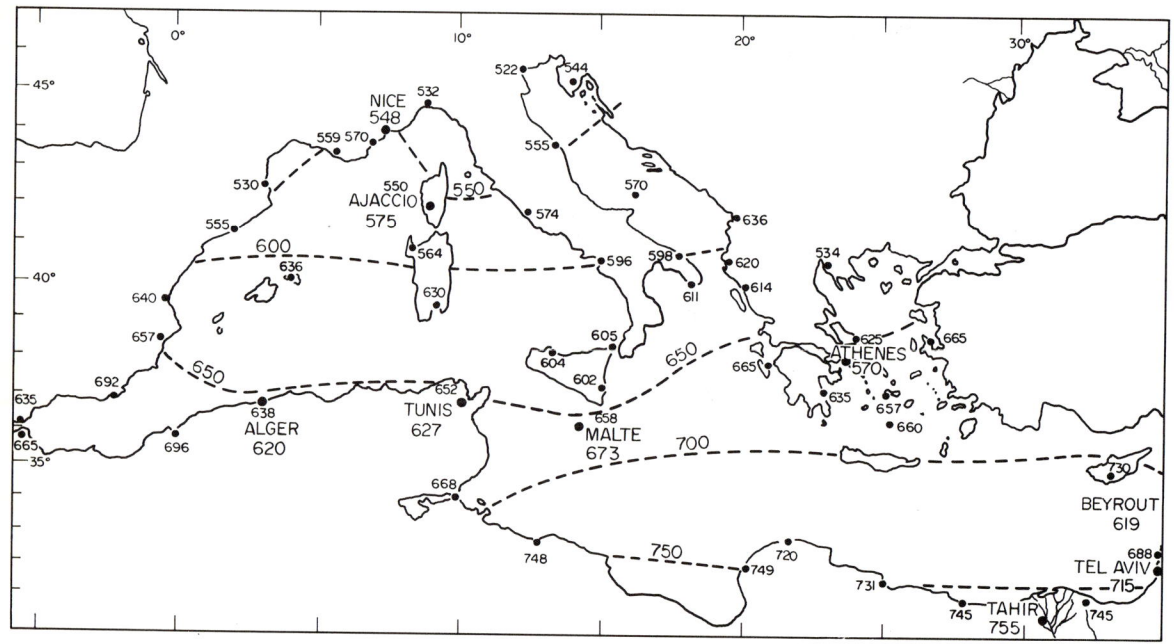

Fig. 5.7. Average solar radiation in the Mediterranean, in kJ cm^{-2} yr^{-1}. Large figures correspond to direct measurements. Small figures are numerical estimates (From Béthoux, 1977).

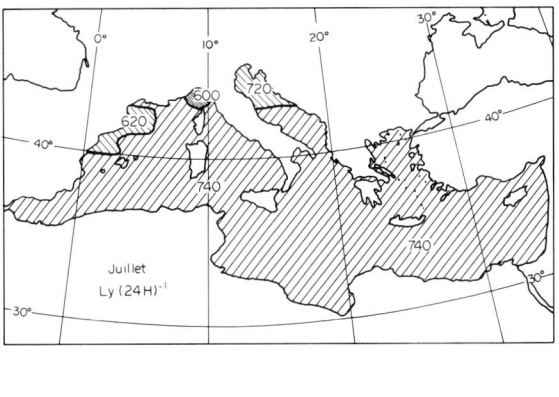
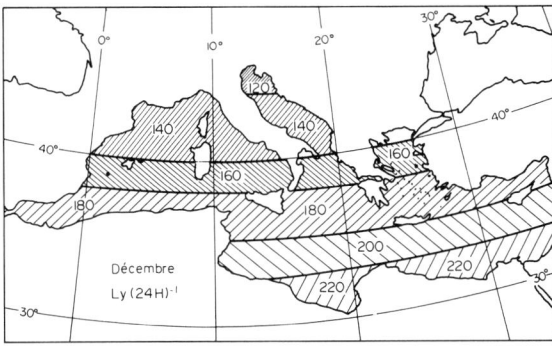

Fig. 5.8. Solar radiation in July and in December, in 1 Ly (24 h)$^{-1}$ = 0.13 kJ cm^{-2} month^{-1}. 1 Ly = 1 calg/cm^2. (From Béthoux, 1977).

evaporation does not seem to take place at high rates in summer, since the wind speed is rather low and the air humidity high.

The Mediterranean region being exposed to cold air masses shows rather low surface temperature during winter though higher than the atmospheric temperature from January through April or May, while from June to December the water is colder than the air.

McGill (1969) estimated the total loss of water through evaporation in 92,000 m^3/s, based on Carter's climatological data and along the lines previously established by Schink (1967). Tixeront (1970) used direct measurements of evaporation rates made in Tunisia, and arrived at a value of about 95,000 m^3/s. This is higher than that used by McGill, but almost 40% below the estimation of 122,000 m^3/s advanced by Béthoux (1977) from thermal considerations. Needless to say that these data refer to the whole Mediterranean. Assuming approximately 100,000 m^3 × 31,536,000 seconds to a year = 3,153 km^3, that spread over a surface of 2.5 millions of km^2 represent a yearly evaporation of 1240 mm, a figure somewhat higher than the estimates of the evaporation averaged over all the oceans, between 930 and 1060 mm/year. This also drives the water in the western basin.

Most of the evaporation takes place during the first half of the year and is linked to the process of deep water formation (explained in more detail in an earlier chapter). The strong continental winds, blowing within directions ranging from north-west to north-east, rapidly reduce the heat content of the surface layer and evaporate water, causing a lowering of the surface temperature to the level of the intermediate and deep waters, with the consequent increase in density. Active vertical convection and thorough mixing of the water of the surface layer, and of this with deeper water are then promoted to a varying degree, depending on the energy of the wind and hydrographic structure.

Although deep water formation is known to take place only in a few locations with peculiar meteorological and hydrographical conditions, strong evaporation takes place during winter all over the basin due to the strong and dry continental winds prevailing. These winds, blowing at velocities of 60 km per hour and even higher at sea level, are of local importance in the Mediterranean region. They are normally of short duration and are commonly restricted to small sections of the basin.

Probably the more important wind in the western part of the Mediterranean region is the Mistral or Mestral of southern France, blowing for days in the lower Rhône valley. The Bora, a north-east wind that blows in the northern Adriatic and the Vardarac in the northern Aegean Sea are also causing deep water formation.

On the other hand, the warm and dry Sirocco that originates over desert tracts and develops maximum intensities during spring, as well as the dry north wind called Meltemi in the Eastern Mediterranean, also contribute substantially to the evaporation of the surface sea-water and formation of intermediate levantine water.

5.4. THE MEDITERRANEAN, AN OLIGOTROPHIC ECOSYSTEM

The biological productivity of the Mediterranean Sea is known to be among the lowest in the world, and has been compared to the Sargasso Sea and other central oceanic gyres (Sournia, 1972). It is true that primary productivity in the central parts of the Mediterranean Sea, and in many of the coastal areas away from the influence of major rivers or urban agglomerations, is rather low and that nutrient concentrations in the deep waters of the Mediterranean, especially of the eastern basin, are also very low. However, fishing activities in the Mediterranean Sea have been going on for centuries, adapting themselves to the local conditions in such a way that a very high ratio catch/primary production exists. Several factors may contribute to this high efficiency, among them, the distribution in time and space of the fertilizing mechanisms.

The sum of biological activity taking place in the water can be represented in rather simple terms

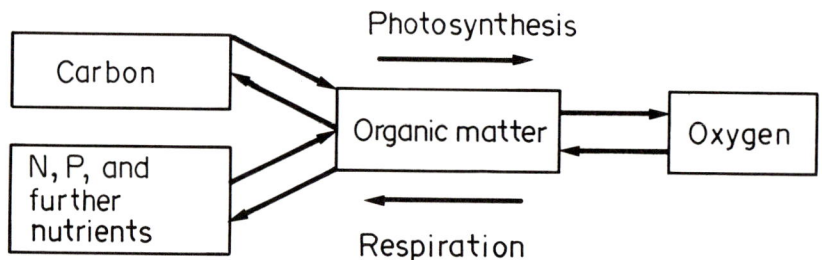

Though light intensity may limit production during winter when vertical turbulence is very high, for a given area in temperate seas the overall biomass production is always limited by nutrient availability. Light entering the marine environment (either directly from the Sun, or through scattering in the atmosphere and by re-radiation from clouds) is reducing intensity by absorption and scattering by water molecules and dissolved materials (Jerlov, 1970), including plants.

The principal nutrients, such as N and P, are often limiting and, together with inorganic C, are the basic elements from which the synthesis of organic matter in the natural environment takes place. It is commonly accepted that the photosynthesis is slowed by the scarcity of a specific nutrient and that there may be a surplus of the others. A great deal remains to be discovered about the role that the various elements play and which is the most important in controlling the rate of primary production. For a number of reasons and for many years, phosphorus has been considered to be the rate-controlling element for marine photosynthesis. Only with the improvement of the analytical techniques for nitrogen compounds, has there been a shift to the belief that they are indeed often controlling this rate. In fact, the various nutrients enter the biological system in variable proportions and there are numerous uncertainties arising from the various factors which affect photosynthesis directly or indirectly (e.g. light intensity, depth, turbulence, species composition and grazing). In view of these complexities it is not important, at least *a priori*, to discuss which element is actually controlling the rate of photosynthesis.

Taking the marine system as a whole it is found that processes which control fertilization are based on the concept of source and sink within the various layers of water masses, since for a given process, a sink in one layer may constitute a source in the adjacent layer.

The sources and sinks of *inorganic nutrients* and the processes connecting them for the three layers mentioned above (Fig. 5.1) are:

SOURCES	SINKS
Surface layer	
(Cycling mechanisms) Transport by advection and diffusion from the intermediate and deep layers. Since it depends on the sign of the gradient, it is generally directed upwards.	Uptake by photosynthetic organisms. This process is controlled by the availability of nutrients and light and it shows a high rate even when the nutrient is in short supply, due to the adaptation of many organisms to extreme conditions.
'*In situ*' regeneration of nutrients by pelagic organisms. It is associated with respiration, especially in connection with grazing and preying. It may take place everywhere.	
(Genuine sources of new nutrients) Run-off of dissolved nutrients from the continents at the sea-land-air boundary.	
Fall-out of soluble inorganic nutrients from the atmosphere at the sea-air boundary. It is especially relevant for nitrogen compounds.	
Nitrogen fixation by cyanobacteria. It takes place in connection with photosynthesis.	
Intermediate and deep layers	
Diffusion from the sediments where the organic matter deposited is degraded by bacterial action. This is the most important process for the regulation of the nutrient concentration in deep sea-water.	Denitrification. It occurs only under anaerobic conditions and mostly at the water-sediment boundary. It does not seem to be important in the Mediterranean.
Erosion of phosphate deposits. It has never been reported so far in the Mediterranean Sea.	Deposition of phosphate at the bottom in certain areas. This hypothesis is being strongly questioned, since phosphate in oceanic deposits seems now to be eroded rather than accumulated. Deposition may happen only beneath the world major upwelling areas.

The sources and sinks of *organic matter* and the processes connecting them in the Mediterranean are:

SOURCES	SINKS
Surface layer	
Photosynthetic production by phytoplanktonic organisms.	Biochemical oxidation associated with respiration.
	Vertical migration of planktonic organisms.
	Settling of particulate living and detrital material.
	Spray to the atmosphere at the sea-land-air boundary.
Intermediate and deep layers	
Chemosynthetic bacterial production.	Biochemical oxidation associated with respiration.
Vertical migration of planktonic organisms.	Progressive burial in the accumulating sediments.
Settling of particulate organic material	

5.4.1. Eutrophication

When organic matter and/or nutrients are introduced into the ecosystem, a self-purification cycle with assimilation of foreign materials, tends to re-establish ecological equilibrium. If the introduction of foreign materials were to be discontinued, then the system would return to its former state, but if discharges persist it would change into a new more eutrophic state. Such change may be advantageous for individual species or populations and may bring the ecosystem to an optimum level. However, beyond a particular threshold any increase in the concentration of the added substances merely produces decline of the most sensitive species or populations.

The self-purifying capacity of the marine waters is directly related to the processes that control the oxygen balance. This balance may be easily broken, giving rise to an oxygen-producing surface layer and an oxygen-consuming bottom layer. This is further complicated by the fact that the oxygen produced in the shallow euphotic zone is greatly reduced by turbidity and tends to escape to the atmosphere while the organic matter produced tends to settle, increasing the oxygen consumption of the deeper layers. Downward transport of oxygen may be severely limited by restricted diffusion due to generally concurrent salinity gradients.

When sewage is continuously discharged, in excess of the self-purification capacity, into a coastal zone with restricted circulation the oxygen cycle is broken. The oxidation of organic matter then proceeds through anaerobic pathways and the zone rapidly becomes a nuisance, turbid and foul-smelling and devoid of natural life. Such a nuisance can be prevented by limiting the quantities of organic matter and nutrients discharged to well below the self-purification capacity of the recipient water body.

Mediterranean waters are oligotrophic except perhaps in the neighbourhood of the large rivers, and sediments have, in general, a low organic carbon content due to the low biological production of the waters and to the presence of high oxygen concentrations in deep waters. Therefore, local oxygen deficiencies are always connected with eutrophicating sources, mostly discharges of raw or treated urban effluents. Sources of eutrophicants in the Mediterranean Sea have been identified and their amounts and effects estimated (UNEP, 1977). Their distribution around the region is uneven, with a maximum in the north-west and in the Adriatic Sea and a minimum on the southern shores. Owing to the strong stratification of the surface waters, eutrophication is more acute in summer, when ambient natural nutrient concentrations are low and the oxygen transport through the thermocline is strongly reduced. Winter mixing allows for the required vertical transport of oxygen to keep the deep waters and the sediments most oxidized all over the Mediterranean Sea.

REFERENCES

Allain, C. (1960) Topographie dynamique et courants généraux dans le bassin occidental de la Méditerranée. *Rev. Trav. Inst. Pêches marit* 24(1), 121–145.

Berkoloff, E. (1952) Evaporation d'eau dans le lac de Tunis. Archives du Bureau d'inventaire des Ressources hydrauliques. Sécrétariat d'Etat à l'Agriculture à Tunis. (Quoted in Tixeront, 1970).

Béthoux, J. P. (1977) Contribution à l'étude thermique de la mer Méditerranée. Rapport n°20, Laboratoire de Physique et Chimie Marines, Université Pierre et Marie Curie, Paris, 199 pp.

Blasco, D. (1972) Acumulación de nitritos en determinados niveles marinos per la acción del fitoplancton. Ph.D. Dissertation, University of Barcelona, 223 pp.

Carter, D. B. (1956) The water balance of the Mediterranean and Black Seas. *Publications in Climatology* 9(3), 127–174.

Cruzado, A. and Kelley, J. C. (1974) Continuous measurements of nutrient concentrations and phytoplankton density in the surface water of the western Mediterranean Sea, winter 1970. *Thalassia Jugosl* 9(1/2), 19–24.

Dugdale, R. C. and Goering, J. J. (1967) Uptake of new and regenerated forms of nitrogen in primary productivity. *Limnol. Oceanogr.* 12, 685–695.

Font, J. (1978) Courants généraux dans la Mer Catalane en automne. Proceedings Workshop on Pollution of the Mediterranean, Antalya (Turkey) 24-27 Nov., 559–562.

Goering, J. J., Dugdale, R. C. and Menzel, D. V. (1966) Estimates of *in situ* rates of nitrogen uptake by *Trichodesmium* sp. in the tropical Atlantic Ocean. *Limno. Oceanogr.* 11, 614–620.

Grill, E. V. and Richards, F. A. (1964) Nutrient regeneration from phytoplankton decomposing in sea water. *Journal of marine Research* 22(1), 51–69.

Harris, R. C. (1966) Biological buffering of oceanic silica. *Nature* 212, 275–276.

Henry, M. C. (1977) The Mediterranean sea, a threatened microcosm. *Ambio* 6, 300–307.

Hopkins, T. S. (1978) Physical processes in the Mediterranean basins. In *Estuarine Transport Processes*, Ed. B. Kjerfve, pp. 269–309, Univ. South Carolina Press, Columbia (S.Ca.).

Horne, R. A. (1969) *Marine Chemistry: the Structure of Water and the Hydrosphere*. Wiley-Interscience, New York, 568 pp.

Jerlov, N. G. (1970) Light, general introduction. In *Marine Ecology*, Ed. O. Kinne, pp. 95–102, Wiley-Interscience, London.

Lacombe, H. and Tchernia, P. (1974) Hydrography of the Mediterranean. Consultation on the protection of living resources and fisheries from pollution in the Mediterranean, FAO, Rome, 19–23 febr. 12 pp.

MacKenzie, F. T., Garrels, R. M., Bricker, O. P. and Bickley, F. (1967) Silica in sea water: control by silica minerals. *Science* 155, 1404–1406.

McGill, D. A. (1969) A budget for dissolved nutrient salts in the Mediterranean Sea. *Cahiers Océanographiques* 21(6), 543–554.

Medoc Group (1970) Observation of formation of deep water in the Mediterranean Sea, 1969. *Nature* 227, 1037–1040.

Merz, A. and Moeller, L. (1928) Hydrographische untersuchungen in Bosphorus und Dardanellen. Institut für Meereskunde am der Universität Berlin. Berlin. (Quoted in McGill, 1969).

Morel, A. (1971) Caractères hydrologiques des eaux échangées entre la bassin oriental et le bassin occidental de la Méditerranée. *Cahiers Océanographiques* 23(4), 329–342.

Ovchinnikov, I. M. (1966) Circulation in the surface and intermediate layers of the Mediterranean. *Oceanology* 6, 48–59.

Packard, T. T., Minas, H. J., Owens, T. and Devol, A. (1977) Deep-sea metabolism in the eastern tropical North Pacific Ocean. *Oceanic Sound Scattering Prediction*, pp. 101–115, Plenum Press, New York.

Perkins, E. J. (1974) *The Biology of Estuaries and Coastal Waters*. Academic Press, London, 678 pp.

Perry, A. H. and Walker, J. M. (1977) *The Ocean–Atmosphere System*. Longman, London, 160 pp.

Redfield, A. C., Ketchum, B. K. and Richards, F. A. (1963) The influence of organisms on the composition of sea water. In *The Sea*, Ed. M. N. Hill, Vol. II, pp. 26–77, Wiley, New York.

Riley, G. A. (1955) Oxygen, phosphate and nitrate in the Atlantic Ocean, *Bingham Oceanographic Coll. Bull.* 13(1).

Rumney, G. R. (1968) *Climatology and the World's Climates*. MacMillan, New York, 656 pp.

Salat, J. and Cruzado, A. (1980) Water masses in the Catalan Sea and adjacent waters. XXVIIᵉ Congrès-Assemblée Plénière CIESM, Cagliari.

Schink, D. R. (1967) Budget for dissolved silica in the Mediterranean Sea. *Geochim. et Cosmochim. Acta* 31, 987–999.

Sournia, A. (1972) Essai de mise à jour sur la production primaire planctonique en Méditerranée. *C.I.M. Operational Unit, Bulletin*, no.7.

Sverdrup, H. V., Johnson, M. W. and Flemming, R. H. (1943) *The Oceans: Their Physics, Chemistry and General Biology*, Prentice-Hall, New York.

Tixeront, J. (1970) Le bilan hydrologique de la Mer Noire et de la Mer Méditerranée. *Cahiers Océanographiques* 22(3), 227-237.

UNEP (1977) Pollutants from land-based sources in the Mediterranean Sea.

UNESCO (1981) International Oceanographic Tables, Vol. 3. Unesco technical papers in marine science 39. Unesco, Paris, 111 pp.

Voorhis, A. D. and Webb, D. C. (1970) Large vertical currents observed in a winter sinking region of the northwestern Mediterranean. *Cahiers Océanographiques* 22, 571–580.

Whitledge, T. E. (1972) The regeneration of nutrients by nekton in the Peru upwelling system. Ph.D. dissertation, University of Washington, Seattle, Wa., 114 pp.

Wyrtki, K. (1962) The oxygen minima in relation to ocean circulation. *Deep-Sea Research* 9, 11–23.

CHAPTER 6

Life and the Productivity of the Open Sea

MARTA ESTRADA, FRANCISCO VIVES and MIGUEL ALCARAZ*

Instituto de Investigaciones Pesqueras, Paseo Nacional, s/n, Barcelona, 3, Spain

CONTENTS

6.1. The large scale distribution of fertility	149
6.1.1. Consequences of the winter mixing and exchange of water with the Atlantic	149
6.1.2. Local mixing and other local sources of fertility	151
6.1.3. River discharge in coastal zones	153
6.2. Vertical organization of phytoplankton and position of the chlorophyll maxima	154
6.3. Seasonal sequences of populations and their taxonomical composition	159
6.4. Spatial heterogeneity of populations	165
6.5. Some figures about primary production	168
6.6. The zooplankton of the Western Mediterranean	170
6.7. A closer view of the pelagic populations of zooplankton	171
6.7.1. Alboran Sea	171
6.7.2. North-African coast	173
6.7.3. Tyrrhenian Sea	173
6.7.4. Ligurian Sea	174
6.7.5. Gulf of Lions	174
6.7.6. Balearic Sea	175
6.7.7. Central Zone	176
6.8. The neritic populations of zooplankton	177
6.9. The populations of interior waters (harbours and small bays)	183
6.10. Planktonic indicators	183
6.11. The trophic position of zooplankton	185
6.12. Vertical distribution and nictemeral migrations	188
6.13. Problems in estimating the secondary production	190
6.14. How much zooplankton in the Mediterranean Sea?	191

* The three authors have contributed different parts to this chapter, viz: Marta Estrada: The primary producers (6.1 to 6.5); Francisco Vives: The zooplankton (6.6 to 6.10); and Miguel Alcaraz: Secondary production (6.11 to 6.14).

6.1. THE LARGE SCALE DISTRIBUTION OF FERTILITY

The Mediterranean is not a fertile sea either for plankton or fish. The comparison of most of it with the rich upwelling areas of the Atlantic or of the Pacific is like comparing an arid plain with a field of maize. However, the Mediterranean, and especially the Western Mediterranean, is far from uniformly poor, and the study of the factors contributing to the local fertility is now an open field of research which poses many interesting questions.

There are several ways to estimate the fertility of a marine area. One of them is to measure the photosynthetic activity of the phytoplankton, which, effectively, is the only input of fresh organic matter into the marine ecosystem. The amount of organic matter fixed per unit time is called the primary production. In oceanographical work, the measurement of primary production usually involves the determination of either oxygen output or carbon uptake (using $H^{14}CO_3$) by plankton samples in natural or artificial light. The experimental bottles are maintained under standard conditions of temperature and other environmental parameters. Such methods have been subject to increasing criticism. When checked against other indicators of production (for example, the oxygen used during a certain period in the oxidation of the sedimented fraction of the produced organic material) the conventional measurements of primary production often appear to be too low. Nevertheless, local and seasonal differences are apparent in the relative values obtained by this technique. For practical purposes, the amount of phytoplankton biomass (measured as number of cells per standard volume of water or as chlorophyll concentration) is often used as an approximate measure of primary production.

As in higher plants, the production of organic material by algae is not only dependent on water, light, and CO_2 (which is plentiful in the ocean, and never limiting, for all practical purposes), but also on the availability of elements such as nitrogen and phosphorus, which in the aquatic environment appear basically as nitrates and phosphates. Although other forms of nitrogen may be important in some situations, this is probably not the case in the Mediterranean. The photosynthetic activity of the phytoplankton depletes the water of these nutrients in the superficial illuminated (euphotic) layers. The maintenance of a high primary production thus requires the return to the surface of the nutrients that slowly become stored in the deeper and dark water layers. This return may be produced by the rise to the surface of deep-water masses (upwelling) or by turbulent mixing of the water column. In both cases, the movement of water requires an expenditure of energy to do the work. In broad terms, the energy is supplied by the interaction between the atmosphere and the sea (wind, currents, etc.). The available external energy is thus the ultimate factor controlling the fertility of marine ecosystems.

6.1.1. Consequences of the winter mixing and exchange of water with the Atlantic

Several processes control the supply of nutrients to the euphotic layers in the Western Mediterranean. Some of them, such as the winter mixing of the water column or the exchange with the Atlantic, operate at a global scale; others, like coastal upwelling or the effects of river discharge, are of a much more local character.

The Mediterranean has been compared with a miniature ocean or a very large negative estuary. Both are relevant to the understanding of its large scale production patterns. As in large oceans, winter formation of deep water occurs in some parts of the basin. In the Western Mediterranean this occurs in a region of the Gulf of Lions, characterized all year round by the presence of a large cyclonic gyre. The formation of deep water includes several successive processes (*see* Chapters 3 and 4). Colder years tend to be more productive, partly because mixing may reach greater depths and incorporate more nutrients, and in part because the formation of deep water may occur over a larger area. The subsequent spreading

TABLE 6.1. Approximate water budgets of the Mediterranean, according to Schink (1967) and other authors. The volume of the Mediterranean is 4.24×10^6 km^3

	Figures in millions of m^3 per second	
	Inputs	Output
Atlantic	0.73 – 1.7	0.69 – 1.6
Black Sea	0.012	0.006
Rivers	0.014	
Rain	0.033	
Evaporation		0.092

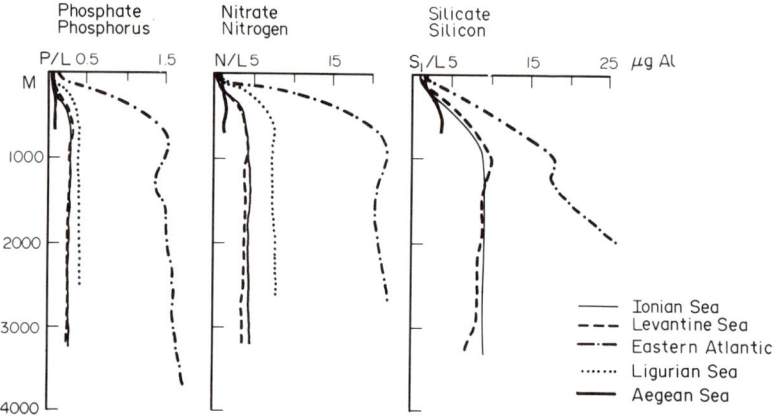

Fig. 6.1. Summary of the nutrient concentrations in several areas of the Mediterranean Sea and in the Western Atlantic. Redrawn from McGill (1965).

of the dense water favours the rise to shallower levels of layers which contain relatively high nutrient concentrations. The year-to-year fluctuations in the vigour of deep water formation are reflected in distant parts of the basin. These provide a possible explanation for the interannual changes in the salinity of subsurface water (a higher salinity indicating a higher proportion of deep water) observed in a 15-years-time series of data taken off Castelló (NE of Spain). High salinity years tend to be more productive due to the higher concentration of nutrients in the water (San Feliu and Muñoz, 1971).

The functioning of the Mediterranean as a negative estuary results from the excess of evaporation over precipitation (Table 6.1) which causes increased salinities in the Mediterranean waters as compared with those of the Atlantic. Surface Atlantic water is relatively low in nutrients. It enters through the Strait of Gibraltar and overcompensates for evaporative losses. At the same time the Mediterranean loses deep water, which is rich in nutrients, into the Atlantic. This exchange may have been one of the main causes of the low nutrient content of the Mediterranean deep water as compared with those of the major oceans (Fig. 6.1). The detailed nutrient budgets of the Mediterranean basins are far from clear. According to Béthoux (1981) the output from the deep Mediterranean water flowing out through the Strait of Gibraltar contains 443,000 tons of phosphorus per year. This would be balanced by an input with the surface Atlantic water of 86,000 tons, and a total run-off from land amounting to 357,000 tons of

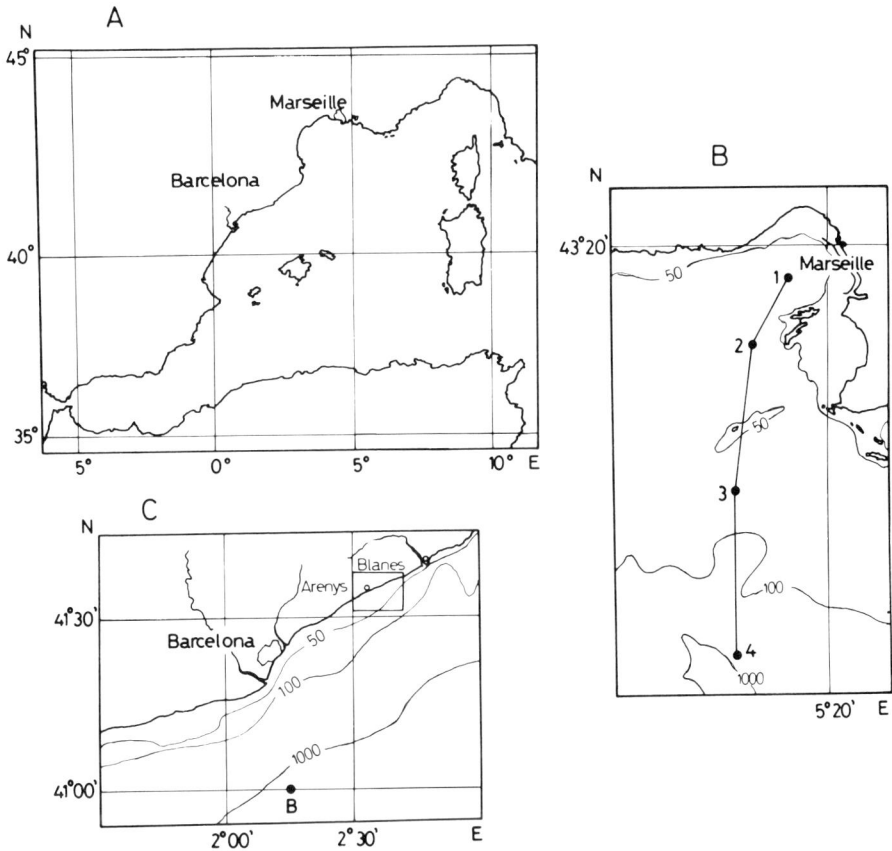

Fig. 6.2. Position of the stations appearing in Fig. 6.3 (C) and 6.4 (B). The rectangle in C indicates the situation of the zone shown in Fig. 6.8.

phosphorus per year. This amount seems large, considering the volume (600 km^3/year) and quantity of dissolved phosphate contributed by the rivers (usually below 100 mg/m^3). Significant amounts may thus also be contributed in organic and particulate form along all the coasts.

6.1.2. Local mixing and other local sources of fertility

Beyond the region of the Gulf of Lions, winter mixing is less intense, but leads also to the enrichment of the surface layers (Fig. 6.2). The alternation of stratified (spring – summer) and mixing periods (autumn – winter) imposes a strong seasonality on primary production in most of the Mediterranean (Fig. 6.3). At some regions of the coast coastal upwelling may bring up deep water which is relatively rich in nutrients. Wind-driven upwelling occurs in the littoral of the Gulf of Lions (a region with a favourable coastal profile) under strong NW (Mistral) winds (Fig. 6.4). Studies in other areas of the Western Mediterranean, such as the Catalano-Levantine coast of Spain, suggest that upwelling may have been an important contribution to the local fertility at some times. However, insufficient data is available to confirm the existence of true upwelling as well as vertical mixing. During summer, doming

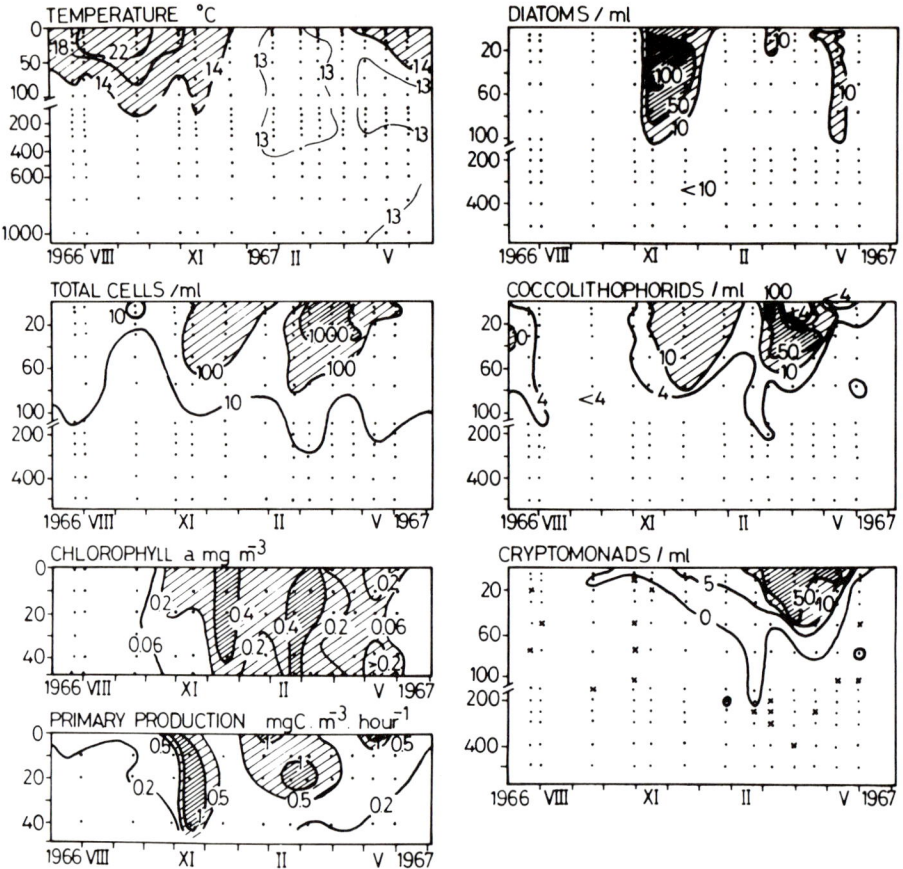

Fig. 6.3. Seasonal distribution of temperature, chlorophyll, primary production and several groups of phytoplankton in a station off the Catalan coast (station B of Fig. 6.2.c). From Ballester et al. (1967) and Margalef and Castellví (1967).

(rising without reaching the surface) of intermediate waters may increase the availability of nutrients in the euphotic layers. Domes and fronts are known in the Ligurian Sea (Jacques et al., 1976) and in other Western Mediterranean areas (Cahet et al., 1972; Margalef et al., 1966; Minas and Blanc, 1970; Estrada, 1985).

Along the coast of Valencia, the advection of water of relatively lower salinity from the south and south-west and probably including Atlantic inputs, has been associated with a change in the planktonic assemblages and with a secondary peak in primary production (Margalef and Castellvi, 1967). In the south-western Mediterranean, relatively high primary production values have been attributed to the 'Atlantic current', which brings surface Atlantic waters into the Mediterranean. The concentration of nutrients in the Atlantic inflow is low (on a world scale) but higher than that of the surface Mediterranean water and may be responsible for the relative fertility of some regions. However, other effects of Atlantic waters must be taken into account. Upwelling in the Alboran Sea off the Spanish coast (Figs. 6.5-6.6) is associated with an anticyclonic gyre, caused by the flow of Atlantic water through the Strait of Gibraltar. In addition, recent studies of the French survey MEDIPROD IV (Minas et al., 1983) have shown that intense turbulent mixing occurring in the Strait drags nutrients from deep layers of the Mediterranean water into the euphotic zone. During the cruise Mediterraneo I (Fig. 6.7),

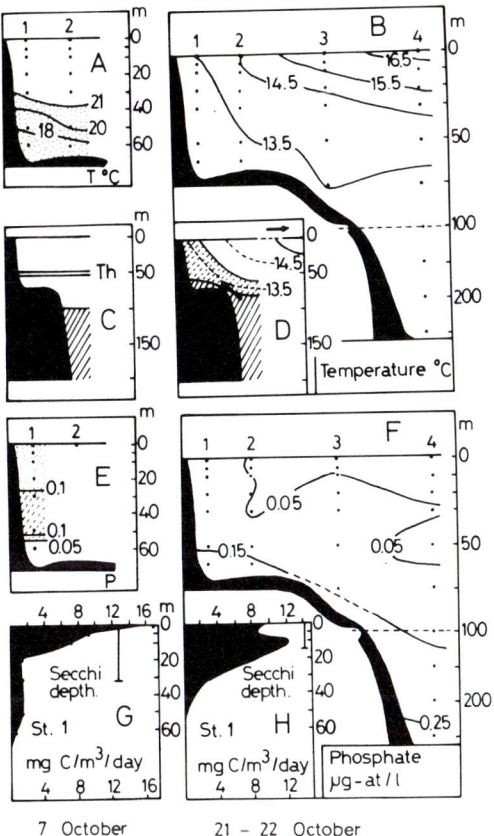

Fig. 6.4. Distribution of temperature (A,B) and phosphate (E,F) and vertical profiles of primary production (G,H) before and during an upwelling event in the Gulf of Marseille (October, 1964). The graphs C and D are schemas of the hydrological structure. Th = Thermocline. See Fig. 6.2, B for the situation of the stations. Redrawn from Minas (1968).

the relatively high fertility of the south-western area (where Atlantic water dominates in the upper levels) was associated with an elevation of the level of the thermocline. In this particular case, the distributions of nutrients and primary production, could be related to the general pattern of circulation of the water masses, rather than to the direct effects of putative Atlantic waters.

6.1.3. River discharge in coastal zones

A significant factor in coastal fertilization along the northern shores of the Western Mediterranean basin is the discharge of rivers. Besides being the main source of contaminants (Chapter 11), they contribute large amounts of phosphates, nitrates and other nutrients. The influence of the Rhône on the hydrographic and biological regimes of neighbouring and distant areas (such as Banyuls, near the French–Spanish border) has been noted by several authors (Coste and Minas, 1967; Jacques et al., 1967). In Spain, the delta of the Ebro is one of the most productive zones of the littoral. Freshwater outflows play an important role in the fertility of enclosed harbours or bays (Palermo, Gulf of Naples,

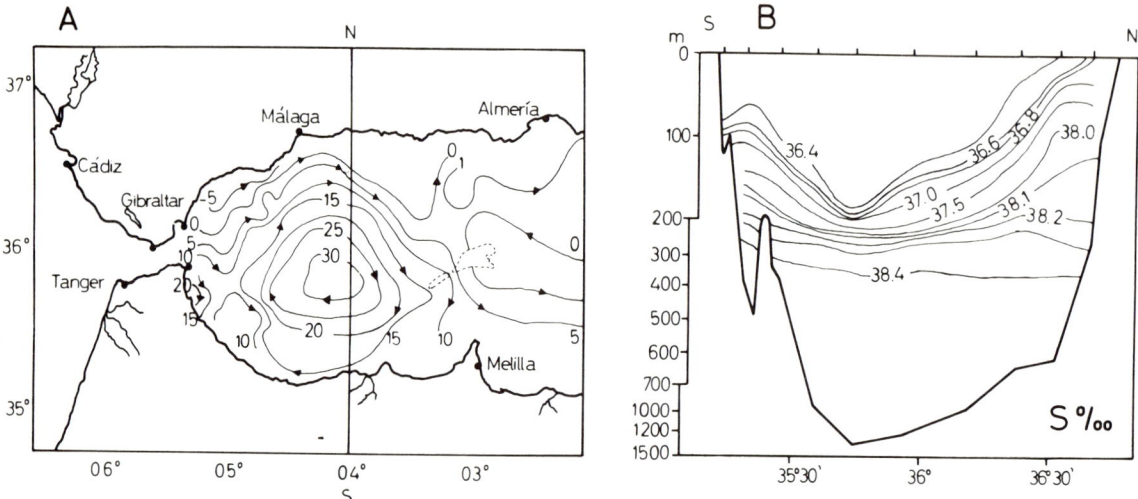

Fig. 6.5. A: Dynamic topography of the Alboran Sea (summer 1962) referred to the 200 dbar surface. Redrawn from Lanoix (1974). B: Salinity profile along the S-N line of the left graph. Observe the presence of upwelling in the Spanish coast of the Alboran Sea. The marks on the upper line indicate the position of the stations. Redrawn from Lanoix (1974).

Tunis, etc.). At some times of the year, temporary run-off due to storms may have powerful effects on the primary production of limited coastal areas (Fig. 6.8).

According to recent estimates (Béthoux, 1981), terrestrial inputs of phosphorus account for one third of the potential new production (i.e. that which is not linked to nutrients released by biological recycling processes) in the Tyrrhenian Sea and in the northern part of the Ligurian basin. The remaining two thirds are contributed mainly by vertical nutrient transfer (through mixing or upwelling) from deep to shallow water layers. With the possible exception of the run-off of industrial phosphates, the terrestrial-nutrient input is much smaller in the southern part of the Western Mediterranean basin, due to the much lower rainfall over the coastal countries.

Freshwater inflows also have fertilizing effects which enhance primary production. This results not only from the nutrient content of the water, but also from the entrainment of relatively rich underlying marine waters, caused by the outflow of freshwater. This is seen from the distribution of freshwater and quantity of nutrients around a delta.

6.2. VERTICAL ORGANIZATION OF PHYTOPLANKTON AND POSITION OF THE CHLOROPHYLL MAXIMA

Turbulent mixing and upwelling promote and modulate phytoplankton growth. If turbulence or advection velocity become too high, then more cells are lost than grown and the populations do not increase even in favourable zones. The relationship between the depth of the mixed layer and that of the euphotic zone is relevant. In winter, light intensity is at a seasonal minimum, the water is cold and mixing of the water column may reach from the surface to far below the euphotic zone. The phytoplankton cells spend on average so much time in inadequately illuminated water that no net growth occurs. Appropriate conditions occur in late winter, or in spring, when stratification of water starts, or in autumn, at the onset of mixing.

The gradients at a zone of mixing or upwelling produce distributions in space comparable to the seasonal sequences in time. The maximum biomass is not located at the centres of mixing, or upwelling, but at the periphery, where increasing stability limits dispersal and favours population accumulation and growth. A typical example of such spatial patterns, after vertical mixing in the Gulf of Lions, is shown in Figs. 6.9 and 6.10. The highest production occurred in the intermediate zone between the stable coastal water and the region of intense mixing.

Summer is the least productive season in the Mediterranean. The stratification of the water reduces the transfer of nutrients from deeper layers. In the upper illuminated waters, a low primary production is maintained mainly by nutrient regeneration associated with bacterial activity and zooplankton excretion. Under these conditions, appreciable rates of photosynthetic carbon fixation may be recorded while the nutrients are undetectable. In the lower layers of the euphotic zone, sporadic proliferations of phytoplankton may follow local enrichments during the period of stratification, due to hydrographic phenomena such as the rise of subsuperficial water layers or the appearance of internal waves which increase turbulence (Castelló, end July).

The vertical distribution of primary producers is determined by the opposing gradients of light and nutrients and by the tendency of the phytoplankton cells to sink. When nutrients are available in the upper parts of the water column, the phytoplankton populations develop in the surface layers and the

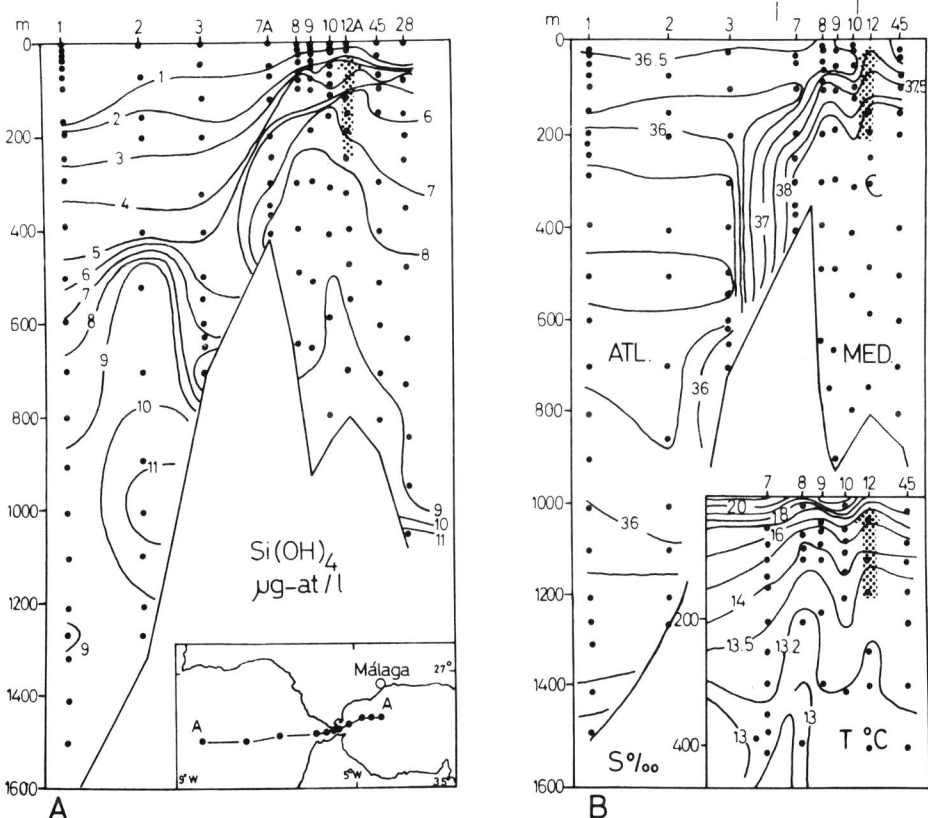

Fig. 6.6. Alboran Sea, cruise MEDIPROD IV (October-November 1981). A: Silicate section through the Strait of Gibraltar. The arrow indicates the zone of upwelling. The graph in the lower right part shows the position of the stations. B: Salinity and temperature sections through the Strait of Gibraltar. The arrow indicates the zone of upwelling. Redrawn from Minas *et al.* (1982).

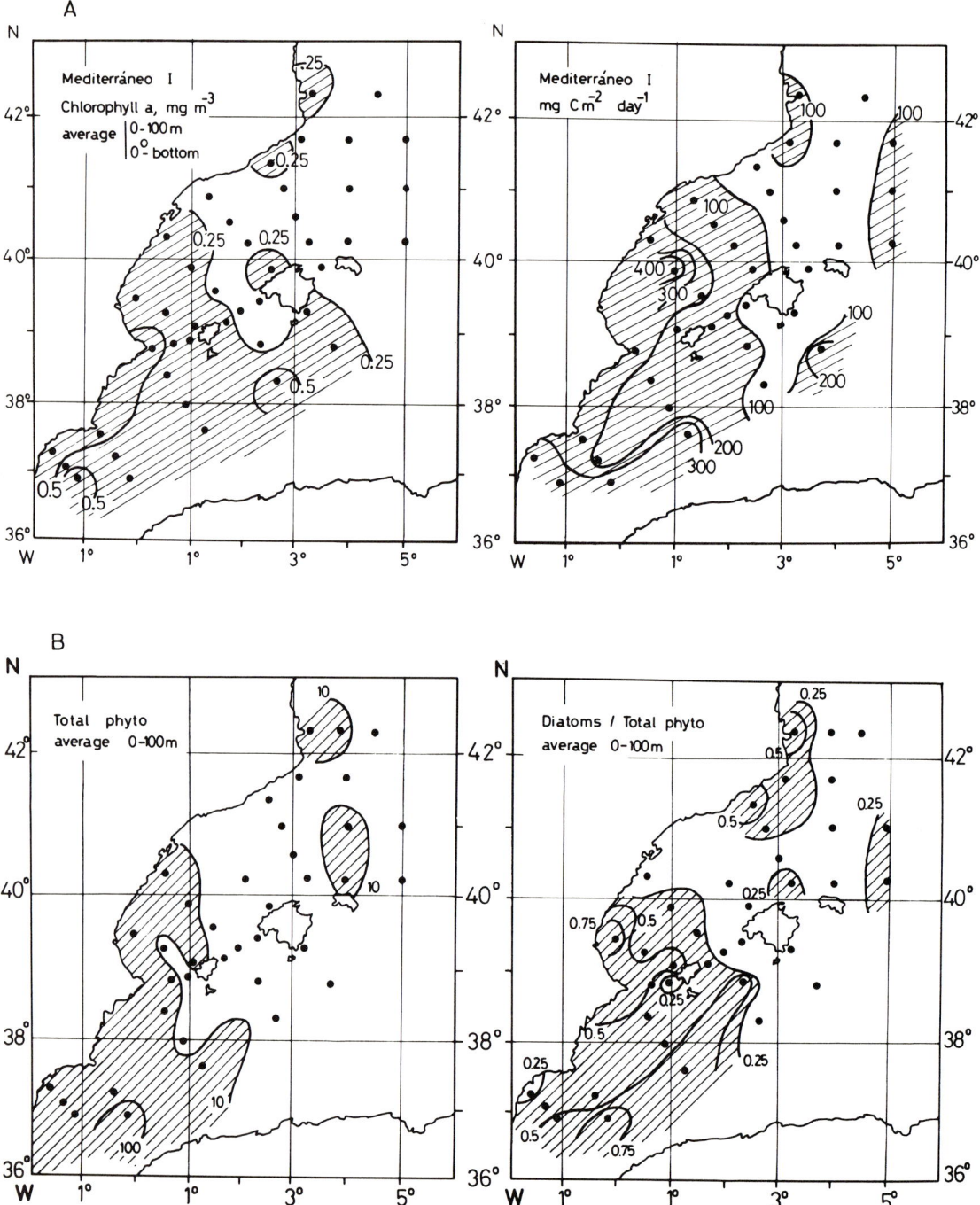

Fig. 6.7. Cruise Mediterráneo I (October, 1976). A: Weighted averages of chlorophyll concentration (left) and primary production (right) for a column of water between 0 and 100 m depth (or between 0 m and the bottom if shallower than 100 m). From Estrada (1981). B: Weighted averages of the phytoplankton concentration (left) and of the relation between the concentration of diatoms and the total phytoplankton concentration (right) for a column of water between 0 and 100 m depth (or between 0 m and the bottom if shallower than 100 m). Data from Estrada.

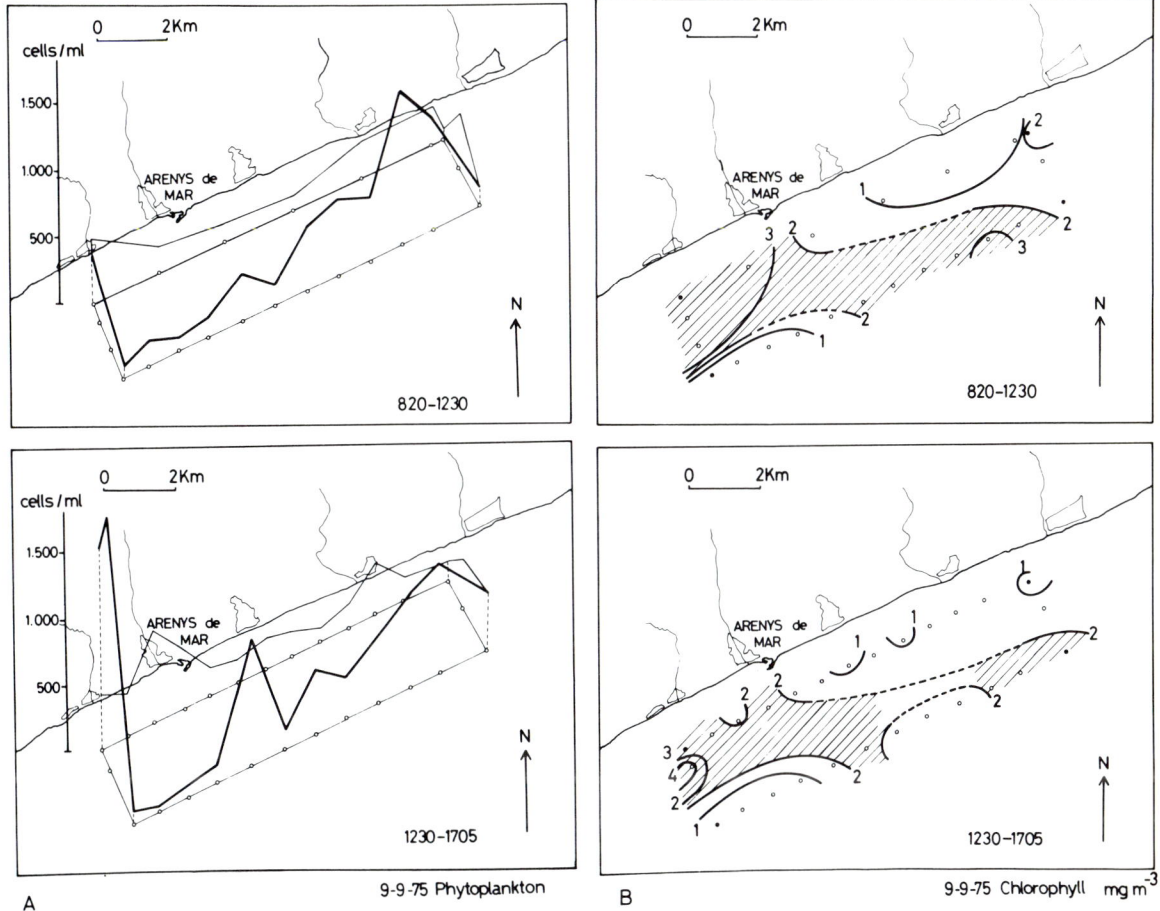

Fig. 6.8. Development of a dense phytoplankton population in the catalan coast, north of Barcelona (see Fig. 6.2) after a heavy rainstorm. The circles indicate the sampling points; the samples were taken between 8:20 and 12:30 h (upper graphs) and between 12:30 and 17:05 h (lower graphs) in September 9, 1975. A: Graphs of the total number of phytoplankton cells per ml. B: Distribution of chlorophyll a (mg/m³). Modified from Estrada (1979).

total phytoplankton biomass declines with increasing depth. The usual decrease towards the surface is a boundary effect; all layers get phytoplankton from overlaying ones, except at the top. During the period of stratification, when nutrients disappear from the surface layers, the vertical distribution of phytoplankton typically shows distinct chlorophyll and cell maxima at the thermocline (at depths close to 30 – 40 m) or below it (down to 100 and even more metres). Such distinctive features in the vertical distribution are characteristic of oligotrophic marine areas and are associated with maxima in the distributions of nitrite and oxygen (Figs. 6.12, 6.13, 6.14). The chlorophyll maximum does not always coincide with the cell maximum, which may be located above the chlorophyll peak. A plausible explanation for this shift is the higher chlorophyll content of the cells lying in the lower layers, where they are exposed to dim light and higher concentration of nutrients. Other possible reasons include the presence of chlorophyll in detritus or the failure to give due account to the smaller chlorophyll-containing organisms (cyanobacteria).

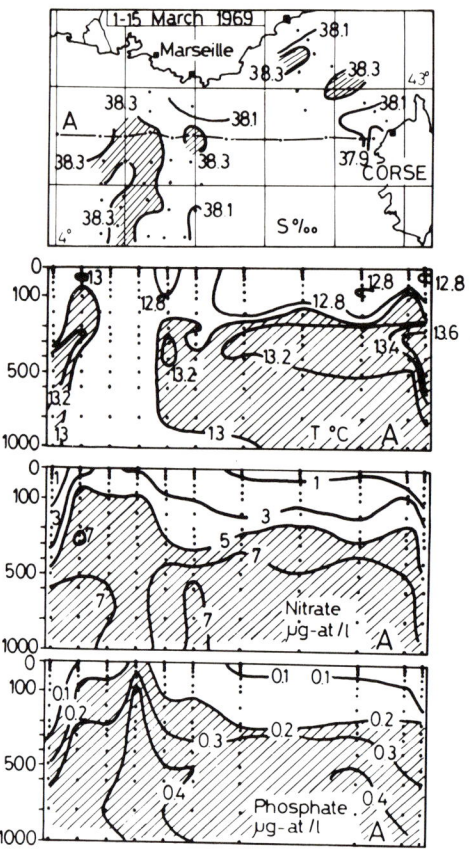

Fig. 6.9. Above: Surface distribution of salinity. The zones of intense vertical mixing are characterized by high salinities (38.30°/∞). Below: East—West section along the line A of the upper graph) of temperature, nitrate and phosphate, during a period of winter mixing in the Gulf of Lions. Depth in m. Redrawn from Coste et al. (1972).

Several mechanisms have been advocated to explain the deep phytoplankton maxima. Their position in the Mediterranean can be related to the availability, in the subsurface layers, of nutrients which diffuse slowly towards and across the thermocline, in low light intensities, barely sufficient for photosynthesis. The scarcity of phytoplankton in the surface layers contributes to a high transparency of the water. In some cases, the deep phytoplankton maxima have been attributed to accumulation of sinking cells, but in others (Estrada, 1985) *in situ* growth combined with low diffusion losses appeared to play a significant role. Another possibility, at least in some cases, is that the phytoplankton populations of the deep maxima originate in well-illuminated layers after periods of mixing, or in relatively fertile littoral waters, from where they would be transported or confined in subsurface levels with appropriate water densities. Photosynthetic rates (per unit of phytoplankton biomass) are low at the level of the chlorophyll maxima because of light limitation. However, the primary production in these maxima may represent an appreciable contribution to the fertility of the Mediterranean. This is particularly important during the summer because it is a new production based largely on inputs from outside the euphotic zone.

The peak in the concentration of oxygen found at the level of the maximum of phytoplankton biomass can be related to both physical and biological processes. Photosynthetic activity increases

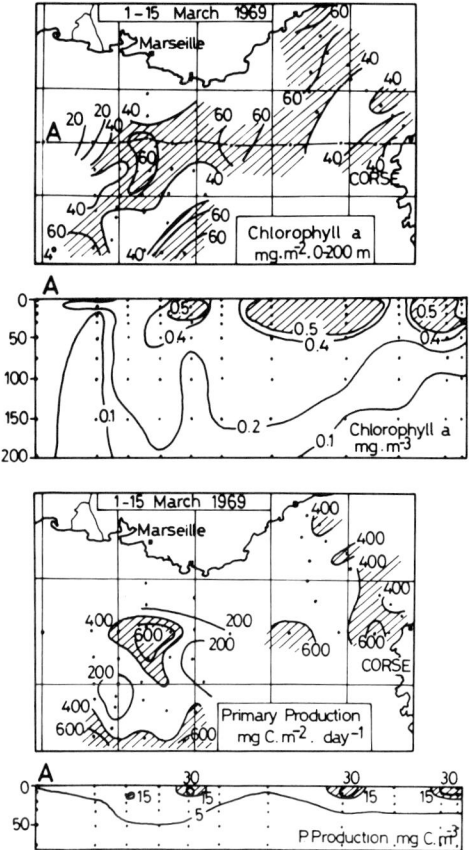

Fig. 6.10. Distribution of chlorophyll *a* and primary production during a period of winter mixing in the Gulf of Lions (*see* Fig. 6.9). The profiles correspond to section A in the upper graph; depth is given in m. Redrawn from Jacques *et al.* (1973).

oxygen concentration in the thermocline, while, at the same time, the consumption of oxygen by regeneration processes and the diffusion of oxygen to the atmosphere are more intense in the warmer upper layers (Minas, 1970).

A nitrite maximum that is commonly observed close to the deep chlorophyll maximum is generally referred to as the first nitrite maximum (to distinguish it from other deeper maxima which appear in certain marine areas). Its formation poses many interesting questions. Nitrite may be produced by bacteria (from the oxidation of ammonia or the reduction of nitrate) and by phytoplankton (from nitrate uptake and reduction to nitrite) when conditions such as light limitation slow down further reduction of nitrite to ammonia inside the cells and the excess nitrite is excreted. Field data and laboratory experiments (Blasco, 1971; Herbland and Voituriez, 1979) suggest the dependence on phytoplankton activity of the first nitrite maximum.

6.3. SEASONAL SEQUENCES OF POPULATIONS AND THEIR TAXONOMICAL COMPOSITION

Studies on the annual phytoplankton and primary production cycles are available for several coastal zones of the Western Mediterranean basin, but information on offshore areas is scarce and based on sporadic surveys (*see* Sournia, 1973, for a revision of primary production data up to 1973). In the

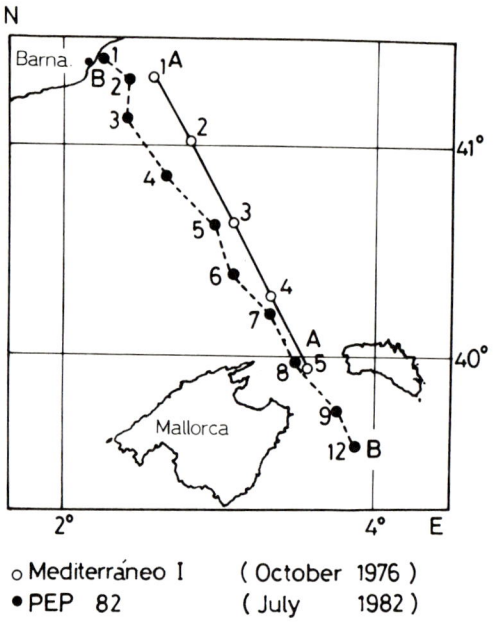

Fig. 6.11. Position of the stations shown in Figs. 6.12, 6.13 and 6.14.

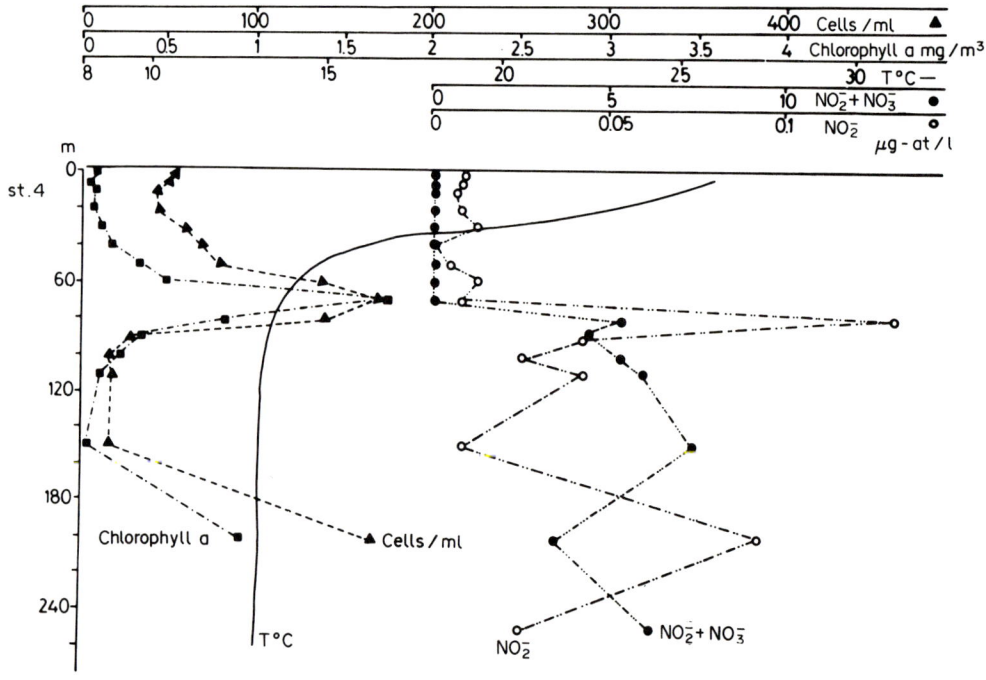

Fig. 6.12. Cruise PEP 82 (Instituto de Investigaciones Pesqueras of Barcelona, July, 1982). Vertical profiles of temperature (T°C), nitrite (NO_2^-, μg-at N/l), nitrate + nitrite ($NO_2^- + NO_3^-$, μg-at N/l), chlorophyll a (mg/m^3) and number of cells per ml (cells/ml) at station 4 of section B in Fig. 6.11.

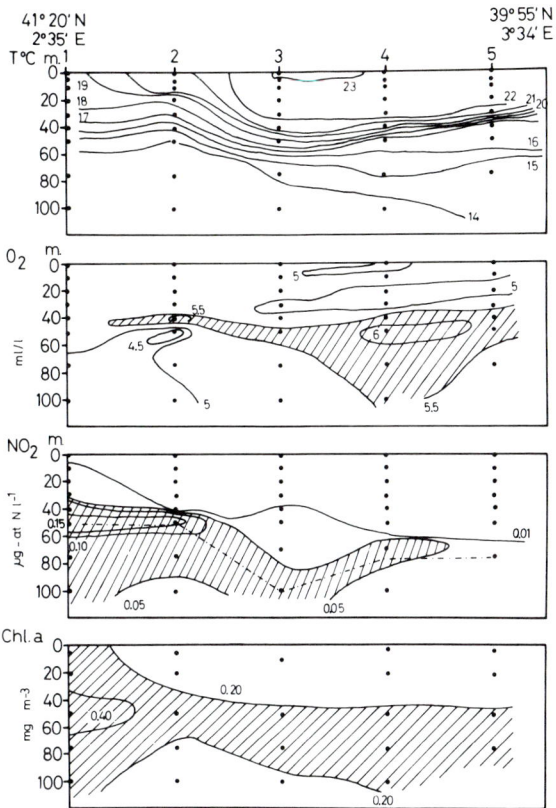

Fig. 6.13. Cruise Mediterráneo I (October, 1976). Vertical distributions of temperature (T°C), oxygen concentration (O$_2$, ml/l), nitrite (NO$_2$, μg-at n/l) and chlorophyll *a* (mg/m^3) for section A – A of Fig. 6.11. Depth in m. The broken line in the nitrite profile shows the position of the maximum. From Estrada (1981).

Western Mediterranean, as in many temperate zones, there are generally two maxima of phytoplankton biomass and production during the year: one in autumn, when the thermocline disappears and another in late winter and spring, after nutrients have been supplied by upwelling or by mixing and stability of the water column is sufficient to allow the development of phytoplankton populations (*see* page 163). The effects of the seasonal variations in irradiance are difficult to separate from those depending on nutrient availability and turbulence. In the Gulf of Marseille (at depths shallower than 75 m) light did not appear to be a limiting factor, even in winter (Travers and Travers, 1973), although, the situation could be different in other areas. Sometimes, secondary population peaks appear (due to essentially local factors) before or after the winter-spring maximum.

The seasonal variation of the phytoplankton populations in the Mediterranean is tightly linked to the annual alternation of periods of stratification and of mixing of the water column. The change in the composition of the phytoplankton community, following enrichment, can be broadly considered as an ecological succession, although the fluctuating characteristics of the aquatic environment do not often allow the high diversity and the low production to biomass ratio which are typical of advanced stages in many successions. Within a succession, different phytoplankton life-forms (groups of organisms adapted to particular environmental conditions) play, sequentially, a dominant role. This process of substitution is especially obvious after the winter-spring bloom. Diatoms (Fig. 6.15) grow rapidly when provided

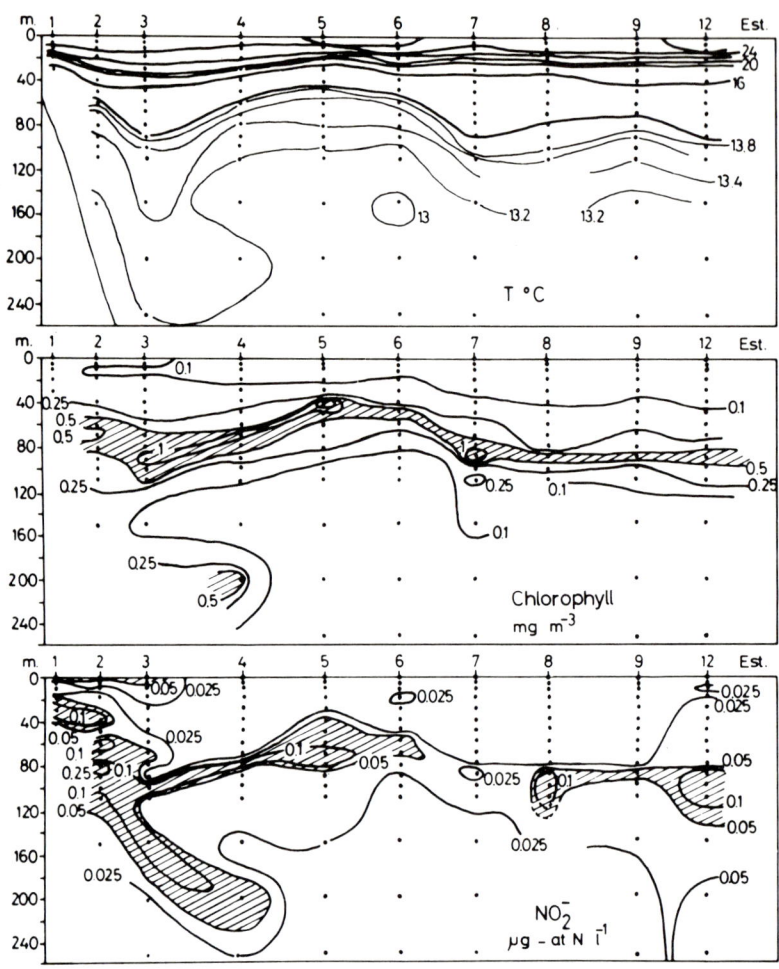

Fig. 6.14. Cruise PEP 82 (Instituto de Investigaciones Pesqueras of Barcelona, July, 1982). Vertical distributions of temperature (T°C), nitrite (NO$_2^-$, µg-at N/l) and chlorophyll *a* (chlorophyll, mg/m^3) along section B – B of Fig. 6.11.

with nutrients in turbulent conditions; they dominate the phytoplankton community during the bloom but become increasingly scarce when the water column stratifies and nutrient concentration decreases. In these new environmental conditions, the dinoflagellates (Fig. 6.15) (which are motile and better adapted to survive under low nutrient concentration) become the dominant organisms. The coccolithophorids (Fig. 6.16), a phytoplankton group characterized morphologically by a cover of calcareous plates, tend to occupy an intermediate position between diatoms and dinoflagellates. The silicoflagellates, a small group of Chrysophyceae possessing a tubular skeleton of silica, reach their maximum concentrations during winter and spring, and remain in subsurface levels during the summer. Small flagellate forms, such as certain haptophytes and cryptomonads, may also play an important role in particular areas, but their study is difficult with conventional methods and there are few accurate identifications of the species involved. Still more uncertain is the distribution of a series of very small photosynthetic microorganisms (i.e. those which pass through 2 µm screens) which belong to different

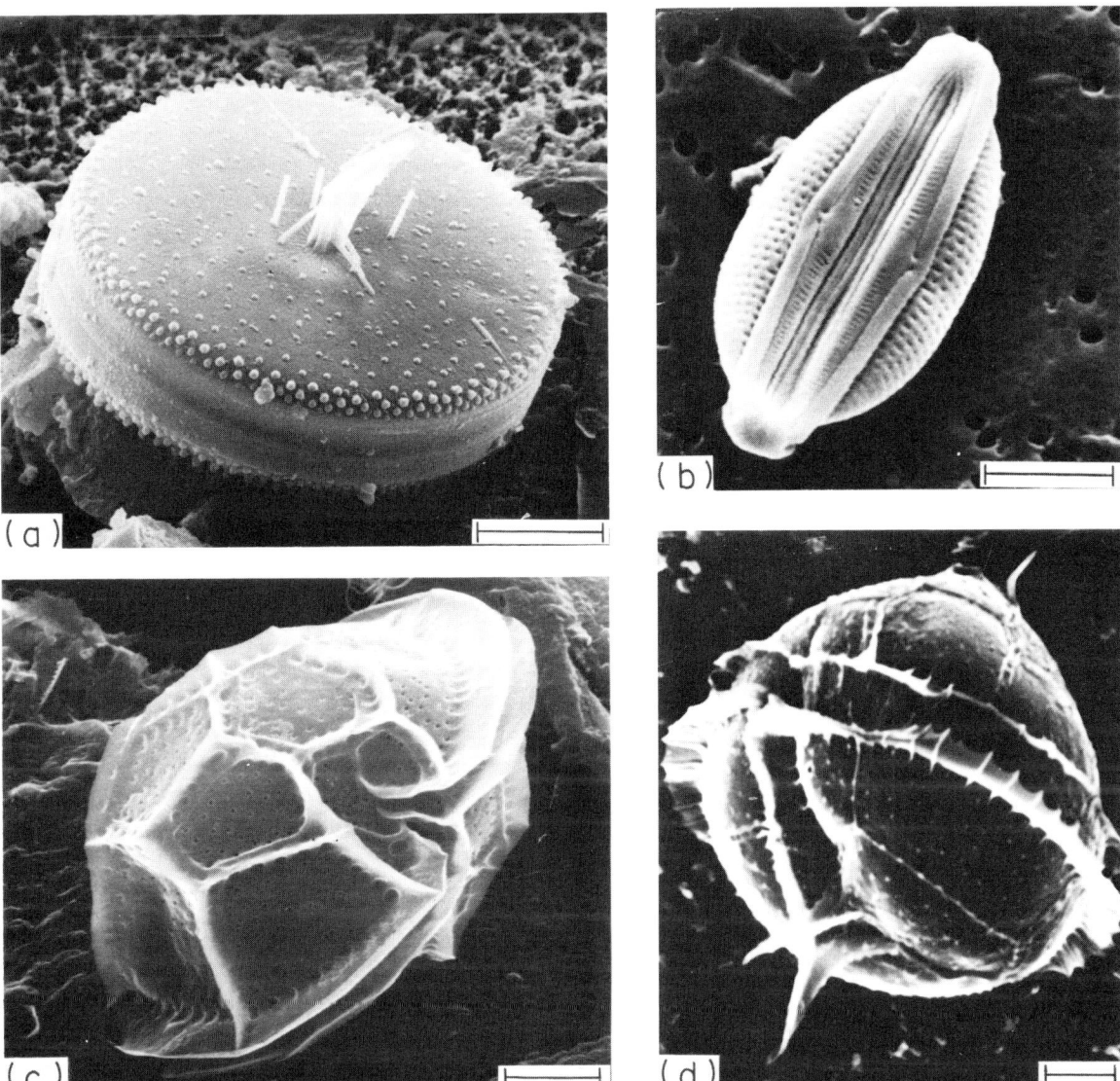

Fig. 6.15. Scanning electron micrographs of diatoms and dinoflagellates. A: *Thalassiosira rotula* (centric diatom). B: Pennate diatom. C: *Goniodoma polyedricum* (dinoflagellate). D: *Peridinium* (dinoflagellate). Scale bars: A, 10 μm; B, 3 μm; C, 10 μm; D. 3 μm.

bacterial (cyanobacteria) and algal groups, and appear to be ubiquitous in marine environments. Recent studies in several marine areas have shown that they may form important fractions of the photosynthetic plankton biomass.

A description of a typical coastal phytoplankton cycle in the Western Mediterranean (based on a series of data extending over several years, from stations off Barcelona and Castellón) is given by Margalef (1969). There was a consistent appearance, in both areas, of three phytoplankton peaks every year: the first due to the breakdown of the thermocline in autumn, the second to mixing and/or moderate upwelling between mid-February and mid-March followed by increased stability and the third to inflow

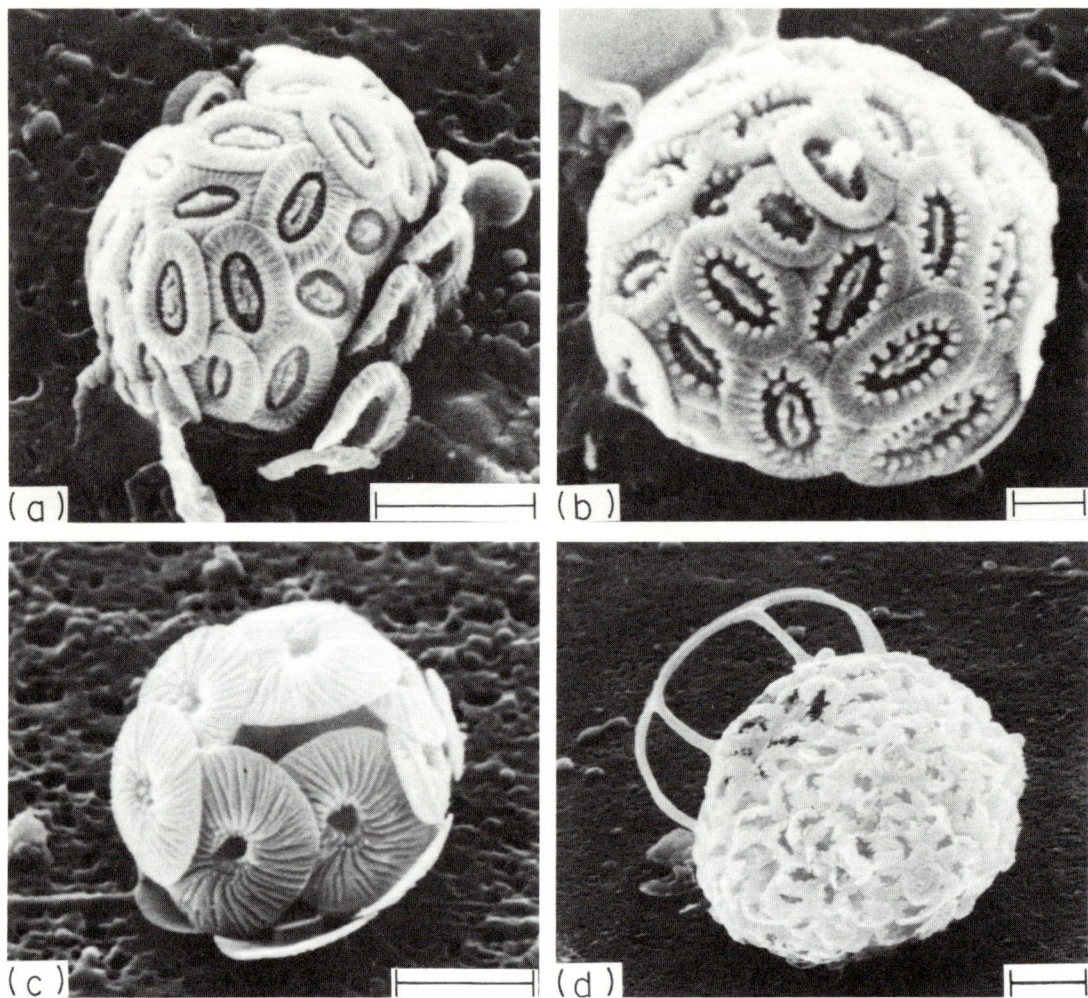

Fig. 6.16. Scanning electron micrographs of coccolithophorids and one tintinnid (loricate ciliate). A: *Caneosphaera molischii* (coccolithophorid), with several external coccoliths. B: *Caneosphaera molischii*; coccoliths with heavily ribbed distal rims. C: *Umbellosphaera tenuis* (coccolithophorid). D: *Dictyocysta* (tintinnid). Scale bars: A, 3 μm; B, 1 μm; C, 3 μm; D, 10 μm.

of water from the south and south-west containing moderate amounts of phosphate and other nutrients in April–May. The principal organisms in the winter-spring maximum were diatoms of the genus *Chaetoceros*, cryptomonads and some green flagellates. The late spring peak was due mainly to diatoms of the genera *Rhizosolenia* and *Nitzschia*. It is remarkable that the same genera are well represented in the deep maximum of chlorophyll in summer. During the summer, dinoflagellates were the predominant group. The most prominent species of the population in the first autumn peak were pennate diatoms such as *Asterionella japonica*, *Thalassionema nitzschioides* and *Thalassiothrix mediterranea*.

The comparison of data gathered by different Western Mediterranean laboratories (Table 6.2) reveals some regularities. Small flagellates and/or a few species of diatoms form the bulk of the phytoplankton population during the main winter-spring, or spring, peak, while the other maxima tend to exhibit a wider variety in species, eventually with an important participation of coccolithophorids. Among these, *Emiliania (Coccolithus) huxleyi* is the most abundant. Dinoflagellates are everywhere the most important group during the summer. Sometimes, the presence of particular phytoplankton species appears to be associated with certain factors, such as the influence of freshwater discharges. This could be the case for the blooms of *Skeletonema costatum* in Banyuls, near the mouth of the Ombrone (Jacques, 1969; Grillini and Lazzara, 1980) and at places along the Spanish coast.

Table 6.2 also includes some figures that indicate the range of the Mediterranean phytoplankton populations expressed in number of cells or in chlorophyll concentration.

Overall, the lists of species recorded from different Western Mediterranean laboratories are similar. However, different areas, or the same area in different years, may present variations both in the timing and intensity of the phytoplankton pulses and in the composition of the phytoplankton populations during the succession. The typical seasonal pattern may be altered in areas which are heavily affected by contamination sources (waste water discharge, river run-off, etc.) but become recognizable at short distances from the pollution sources.

When the supply of nutrients is adequate and the stability of the water high (conditions not commonly found in the Mediterranean, except in enclosed bays and polluted harbours) dense populations of phytoplankton may accumulate, with the result that the water appears reddish or otherwise coloured. Such events, named in a broad sense 'red tides', are generally caused by dinoflagellates whose growth and secondary concentration contribute to this effect. Jacques and Sournia (1978 – 1979) have reviewed the information on the occurrence of red tides in the Mediterranean. The most prominent organisms are dinoflagellates (*Scrippsiella*, *Cochlodinium* and the non-photosynthetic *Noctiluca scintillans*), but *Chattonella subsalsa* and other flagellates have also been recorded. No significant toxic effects have been reported with Mediterranean red tides.

6.4. SPATIAL HETEROGENEITY OF POPULATIONS

The phytoplankton populations exhibit considerable spatial heterogeneity. Large scale variability of the phytoplankton distributions may reflect the presence of different water masses or the effects of large scale enrichment processes (Fig. 6.7). Diatoms tend to dominate in the richer and more turbulent zones, and dinoflagellates in the poorer and more stratified. Coccolithophorids may be important in some regions, although they do not appear to be as abundant in the Mediterranean as was previously accepted.

Smaller scale variations are superimposed on the main trends of the phytoplankton distributions, like peaks and secondary peaks on a mountain range. Some large-scale properties may influence local distributions. For example, the persistent stratification below the summer thermocline, permits the segregation of deep chlorophyll maximum into thin layers with rather distinct phytoplankton populations. Phenomena such as eddies of different sizes, or the breaking of internal waves, can lead to nutrient patchiness, with a pattern that is amplified by phytoplankton growth, to produce characteristic profiles of high-abundance peaks on a low density background (Fig. 6.8).

At the lower end of the variability spectrum, turbulence plays a leading role in generating hydrographic and biological heterogeneity. Fig. 6.8 illustrates examples of small scale variability of phytoplankton populations in the Western Mediterranean. The study of such distributions is important for, among other reasons, it may provide a link between the results of experimental work in the laboratory and the observations made in large field surveys. Experimental research is usually done with small volumes of water in bottles, but in the sea, the water moves up and down and diffuses in all directions. When referring to the heterogeneity of populations, biologists used to speak of 'patches' of plankton. But in reality there are patches of patches of patches in an infinite regression, and only a spectral representation can do justice to the fractal nature of these distributions. Unsophisticated statistical methods cannot cope with the complexity of the situation.

TABLE 6.2. Some examples of seasonal variability of the phytoplankton in the Western Mediterranean. Cell counts based on sedimentation techniques

Source	Zone	Years	Population maxima	Dominant species or groups	Phytoplankton density cells/ml	Chlorophyll concentration mg m^{-3}
Margalef (1963) Herrera and Margalef (1963) Margalef and Herrera, 1964	Catalan coast, from the Ebro river to Castelló	1961-1962	autumn	Pennate diatoms (*Asterionella*, *Thalassionema*, *Thalassiothrix*)	1-600 cells/ml	<0.02-0.40 mg m^{-3}
			February	*Chaetoceros*		
			spring	*Nitzchi delicatissima*, *Ch. compressus*		
Bernhard and Rampi (1967) Bernhard, Rampi and Zattera (1967)	Ligurian Sea	1959	June	*Nitzschia* 'seriata like', *Leptocylindrus danicus*, *Bacteriastrum delicatulum*	1-160 cells/ml (including diatoms, dinoflagellates, coccolithophorids and silicoflagellates)	
			late summer (Sep.)	all groups		
			winter (Nov.-Dec.)	*Emiliania huxleyi*		
		1962	No blooms	More diatoms and coccolithophorids in winter, dinoflagellates all the year	<1-23 cells/ml (including diatoms, dinoflagellates, coccolithophorids and silicoflagellates)	
Margalef and Ballester (1967)	Catalan Coast	1965-1966	Feb.-March	Coccolithophorids Cryptomonads	3-443 cells/ml	<0.05-2.5 mg m^{-3}
			April-May	*Nitzschia* 'seriata like', *Dactyliosolen mediterraneus*, *Rhizosolenia imbricata*, coccolithophorids		
			autumn	*Chaetoceros compressus*, *Chaetoceros* spp., *Asterionella japonica*, *Rhizosolenia stolterfothii*, *Thalassionema nitzschioides*		
Margalef and Castellví (1967)	Catalan Coast	1966-1967	autumn	*Asterionella japonica*	2-400 cells/ml	<0.05-1.2 mg m^{-3}
			Feb.-March	*Skeletonema costatum*, Cryptomonads, coccolithophorids		
			May	*Nitzschia* 'seriata like', *Lauderia annulata*		

Reference	Years	Month	Species	Cells	Biomass
Jacques (1968, 1969)	Banyuls 1965-1968 (diatoms and dino.)	February	Skeletonema costatum, Cryptomonads	<1-5000 cells/ml (diatoms and dinoflagellates)	<0.1-2.5 mg m^{-3}
	1965-1967 (flagellates)	April-May	Chaetoceros spp., Nitzschia spp., Rhizosolenia delicatula	<100-2900 cells/ml (flagellates)	
		autumn	Diatoms		
Travers (1973, 1975)	Gulf of Marseille 1960-1965	February	Skeletonema costatum, Chaetoceros spp., Rhizosolenia stolterfothii	<1-1000 cells/ml (mean values for the water column)	
		March-May	S. costatum, Ch. curvisetus, Lauderia annulata, etc.		
		autumn	S. costatum, Leptocylindrus danicus, Thalassionema nitzschioides, Thalassiothrix frauenfeldii, etc.		
Grillini and Lazzara (1978, 1980)	Parco della Maremma (NW Italy) 1975-1976	February	Skeletonema costatum, Thalassiosira decipiens, Chaetoceros, Lauderia	<1-831 cells/ml	
		November	Thalassiothrix frauenfeldii, Chaetoceros		
Carrada et al. (1980)	Gulf of Naples 1976-1977	May-June	Cilindrotheca closterium, Chaetoceros compressus, Nitzschia longissima	40-241 cells/ml	<0.2-0.7 mg m^{-3}
		October	Dinoflagellates, Emiliania huxleyi		
		winter	E. huxleyi		
Gaumer (1981) Samson-Kechacha (1981)	Alger Bay 1979-1980	Feb.-March	Rhizosolenia delicatula, Chaetoceros	Diatoms: 224-280 × 10^3 cells/ml	<0.1-12 mg m^{-3}
		April-May	R.h. delicatula, Skeletonema costatum, Chaetoceros		
		Jun.-Sept.	S. costatum, Leptocylindrus danicus		

Water moves and with it the phytoplankton populations. Not only turbulence, but at the other end of the scale, advection or transport also have to be taken into account when dealing with plankton distributions. Periodic sampling at a geographically-fixed point reveals populations that belong to separate volumes of water, but to mark a volume of water with a drogue and follow its phytoplankton population also poses difficulties that are generally unsurmountable. Jörgensen, in his study of the phytoplankton collected during the Danish expeditions at the beginning of the century, considered specifically a possible evolution of the populations of *Ceratium* in the surface Atlantic water, from Gibraltar towards the East. Almost nothing positive resulted from the attempt, but it stressed the need to consider simultaneously, the dynamics of populations, the movements of water and the processes of selection.

Another aspect of interest, concerns the temporal and spatial distribution of phytoplankton species, and the need to classify the phytoplankton associations (a term which will be loosely used here to designate groups of species whose abundances are positively correlated). Several studies have demonstrated phytoplankton assemblages which show a coherent distribution, in time or space, that can be related to environmental conditions. Some of these groups contained exclusively diatoms, or dinoflagellates, or coccolithophorids: a finding which reflects the ecological similarities which exist between species of the same major taxonomical units. The resulting segregation of the groups supports the idea that nutrient supply and water turbulence are key factors in the selection of the dominant species in each particular marine environment. An example of this kind of work in Western Mediterranean regions is summarized in Fig. 6.17. So far, the available information is rather limited and no clear opinion can be formed on the temporal persistence and geographical distribution of the described associations.

6.5. SOME FIGURES ABOUT PRIMARY PRODUCTION

Net primary production by plants is the input that supports the rest of life, in the case of phytoplankton: zooplankton, fishes and bacteria. Energy cascades through the different trophic levels and primary production is the first throttle.

The highest production values reported in unpolluted areas (1.2 g C per m^2 per day) have been measured in the area of divergence of the Gulf of Lions (Jacques et al., 1973) in April. In the remaining open sea, primary production values are generally lower than 0.1 $C/m^2/day$, but the presence of enrichment mechanisms, such as raising the level of the thermocline, may result in higher figures for particular areas (Fig. 6.7). Coastal regions tend to be richer than offshore areas for several reasons, notably: the supply of nutrients by run-off, coastal upwelling, and the shallow depth, which limits losses through turbulence during the periods of mixing. The comparison of data from different places is difficult, because of interannual fluctuations and small-scale variability related to local topographic and hydrographic factors, or to the influence of discharges of freshwater. For instance, at a station near Barcelona, the annual averaged primary production was estimated to be 76 g $C/(m^2/year)$ in 1965−66 (Margalef and Ballester, 1967), and 85 g $C/(m^2/year)$ in 1966−67 (Margalef and Castellví, 1967); approximatively 200 km to the north, near Banyuls (France), the evaluated production was 86 g/(m^2/year), in 1967, and 142 g $C/(m^2/year)$ in 1968 (Jacques, 1970). Carbon fixation rates can be higher in polluted areas, such as the harbour of Palermo, where a production of 0.47 g $C/(m^2/year)$ has been recorded (Genovese et al., 1972).

In general, the total annual production of a particular area depends on several fertilization events generated by different mechanisms (upwelling, mixing freshwater run-off, advection of water masses) which fluctuate independently, with the result that total production (integrated over the whole year) is

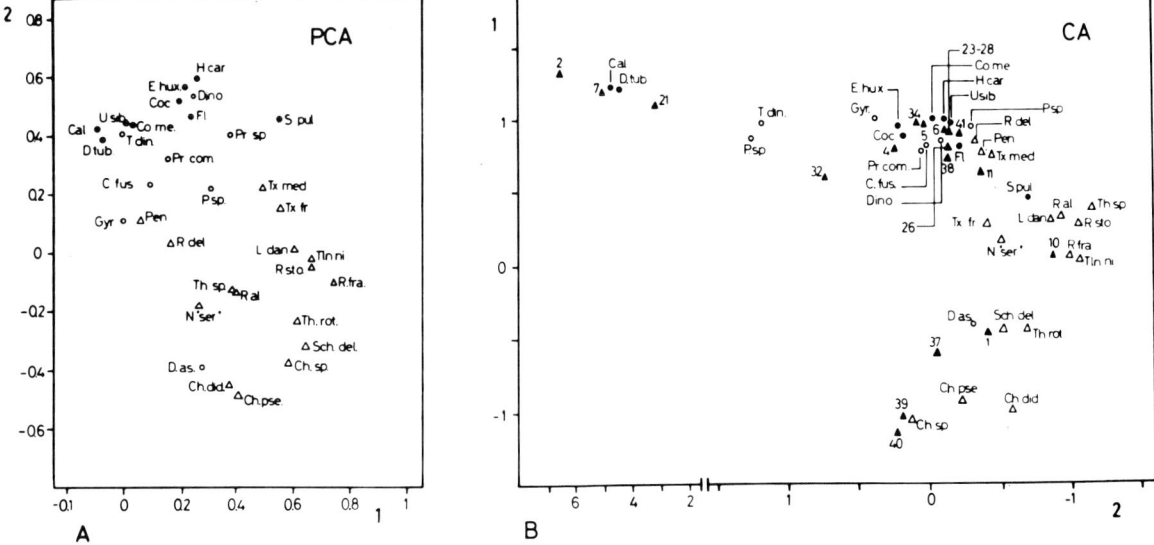

Fig. 6.17. Results of the application of two multivariate statistical methods, principal component analysis (PCA) and correspondence analysis (CA) to the abundance data of 33 species in a set of 187 phytoplankton samples from the cruise Mediterráneo I (see Fig. 6.7 for the layout of the stations; the samples were taken from several depths between 0 and 100 m). From Estrada (1982). A: Position of the 33 species and the surface samples in the space dimensioned by the first and second principal components of the PCA. B: Position of the 33 species and the surface samples in the space of the first two axes of the CA. Note the different position of the axes with respect to A and the change of scale in axis 2.
▲ Surface samples; △ Diatoms; ○ Dinoflagellates; ⊙ Coccolithophorids; ● Flagellates.
Abbreviations:

C. fus. = *Ceratium fusus*
D. as. = *Diplopsalis asymmetrica*
Gyr. = *Gyrodinium* spp.
P. sp. = *Peridinium* spp.
Pr. com. = *Prorocentrum compressum*
Pr. sp. = *Prorocentrum* spp.
T. din. = Unidentified tecate dinoflagellate
Dino. = Small dinoflagellates
Cal. = *Calyptrosphaera* sp.
Co. me. = *Coronosphaera mediterranea*
E. hux. = *Emiliania huxleyi*
D. tub. = *Discosphaera tubifer*
H. car. = *Helicosphaera carteri*
S. pul. = *Syracosphaera* cf. *pulchra*
U. sib. = *Umbilicosphaera sibogae*
Coc. = Unidentified coccolithophorids
Ch. did. = *Chaetoceros didymus*

Ch. pse. = *Chaetoceros pseudocurvisetus*
Ch. sp. = *Chaetoceros* spp.
L. dan. = *Leptocylindrus danicus*
N. 'ser' = *Nitzschia* 'seriata-like'
R. al. = *Rhizosolenia alata*
R. del. = *Rhizosolenia delicatula*
R. fra. = *Rhizosolenia fragilissima*
R. sto. = *Rhizosolenia stoltherfothii*
Sch. del. = *Schröderella delicatula*
Tin. ni. = *Thalassionema nitzschioides*
Th. rot. = *Thalassiosira rotula*
Th. sp. = *Thalassiosira* spp.
Tx. fr. = *Thalossiothrix frauenfeldii*
Tx. med. = *Thalossiothrix mediterranea*
Pen. = Unidentified small pennate diatoms
Fl. = Flagellates

probably maintained within closer ranges than would be expected if total production were dependent on a single, key, fertilizing event (Margalef and Herrera, 1963).

Estimates of the nutrient inputs to the euphotic zone, make it possible to evaluate the potential new production in a particular marine area. Table 6.3 shows an example of calculations applied to the Western Mediterranean basin. The measures of primary production, as is usually done, include also the contribution of recycling by biological processes, that may vary widely. Assuming, as Béthoux does, that 60% of the measured production is based on recycled nutrients, the average total primary production of the Western Mediterranean would be 90 g C/m²/year, which is very close to the estimates deduced from reported measurements (Sournia, 1973).

TABLE 6.3. Potential fertility based on phosphorus budgets. After Bethoux (1981)

Inputs to the surface layers	Nord occidental basin $(3.0 \times 10^{11} m^2)$	Sud occidental basin $(3.0 \times 10^{11} m^2)$	Tyrrhenian $(2.4 \times 10^{11} m^2)$	Occidental basin $(8.6 \times 10^{11} m^2)$
Land run-off	210×10^6 kg y^{-1}	7.5×10^6 kg y^{-1}	17.8×10^6 kg y^{-1}	
Vertical transfers from deep layers	470×10^6 kg y^{-1}	92×10^6 kg y^{-1}	40.9×10^6 kg y^{-1}	
Surface fluxes	12×10^6 kg y^{-1}	28.6×10^6 kg y^{-1}	1.7×10^6 kg y^{-1}	
Potential fertility	86 g C m^{-2} y^{-1}	9 g C m^{-2} y^{-1}	10 g C m^{-2} y^{-1}	36 g C m^{-2} y^{-1}

6.6. THE ZOOPLANKTON OF THE WESTERN MEDITERRANEAN

Despite the numerous studies which have been carried out in the Western Mediterranean during the present century, the current information of the planktonic population of these waters is far from complete and insufficient to recognize a general pattern in the distribution and behaviour of the species throughout the year. The data have generally been obtained during rather short cruises in limited areas. Thus although we know the composition of the populations (in which generally only a few zoological groups, usually the most common, were studied) the time successions and space distribution are not known in detail. Furthermore, the nets employed to collect plankton, and the processing of the samples have not been standardized, and it is obvious that it would be very difficult, if not impossible, to establish comparisons.

Being aware of, but not utterly discouraged by these difficulties, and using the results of the efforts of scores of planktologists (especially those of the second half of this century) we will try to summarize some general characteristics and properties of the animals that form the planktonic community of our waters.

First, we will distinguish those animals that spend their entire life in the plankton (holoplankton) and those which are part of this community only during a certain period of their development (meroplankton). We will also classify them according their habitat: pelagic communities are those of the open sea and neritic communities those which occur above the shelf. A special and extreme case of neritic communities are the populations of confined waters (e.g. harbours or small bays).

From a practical point of view it is important to know the amount of available zooplankton to answer questions about the fertility of the Mediterranean. The next question concerns its diversity. Is the Mediterranean zooplankton very diversified or does it have relatively few species?

Recent geological events and extreme climatic variations (glaciations separated by interglacial periods) has produced a mixture of faunas of very diverse origin which have experienced severe selective pressures. With time the original conditions have changed so much that the primitive composition of the original fauna, or of any of its components, is very difficult to discover. Furthermore, present day Mediterranean plankton (which is as notably diverse) contains a large number of subtropical forms and some northern ones (a minority that would qualify as boreal, together with a large number of species from temperate-cold areas). The prevailing forms are, however, subtropical species, together with some of clearly tropical origin.

The extent of the diversity, can be seen from the recorded holoplanktonic species and the list must be expected to grow longer (Table 6.4).

Among the Protozoa, the Radiolaria are the largest group and it seems likely that many more will be found, since only those captured in the first 500 m of water are included. Below this depth no appropriate data exist. In the second place are the Tintinnida, which, very possibly, could yield a longer list in future reviews of the group, for very few studies of these organisms has been made in the pelagic

TABLE 6.4. Number of holoplanktonic species of different zoological groups

Foraminifera	10	Ostracoda	28
Heliozoa	2	Cladocera	6
Acantharia	83	Copepoda	284
Radiolaria	350	Mysidacea	17
Tintinnida	175	Amphipoda	110
Medusae	116	Euphausiacea	16
Siphonophora	44	Decapoda	14
Ctenophora	17	Pteropoda	30
Chaetognatha	18	Appendicularia	36
Annelida	39	Thaliacea	17

zone. The expected increase in the number of species will be complicated by past taxonomic errors (in which growth stages and other intraspecific differences have been described as separate species). Thus a number of tintinnid and radiolarian species will undoubtedly fall into synonymy. A similar situation exists in the Acantharia, and in planktonic Foraminifera in which there are surprisingly few species, in contrast to the numerous benthic species. The Heliozoa consists of a few parasitic species which have yet to be studied.

Among the Metazoa, the Copepoda, as expected, are by far the most numerous group, followed by Medusae. In third place we find the Amphipoda and next the Siphonophora. The remaining groups are represented by a small number of species, but this does not reflect their actual and considerable importance in the food chain.

We must consider both the size of individuals and their number. Interestingly, the group with the lowest number of species during the summer, the Cladocera, are represented by more individuals than the Copepoda in several coastal areas.

The present Western Mediterranean fauna, except for a few relict endemic species, and those of oriental origin, can thus be considered as a true Atlantic fauna.

6.7. A CLOSER VIEW OF THE PELAGIC POPULATIONS OF ZOOPLANKTON

The biological characteristics of each one of the sectors into which we can divide the Western Mediterranean depend principally on the magnitude of the Atlantic influence and the local hydrography. These factors and the basic production determine the abundance and variety of the animal plankton.

The areas of our sea which can be recognized and separated by their zooplankton assemblages are: Alboran Sea, North African coast, Tyrrhenian Sea, Ligurian Sea, Gulf of Lions, Balearic Sea and Central sector, according to the boundaries proposed by Furnestin (1966) (Fig. 6.18).

6.7.1. Alboran Sea

The Alboran Sea is the entrance to the Mediterranean, and as such represents a transition zone between it and the Atlantic.

Below a depth of 80 – 100 m (off the African coast of this sector and in shallower strata in the Spanish coast) the typical Mediterranean water flows westward towards Gibraltar. Above this level nearly a million cubic metres per second of Atlantic water moves eastward. The saline gradient is maximal at the interface between these waters masses.

Predictably, the zooplankton population in this zone shows a large specific diversity due to the coexistence of Atlantic and Mediterranean species, from both the pelagic and neritic communities.

The high primary production in the area allows the development of large numbers of individuals. In consequence, the biomass of these waters is, on average, equal or superior to that of the Gulf of Cadiz, on the Atlantic side of the Strait.

The populations of Atlantic origin contain some species which are capable of crossing the Strait. These can be divided into (1) Those species found above the 150 – 200 m depth level in the Spanish Sea (or Gulf of Cadiz), and which are able to enter the Alboran Sea. (2) Those living at greater depths that can reach the surface or subsuperficial zone during the night by vertical migration. In the African zone near the Strait, during certain periods of the year, the local upwelling will also contribute to an increase in the Atlantic fauna. This change refers to the adults, since the juvenile forms of many species spend the first stages of their life in superficial levels and can therefore be easily moved horizontally during the early stages of their development (Fig. 6.19).

Surface samples from each side of the Strait reveal populations of different composition. Many Alboran species are not present in the upper levels of the Spanish Sea. (For example, among the Copepoda, *Eucalanus monachus*, *Rhincalanus nasutus*, *Aetideus armatus*, *Pleuromamma abdominalis*, *Oncaea dentipes* and *Vettoria granulosa* are frequent in Alboran, but are not found on the surface in the Atlantic zone). Differences are also seen in the diversity indexes estimated from the numbers of Atlantic and Mediterranean species taken at the surface (at 26 and 45, respectively).

How can these differences be explained? Perhaps the hydrographic characteristics of the Alboran Sea predominate. We know, for example, that the upwelling water in the NW zone advects deep water forms to the superficial zone. The whirlpools that appear in the central area, and the effect of the interface or boundary between Atlantic and Mediterranean water, also acts as a zoological barrier. Thus all the species taken in vertical tows (between 0 – 500 m) in the Spanish Sea (except for those which do not cross the Strait) become concentrated in the top 200 m in the Alboran Sea.

Paracalanus parvus, *Clausocalanus* ssp., *Centropages chierchiae* and especially *Acartia clausi* are the main contributors to the biomass. Altogether, there are 20 batipelagic or mesopelagic, 50 epiplanktonic and fewer than 10 neritic species.

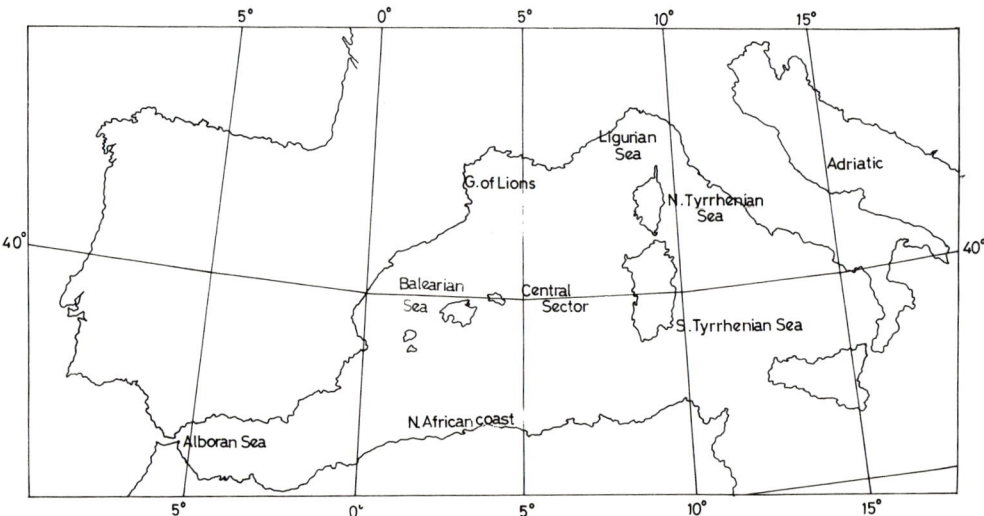

Fig. 6.18. Chart showing the different sectors from the Western Mediterranean with self-defining characteristics from hydrographical and biological point of view.

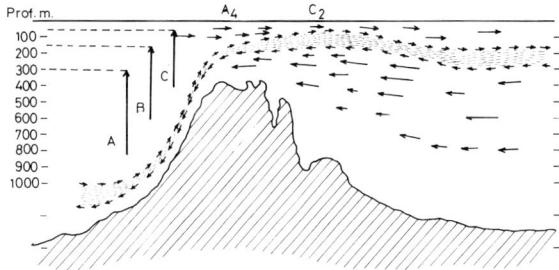

Fig. 6.19. The hydrographical particularities that show the Gibraltar Strait has a high repercussion on the distribution of the pelagic populations (*see* text).

Besides the Copepoda, 13 of the Euphausiacea are found in the Western Mediterranean. *Thysanoessa gregaria* is typical of Alboran. Twelve species of Chaetognatha have been listed; among them, the most abundant are *Sagitta friderici*, *S. tasmanica*, *Pterosagitta draco* and *S. planctonis*. All of these are typical of Atlantic waters. The number of recorded mollusca is twenty-one, principally Pteropoda tecosomata, among which, *Limacina trochiformis*, *Cuvierina columnella* and *Diacria quadridentata* are present and testify to the warm character of these waters.

6.7.2. North-African coast

Hydrographically speaking, the African coast is an extension of the Alboran Sea. The continental platform is narrow, with the exception of the Tunisian coast, and the large proportion of Atlantic water explains the presence of a fauna which resembles closely that of the Alboran Sea. This is seen in the continued presence of Atlantic forms, such as *Sagitta friderici*, *S. planctonis*, *Pterosagitta draco*, etc.

Among the Copepoda we find *Temora longicornis*, *Acartia danae* and *Centropages chierchiae*, which are very abundant. These species, and many others, are also typical of the Alboran Sea.

6.7.3. Tyrrhenian Sea

The Atlantic influence continues in the superficial waters, although the presence of warm and salty oriental waters is important for all the pelagic areas in this sector.

Zooplankton is more abundant in the southern Tyrrhenian, than in the north due to the wider continental shelf and to the influence of Atlantic waters.

Even though typical Mediterranean species of Copepoda are found here, there are major quantitative differences relative to other areas. A number of important species occur here: *Pleuromamma gracilis*, *Lucicutia flavicornis*, *Haloptilus longicornis* and especially *Oithona plumifera*.

Among the Chaetognatha, *Sagitta enflata* and *S. bipunctata* prevail. Among the Euphausiacea, *Euphausia krohni* is typical, as is *Thysanopoda aequalis*, which is more common in western waters. We also find *Stylocheiron suhmii*, an indicator of eastern waters.

On the other hand, species of Atlantic origin are not totally absent and include *Pterosagitta draco*, *Acartia danae*, *Temora longicornis* and *Centropages violaceus*.

The influence of Atlantic water is also manifest in the northern Tyrrhenian, especially to the east of Corsica. In the central region of this sector, the waters of the east are very important.

On the eastern coast of Corsica *Sagitta setosa* and *Penilia avirrostris* are very typical, in waters of relatively low salinity. Certain Atlantic indicator species can also be found: *Acartia danae*, *Centropages violaceus* and others. Species of *Clausocalanus* and *Temora stylifera* are permanent and constant, while other species vary in importance from year to year.

6.7.4. Ligurian Sea

In this zone there is a large whirlpool of Mediterranean water surrounded by a narrow band of Atlantic water. This Atlantic water has undergone many changes throughout its long journey and constitutes what is known as the coastal current of the Ligurian Sea.

In the centre of the whirlpool, a divergence leads deep water to the surface. The eastern water flows at a great depth and in the same cyclonic direction (counterclockwise).

The Copepod fauna is not abundant. Sixty species have been listed, including *Clausocalanus* ssp. and common members of the families *Oithonidae*, *Paracalanidae*, *Centropagidae*, *Temoridae* and *Acartiidae*. The presence of Atlantic water indicators is irregular.

The Chaetognatha do not show a large variety: only 8 – 9 species, among which there are deep water forms such as *Sagitta lyra* and *S. hexaptera*. The presence of high salinity water favours the increase in the surface of typically pelagic species such as *Sagitta bipunctata*, and determines the relative scarcity of semineritic ones like *S. enflata*.

Thirteen species of Euphausiacea known in the Mediterranean have been observed. Eight species of Pteropoda thecosomata have been reported, of which some are considered rare (*Gleba cordata*, *Peraclis apicifulva*, etc.), even though in the surface the abundance of species are *Creseis*.

Many species have been recorded from Villefranche-sur-Mer, some of which are not consistently reported for the Mediterranean. This is a result of careful study of zooplankton of the bay since the end of the last century, rather than a reflection of local enrichment of the fauna.

6.7.5. Gulf of Lions

The hydrography has been well-studied. The surface water (salinity: $35.5 - 37.00\ ^\circ/_{oo}$), mixes with the Rhône river, turns south-west, and spreads through the entire gulf. At Cape of Creus, some of the water turns back to create 'Languedoc's current'. On the eastern side of the Gulf, high salinity water flows in from the Ligurian divergence ($38.2^\circ/_{oo}$), from the deep eastern current and from the Atlantic-provençal branch ($37.8^\circ/_{oo}$). The proportions of the different waters change according to the seasons.

To the south-east of the Gulf and north-east of the Balearic Islands the cold winter winds, blowing from the north, lower the temperature of the surface water, which then sinks to its density level. As explained elsewhere (Chapter 4) two main types of water are formed: (a) an intermediate winter water formed near the coast and with the following characteristics: $12.8 - 13^\circ C$ and $38.2 - 38.5^\circ/_{oo}$ salinity; and (b) deep water, formed at open sea and characterized by a temperature of $12.7 - 12.9^\circ C$ and higher salinity ($38.4 - 38.5^\circ/_{oo}$).

The neritic facies consists of species typical of diluted waters. The diversity of crustaceans is notable. Besides abundant coastal Copepoda there is a diversified population of Mysidacea, and this is rather exceptional since this group is usually scarce in plankton. There are many members of the Gammaridea and Caprellida. Among the Chaetognatha *Sagitta setosa*, typical of diluted waters, is abundant as is *Evadne nordmannii* among the Cladocera. There are also species of medium salinity waters, such as

Doliolum nationalis. Species characteristic of Atlantic waters occasionally show up, for example *Acartia danae, Temora longicornis* and *Centropages chierchiae.*

The pelagic group comprises fewer species and practically all of them are considered as characteristic of high salinity water. The Coelenterata (*Eudoxoides, Sulculeolaria, Abylopsis, Aglaura,* etc.) are abundant. Among the Chaetognatha we find *Sagitta bipunctata* (typically pelagic), sometimes together with *Sagitta setosa*. The Thaliacea *Salpa democratica* and *S. fusiformis* are frequent. The Copepoda fauna is diverse and includes more than a hundred species.

6.7.6. Balearic Sea

The coastal surface waters of the Balearic Sea are of relatively low salinity and come from the Rhône, in the Catalonian zone, and from the Ebro, in the Valencia coasts. At the periphery, and even its centre, the Balearic Sea is influenced by Atlantic waters from the south-west, which flow in at the northern part of the Balearic archipelago. At intermediate depths (300 – 400 m) there is a branch of eastern water current, beneath which is the typical northern water of the Western Mediterranean. This mixture of water masses produces a diversified population, especially in the NE areas, although in the central areas and further south, the populations tend to become increasingly uniform.

The Copepoda has been the most extensively studied group. At the surface there are many neritic species showing similar zonal distribution with reference to the coast, as, for example, *Temora stylifera* (Fig. 6.20), *Paracalanus parvus* and *Oncaea media*. However, below 50 m, typically pelagic species are present. The families which show the larger abundance of species are the *Aetideidae, Euchaetidae, Scolecithridae, Metridiidae,* and *Heterorhabdidae.*

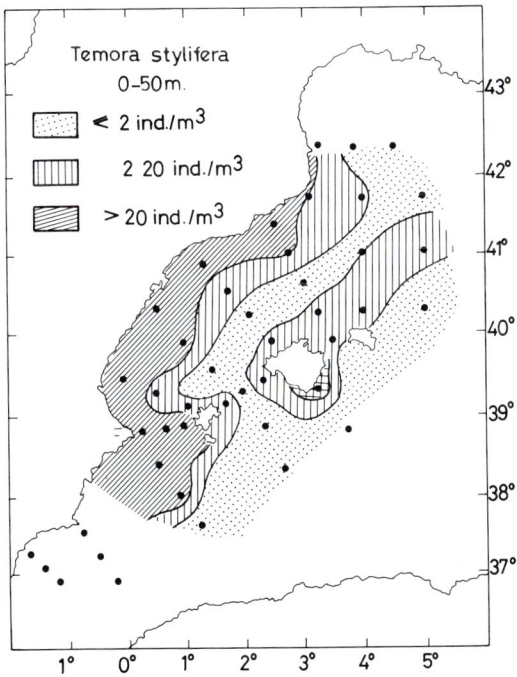

Fig. 6.20. Zonal distribution of the Copepod *Temora stylifera* in the Spanish coasts.

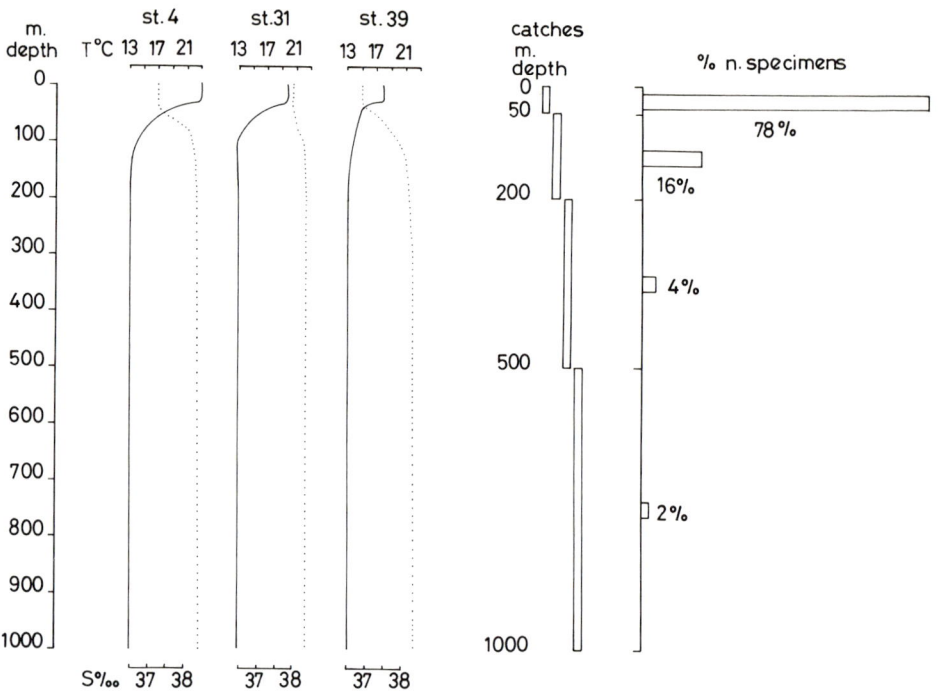

Fig. 6.21. Vertical distribution of temperature, salinity and proportional quantities (%) of the zooplankton specimens in the Western Mediterranean.

The Chaetognatha are relatively scarce. Of the 18 Mediterranean species, only 9 have been found here, almost all above 200 m, except for S. lyra and S. decipiens which abound between 200 and 500 m.

Few studies on the Coelenterata are available. Recent unpublished work reveals 19 species of Medusae. The most abundant are *Aglaura hemistoma*, *Persa incolorata*, *Rhopalonema velatum*, *Phialidium hemisphaericum* and *Lyriope tetraphylla*, which are common.

The Salpae and Doliolidae are very scarce, although some small shoals have been observed between the Balearic Islands and Valencia above depths of more than 1000 m.

6.7.7. Central zone

The central area of the western basin, especially in the first 50 m, influenced by the Atlantic water from Gibraltar (salinities between 36.5 and 37.0 °/$_{oo}$). Beneath this layer flows water from the east and the typical northern water.

In the southern areas, the Argelian Atlantic water moves to the surface and greatly modifies the vertical distribution of zooplankton.

The distribution of certain indicated species in the surface layers reveals the remote origin of local patches of water.

Rather irregular distributions are observed in intermediate and deep waters. The quantitative decrease (Fig. 6.21) clearly shows the effects of food limitation beneath 200 m depth and the uniformity of the

LIFE AND THE PRODUCTIVITY OF THE OPEN SEA

temperatures from 150 – 200 m to the bottom, both in individuals and in species. The water mass between 500 and 1000 m contains only 4 – 5 species of Copepoda typical for this depth. This contrasts strongly with the Atlantic, where, in equivalent depth, 40 – 50 species are commonly recorded.

6.8. THE NERITIC POPULATIONS OF ZOOPLANKTON

More studies are available on the planktonic populations in the neritic zones than in the pelagic ones. Many of these were carried out during the past two decades. The specific composition of the zooplankton populations on the shelf is very diverse, but an attempt at a synthesis is shown in Figs. 6.22 and 6.23.

Nevertheless, for the mixed populations of each group a characteristic quantitative composition is found throughout the year, showing a uniform, regular and specific distribution (Fig. 6.24). A very small number of species which dominate the whole population. In the neritic zone of the Spanish Mediterranean, from a total of 81 species of Copepods, as few as 19 account for 4,583,352 individuals, from 4,751,352 specimens: i.e. 92% of the total. This shows the highly disproportionate participation of the different species in the population (Table 6.5).

Among the main causes which determine the different abundances are: (1) their own community structure, since, according to Margalef (1962), the food chain and the kinds of relationship among associated species lead to an unequal representation of the same; (2) the consequences of the natural evolution: it is known that in nature there are few species which are very numerous, with wide

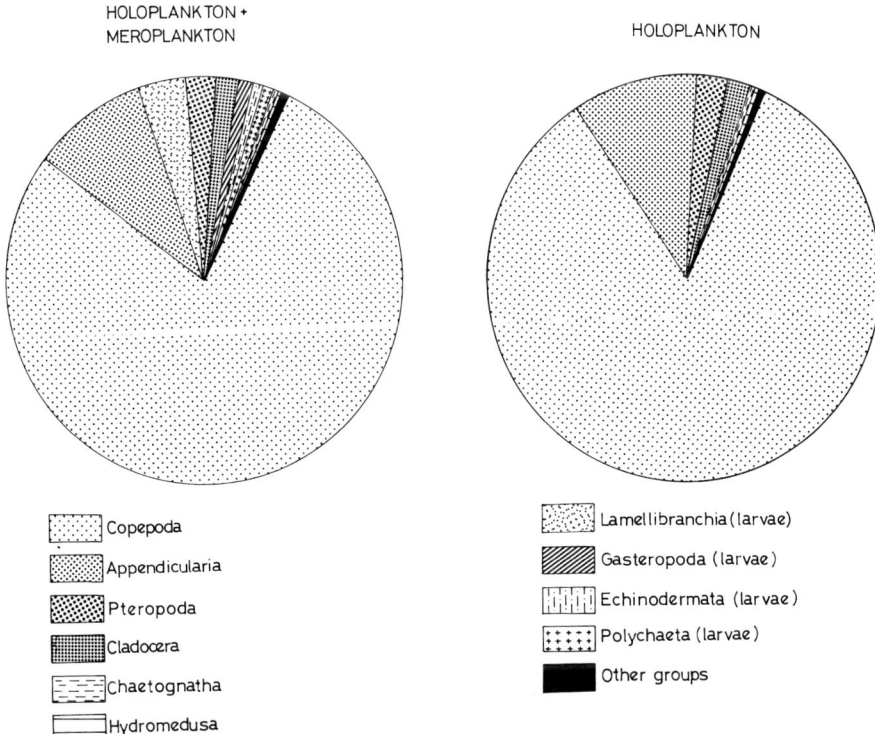

Fig. 6.22. Example of the zooplankton average quantitative composition caught during an annual cycle in the neritic waters from the Catalonian coasts (number of specimens).

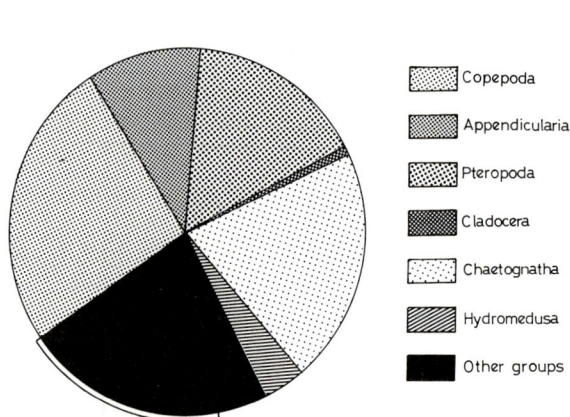

Fig. 6.23. Annual average composition of the neritic holoplankton (in weight).

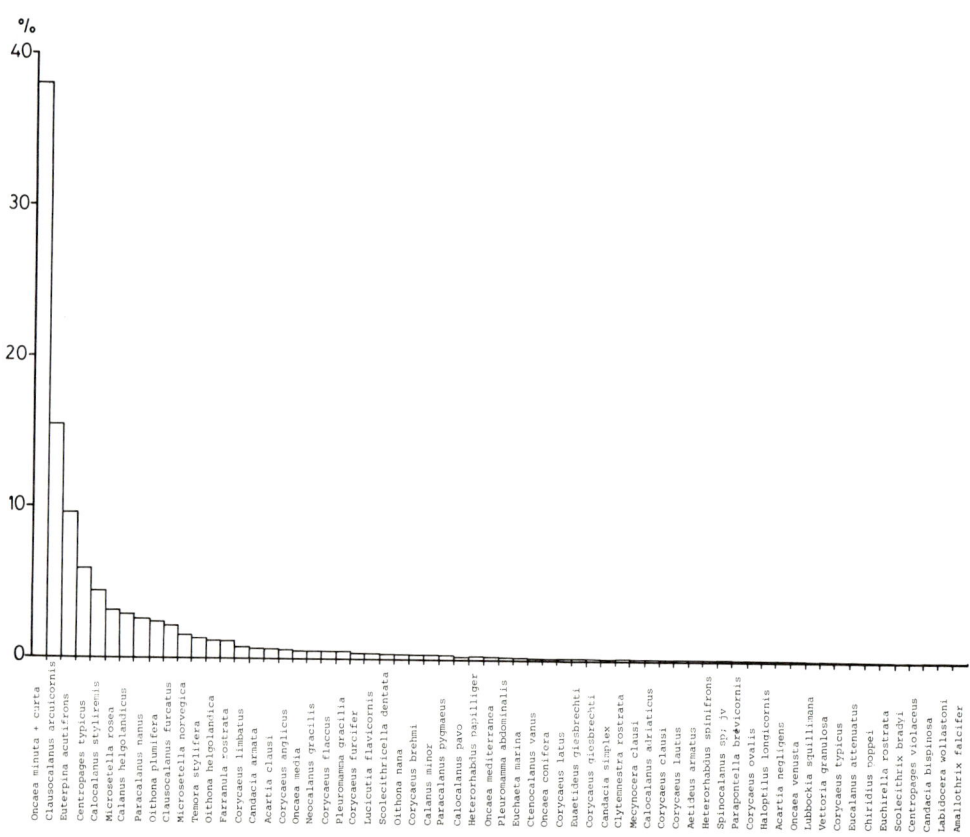

Fig. 6.24. Species of Copepods of the Catalonian neritic waters, ordered after their abundances.

TABLE 6.5. Species of copepods representing 92% of the total neritic populations (N = 4.751.352 specimens)

Calanus helgolandicus	*Oithona helgolandica*
Porocalanus nanus	*Oithona plumifera*
Calocalanus styliremis	*Microsetella rosea*
Clausocalanus arcuicornis	*Microsetella norvegica*
Clausocalanus furcatus	*Euterpina acutifrons*
Temora stylifera	*Oncaea media*
Centropages typicus	*Oncaea minuta + O. curta*
Acartia clausi	*Corycaeus anglicus*
Oithona nana	*Farranula rostrata*

geographical distributions and many other, less abundant ones, with more specific needs and restricted distribution; and (3) the different fecundity of the species: so, in our waters we encounter certain species which, during the year, present 5 or 6 generations (polycyclic species) whilst others only have one generation (monocyclic species).

On the other hand, the population from offshore or the oceanic zone studied with wide time (from hours to years apart) and spatial scales (from hundreds of metres to over 1000 km apart) shows some shifts in ranking. However, rare species never become abundant. McGowan and Walker (1982), working in the North Pacific, concluded that 'on spatial scales of 100's of metres to about 1000 km, we do not see the expected shifts in dominance'.

The situation may be different in the Mediterranean, especially in the neritic zone, where environmental variables change noticeably. Temperature, for instance, fluctuates in a yearly range of 10 – 12 °C, in surface waters, and 3 – 4 °C in deeper waters on the shelf. Salinity also differs depending on terrestrial run-off, and its variability may change from one area to another. Seasonality is important in defining and reorganizing the composition of local populations, since summer stratification and winter mixing act as selective filters on neritic populations. Table 6.3 compares mixed populations of Copepods in winter and in summer. Figs. 6.25 and 6.26 show the relative representation of the more important neritic species in the respective assemblages.

An example of the sequence of neritic zooplankton populations is provided by studies of the Copepod populations along the French and Spanish coasts. These are characterized by two species which single out respectively the two most important periods of the year: *Centropages typicus*, which is frequent and abundant during the first semester of the year, and *Temora stylifera*, a thermophilic species, which prevails throughout the second semester, although strictly both are perennial in Spanish waters.

The populations of *Centropages* exhibit a characteristic *facies*. This is especially obvious in winter and at the beginning of spring, when temperature is uniform from surface to the bottom. At this time deep water forms abound on the shelf, and also on the slope, and the surface Copepods (e.g. *Acartia*, *Oithona*, *Clausocalanus*, *Ctenocalanus*) are joined by sub-surface forms including representatives of the families Aetideidae, Euchaetidae, Metridiidae, Heterorhabdidae and others (Table 6.6).

Simultaneously the Siphonophora, *Muggiaea kochi* and *M. atlantica*, and the Medusae (*Trachymedusae* and *Narcomedusae*) become frequent, and the *Fritillaridae* also abound. After the phytoplankton blooms at the end of the winter, the *Thaliacea* can multiply. The large shoals of *Salpa democratica* are sometimes so dense that the operation of purse seines becomes difficult.

Although Chaetognatha might be scarce in midwinter; both *Sagitta setosa* (near river mouths) and *S. bipunctata* (typical of the open sea) are extraordinarily abundant with the arrival of spring.

The meroplankton also increases considerably in spring and the larvae of Brachyura which have been present throughout the winter, are outnumbered by those of the Macrura and Anomura (*Pandalina*, *Alpheus*, *Jaxea*, *Upogebia*, etc.).

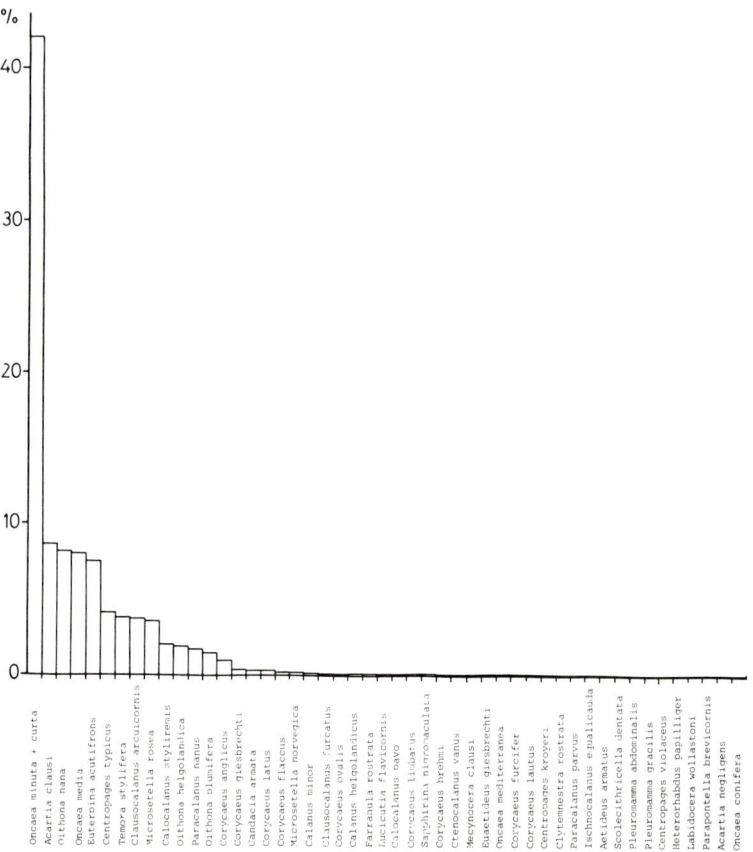

Fig. 6.25. Summer Copepod populations of the Spanish neritic zones (%).

The larvae of Gasteropoda (*Bittium*, *Eulima*, *Triphora*, etc.) increase together with those of Bivalva (*Venus*, *Cardium*, *Venerupis*, etc.).

There are two very abundant species of pelagic fish: the sardine and the anchovy. While the former spawns principally in December-January, producing a large number of eggs and larvae, the anchovy begins its spawning in May, and the larvae appear in abundance in the meroplankton at the end of spring.

In May large populations of *Doliolidae* appear to substitute for the Thaliacea (*Salpa*) and, coinciding with these, large populations of Cladocera appear. May and June mark the beginning of the concentrations of *Evadne spinifera*, followed by *Penilia avirrostris* and *Evadne tergestina*. Beneath the thermocline large swarms of *Acartia clausi* appear and throughout the water column *Paracalanus parvus* is the most common species.

At midsummer (July), the population of *Temora stylifera* (which during the winter and spring was present only at low concentrations) increases considerably and becomes one of the most important components of the neritic zooplankton on the Spanish and French coasts. At the same time, a population of small-sized Copepoda such as the *Oncaeidae* and the *Corycaeidae* also appear. By this time the *Aetideidae*, *Metridiidae* and *Heterorhabdidae* have disappeared completely. Members of this group reappear from October-November, when the stratification is broken and the temperature becomes more uniform through the whole depth of water.

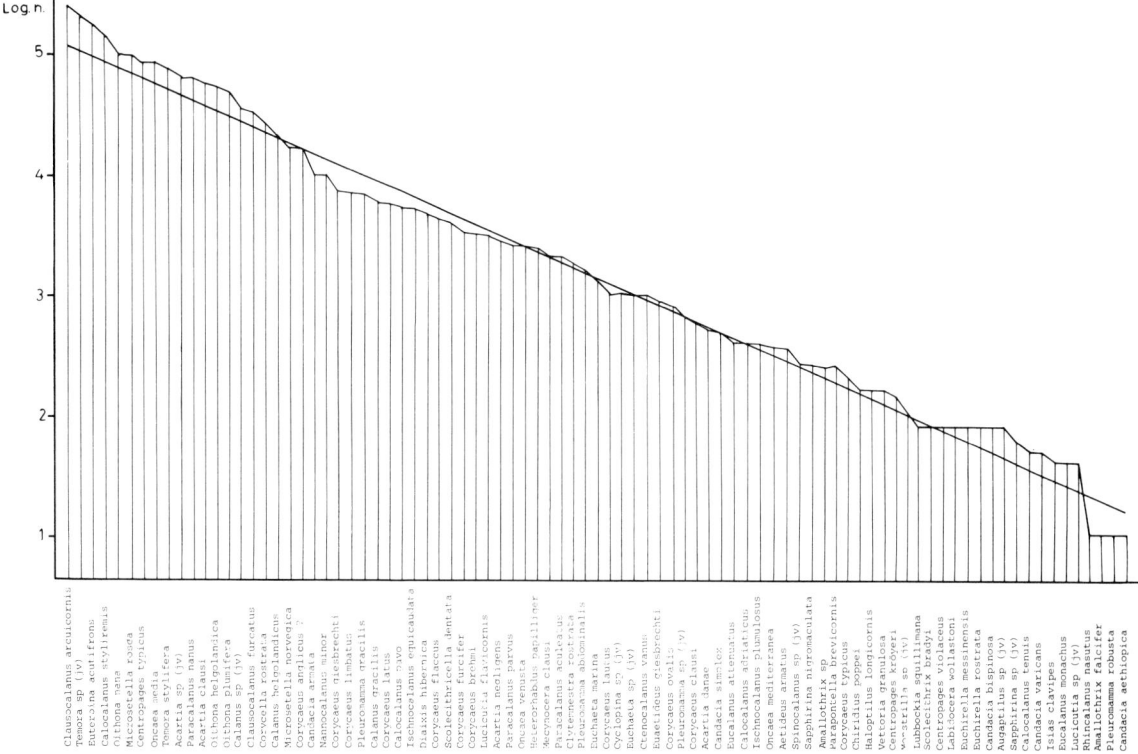

Fig. 6.26. Winter Copepod populations of the Spanish neritic zones (%).

Other groups are associated with the Copepod population during the second half of the summer. The Medusae, *Liriope*, *Phialidium*, *Sarsia* and *Rhopalonema* are common, *Muggiaea kochi* and *M. atlantica* are frequent (although not as abundant as they were in winter and spring). *Chelophyes appendiculata* is very abundant.

Among the Chaetognatha the unmistakable *Sagitta enflata* prevails, while *Evadne spinifera* and *Penilia avirrostris* continue in high concentrations, especially in areas of lower salinity, such as those near the mouths of important rivers (Rhône and Ebro).

In the meroplankton, the larvae of Decapoda remain abundant throughout the second half of the summer (*Anapagurus*, *Upogebia*, *Callianassa*, etc.) while larval Molluscs (both Gasteropoda and Lamellibranchiata) are noticeably reduced.

The number of fish eggs increases once again, at the end of autumn and beginning of winter, as pelagic species (especially the sardine) start to spawn.

The *Thaliacea* (Salps) appear again, possibly following the autumnal blooms of phytoplankton. Later, their number are gradually reduced, simultaneously with the Cladocera, giving way to the winter scarcity, in which no group (except for the gelatinous animals) is really important.

The coexistence of the populations of *Temora* and *Centropages* poses some intriguing problems as emphasised by Razouls (1972) and Gaudy (1970). Razouls has discovered a close correlation between the populations of *Temora* and the temperature, but not for those of *Centropages*. Although not rigorously confirmed, this observation suggests that there is a degree of competition between both species and their non-coincidence can be considered as an example of ecological segregation. The neritic zooplankton of the Alboran Sea exhibits some pecularities which are worthy of mention. Large densities of *Centropages*

TABLE 6.6. Copepods populations of the neritic waters during winter and summer stations (in %)

species	winter %	summer %	species	winter %	summer %
Calanus helgolandicus	2.81	0.09	Candacia armata	0.67	0.27
Calanus gracilis	0.55	—	Candacia bispinosa	0.01	—
Calanus minor	0.25	0.14	Candacia simplex	0.07	—
Eucalanus attenuatus	0.01	—	Labidocera wollastoni	0.01	+
Mecynocera clausi	0.06	0.02	Parapontella brevi	0.03	+
Paracalanus parvus	—	+	Acartia clausi	0.63	8.66
Paracalanus pygmaeus	0.24	+	Acartia negligens	0.02	+
Paracalanus nanus	2.60	1.78	Oithona nana	0.26	8.12
Calocalanus pavo	0.18	0.06	Oithona helgolandica	1.04	1.90
Calocalanus styliremis	4.40	2.00	Oithona plumifera	2.41	1.42
Calocalanus adriaticus	0.06	—	Microsetella rosea	3.13	3.58
Ischnocalanus equalicauda	—	+	Microsetella norvegica	1.56	0.15
Clausocalanus arcuicornis	15.43	3.79	Euterpina acutifrons	9.49	7.46
Clausocalanus furcatus	2.12	0.13	Clytemnestra rostrata	0.07	0.01
Ctenocalanus vanus	0.12	0.04	Oncaea venusta	0.02	—
Spinocalanus sp. jv.	0.04	—	Oncaea mediterranea	0.18	0.02
Aetideus armatus	0.05	+	Oncaea media	0.57	8.03
Euaetideus giesbrechti	0.09	0.02	Oncaea conifera	0.11	+
Chiridius poppei	0.01	—	Oncaea minuta + curta	38.22	42.00
Euchirella rostrata	0.01	—	Lubbockia squillimana	0.02	—
Euchaeta marina	0.14	—	Sapphirina nigromaculata	—	0.05
Scolecithrix bradyi	0.01	—	Vettoria granulosa	0.02	—
Amallothrix falcifer	+	—	Corycaeus clausi	0.06	—
Scolecithricella dentata	0.32	+	Corycaeus limbatus	0.70	0.06
Temora stylifera	1.28	3.80	Corycaeus typicus	0.02	—
Pleuromamma abdominalis	0.16	+	Corycaeus flaccus	0.47	0.18
Pleuromamma gracilis	0.46	+	Corycaeus giesbrechti	0.09	0.38
Centropages typicus	5.84	4.07	Corycaeus latus	0.10	0.25
Centropages kroyeri	—	0.01	Corycaeus ovalis	0.03	0.10
Centropages violaceus	0.01	+	Corycaeus anglicus	0.59	0.90
Lucicutia flavicornis	0.37	0.07	Corycaeus brehmi	0.26	0.05
Heterorhabdus papilliger	0.18	+	Corycaeus furcifer	0.38	0.02
Heterorhabdus spinifrons	0.05	—	Corycaeus lautus	0.06	0.02
Haloptilus longicornis	0.02	—	Farranula rostrata	1.04	0.09

typicus are observed, as in other areas, but here only occasionally; usually this species is substituted by *Centropages chierchiae*, which is very abundant and frequent in the Portuguese and North African coasts. Curiously, its increases in number parallels those of *Temora stylifera* (according to Rodriguez, 1979). This does not accord with the hypothesis of competition between both genera.

In summary it can be seen that the neritic zone presents relatively prominent populations with rather low values of diversity; the species are tolerant (euryhaline and eurythermal) and most components are phytophagous. These collective properties characterize the facies of *C. typicus*, Cladocera (*Evadne*, *Penilia*) and *Temora stylifera*.

During the period of isothermy we find, together with *Centropages typicus*, representatives of bathypelagic species (winter, beginning of spring).

Other groups, like the Thaliacea, follow the large phytoplankton blooms. The Doliolida increase extraordinarily when the population of Thaliacea decreases. In general the neritic zone is characterized by many individuals and few holoplanktonic species.

The larvae of benthic and nektonic animals contribute to meroplanktonic populations, which generally show a considerable increase during the second half of spring and the beginning of summer. The allochthonous species depend on the hydrographic conditions: homothermy or stratification, and on the direction of the currents, which affect the shelf with more or less intensity.

6.9. THE POPULATIONS OF INTERIOR WATERS (HARBOURS AND SMALL BAYS)

Interior waters are notoriously heterogeneous and samples are usually taken separately from study areas judged, *a priori*, as having different characteristics. Some are placed in more or less inner areas, while others are established in the mouth of the port, and in exterior waters.

The magnitude of exchange between the water of the exterior and the inner stations is often reflected in the gradient of diversity. Diversity is usually low in the relative isolation in the inner areas.

The zooplankton is basically composed of Copepoda belonging to the families *Acartidae* and *Oithonidae*. Among the representatives of the former are *Acartia clausi*, *A. margalefi* (Alcaraz, 1976), and *A. latisetosa*, and *A. granii* which is quite common. Among the second family *Oithona nana* and *O. helgolandica* are common. The remaining Copepoda come more or less from outside. Species of *Acartia* and *Oithona*, together with *Tisbe* sp., *Sapphirella* and *Euterpina acutifrons* can be considered as characteristic, since their numbers are higher in the inner stations and diminish at the outer ones. This is exactly the opposite of the situation with the rest of the species.

Other groups from this fauna can be considered typical of the interior zone. For example, the Tintinnida form large swarms, principally composed of *Tintinnopsis campanula*, *Codonella galea* and species of *Favella*. The Cnidaria are not very important, some Medusae are found usually in their juvenile forms. Among these, those of the genus *Obelia* are prominent. The larvae of Polychaeta are common in interior waters. In the harbour of Barcelona, according to Rubies (1974) at times they make up to the 40% of the total zooplankton.

The Cladocera are scarce, the genus *Podon* being an exception; other genera behave rather as allochthonous. The juvenile forms of *Podon intermedius* are sometimes present (especially in May) in large quantities.

Among the Cirripeda, *Balanus* usually covers the rocks and jetties of the harbours completely. Their *nauplia* and *metanauplia* are very abundant in summer, and late spring (May).

Among the Appendicularia, *Oikopleura* is perennial and autochthonous in the interior waters.

All these species are eurythermic and euryhaline, and thus capable of resisting sudden physicochemical changes. The same species are found in the open coastal waters and in relatively sheltered areas.

6.10 PLANKTONIC INDICATORS

The concept of indicator organisms is supported by two factors: (1) all water masses harbour characteristic populations of animals (because of the precise environmental conditions that makes life possible for these animals and exclude others), (2) by definition, movement of the planktonic organisms cannot oppose the displacements of water masses.

The concept of indicators is, therefore, based on the relation between the hydrological conditions of the water mass and the planktonic forms that inhabit it. Among these, the tolerant species are easily dispersed and will adapt to the properties of other or of mixed water masses. These cannot be used as indicator species. However, there will be other planktonic forms which cannot survive changed conditions, and will therefore, not survive in non-favourable environments. These species can be considered true hydrological indicators.

We find good examples of indicator species in different groups; for example, in Chaetognatha (which were selected as indicators for the first time in the NE Atlantic) Copepoda, Medusae, Siphonophora and Pteropoda.

Taken separately, the distribution of alleged indicators is often confusing, as they are swamped by populations of rather cosmopolitan species. It may be advisable to resort to indicators belonging to

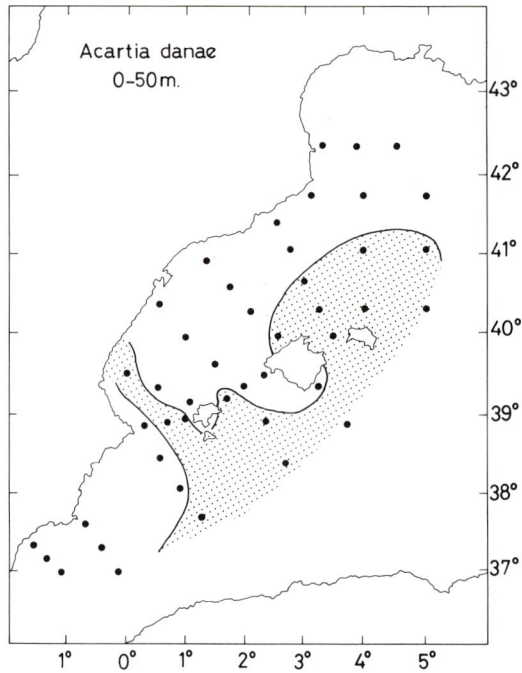

Fig. 6.27. The horizontal distribution of *Acartia danae* concurs with the more influenced areas by the Atlantic water in the Western Mediterranean.

diverse zoological groups, in other words, to *indicator communities*. But such procedure requires a complete knowledge of the zooplankton of the different water masses involved.

The *hydrological indicator* must be a relatively long-lived species whose life span falls between several weeks and a few months. This is the case of most of the zooplankton organisms (Medusae, Siphonophora, Copepoda, Eupausiacea, etc.).

On the other hand, the presence of the *ecological indicator* will depend on the physicochemical characteristics of the water mass. It is easy to predict the season from a sample of neritic plankton; the high frequency of certain species of Cladocera (for example *Evadne spinifera*, *E. tergestina*, etc.) is characteristic of the warmer waters of the summer.

The phytoplankton is much more useful than the animal plankton as an indicator for the diagnosis of physicochemical changes because their life cycles show fast reactivity. This can be observed simply by incubating a sample of water with its phytoplankton at a different temperature. In one day the dominant species will have changed completely.

We can conclude, in general, that the zooplankton is a good hydrological indicator while the phytoplankton is a better ecological indicator. This is exemplified by the appearance of Medusae in sheltered ports and beaches at midsummer or at the end of summer (August — September). Usually their presence is associated with cleaner and more transparent water. These Medusae show that an exchange of waters has occurred, for, being pelagical animals, reached the coast in their own water mass.

Another example is provided by the Copepoda. For the last few decades, Mediterranean planktologists have tried to recognize species which can be used as indicators of Atlantic waters. Extensive lists have been compiled, but many species have to be deleted, since their juvenile forms are widely found, even in areas far from any Atlantic influence. However, some remain as potential Atlantic indicators, notably: *Acartia danae* (Fig. 6.27), *Temora longicornis*, *Centropages chierchiae* and *Corycaeus latus*. Other useful Atlantic indicators can be found, in the Euphausiacea, *Thysanoessa gregaria* (Fig. 6.28), *Euphausia krohni*, *E. brevis* and *E. hemigibba*. Among the Pteropoda there is *Spiratella lesueuri* (Fig. 6.29) which is found precisely over all the areas covered by the Atlantic current.

The so-called relicts or 'vestige' indicators are characteristic of boreal regions. It is well known that during colder periods of the Pliocene, northern forms could replace other lineages of tropical distribution. Posterior warming, after the last glacial period, considerably restricted the existing fauna of boreal origin. However, not all the species disappeared and some of them continue to live in our sea. These are true relict species. Even though a precise definition is very difficult today, the following are considered as such. Among the Chaetognatha there is *Sagitta setosa*, which is not found on the Portuguese coast and is very clearly present in the NE Atlantic. Among several other Copepoda, *Ctenocalanus vanus* and *Pseudocalanus elongatus* are very frequent in the north Adriatic where the winter temperatures are relatively low.

6.11. THE TROPHIC POSITION OF ZOOPLANKTON

In the open sea, the primary producers (phytoplankton) are unicellular algae which are more or less dispersed throughout the illuminated water layers; their concentration is in general very low, except in particular spots of enhanced fertility such as upwelling areas, bays with a continuous nutrient supply, or when the circulation patterns and the density gradients provided special conditions for their accumulation.

Small size, extreme dilution and low energetic value are characteristics of marine phytoplankton which impose certain restrictive conditions for its efficient exploitation by pelagic herbivores, which are the first level of consumers, and a fundamental knot in the pelagic food web. Most of them are small animals known as *zooplankton*. Classically they are grouped together because they share a common characteristic of being at the mercy of the currents; their swimming capacity is insufficient to prevent them from being carried by water movement. From a taxonomic point of view, the zooplankton includes many different phyla, from protozoans to vertebrates. Crustaceans are, in general, the most abundant, and among them, Copepods are considered to play a basic role in the transfer of matter and energy in marine ecosystems.

Not all the organic matter present in the sea is in the form of living organisms; besides phyto and zooplankton, there is a considerable proportion of dissolved and particulate organic matter, as a

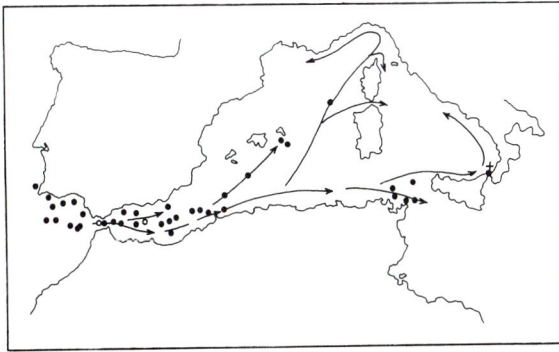

Fig. 6.28. Distribution of *Thysanoessa gregaria* in the Western Mediterranean and directions of the Atlantic current (after B. Soulier).

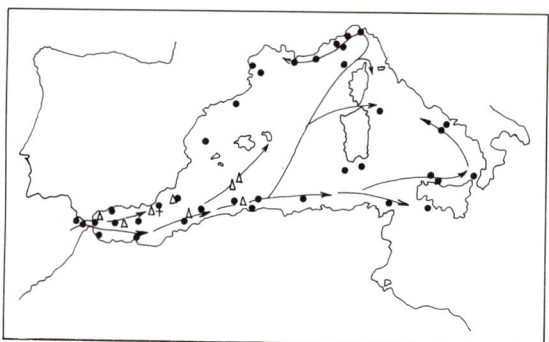

Fig. 6.29. Distribution of Pteropoda coming from the Atlantic Ocean: (*Spiratella lesueuri* (+), *Cuvierina columnella* (△), and *Diacria trispinosa, D. quadridentata, Peraclis bispinosa, Cavolinia longirrostris* and *Spiratella bulimoides* (●)), in the Western Mediterranean (after J. Rampal).

consequence of the metabolic activities of both producers and consumers. Dissolved organic matter can be actively excreted during the photosynthetic activity of phytoplankton, or may be a product of lysis of dead organisms and faeces. Its persistence depends on the impossibility of it being reconverted, so a great proportion of dissolved organic matter consists of polymerized compounds which are free of nitrogen and phosphorus.

Particulate detritic material can be formed from dissolved organic matter, through such agents as air-bubble condensation and bacterial activity. However, probably the greatest contribution comes from dead phytoplankton cells and zooplankters, their faecal pellets and moults, etc. The contribution of detritic material to marine food webs is poorly understood. It seems to be of little use for suspension feeders in the pelagic system, for detritic material would not persist at concentrations that are many times greater than living phytoplankton. However, in relatively shallow areas fresh detritic material seems to be the main source of matter and energy for benthic organisms (Strickland, 1970).

Apart from the taxonomic categories, size range is also used in the classification of zooplankton, although this feature obviously depends on the sampling technique (mesh sizes and the pore size of filters, etc.). These size classes may coincide with the traditional plankton classification terminology (Fig. 6.30, Sieburth et al., 1978).

The trophic habits of planktonic animals are also far from uniform: besides a dominance of herbivores, many zooplankters are carnivores of the first and second order (depending whether they feed on

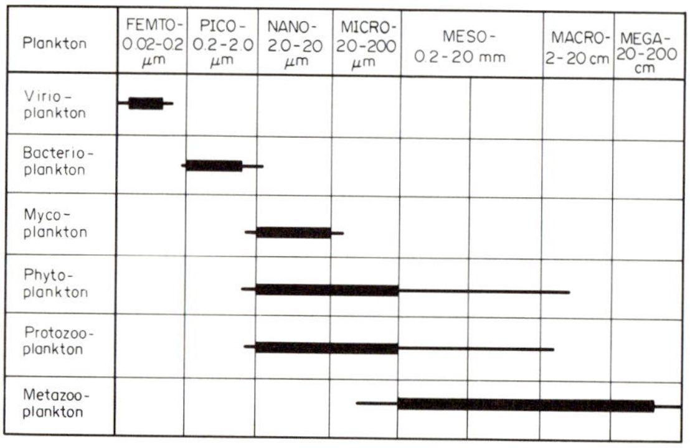

Plankton	FEMTO- 0.02-0.2 μm	PICO- 0.2-2.0 μm	NANO- 2.0-20 μm	MICRO- 20-200 μm	MESO- 0.2-20 mm	MACRO- 2-20 cm	MEGA- 20-200 cm
Virio-plankton	▬						
Bacterio-plankton		▬▬					
Myco-plankton			▬▬▬				
Phyto-plankton			▬▬▬▬▬				
Protozoo-plankton			▬▬▬▬▬				
Metazoo-plankton				▬▬▬▬▬▬▬			

Fig. 6.30. Classification of the different components of the pelagic ecosystem based in size fractions, compared with the additional terminology of plankton classes (from Sieburth et al., 1978, modified).

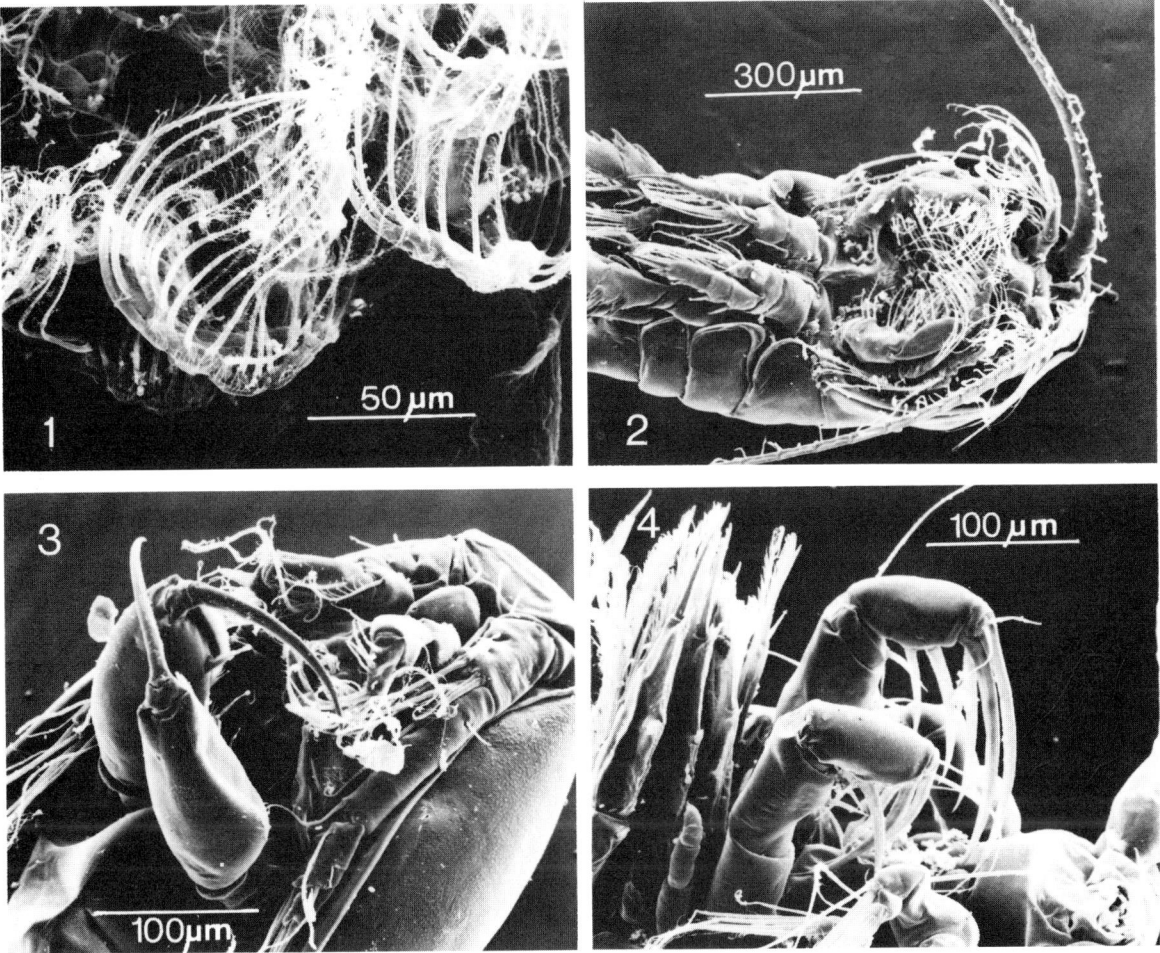

Fig. 6.31. Feeding appendages of marine microcrustaceans. 1. *Penilia avirrostris*, Cladoceran, microphage; 2. *Temora stylifera*, Copepod, microphage; 3. *Oncaea* sp., copepod, macrophage; 4. *Candacia* sp., Copepod, macrophage. Scanning microscopy (Alcaraz, unpubl.).

herbivores or other carnivores) as well as detritivores, and omnivores (Fig. 6.31). Classification can also be made according to the relative size of the animals and their preys. In this sense, two major groups can be reorganized: *microphagous* and *macrophagous*. The first consists mainly of herbivores and detritvores which feed on small particles, continuously filtering the water through net or basket-like morphologic structures, which were formerly considered to retain the food passively. This is true for some organisms, even those using mucous secretions to trap the particles (as do pelagic tunicates and other zooplankters) but in some cases (crustaceans), they are able to detect, capture and ingest or reject the food (Alcaraz *et al.*, 1980).

Macrophagous zooplankton are predators, carnivores or omnivores which feed on relatively large sized organisms. Prey capture requires developed sense organs and additional energy expenditure. As a compensation the macrophagous diet has a higher nutritive value than the microphagous one.

As all the energy in the form of organic matter which enters into the pelagic systems is exclusively supplied by the primary producers, any factor which affects the photosynthetic activity will also change the dynamics of the successive trophic levels. Zooplankton, the main representative of the primary consumers, is thus directly influenced by changes in phytoplankton production, and conversely, zooplankton can modify the characteristics of phytoplankton populations. Selective feeding can produce important qualitative changes both in the size spectrum and in the specific composition of phytoplankton. Zooplankton activity also has remarkable quantitative consequences, which can have negative effects, such as reduction of phytoplankton biomass by grazing, or positive ones, resulting from excretion of metabolites (mainly compounds rich in nitrogen and phosphorus) which tend to shorten the cycle of nutrient regeneration.

The interaction between phyto and zooplankton is essentially a predator-prey relationship which tends to generate a negative feed-back circuit, although external factors usually modify the expected development of both trophic levels. As a consequence, the characteristic sequence (of zooplankton following the evolution of phytoplankton biomass with some time lag, and the subsequent reduction of phytoplankton as zoo increases) can only be seen in some parts of their cycles (Fig. 6.32).

6.12. VERTICAL DISTRIBUTION AND NICTEMERAL MIGRATIONS

The vertical distribution of zooplankton shows similar trends in different marine areas. The biomass decreases with depth following a pattern which conforms to a negative exponential function (Vinogradov, in Bougis, 1974),

$$y_z = a \cdot e^{-kz}$$

being y_z the zooplankton biomass at the depth z; a is a constant, the surface biomass, and k is the decreasing coefficient.

In the Mediterranean Sea, the values of the constant a (the surface biomass), are lower than in the Atlantic areas, while the absolute values of the decreasing coefficient, k, are greater, implying a higher rate of biomass reduction with depth (Fig. 6.33). This, of course, reflects a situation of more light and less food. A similar distribution is also found in the case of several taxonomic groups, as in Copepods, in the Western Mediterranean basin. The distribution of Copepods from surface to a depth of 1000 m is as follows: 78% of the population, from the surface to 50 m depth; 16%, from 50 m to 200 m depth; 4%, from 200 m to 500 m; and only 2%, from 500 m to 1000 m depth (Vives, 1978). In this case, the mean calculated value of the diminishing coefficient is k = 0.011, taking the depth, z, in metres.

Such regularities make more sense when the distributions over great depths are considered, because the vertical distribution of zooplankton in the upper 200 m is regularly modified by one of the most

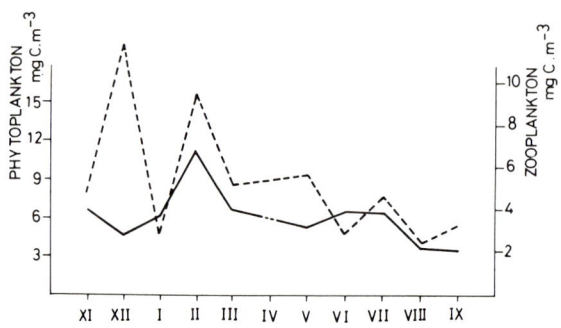

Fig. 6.32. Seasonal changes of phytoplankton (dashed line) and zooplankton (continuous line) biomass near Castelló (from Vives, 1966, modified).

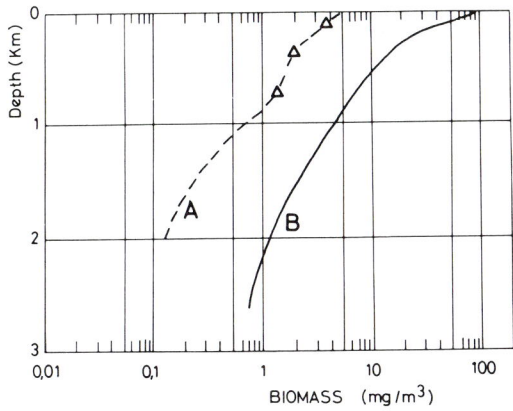

Fig. 6.33. Vertical distribution of zooplankton biomass (wet weight) in the Western Mediterranean basin (dashed line), and in the Canary Current (Continuous line). From Vinogradov, 1970, cited in Bougis, 1974, modified).

striking aspects of the zooplankton behaviour: the vertical migration displayed in a nictemeral (day-night) rhythm by almost all the zooplankters. The pattern of diel changes in the mean depth distribution of the organisms is in general very similar. At dusk there is a movement towards the surface from the so-called 'day depth'. At around midnight, the zooplankton descends from the surface (midnight sinking), and subsequently there is a regrouping near the surface at dawn (dawn rise), followed by the descent in the early post-dawn hours to the 'day depth'.

This phenomenon involves a series of complex factors which trigger the behaviour. Among these are the close relationship with the alternation of day and night (indicating the primary role of light in controlling the movements) and other factors which can modify the basic pattern, such as temperature (strong gradients can interrupt or shorten the range of movements), pressure and gravity.

Many theories have been developed to explain light-migration. It was generally assumed to involve negative phototactism. But this view has been recently challenged by the theory of optimal lighting, which proposes that migration is a consequence of the organisms choosing an optimal light intensity. According to Rudjakov (1970) vertical migration is a consequence of zooplankton swimming, while the descent is due to passive sinking.

Apart from the mechanisms involved, the ecological consequences of this migration are of great importance from the trophic point of view. Migrating herbivores feed at night, on phytoplankton produced in the surface waters during day hours (when the zooplankters are in the deep layers). This should accelerate the rate of vertical transfer of material and energy. In carnivores on the other hand, nictemeral migration is only a matter of following their food.

Migrating organisms can also gain advantages from feeding in warm surface waters and resting in colder deep layers, because fecundity and size are inversely related to temperature (McLaren, 1963). Nevertheless, if the energy expended by the organisms in their vertical displacements was too high, it would be difficult to explain its ecological advantage. According to Petipa (1966) the contribution of swimming to total energy consumption is about 95%. Even the more conservative estimations of 40% total metabolism (Klyashtorin and Yarzhombek, 1973) seems to be very high. It is more likely the cost of swimming at a 'cruising speed', without escape reactions, may be as low as 0.1% of total metabolism (Vlymen, 1970; Strickler, 1975), for otherwise, vertical migration would not make evolutionary sense.

The differences in the day-night vertical distribution of zooplankton in the Western Mediterranean basin (Alcaraz, 1981) are shown from data for the upper 100 m which show an homogeneous distribution during the day, while in the night the maximum concentration is found at the surface (Fig. 6.34).

6.13. PROBLEMS IN ESTIMATING THE SECONDARY PRODUCTION

Secondary production is the rate at which new organic matter is produced by consumers using the energy supplied by the primary producers.

Techniques for measuring feeding rate, assimilation, growth and transformation of food by zooplankton are complex in comparison to those used for determining primary production, which are based, for example, on the estimate of photosynthesis via the rate of radiocarbon uptake or the oxygen release. The great variety of feeding systems and the wide range of sizes prevent any simple estimate being made of secondary production in the pelagic community. The problem is, therefore, usually approached in two ways: studying metabolic activities (feeding rates and feeding efficiencies), or population dynamics (birth and death rates and the changes in the structure and biomass of the populations).

Temporal changes in zooplankton biomass (standing stock) can similarly reflect the general trends of secondary production in marine areas. Nevertheless, even this simple measure has only relative value, for zooplankton biomass is usually expressed as number of individuals, displacement volume, weight of dry organic matter or, more exactly, as organic carbon or nitrogen by unit volume. These are not estimates of total zooplankton biomass, but only of certain size classes of captured organisms which depend on the dimensions of the filter (mesh) used to concentrate or to catch the sample.

Thus only partial estimates of biomass can be obtained with the plankton nets (the usual sampling method for zooplankton) for only a fraction of the total population is caught, and many taxonomic groups or development stages are drastically under-represented or even totally absent due to the selectivity imposed by mesh size. Net sampling also involves integration and mixing of zooplankton populations over a large vertical or horizontal space, frequently through environmental gradients, so that they are virtually useless for the study of the fine grain distribution patterns.

In addition, the available data on zoo and phytoplankton dynamics are difficult to compare due, in many cases, to lack of synchronization or continuity in the sampling, or to methodological differences (vertical, oblique or horizontal net tows, discrete bottle samples, different mesh sizes on nets and filters, etc.). However, the main problem arises from the various strategies adopted in the oceanographic

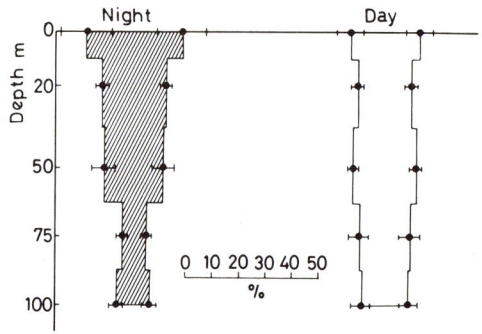

Fig. 6.34. Night and day vertical distribution of mesoplankton biomass in the upper 100 m in the Catalan Sea in October-November 1976, Mediterráneo I cruise, (see Fig. 6.11) expressed as mean percentage values of organic nitrogen (from Alcaraz, 1981).

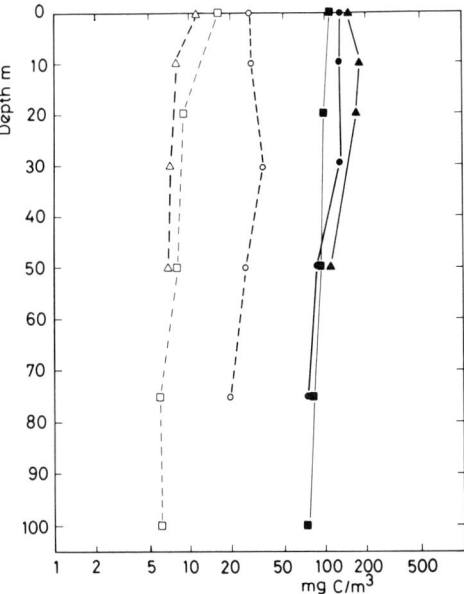

Fig. 6.35. Mean vertical distribution of total particulate organic matter (continuous line) and mesoplankton biomass (dashed lines). White triangles: coastal-urban-polluted area near Barcelona; white and black squares: Catalan Sea; white and black circles and black triangles: NW Africa upwelling region (from Alcaraz, 1980, 1981; Alcaraz et al., 1982, modified).

surveys. The intensive temporal studies are generally restricted to coastal waters and bays, while in the open sea only extensive studies have generally been done, giving synoptic descriptions without dynamic significance.

6.14 HOW MUCH ZOOPLANKTON IN THE MEDITERRANEAN SEA?

The oligotrophic character of the Mediterranean Sea determines its low zooplankton biomass, compared with similar Atlantic areas (Fig. 6.35). In the Mediterranean, three different, and relatively isolated, subsystems can be identified (in terms of the structure of phyto and zooplankton communities, as well as their production and fertilization characteristics): the *open sea*, mainly influenced by enriching mechanisms which operate at a global scale, like the winter mixing and deep water formation intensity; *coastal waters*, which are more affected by wind-induced upwelling, rivers and land run-off; and *embayments*, in which the planktonic production is strongly influenced by urban effluents and stability conditions.

The general trends of zooplankton distribution (in Mediterranean areas not affected by continental enrichment) show an increasing abundance towards the SW end of the western basin. In the Alboran Sea the abundance of zooplankton contrasts with the low values of biomass observed on the Atlantic side of the Gibraltar Strait (Fig. 6.36), so Alboran fertility should depend more on local upwelling and the effect of the cyclonic gyre than on the influence of the Atlantic waters entering into the Mediterranean. In the coastal upwelling area of the northern part of the Alboran Sea, the recorded mean annual zooplankton biomass in the upper 50 m can be as high as 36 mg $C.m^{-3}$. In open sea, the mean values are of the same order of magnitude as those of other Mediterranean regions, with relative maxima in Spring, Summer and Autumn (Camiñas, 1981).

The fertilizing effect of river discharges on phytoplankton populations, has led to an increase of zooplankton in the areas affected. In the coastal region between Castelló and the delta of the Ebro river, in the Spanish Mediterranean, there is a continuous increase of zooplankton biomass from the South to the North; the annual mean values range from 2.78 mgC.m^{-3} in the southern part, off Castelló, to 5.08 mgC.m^{-3} at a station near the Ebro delta (Vives, 1966). Other nutrient supplies of continental origin, such as human waste waters, generally have important but strictly local effects, with the associated pollution as an undesirable consequence. The maximal values of zoo and phytoplankton biomasses may not coincide in such areas. In many cases there is a greater contribution of small, high productive zooplankton in the eutrophic zone, while in the external waters around (where biomass can be higher), larger organisms dominate the zooplankton populations, as in the Gulf of Naples (Carrada et al., 1980).

In a coastal, polluted area near Barcelona, the annual mean of zooplankton biomass was 8.53 mgC.m^{-3} (Alcaraz, 1980), more or less twice the corresponding average of a neighbouring, but unpolluted zone (4.86 mgC.m^{-3}, Dinofrio, 1979). This development coincides with general trends in other Mediterranean regions, with two annual maxima corresponding to spring and autumn.

When circulation is restricted (as in bays, harbours and other semi-landlocked areas) meteorological events which modify the hydrographic conditions (winds, insolation, etc.) can have a disrupting effect, altering the relative distributions of phyto and zooplankton and accelerating their changes.

The scarcity and the differences in sampling methods of surveys made in the open sea of the Western Mediterranean basin make it difficult to interpret the available data. The annual cycles demonstrate that average zooplankton biomass is generally lower in offshore stations than in coastal ones (2.8 mgC.m^{-3} off Barcelona, against 4.85 mgC.m^{-3} in a coastal station, mean annual values, Dinofrio, 1979). Synoptic data from larger areas can give valuable information on the spatial distribution of zooplankton. However, the time lag between surveys is often too long, so that temporal variability and even average abundance of zooplankton in the different geographic zones cannot be compared. In Table 6.7, the values of zooplankton biomass from several zones of the Western basin are shown, as well as their integrated depths, collecting methods and period of the year.

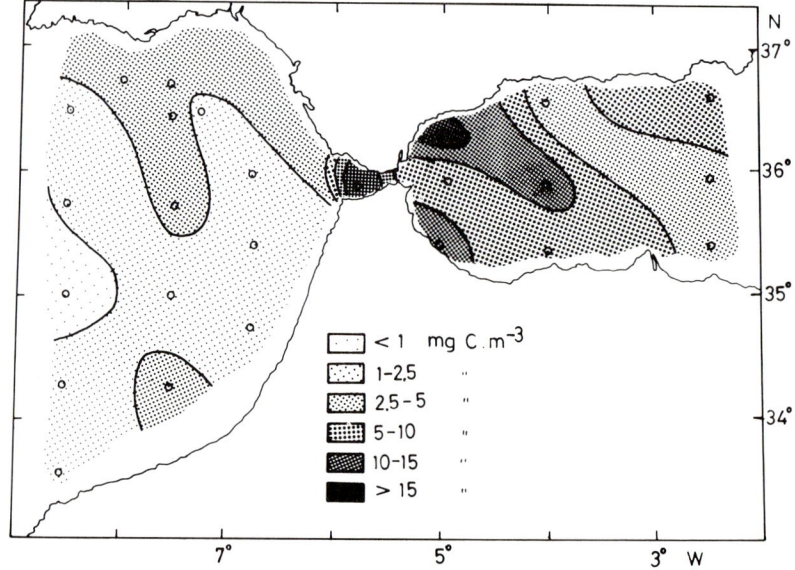

Fig. 6.36. Spatial distribution of mesoplankton biomass in the upper 200 m at the two sides of the Gibraltar Strait (June – July 1972, Maroc-Iberia I cruise).

TABLE 6.7. Zooplankton biomass expressed as mg C. m^{-3} in different areas of the Western Mediterranean basin. Values with asterisk have been calculated from data on displacement volume after Cushing et al. (1958)

Area	Month	Biomass (mg C. m^{-3})	Integrated depth (m)	Sampling device and mesh size	Author
Balearic Sea	March	7.72*	500-0	Bongo Net, (333 μm)	Suau, 1977
Balearic Sea	October-November	9.36	100-0	Discrete bottle samp. (200 μm)	Alcaraz, 1981
Balearic Sea	October-November	6.31*	200-0	WP2 Net (200 μm)	Vives, 1979
Balearic Sea	August	2.02*	300-0	Bongo Net (333 μm)	Palomera (unpubl. data)
Barcelona, coastal area	July-March	8.53	50-0	Discrete bottle samp. (200 μm)	Alcaraz, 1980
Barcelona, coastal area	June-March	4.85*	200-0	WP2 Net (200 μm)	Dinofrio, 1979
Barcelona, open sea	June-March	2.80*	200-0	WP2 Net (200 μm)	Dinofrio, 1979
Castelló	November-September	3.68*	15-0, 60-0	WP2 Net (200 μm)	Vives, 1966
Alboran Sea	July	8.35*	200-0	WP2 Net (200 μm)	Maroc-Iberia I (unpubl. data)
Villefranche-Calvi	September	0.63*	200-0	Juday-Bogorov; (160 μm)	Razouls, 1972
Alguero-Tunisian area	April	1.09*	200-0	Juday-Bogorov (160 μm)	Razouls, 1972
Ligurian Sea	March	0.66*	200-0	Juday-Bogorov net (160 μm) WP2 net (200 μm)	Razouls, 1972
Ligurian Sea	April	9.14*	200-0	Juday-Bogorov net (160 μm) WP2 net (200 μm)	Razouls, 1972

The recorded biomass is that of the *mesoplankton* (i.e. organisms larger than 200 μm) (Fig. 6.30). As a consequence, any quantitative approach to zooplankton dynamics using such data must be very conservative. The unrepresented fraction consists of small ciliates and other microplankton whose participation, in terms of biomass, is around 20% of mesoplankton (Margalef and Vives, 1967; Rassoulzadegan, 1982), but in terms of production is similar to that of mesoplankton.

Little is known about the efficiency of pelagic food webs, and the above mentioned difficulties prevent precise quantification of the phytoplankton-zooplankton relationships. Nevertheless, a model has been attempted for the northern part of the Western Mediterranean basin (Gulf of Lions and Ligurian Sea). After the formation of deep water in winter, the potential primary production and the zooplankton growth are reduced by the excess of turbulence in the central part of the cyclonic gyre. Only in the peripheral zones does the stability allow the almost simultaneous development of phytoplankton and zooplankton populations. These invade the central zone when there is a reduction in the turbulence. The estimated secondary production increases from 18 mgC.m^{-2}.day^{-1} in March, to 230 mgC.m^{-2}.day^{-1} in April, with an efficiency in the energy transfer phyto-zooplankton ranging from 7% to 26% (Nival, 1976). The grazing pressure, which is higher in the peripheral zone, represents an average 50% of all the phytoplankton losses, while during springtime zooplankton excretion has only a secondary role as nutrient supply.

During the summer, the diffusion of nutrients from the deep layers to the photic zone is restricted by the density gradient. However, some development of phyto- and zooplankton at the pycnocline level can occur, due to the effect of the cyclonic gyre, which pushes up the nutrient-rich intermediate water to form a typical dome (Gostan and Nival, 1967). In the Catalan Sea, appreciable concentrations of phytoplankton have been found in thin layers below the thermocline (at 60 to 90 m depth). In these deep phytoplankton-rich layers, there was a greater proportion of small, (less than 2 μm) (picoplankton)

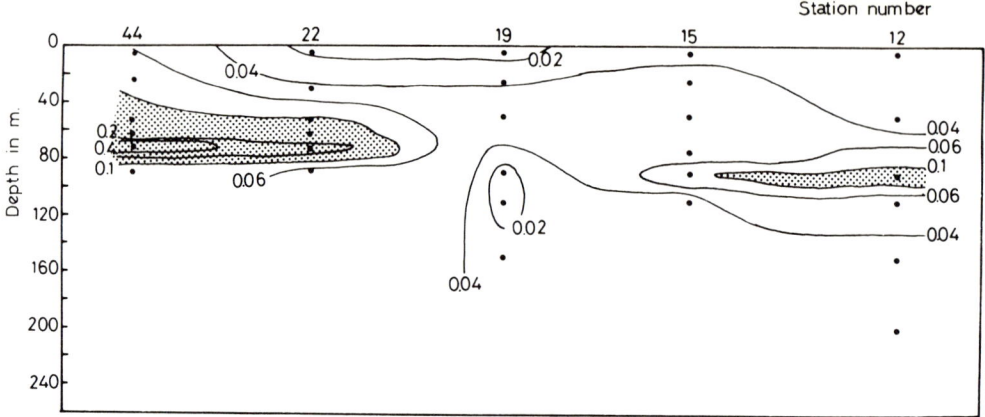

Fig. 6.37. Vertical distribution of mesoplankton biomass (particulate organic nitrogen, μ at.gN.l^{-1}) in the upper 200 m in the Catalan Sea in July, 1982, PEP 82 cruise. Compare with Fig. 6.12 (Alcaraz, in press).

cells, which are supposed to be blue-green algae or similar (Estrada, 1985; Vallespinos, in prep.). The vertical distribution of zooplankton was similar to that of phytoplankton; 70% of the mesoplankton biomass of the upper 200 m was frequently concentrated in a 10 – 20 m thick layer (Fig. 6.37, Alcaraz, in press') coinciding with the chlorophyll maximum (*see* Fig. 6.12).

ACKNOWLEDGEMENTS

We thank Dr. H. J. Minas for his comments on the primary production sections. Mr. A. Fauquet operated the Scanning Electron Microscope and obtained the micrographs. Mrs. A. Cruz and Miss C. Bas provided technical assistance.

REFERENCES

Alcaraz, M. (1976) Description of *Acartia margalefi*, a new species of pelagic Copepod, and its relationship with *A. clausi*. *Inv. Pesq.* 40(1), 59 – 74.
Alcaraz, M. (1980) Evolución y distribución vertical de la biomasa de zooplancton, expresada en carbono y nitrógeno orgánicos, relación C/N y carbono detrítico en una zona marina afectada por aguas residuales procedentes de la ciudad de Barcelona. *Inv. Pesq.* 44(2), 265 – 274.
Alcaraz, M. (1981) Carbono y nitrógeno orgánicos del mesozooplancton en el Mediterráneo Occidental en octubre de 1976: distribución espacial y relación C/N. *Res. Exp. Cient.* 9, 129 – 141.
Alcaraz, M. (in press) Vertical distribution of zooplankton biomass during summer stratification in the Western Mediterranean. In *19th EMBS*.
Alcaraz, M., Estrada, M., Floss, J. and Fraga, F. (1982) Relationships between particulate organic carbon and nitrogen and phytoplankton biomass in oligotrophic and upwelling systems. JOA Poster abstracts 1, 81.
Alcaraz, M., Paffenhöffer, G. A. and Strickler, J. R. (1980) Catching the algae: a first account of visual observations on filter-feeding calanoids. In *Evolution and Ecology of zooplankton communities*. Edit. W. C. Kerfoot, pp. 241 – 248. Univ. Press of New England.
Ballester, A., Arias, E., Cruzado, A., Blasco, D. and Camps, J. M. (1967) Estudio hidrográfico de la costa catalana de junio de 1965 a mayo de 1967. *Inv. Pesq.* 31(3), 621 – 662.
Bernhard, M. and Rampi, L. (1967) The annual cycle of the Utermöhl-phytoplankton in the sea in 1959 and 1962. *Publ. Staz. Zool. Napoli* 35(2), 137 – 169.
Barnhard, M., Rampi, L. and Zattera, A. (1969) La distribuzione del fitoplancton nel Mar Ligure. *Publ. Staz. Zool. Napoli* 37 (Suppl. 2°), 73 – 114.

Béthoux, J. P. (1981) Le phosphore et l'azote en Mer Méditerranée, bilans et fertilité potentielle. *Marine Chemistry* 10, 141–158.

Blasco, D. (1971) *Acumulación de nitritos en determinados niveles marinos por acción del fitoplancton*. Ph.D. Thesis, University of Barcelona, 223 pp.

Bougis, P. (1974) *Ecologie du plancton marin. II Le zooplancton*. Masson et Cie, Paris, 200 pp.

Cahet, G., Fiala, M., Jacques, G. and Panousse, M. (1972) Production primaire au niveau de la thermocline en zone néritique de Méditerranée Nord-Occidentale. *Mar. Biol.* 14, 32–40.

Camiñas, J. A. (1981) Distribution spatiale et temporelle de la biomasse zooplanctonique superficielle dans le secteur nord-occidentale de la mer d'Alboran. *Rapp. Proc. Ver. Réun. CIESM* 27(7), 125–127.

Carrada, G. C., Hopkins, T. S., Bonaduce, G., Ianora, A., Marino, D., Modigh, M., Ribera D'Alcala, M. and Scotto di Carlo, B. (1980) Variability in the Hydrographic and Biological Features of the Gulf of Naples. *Mar. Ecol.* 1, 105–120.

Casanova, B. (1974) Les Euphausiecés de Méditerranée. Thése Doctorat, pp. 1–380, Marseille.

Casanova, J. P. (1977) *La Faune Pèlagique Profonde (zooplancton et Micronecton) de la Province Atlanto-Méditerrenne*. Thése Doctorat, pp. 1–455, Marseille.

Coste, B. and Minas, J. H. (1967) Premières observations sur la distribution des taux de productivité et des concentrations en sels nutritifs des eaux de surface du golfe du Lion. *Cah. Océanogr.* 19(5), 417–429.

Coste, B., Gostan, J. and Minas, H. J. (1972) Influence des conditions hivernales sur les productions phyto- et zooplanctoniques en Méditerranée nord-occidentale. Estructures hydrologiques et distribution des sels nutritifs. *Mar. Biol.* 16(4), 320–348.

Cushing, D. H., Humphrey, G. F., Banse, K. and Laevastu, T. (1958) Report of the committee on terms and equivalents. *Rapp. Proc. V. CIEM* 144, 15–16.

Dinofrio, E. O. (1979) *Estudio de las variaciones espacio-temporales de los copépodos y especies acompañantes en las costas catalanas*. Tesina Universidad de Barcelona, 96 pp.

Estrada, M. (1979) Observaciones sobre la heterogeneidad del fitoplancton en una zona costera del mar Catalán. *Inv. Pesq.* 43(3), 637–666.

Estrada, M. (1981) Biomasa fitoplanctónica y producción primaria en el Mediterráneo occidental a principios de otoño. *Inv. Pesq.* 45(1), 211–230.

Estrada, M. (1985) Deep phytoplankton and chlorophyll maxima in the Western Mediterranean. In *Mediterranean Marine Ecosystems*, eds. M. Moraitou-Apostolopoulou and V. Kiortsis, pp. 247–277. Plenum.

Furnestin, M. L., Maurin, C., Lee, J. Y. and Raimbault, R. (1966) Eléments de Planctonologie Appliquée. Inst. Scient. Téch. Pêches marit, pp. 1–166, Paris.

Gaudy, R. (1970) *Contribution à la connaissance du cycle biologique et de la physiologie des copépodes du golfe de Marseille*. Thése doctorat, pp. 1–294, Marseille.

Gaumer, G. (1981) Evolution annuelle des communautés phytoplanctoniques de la baie d'Alger. *Variations de la composition spécifique liées à la nature de facteur nutritionnel limitant la biomasse algale*. Thése 3e cycle. Université Pierre et Marie Curie, 91 pp.

Genovese, S., Gangemi, G. and De Domenico, E. (1972) Campaña estival 1970 della n/o"Bannock" nel Mar Tirreno. Misure di produzione primaria lungo la transversale Palermo-Cagliari. *Boll. Pesca Piscic. Idrobiol.* 27(1), 139–147.

Gostan, J. and Nival, P. (1967) Relations entre la distribution des phosphates mineraux dissous et la répartition des pigments dans les eaux superficielles du Golfe du Lion. *Cah. Océanogr.* 19(1), 41–52.

Grillini, C. L. and Lazzara, L. (1978) Ciclo annuale del fitoplancton nelle acque costieri del Parco Naturale della Marenna. I. Variazioni quantitative. *Giornale Botanico Italiano* 112(3), 157–173.

Grillini, C. L. and Lazzara, L. (1980) Ciclo annuale del fitoplancton nelle acque costieri del Parco Naturale della Marenna. II. Flora e variazioni delle comunità. *Giornale Botanico Italiano* 114, 199–215.

Herbland, A. and Voiturez, B. (1979) Hydrological structure analysis for estimating the primary production in the tropical Atlantic Ocean. *J. Mar. Res.* 37, 87–101.

Herrera, J. and Margalef, R. (1963) Hidrografía y fitoplancton de la costa comprendida entre Castellón y la desembocadura del Ebro, de julio de 1960 a junio de 1961. *Inv. Pesq.* 24, 33–112.

Jacques, G. (1968) Aspects quantitatifs du phytoplancton de Banyuls-sur-Mer (Golfe du Lion) II. Cycle des Flagellés nanoplanctoniques (juin 1965 – novembre 1965 – décembre 1967). *Vie et Milieu, sér. B.* 19, 17–33.

Jacques, G. (1970) Aspects quantitatifs du phytoplancton de Banyuls-sur-Mer (Golfe du Lion). - Biomasse et production, 1965–1969. *Vie et Milieu*, sér. B. 21, 37–102.

Jacques, G. (1971) Floraison printanière du phytoplancton à Banyuls-sur-Mer (Golfe du Lion) en 1968. *Rapp. Com. int. Mer Médit.* 20, 311–313.

Jacques, G., Minas, H. J., Minas, M. and Nival, P. (1973) Influence des conditions hivernales sur les productions phyto- et zooplanctoniques en Méditerranée Nord-Occidentale. II. Biomasse et production phytoplanctonique. *Mar. Biol.* 23, 251–265.

Jacques, G., Minas, M., Neveux, J., Nival, P. and Slawik, G. O. (1976) Conditions estivales dans la divergence de Méditerranée nord-occidentale. III Phytoplancton. *Ann. Inst. océanogr. n.s.* 52(2), 141–152.

Jacques, G. and Sournia, A. (1978-1979) Les "eaux rouges" dues au phytoplancton en Méditerranée. *Vie et Milieu* sér. AB 28-29(2), 175–187.

Klyashtorin, L. and Yarzhombek, A. A. (1973) Energy consumption in active movements of planktonic organism. *Oceanology* 13, 575–580. (in Russian).

Lanoix, F. (1974) Project Alboran. Etude hydrologique et dynamique de la mer d'Alboran. *Rapport Technique OTAN* No. 66, 39 pp. + 31 fig.
Margalef, R. (1962) *Comunidades Naturales*. Inst. Biol. Mar. Univ. Puerto Rico. Publ. esp. 1 — 469.
Margalef, R. (1963) El ecosistema pelágico de un área costera del Mediterráneo occidental. *Mem. Real Acad. Cienc. Art. Barcelona* 35, 3 — 48.
Margalef, R. (1969) Composición específica del fitoplancton de la costa catalano-levantina (Mediterráneo occidental) en 1962 — 1967. *Inv. Pesq.* 33(1), 345 — 380.
Margalef, R. (1971) Distribución del fitoplancton entre Córcega y Barcelona, en relación con la mezcla vertical del agua, en marzo de 1970. *Inv. Pesq.* 35(2), 687 — 698.
Margalef, R. (1977) *Ecología*. Ediciones Omega: 1 — 951, Barcelona.
Margalef, R. and Ballester, A. (1967) Fitoplancton y producción primaria de la costa catalana, de junio de 1965 a junio de 1966. *Inv. Pesq.* 31(1), 165 — 182.
Margalef, R. and Castellví, J. (1967) Fitoplancton y producción primaria de la costa catalan, de julio de 1966 a julio de 1967. *Inv. Pesq.* 31(3), 491 — 502.
Margalef, R. and Herrera, J. (1963) Hidrografía y fitoplancton de las costas de Castellón de julio de 1959 a junio de 1960. *Inv. Pesq.* 22, 49 — 109.
Margalef, R. and Herrera, J. (1964) Hidrografía y fitoplancton de las costas comprendidas entre Castellón y la desembocadura del Ebro, de julio de 1961 a julio de 1962. *Inv. Pesq.* 26, 49 — 90.
Margalef, R., Herrera, J., Steyaert, M. and Steyaert, J. (1965) Distribution et caractéristiques des communautés phytoplanctoniques dans le bassin Tyrrhenien de la Méditerranée en fonction des facteurs ambiants et à la fin de la stratification estivale de l'année 1963. *Bull. Inst. r. Sci. nat. Belg.* 42(5), 1 — 56.
Margalef, R. and Vives, F. (1967) La vida suspendida en las aguas. In *Ecología Marina*, Ed. La Salle, pp. 439 — 562, Ciencias Naturales, Caracas.
Massuti, M. and Margalef, R. (1950) *Introducción al estudio del plancton marino*. Patr."Juan de la Cierva" Inv. Téc. Sec. Biol. Mar. 1 — 182, Barcelona.
McGill, D. A. (1965) The relative supplies of phosphate, nitrate and silicate in the Mediterranean Sea. *Rapp. Proc. Ver. CIESMM* 18(3), 737 — 744.
McGowan, J. A. and Walker, P. W. (1982) Stability and Resilience in an Oceanic Ecosystem. *Scripps. Inst. Ocean. La Jolla* (Mimeo), 1 — 77.
McLaren, I. A. (1963) Effects of temperature on growth of zooplankton and the adaptive value of vertical migration. *J. Fish. Res. Bd. Canada* 20, 685 — 727.
Minas, H. J. (1968) A propos d'une remontée d'eaux"profondes" dans les parages du golfe de Marseille (octobre 1964). Conséquences biologiques. *Cah. Océanogr.* XX(8), 647 — 674.
Minas, H. J. and Blanc, F. (1970) Production organique primaire au large et près des côtes méditerranéennes françaises (juin-juillet 1965), influence de la zone de divergence. *Téthys* 2(2), 299 — 316.
Minas, H. J., Coste, B., Gascard, J. C., Le Corre, P. and Richez, C. (1982). Proprietés chimiques et circulation des masses d'eau dans le détroit de Gibraltar et en mer d'Alboran (Campagne MEDIPROD IV du N.O.JK. Charcot, oct.-nov. 1981). *Rapp. Proc. verg. Réun. Comm. int. Mer Médit.* 28(2), 129 — 130.
Nival, P. (1976) *Rélations phytoplancton-zooplancton; essai de modellisation*. Thèse de Doctorat d'Etat. Univ. Pierre et Marie Curie, Tome I, 219 pp.
Petipa, T. S. (1966) On the energy balance of *Calanus helgolandicus* (Claus) in the Black Sea. In *Physiology of Marine Animals*. Akad. Nauk. SSSR, pp. 60 — 81. (in Russian)
Rampal, J. (1975) *Les Thecosomes (Mollusques pélagiques) systematique et Evolution-Ecologie et Biogeographie Méditerranéennes*. Thèse de doctorat, pp. 1 — 485, Marseille.
Rassoulzadegan, F. (1982) *Le rôle fonctionnel du microzooplancton dans un écosystème méditerranéen*. Thèse de Doctorat d'Etat Univ. Pierre et Marie Curie. Tome I, 138 pp.
Razouls, C. (1972) *Estimation de la production sécondaire (copépodes pélagiques) dans une province néritique méditerranéenne (Golfe du Lion)*, Thèse Doctorat d'Etat, Fac. Sciences de Paris, 301 pp.
Rodríguez, J. (1979) *Zooplancton de la bahía de Málaga*. Tesis doctoral, 1 — 147, Málaga.
Rodríguez, V. (1981) *Estudio de un ecosistema portuario*. Tesis doctoral, pp. 1 — 228, Málaga.
Rubiés, P. (1974) *Contribución a la ecología del Puerto de Barcelona: Zooplancton y polución*. Tesina, pp. 1 — 169, Barcelona.
Rudjakov, J. A. (1970) The possible causes of diel vertical migrations of planktonic animals. *Mar. Biol.* 6, 98 — 105.
Samson-Kechacha, F. L. (1981) *Variations saisonnières des matières nutritives de la baie d'Alger. Recherche des facteurs contrôlant le developpement du phytoplancton*. Thèse 3ᵉ cycle. Université Pierre et Marie Curie, Paris, 98 pp.
San Feliu, J. M. and Muñoz, F. (1971) Fluctuations d'une année à l'autre dans l'intensité de l'affleurement dans la Méditerranée occidentale. *Inv. Pesq.* 35(1), 155 — 159.
Schink, D. R. (1967) *Geochim. et Cosmochim*. Acta 31, 987 — 999.
Sieburth, J. Mc N., Smetacek, V. and Lenz, J. (1978) Pelagic ecosystem structure: heterotrophic components of the plankton and their relationship to plankton size-fractions. *Limnol. Oceanogr.* 23, 1256 — 1263.
Soulier, B. (1965) Euphausiacés des bancs de Terre-Neuve, de Nouvelle Ecosse et du Golfe du Maine. *Rev. Trav. Inst. Pêches marit.* 29(3), 173 — 189.
Sournia, A. (1973) La production primaire planctonique en Méditerranée. Essai de mise à jour. *Bulletin de l'étude en commun de la Méditerranée, 5, Num. spéc.* 128 pp.
Strickland, J. D. H. (1970) Recycling of organic matter. In *Marine Food Chains*: 3-5, Oliver & Boyd, Edinburgh, 552 pp.

Strickler, J. R. (1975) Swimming of planktonic *Cyclops* (Copepoda: Crustacea): Pattern, movements and their control. In *Swimming and Flying in Nature*, Eds. T. Y. T. Wu et al., V.II, pp. 599–613, Plenum.

Suau, P. (1981) Campaña "Mediterráneo II" (marzo, 1977). *Datos Informativos* 8, 239 pp.

Thiriot, A. (1970) *Cycle et Distribution des crustacés planctoniques de la Région de Banyuls-sur-Mer (Golfe du Lion) Etude spéciale des cladocéres*. Thèse Doctorat, pp.1–307, Marseille.

Travers, M. (1973) Le microplancton du golfe de Marseille: variations de la composition systématique et de la densité des populations. *Téthys* 5(1), 31–53.

Travers, M. (1975) Le microplancton du golfe de Marseille. Schéma du cycle annuel, répartitions horizontale et verticale. *Téthys* 6(4), 713–726.

Travers, A. and Travers, M. (1973) Données sur quelques facteurs de l'écologie du plancton dans la région de Marseille. 3. La lumière. *Téthys* 5(1). 7–30.

Tregouboff, G. and Rose, M. (1957) *Manuel de Planctonologie Méditerranéenne*. C.N.R.S., pp. 1–587, Paris.

Vives, F. (1966) Zooplancton nerítico de las aguas de Castellón. (Mediterráneo occidental). *Inv. Pesq.* 30, 49–166.

Vives, F. (1970) *Contribución al estudio de los crustáceos planctónicos del Mediterráneo occidental*. Tesis Doctoral, pp. 1–287, Barcelona.

Vives, F. (1978) Distribución de la población de copépodos en el Mediterráneo occidental. *Res. Exp. Cient. B/O Cornide* 7, 263–302.

Vives, F. (1979) Campaña "Mediterráneo I" (octubre-noviembre 1976). *Datos Informativos* 7, 164 pp.

Vives, F., Santamaria, G. and Trepat, I. (1975) El zooplancton de los alrededores del estrecho de Gibraltar en junio-julio de 1972. *Res. Exp. Cient. B/O Cornide* 4, 7–100.

Vlymen, W. J. (1970) Energy expenditure by swimming copepods. *Limnol. Oceanography* 15, 348–356.

CHAPTER 7

History of the Mediterranean Biota and the Colonization of the Depths

J. M. PÉRÈS

Station Marine d'Endoume, Marseille, France

CONTENTS

7.1. The History	198
7.1.1. Ancient times	198
7.1.2. The salinity crisis at the Late Miocene	199
7.1.3. Pliocene and Pleistocene	202
7.1.4. The 'Quaternary'	203
7.2. The Mediterranean Benthos and its actual and recent changes	206
7.3. The problem of the Mediterranean endemics	207
7.4. The circalittoral soft bottoms	209
7.4.1. Coastal detritic assemblage	210
7.4.2. Muddy-detritic assemblages	213
7.4.3. Terrigenous mud-shelf assemblages	213
7.4.4. Shelf-edge detritic assemblage	215
7.5. Offshore rocky-bottom assemblages	217
7.5.1. Offshore rocky-bottom assemblage	217
7.5.2. Bottoms with large brachiopoda	218
7.6. The Bathyal assemblages	219
7.6.1. Deep-sea coral assemblage	219
7.6.2. The bathyal mud assemblage	223
7.7. Is there an Abyssal Zone in the Mediterranean Sea?	227

7.1. THE HISTORY

7.1.1. Ancient times

To recognise a sea that could properly be identified as an ancestor of the present Mediterranean, it is necessary to go back to the Oligocene. Then, a sea is recognisable between the Euro-asiatic and the African shields, ranging approximately from 35° to 45°N of latitude, opening to the Atlantic Ocean at approximately betweeen 30 − 40°N and 05°E and to the Indian Ocean possibly between 10 − 20°N and 50 − 60°E. This oligocene Mediterranean was a part of the mesogean sea that Suess named the Tethys Sea, which existed since the Lower Cambrian and continued until the late Tertiary, perhaps with some brief local interruptions (Ekman, 1953).

During the Miocene, the alpine geosynclinal was active in an orogenetic multiple event which gave birth to the Caucasus, the Dinarids, the Appenins, the Betico-Riffan range, the Pyrenees and, finally, the Alps. By the later Miocene (Pontian) the communication with the Indian Ocean was definitively interrupted.

It is still a matter of discussion what happened at this time, the Tortonian, when the Mediterranean Sea was certainly still in communication with the Atlantic Ocean. It seems probable, however, that this communication was mostly restricted to shallow waters, if not interrupted altogether. This stage is the Messinian, as interpreted recently by Rouchy (1982).

7.1.2. The salinity crisis at the Late Miocene

By the Late Miocene and especially since the Messinian (6.5 to 5 million years B.P.) the shape of the Mediterranean basin resembled its present form, and the period was marked by important geodynamic and hydrodynamic events. The result was a succession of non-compensated evaporations which led to saline deposits ('halines'), with a thickness of up to 1000 − 1500 m in the Western basin, and up to 3500 m in the Eastern basin (Hsü et al., 1973). The thick evaporites were mainly limited to the abyssal plains of the present Mediterranean, for example in the Algéro-Provençcal basin, which for about five sixths of its total surface is covered by such layers. After some pre-evaporitic formations, sometimes made of marls, the true evaporites consisted of two superimposed layers, namely: a lower one consisting of salt, and an upper one exhibiting alternative layers of marls and gypso-anhydrites. It appears that open sea conditions suddenly overtook and replaced the brackish or lagunar conditions somewhere between the Miocene (Tortonian) and the Pliocene.

Different estimates have been advanced to account for the maximum depth of the Mediterranean before the deposition of the evaporites, and for the pattern of their deposition. A depth of several thousand metres was advanced by K. J. Hsü et al. (1973), whereas observations made by W. D. Nesteroff (1973) suggested not more than a few hundred metres. Probably the truth lies somewhere between these estimates, closer to more recent ones made by Genesseaux and Lefebvre (1980) and Rehault (1981), using geophysical methods, who suggested that, just prior to the evaporitic phase, the depth was about of one thousand metres, and no more, a value with which Rouchy (1982) agrees.

Different views have been proposed concerning the pattern of the deposition of the evaporites, for example: either total dessication, leading to residual environments of the type of sebka or playa (Hsü et al., 1973), or the persistence of a small amount of water, including some dynamic processes which took place in the shallow basins (Rouchy, 1976, 1980, 1981) (see also Chapter 2).

The origin and pattern of deposition of the evaporites indicates a marine origin, with, possibly, washing of the substrate and dilution of the brine by meteoric water. In fact, the evaporites contain the remnants of planktonic organisms: nannoplankton, diatoms, foraminifera, including benthic species, and both within the evaporites themselves and in the included layers of marls. The diatom populations closest to the evaporites show an increase in the proportion of brackish or even freshwater species, testifying to 'residual' environments, in which the physical and chemical parameters changed erratically, in accordance to the climate. In the three sub-basins of the Western Mediterranean (Fig 7.1) the geological facies display, from the lower part to the top, a sequence of marine marls, pre-evaporitive laminites, and evaporites. After the interruption of the active marine conditions, in increasingly confined basins, the microflora and microfauna reveal a diminution in the height of the water column, and, at the

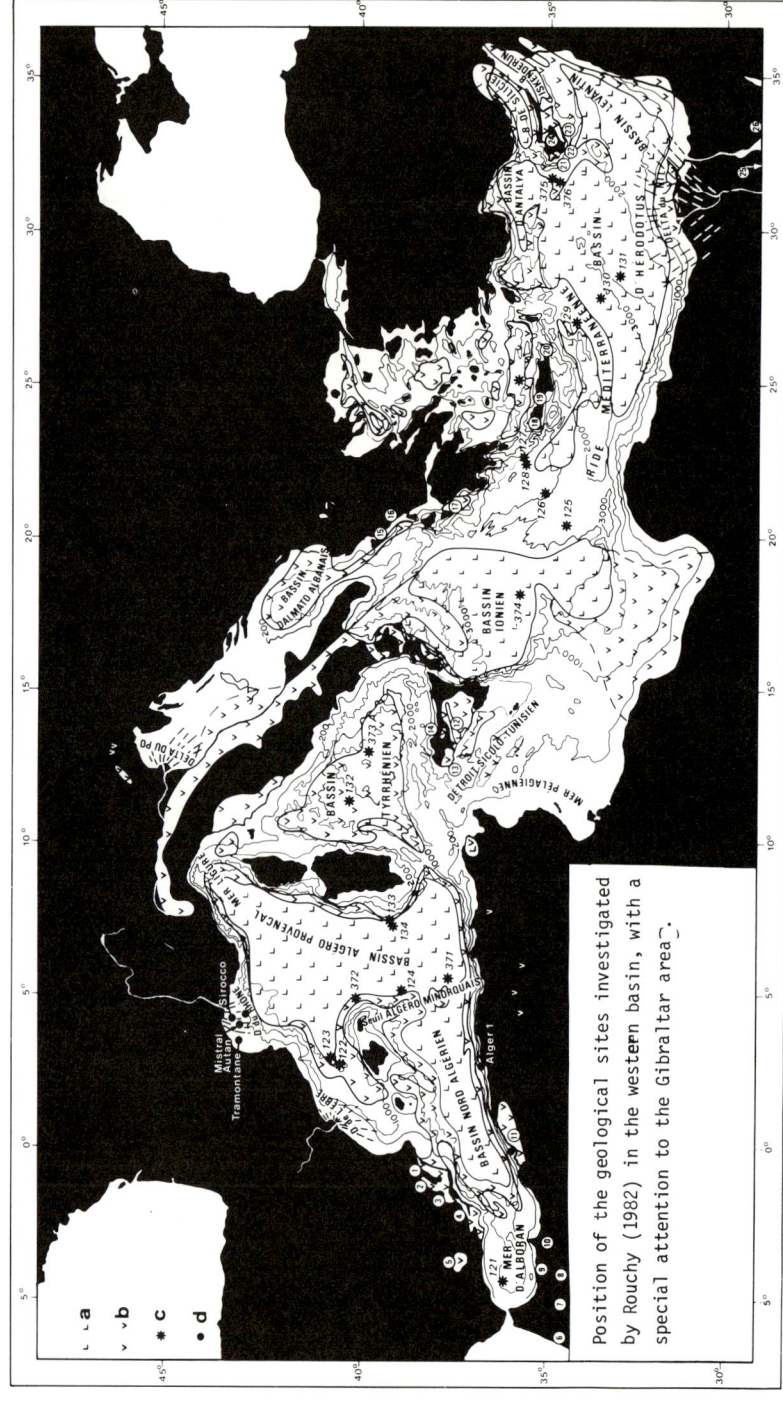

Fig. 7.1. Distribution of Messinian evaporites in the Mediterranean basins as summarized by Rouchy (1982). In spite of global coincidence between boundaries of the massive salt bodies and the present abyssal plain, the obliquity is a frequent situation. a, thick evaporite formations including a saliferous term. b, evaporite formations chiefly of gypsoanhydritic nature. c, legs of the Deep-Sea Drilling Project. d, cores extracted for surveying the potential oil-bearing areas. (After Rouchy (1982), simplified; reproduced by permission of the author).

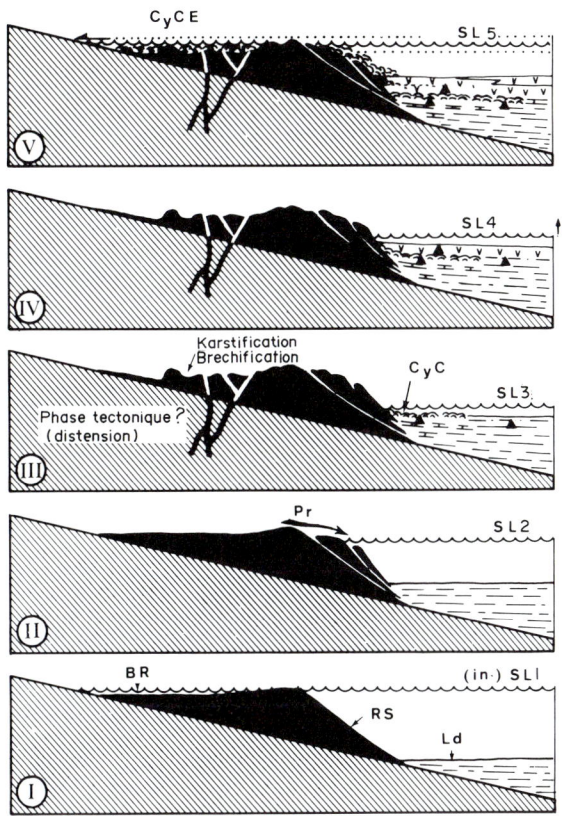

Fig. 7.2. Tentative reconstruction of the different evolutionary stages of a reef complex (coral and stromatolitic reefs) and its relations with pre-evaporitic laminates and evaporites (composite sections obtained from Morocco, Algeria, Spain and Sicily). Several fundamental processes are shown: downward prograding coral lenses; episodic drop of the sea level during concentration stages and subsequent karstification and erosion of the emerged part of the reef; precedence of coral build-up with respect to evaporitic phase; colonization of restricted sedimentary environments by cyanobacteria, after the definitive interruption of coral build-up at the end of the pre-evaporitic phase — Initial (in) 1, 2, 3, 4, 5; SL, sea-level; Ld: deposit of laminits; BR: reef-flat and/or back-reef; RS: reef slope - (II); Pr: progradation of coral patches - (III). Beginning of colonization by a cyanophyte 'carpet' — (IV) This 'carpet' extends more and more as (V) the sea level is rising : Cy C E — (After Rouchy (1982); reproduced by permission of the author).

same time, indications of a decreasing salinity appear. Therefore, the whole Messinian event may be analysed in three phases: a pre-evaporitic one, an evaporitic one, and a post-evaporitic phase leading to the beginning of the lower Pliocene. This does not imply a complete dessication of the Mediterranean area during the Messinian, but a persistence of local and residual aquatic environments; restriction of the sedimentary area involves repetitive emersion more or less developed in the marginal parts of basins (Rouchy, 1982).

According to Rouchy (1982) the building of coral reefs occurred mainly along the margins of the Western basin in small and shallow openings to the Atlantic. These reefs were from a very small number of species, sometimes only a single one, and were dominated by the genus *Porites* (some species of which still colonize the upper levels of many Indian and Pacific Ocean reefs, where they form microatolls). In fact, the impoverishment of the Scleractinian fauna began earlier, at the Tortonian, whereas at the Messinian relatively more diversified assemblages were still present in the eastern basin.

Fig. 7.2 (Rouchy, 1982) summarises several fundamental processes, namely: downward centrifugal prograding coral lenses (as colonies expanded towards better conditions for growth and food supply), episodic drop of the sea level, during the stages of concentration of water and subsequent karstification and erosion of the emerged part of the reef, coral building prior to the evaporitic phase and colonization by cyanobacteria of the restricted sedimentary environments, after the definitive interruption of coral growth at the end of the pre-evaporitic phase. These processes resulted in the superposition of two units: (1) a lower one composed of Scleractinians, and (2) an upper level of stromatolites.

The stromatolitic formations occur from sea level to a depth of about 3 m. The name Stromatolite may be applied to the formations of different origins, for example: the growth of encrusting red algae of the family Corallinaceae, or of mats of Cyanobacteria, which precipitate calcium carbonate and trap other mineral sediments; some other Cyanobacteria dissolve and penetrate the limestones and the skeletons and shells of invertebrates (Fairbridge, 1968). Nevertheless, geologists prefer to restrict the name stromatolites to the formations originated by filamentous cyanophytes.

Subsequent partial filling prevented further colonization of the Western Mediterranean by hermatypic Scleractinians after the Messinian; and the same is also true for the eastern basin. The two species of hermatypic corals at present in the Mediterranean cannot be considered as relicts from the Tethys (Zibrowius, 1974, 1983a,b). The nearshore shallow areas, from 0 to 2 or 5 m depth, were occupied only by cyanobacterial mats. During the whole period (which corresponds mainly with the long evaporitic phase) Rouchy (1982) assumed that fluctuations in the water level occurred, extending down to about 200 m in depth. This resulted in retraction in the sedimentation area, which tended to migrate towards the deepest part in the centre of the sub-basins of the Western Mediterranean.

Thus, the increasing concentration of salt, the depression of the level of water, the migration of the most concentrated brines, and the general horizontal displacement of the sedimentation areas, were all associated with the entrapment of salt.

The distribution of the evaporite masses emphasize the leading role of the morphology of the so called abyssal 'plain'. The depth of the Messinian Mediterranean, at the end of the Miocene, is rather poorly documented, but the most reliable estimations range from about 1000 m (Rouchy, 1982), 1200 to 1400 (Genesseaux and Lefebvre, 1980; Rehault, 1981) to 1500 m, suggested by Wright (1978) from the study of benthic microfauna. As a consequence of tectonic activity (first in the Eastern basin, and somewhat later in the Western basin) the Messinian-Pliocene subsidence attained 1500 – 2500 m beyond the shelf, and one of the results was the formation of submarine canyons (Montadert et al., 1978).

7.1.3. Pliocene and Pleistocene

At the dawn of the Pliocene the waters returned: a sudden transgression that was something of a catastrophic event. This stage corresponds to geological formations which are attributed to Astian and Plaisancian. The Astian is mainly related to coastal formations, and the later Plaisancian shows largely marine muds, possibly deposited beneath a higher water column.

It is still uncertain whether this event results from two different transgressions or a single, progressive one. During the Plaisancian transgression (about 5 to 3.1 million years B.P.) the climatic conditions remained relatively constant and the influence of warm waters predominated. The opening through the Gibraltar area allowed the Atlantic fauna from the Ibero-Moroccan Bay to enter the Mediterranean Sea and to re-populate it. Some belated Miocene species were re-introduced again.

In the Late Pliocene (2.0 – 1.9 million years B.P.) the sea level dropped slightly. Some post-alpine motions tended to reactivate the Betico-Riffan cordillera and the Calabro-Sicilian arc, and sliding occurred along the continental flexure.

To describe the history of the Mediterranean biota, it is convenient to consider the Pliocene and Pleistocene together, to understand the events which occurred during its long and eventful history, and to relate the changes in the biota and those of the surroundings.

It is possible to recognize several different marine assemblages from the Plaisancian transgressive period. Some of these are similar to those of the present time, for example: coralligenous limestones with bryozoans, tubicolous polychaetes and foraminiferans; muddy bottoms at the lower levels of the shelf (e.g. the biocenosis of the coastal terrigenous mud); assemblages of colloidal muds, possibly deposited in offshore areas and displaying a bathyal microfauna.

By the Late Plaisancian, the climatic conditions would be subtropical. However, some faunal elements, possibly from assemblages of the bathyal zone (deeper than 300 m) inhabiting North Atlantic temperate waters, could enter the Mediterranean Sea. During the regression following the Plaisancian, cooling occurred and the temperate species of North Atlantic origin increased. Some of these evolved *in situ* into 'vicariant' species of the original Atlantic stock and generated a part of the true endemic element of the Mediterranean fauna. From this time the Mediterranean Sea became a temperate sea and received from the Atlantic Ocean a large number of species, which constituted the predominant elements of its fauna (the 'atlanto-mediterranean' stock) and which were preserved almost to present time. The surface waters temperature tended to decrease slightly, as indicated by the upward movement of species such as the pectinid *Amussium cristatum* (Mars, 1963) which were restricted to deeper bottoms during the Pliocene.

7.1.4. The 'Quaternary'

At the dawn of the 'Quaternary' Period, the temperature of the surface water decreased by several degrees, as calculated from $0^{18}/0^{16}$ isotope measurements of pelagic foraminifera and on the basis of many biological indicators (Fairbridge, 1968). There was also an alternating displacement of the trade-winds deduced by Flohn (1953, after Fairbridge, 1968) from meteorological data.

During the maximum of a 'glacial period', the anticyclonic gyre of the North Atlantic hemisphere, was enhanced and the northern trade-winds were blowing from the north-northwest to about 10°N. The climatic conditions tended to be dry and desertic areas developed. The south-southwest trade-winds tended to turn to westerlies (like 'monsoon' of equatorial regions). In such periods the offshore waters tended to flow to the right and finally give the north-equatorial current. Some westerlies entered the Mediterranean Sea, and a coastal current tended to bring species from the Senegalian subtropical Province into the Mediterranean fauna.

During the maximum of a 'non-glacial' period, the climatic conditions became wetter and the south-west monsoon predominated on the African coasts (up to 23 – 25°N), a situation similar to that of the present summer season of the boreal hemisphere.

Moreover, during a period of glacial decline, the melting ice increased the amount of freshwater that the Mediterranean Sea received, whereas during an increase in the glacier ice-cap the water balance of the Mediterranean tended to be reversed, depending on the depth and width of the communication between the Atlantic Ocean and the Mediterranean.

From the hinge Miocene – Pliocene, it is generally agreed, as emphasized above, that the climatic conditions remained almost constant during the Plaisancian, despite some fluctuations of the global oceanic level, in accordance with the atmospheric circulation which will be considered later.

However, from this time (1.9 million years) the Antarctic ice-cap and the Arctic glaciers increased, thus initiating the beginning of the glacio-eustatic mechanism. An opposing theory was proposed by Ryan (1973) and Cita (1979) who postulated that the evaporitic Messinian crisis itself induced the glacio-eustatic process.

Whatever the origin of this process, it is certain that it dominated the whole of the Pleistocene and Quaternary periods. The glacio-eustatic oceanic lowering, together with the tectonic factors, contributed to the morphologic isolation of the Mediterranean basin up to present time, with the narrow Gibraltar Strait whose sill does exceed 298 m in depth.

Except for changes in the height of the water column resulting from the glacial-eustatic mechanism, it seems likely that the configuration of the depths in the western basin of the Mediterranean Sea was similar to that at the present time.

By the Calabrian period (1.9 to 0.9 million years B.P.) a third stock (which represents a more septentrional element) was added to the Atlanto-Mediterranean stock and the surviving species of the palaeomediterranean stock. This element corresponds to a surviving fauna at present in the north-eastern areas of the Atlantic Ocean (sometimes named the 'Celtic' fauna), the most characteristic species of which are: the bivalves *Cyprina islandica* and *Mya truncata*, the gastropod *Buccinum undatum*, and other bivalves, as *Panomya arctica*, *Modiolus modiolus* and *Chlamys islandica* (Pérès and Picard, 1964). Possibly the flatfish, *Platichthys flesus flesus* and *Raia clavata*, came to the northern shore of the Calabrian Mediterranean together with these molluscs (Quignard, 1978).

At the maximum of Günz glaciation, the Late Calabrian shows a small increase of the sea level (i.e. a small transgressive phase). At this time the 'Celtic' fauna declined. When *Cyprina islandica* and *Buccinum undatum* were still present, surviving species of the paleomediterranean stock are also found. Many of these are typical of temperate waters, notably: *Pecten jacobaeus*, *Clamys varia*, *Ch. flexuosa*, *Arca diluvii*, *Cardium tuberculatum*, *Nassa gibberula*, *Murex trunculus* var. *conglobata*, and others. Mars (1963) pointed out that the simultaneous presence of *Nucella lapidus* (a cold-temperate species) and *Turris indatiruga*, (a more meridional species) gives some indication of the hydrological and climatic conditions at the time. Both co-exist in the Gibraltar area, where the surface temperature is 21°C in summer and 15°C in winter. The 'celtic' fauna inhabited waters of salinity below $35.0 - 35.5\ ^o/_{oo}$ and certainly occupied the middle Atlantic Ocean at some periods during the Quaternary. The record of *C. islandica* and *Mya truncata* in a sample from the vicinity of Cadiz, shows how they could enter into the Mediterranean. A lowering of $1\ ^o/_{oo}$ in the present Atlantic Ocean salinity is sufficient to drive the $35.5\ ^o/_{oo}$ isohaline to the south of Gibraltar and along the Senegal shelf; however, because of the higher temperature *C. islandica* cannot develop here. *C. islandica* could not survive longer than Late Calabrian, because it spawns only when the temperature exceeds 13.5°C and is found at depths of more than 100 m. Thus, although it would have been able to survive the Mediterranean homothermal (12.5 − 13.0°C) the salinity would have been too high, except close to deltas of the septentrional shores of the Mediterranean Sea, influenced by the heavy rains in the basin itself or by deglaciation during the decline of Günz glacial period. The *Modiolus modiolus* population, recorded within the dilution area of the Loire estuary, must be a 'relict' of the former distribution of this species, which is present in the English Channel and more widespread in the North Sea and Norwegian Sea (Mars and Picard, 1958).

The second Quaternary glacial period (Mindel glaciation) initiated the Lower Sicilian which coincides with a fresh inflow of rather cold − temperate Atlantic waters. The return of the Calabrian fauna brought back the cold-loving species (e.g. *Trichotropis borealis* and *Chlamys tigrina*). At this time, the stocks of palaeomediterranean species and recent northern immigrant species were equally well represented (both groups constituting about 4% of the Sicilian fauna). This implies that the temperature — at least of the surface waters — was much colder than at present. By Late Sicilian, the species of septentrional origin became scarce and disappeared mainly from littoral bottoms, except in areas which were exposed to dilution by fresh waters from the melting alpine glaciers. This resulted in a temperate fauna at about 20 − 30 m above the present level. However, this rise was probably not of eustatic origin but was generated by the final after-effects of alpine foldings.

Much more important, was the next transgressive period, corresponding to the maximum of the Riss glaciation (i.e. Tyrrhenian, or Eutyrrhenian, according to Mars, 1963).

This true Tyrrhenian or Eutyrrhenian did not seem to have risen more than 12 metres above the present sea level. In the shallow waters a new element, termed 'Senegalian', is added to the earlier ones (i.e. Northern Atlantic and endemic elements arising from the palaeomediterranean fauna). This element included warm temperate species, but not a truly tropical fauna, and was not very different from that existing today on the shores of Senegal. Among the most conspicuous, but now extinct mollusc species, were: *Strombus bubonius, Natica turtoni, Cantharus viverratus, Cymatium trigonum, Conus testudinarius, Acteocina knockeri, Arca plicata, Brachyodontes senegalensis, Mactra largillierti* and *Tugonia anatina*. Several of the Senegalian species are, however, still present in the Mediterranean Sea, particularly in the warmest parts. These include: *Fissurella nubecula, Pirenella conica, Cypraea lurida, Purpura haemastoma, Mitra fusca, Clavatula nifat*, some ascidians, such as *Eudistoma planum, E. paesslerioides, Polysyncraton amethysteum, Ecteinascidia turbinata*, and others. These are known only from the warm temperate Atlantic and eastern Tunisian coast and are probably also Tyrrhenian relicts. Similarly, the present discontinuous distribution of many fish species can, according to Postel (1956), be explained. The fishes *Dasyatis marmorata* and *Taeniura grabata* are, for example, totally absent in Algeria but present in the Gulf of Gabès. Several other fishes are common both in this area and on the sub-tropical shores of western Africa, yet are very rare on the Algerian coasts; these are: *Sphyrna zygaena, Rhinobatus cemiculus, Pteroplatia altavela, Morone punctata, Orthopristis bennetti, Dentex filosus, Pagrus ehrenbergi*, and *Stromateus fiatola*. According to Quignard (1978) the benthic or nektobenthic fishes of the genera *Epinephelus, Serranus* and *Crenilabrus* represent a part of the coastal fish fauna in the Mediterranean possibly introduced during the interglacial phases. Holthuis and Gottlieb (1958) mention four species of decapods with comparable distributions: *Athanas amazone* and *Salmoneus jarli* (Nigeria — Israel); *Micropanope rufopunctata* (from Ghana up to the Azores in the Atlantic; Alexandria, Israel), and *Maja goltziana* (from Congo up to Portugal in the Atlantic; Israel). The case of *Pachygrapsus transversus* is different, since this crab is known from both the Pacific and Atlantic coasts of America, and also from the coasts of Africa, Madeira, the Canaries and Angola. In the Mediterranean it was recorded from Port Said, Alexandria, the coast of Israel, the southern coast of Turkey, in the Dodecaneses (Karpathos) and a single individual has been found at Marseilles.

According to Mars, the Eutyrrhenian deep sea bottom fauna is not well described, but there are signs of a more recent shoreline (Neotyrrhenian) at about 2 metres above the present sea level, where the Senegalian element becomes impoverished (Pérès and Picard, 1964; Pérès, 1967).

The displacement of the jet streams towards the equator resulted in a catastrophic drop in precipitation not only in Africa, but also in the southern hemisphere. However, 'the drop in rainfall was not worldwide, because an increase of wind velocities (e.g. of the main westerlies) in spite of lowered temperatures, would also raise evaporation and, therefore precipitation, in areas of marine climate thus ''feeding'' the growing ice caps. On the other hand, the expansion of the mid-latitude continental ice-sheets would set up cold high-pressure cells, which would lead to anticyclonic wind resulting in cold dry deserts' (Fairbridge, 1968).

The successive inflow of invaders, including sometimes boreal, sometimes southern species, were probably related to the current system in the Strait of Gibraltar, itself controlled by the input of meteoric and continental waters that this basin received.

The mechanism of change, proposed by Mars (1963), may be summarized as follows. In the conditions of a characteristic interglacial period, the Mediterranean region had a warm and arid climate and a deficient water balance. There was, therefore, an important penetration of Atlantic water into the Mediterranean through the Strait of Gibraltar, in which the depth of this current from west to east was clearly augmented. The Atlantic was itself in a period in which tropical and warm temperature waters extended considerably further towards the north (thus modifying the biogeographical provinces). All the conditions were therefore present which would allow both the introduction and maintenance of tropical littoral faunas in the Mediterranean. This sea with its relatively high salinity and temperature was a very favourable environment for such species. The temperature of the homothermal deep layer was at the same time higher than in present times, because the temperature of the surface waters during the coldest months of the year (which determine the temperature of the deep water) was also higher. Such conditions were, therefore, inimical to the survival of previous boreal forms.

During a glacial period, eustatic regression would reduce the water level everywhere, including at the Strait of Gibraltar. Further, the displacement of the arid zone towards the south implies that the humid temperate and cool zone would come to occupy the Mediterranean regions. Under these conditions precipitation and run-off would increase, so that the water balance would be reversed, and a surface current of Mediterranean water would pass outwards towards the Atlantic, while a bottom countercurrent of Atlantic water would be directed inwards, towards the Mediterranean. At the same time, the shelf and slope would support large benthic populations, the biomass of which would be quite comparable with those of the present benthic biocoenoses of the boreal Atlantic. The flourishing Quaternary population of deep water ahermatypic corals of the bathyal zone (Blanc, Pérès and Picard, 1959) or the very rich populations of pelecypods, present in the northern fauna of the canyons of the eastern coast of the Pyrenees are impressive examples.

The generally accepted hypothesis of the inversion of currents, also explains the successive waves of littoral immigrant faunas during interglacial periods and of deep water fauna (circalittoral or bathyal) during glacial periods. As Mars (1963) emphasized, these variations cannot have resulted from a succession of distinct glacial and interglacial maxima, but can be interpreted as successive advances and retreats of smaller amplitude. The maximum effects would, of course, more or less, coincide with the maxima of the glacial or interglacial periods. However, the secondary oscillations would, nevertheless, have had marked influences. There were also retarding processes, the most evident of which is associated with the melting of the glacial mass during a period of deglaciation and which would have been prolonged in the Mediterranean during a part of the subsequent glacial eustatic transgression, when the conditions are favourable to the survival of northern forms (Pérès, 1967).

7.2. THE MEDITERRANEAN BENTHOS AND ITS ACTUAL AND RECENT CHANGES

The scope of this book is restricted to the Western basin of the Mediterranean Sea. We can thus disregard the 'Lessepsian' migration which began in the forties and augmented the number of Red Sea species invading the Mediterranean Sea, both planktonic and benthic. However, the latter tend to expand their geographical areas less easily than the planktonic ones.

In contrast, the obvious contemporary penetration of Atlantic species should be compared with the present populations and with that of the most recent fossils, taking into account also the extinct faunas from the Miocene, Pliocene and Low Pleistocene discussed above (p. 201).

During recent times a certain number of species have entered the Mediterranean with the Atlantic current through the Strait of Gibraltar. The gradual modification of hydrological factors (slowly along the coast of North Africa and more rapidly along the coasts of Spain) means that along North Africa these species are more extended; for example: *Laminaria ochroleuca* (from Britanny to the Canaries) which goes as far as Algiers and the Sea of Alboran; *Patella safiana* (from Madeira and the Canaries) as far as the coasts of Oran; *Siphonaria pectinata* (from Senegal) as far as Cap Matifou (Algeria) and to Motril (Spain); *Lithophaga aristata* (from Senegal to Biarritz) as far as Malta and the Balearic Islands; *Mytilus perna* (the coasts of western Africa) as far as Philippeville (Algeria) and Motril (Spain); *Eastonia rugosa* (Portugal, Canary I., Morocco) as far as Oran and Malaga; *Mesalia brevialis* (Senegal to Algiers); the small sea-urchin *Arbaciella elegans* (Angola) as far as Majorca and the Sicilian – Tunisian sill (Pérès and Picard, 1964; Regis, 1983) (this will be considered later as a 'senegalian' species). Some benthic fishes certainly entered with Atlantic waters and extend from the Strait of Gibraltar along the coasts of northern Africa. There is for example, *Diagramma mediterraneum*, which is known from Senegal and Morocco and is also present on the coasts of Algeria, but absent along those of Tunisia (Postel, 1956), and *Scorpaena loppei*, known from Senegal to the north of Gibraltar and present also in the region of the Balearic Islands.

To summarize, the number of present-day immigrants from the Atlantic appear to be rather restricted and localized in the Alboran Sea, but somewhat more numerous on the coasts of north-west Africa which are more directly bathed by the Atlantic current (a stream passing some kilometres from the coast and not a vast sheet which directly touches it) (Pérès, 1967).

Some recent Mediterranean introductions certainly result from the rapidly increasing maritime traffic. This has occurred since the early '50s and is mainly concerned with oils and ores for industry, especially within the western basin, where the largest harbour and industrial complexes are developed.

Certainly, the most striking example is the hermatypic scleractinian *Oculina patagonica*. This was recorded for the first time in 1966 in the infralittoral rocky assemblage near the Savona harbour (Gulf of Genoa) and was found, prosperous and spreading, when surveyed again in 1971 and 1972 (Zibrowius, 1974). 'The water temperature in the area was found to vary from about 11°C to about 26°C but for the greater part of the year, remains far below those generally considered necessary for growth and even survival of hermatypic scleractinians. In spite of the local pollution — both domestic and industrial — which favours dense and prosperous populations of barnacles and ascidians, the encrusting colonies of *O. patagonica* (now measuring 1.2 m × 1.5 m) tend to cover other organisms. The Savona record of *O. patagonica* is the first one of a living colony, since nothing was known on the actual range of the species, which probably arose on the Atlantic coast of South America. Previously, *O. patagonica* was known only from worn fragments, fossil or subfossil, from Argentina. Liberation of larvae from a mature colony that came to Savona waters on the hull of a ship is the only acceptable explanation for the unexpected arrival of this exotic Scleractinian in the north-western Mediterranean. Transplantation of samples of *O. patagonica* broken from the Savona colony to clean waters of Marseilles has been very successful (Zibrowius, 1974). Recently, Zibrowius and Ramos (1983) recorded it on the south-east of Spain, where it is particularly abundant in extreme environments, which are unusual for a scleractinian (e.g. harbour pollution, waters charged with wastes from mining, on exposed rocks among sand at shallow depth, etc.). In the Alicante harbour, this coral outgrows mussels, oysters, barnacles and ascidians. In the dark, colonies survived for 29 months, losing their zooxanthellae; they recovered their symbionts when placed back into full light close to normal colonies, and then seem to be very well-adapted to its new habitat on Mediterranean temperate shores.

In fact, *O. patagonica* is not the sole hermatypic scleractinian in the present Mediterranean Sea. *Cladocora caespitosa*, which also harbours zooxanthellae is very widespread in the infralittoral, mainly on the Tunisian eastern coast and in Greece (Duclaux and Lafargue, 1973). Zibrowius (1974) also mentions subfossil domes of *Cladocora* near Nabeul (Tunisia), which would be possibly of Tyrrhenian age.

Among polychaetes, the serpulids, *Spirorbis marioni* and *Pileolaria berkeleyana*, are common on shallow rocky bottoms, at least from the Costa Brava (Spain) via the French coast and northern Corsica up to the Ligurian coast; both originated from the eastern Pacific. Both are very abundant in the outlet of the reception basin of a fuel-power plant, cooled by coastal waters, on the northern shore of Gulf of Marseilles. During a very hot summer the operation of the plant was completely stopped by obstruction of the whole circulation system by myriads of hydroids of a species of *Tubularia* from the Persian Gulf.

7.3. THE PROBLEM OF THE MEDITERRANEAN ENDEMICS

During the Messinian salinity crisis almost all the true marine benthic species were extinguished (except possibly within small areas, with some river input). At this time, the Mediterranean Sea was probably completely isolated from both the Atlantic and Indo-Pacific Oceans.

However, Pérès and Picard (1964) and Pérès (1967) mention a paleomediterranean stock, which could have participated in a fresh re-population of this sea from the Pliocene, the history of which has already been outlined (p. 202).

It seems, in fact, that the paleomediterranean benthic fauna came from the Ibero-Moroccan gulf.

It is impossible to say whether some species which exist throughout the whole southern Mediterranean are Senegalian relicts, which have existed with a continuous distribution since the Eutyrrhenian or whether they are recent immigrants that have travelled faster and further than those mentioned above. The first assumption is probably the most acceptable and would also account for the distribution of the two decapods, recorded by Holthuis and Gottlieb (1958): *Albunea carabus*, which occurs from Nigeria to Israel, is also present in Algeria, Minorca and Sicily while *Ocypode cursor*, known from Angola to Turkey and Greece, is passing along the western and northern coasts of Africa.

With the exception of relatively recent immigrations via the Suez Canal (which are out of place in a book devoted to the Western Mediterranean) this whole sea contains only very few species — if any — whose distributions encompass the Mediterranean and the Indo-Pacific. Some of these (such as the opisthobranch *Caloplacamus ramosus*, the ascidian *Amaroucium profundum*, and the polychaete *Pallasia porrecta*) probably exhibit a circumtropical or cosmopolitan distribution which is, as yet, incompletely known.

The species which are at present strictly localized in the eastern basin (mainly in areas of high temperature and salinity) probably disappeared from the western basin after the maximum of the Tyrrhenian transgression.

Among existing species in the extreme east of the eastern basin (on the coasts of Syria, Lebanon, Israel and east coast of Tunisia) it is very difficult to decide whether they came into the Mediterranean from the Red Sea before the massive input since the 1940s, or if they entered it during the Tyrrhenian and survived only in the warmest and more saline areas. This is the case of the alga *Caulerpa scalpelliformis*, the hydroid *Sertularia marginata* and the ascidian *Symplegma viride*, which exist in the Red Sea, but also occur along the subtropical western Africa coast, and probably therefore must be considered as circumtropical species.

'Relicts' from the Tethys Sea, are represented only by examples of 'vicariant' species such as *Lysmata seticaudata* (Mediterranean) and *L. ternatensis* (Indo-Pacific), *Dromia vulgaris* (Mediterranean) and *D. dromia* (Indo-Pacific), *Octopus macropus* (Mediterranean) and *O. variabilis* (Japan), *Posidonia oceanica* (Mediterranean) and *P. australis* (South Australia). A whole series of animal genera from the Atlanto-Mediterranean Region are also found in the western Indo-Pacific. These include *Pteroides*, *Maia*, *Eurynome*, *Amphipalaemon*, *Lysiosquilla*, *Echinothrix*, *Pseudocucumis*, the whole family of the Sepiolidae and several genera of fishes. Some benthic and nektobenthic fishes including some species of the genera *Pagrus*, *Sparus*, *Trachinus*, *Cepola*, may also be considered as Tethys 'relicts' (Quignard, 1978).

This endemism may be partially influenced by reproduction in a restricted space which leads to genetic isolation.

If the locally distinct populations of a given species mix during their reproductive period, the 'species' conserves its unity, whereas, if the different populations prosper independently of one another, then relatively rapid morphological differences may be established between them.

I have, so far, listed extra-Mediterranean species from only four animal groups: hydroids, decapoda Reptantia, echinoderms and ascidians (Pérès, 1967). These lists are now out-of-date and should be reviewed and extended, according to the fresh taxonomic and world distribution data. However, it seems that Mediterranean ascidians, which are all sessile, have the highest percentage of endemics (possibly about 40%), because the planktonic free stage is very short. In hydroids and echinoderms the percentage is about 26 – 27%. Hydroids are also sessile but have a planktonic stage which is usually longer than that of ascidians. Echinoderms often have a relatively long planktonic stage (e.g. 1 – 3 weeks), but are able also to walk or crawl slowly, the result being a comparable range of Mediterranean endemics. Finally, crawling decapods (13.2%) have the lowest percentage of endemics, since the larval planktonic stage is usually rather long and, in addition, some crabs (e.g. Portunids) and Anomura can combine walking and swimming.

Ekman (1953) has suggested in a very general way, that the endemic Mediterranean species may have three possible origins, namely: (i) appearance during the known isolation of the Mediterranean at the end of the Miocene and subsequent evolution by speciation from other Miocene relicts or from immigration during the Pliocene, Pleistocene and Quaternary times; (ii) they have been maintained, only in the Mediterranean, from an early epoch and have disappeared in all other regions of the World Ocean because of hydrological changes of climatic origin or (iii) they may have persisted since the fragmentation of the Mesogean Sea or Tethys. Ekman has correctly emphasized that the true species of the Tethys must not be confused with those which have simply descended from species of the Mesogean fauna. To affirm that a species is a Tethys relict it is necessary to be sure that it existed in the Mediterranean, without interruptions since the dislocation of that Sea. Also that it is not a species which temporarily disappeared from the Mediterranean during the period of marked variations in abiotic factors, to subsequently re-enter, for example, from the Senegalian Province. As shown above it is, in fact, generally quite impossible to be sure in such cases.

It thus seems certain that there are very few species (possibly not even one) which are true Mesogean relicts. It is, as we have seen above, only the *genera* which are relicts and these have spawned endemic species (Pérès, 1967).

After the maximum of the last glaciation (Würm, 40,000 years B.P.) the sea level probably fell by -120 m (as compared with present level). During the subsequent deglaciation phase, the surface water outflow largely exceeded the potential inflow from Atlantic waters and thus, probably preventing entry of species into the Mediterranean. The last transgression took place from about 32,000 to 8,500 years B.P. and was not followed by any marked change of water level, except the small Flandrian transgression (from $+1$ to $+3$ m).

From this the composition of the Mediterranean benthos could not change, except for the minor additions which have been mentioned.

7.4. THE CIRCALITTORAL SOFT BOTTOMS

The whole of a shelf can be arbitrarily divided into hard and soft bottoms. In some places, especially on young coasts, it constitutes a syngenetic series from the shore line down to the greatest depth. This is because the categories of bottom and their assemblages may intermix within a given shelf (and upper slope) according to the geomorphological peculiarities of the shoreline.

However, the biotopes of the lower shelf in the Western Mediterranean usually consist of rather fine terrigenous sediments, and organogenous remnants. The latter, which can mix with the terrigenous sediments, arise from benthic organisms and are frequently coarse. The distribution of the different assemblages depends mainly on the granulometric composition of sediment. This, in turn, is a function of the sedimentation rate and water movement and also governs the food supply of the biota.

It is a well-documented fact that the small particles of the terrigenous elements tend to be carried further from the shoreline, to an extent which is inversely related to their size. The average particle size on shelf soft bottoms should thus decrease offshore. However, this theoretical prediction is not confirmed. For example, in the vicinity of an estuary, where a large amount of fine particles is brought into the sea, considerable precipitation and floculation results from the mixing of fresh and sea-waters. Furthermore, at any given point on the shelf, many changes of depth have occurred from the Pliocene to present time, according to the particular history of transgression — regression cycles (Pérès, 1967). Thus, a bottom now at a depth of 100 m, and receiving fine sediments, may have been submerged under only

20 m of water during one of the Quaternary regressions when it received coarse terrigenous sediments, or, perhaps, even supported organogenous formation. With the most recent transgressions these coarse detritic bottoms were exposed to varying conditions. For example, a shelf ceases to receive fine sediments from rivers. Then, from the edge of the shelf up to the present shallow waters there will be, an uninterrupted range (both in time and space) of this kind of bottom. The narrow shelf of the French coast to the east of Marseilles shows features usually associated with the 'detritic bottoms'. In contrast the present deposition of fine sediments (silt, clay and mud) in an estuary, may cover all the older detritic sequences, thus forming a more or less uniform strip of very soft bottoms, down to the commencement of the slope, which, may also be covered by fine deposits. When the input of terrigenous material is small, or the fine particles are carried away by strong currents, there may be little accumulation at the shoreline (Pérès, 1967). In such cases, a strip of muddy bottom (shelf terrigenous mud) is formed, more or less parallel to the coast, between two strips of detritic bottoms. Closer to the coast, the 'coastal detritic' contains organic remains, almost always calcareous, which are mainly derived from present or recent benthic biota. Some of these remnants may be produced by the local assemblage, while others are formed by bottom currents from neighbouring ones. Their nature depends on the existing, or recent, benthos assemblages: broken shells, coral fragments, calcareous algae, pieces of bryozoan zoaria, etc. On the outer side of the terrigenous shelf mud, there is another strip of detritic bottom called 'self-edge detritic'. This deeper detritic bottom is correlated with the stronger currents in the vicinity of the shelf edge; calcareous remnants of local benthic organisms arising from presently living species often mix with those from thanatocoenoses. Finally, the coastal detritic may be mixed with some mud. In such cases, first recognised by Picard (1965) on the Mediterranean shelf of France, a fourth type of circalittoral soft bottom, the 'muddy detritic' assemblage, may be observed (Pérès, 1982).

7.4.1. Coastal detritic assemblage

On the French Mediterranean-shelf the coastal detritic assemblage mainly occupies the upper part of the circalittoral zone, just below the last *Posidonia* beds or well-sorted infralittoral sands. In areas with small deposition of fine-sediment, it may be as deep as 90 – 95 m at the edge of the shelf-edge detritic

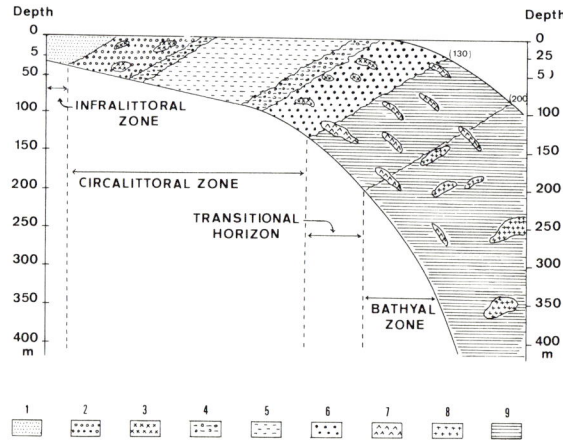

Fig. 7.3. Generalized diagram of the main benthic assemblages of the lower shelf and upper slope of the north-western Mediterranean. 1. Fine well-sorted sand assemblage or *Posidonia* meadow assemblage. 2. Coastal detritic assemblage with (3) concretioned masses displaying enclaves of the corralligenous assemblage. 4 and 6, Muddy-detritic assemblage. 5. Terrigenous mud-shelf assemblage. 6. Shelf-edge detritic assemblage. 7. Off-shore rocky assemblage. 8. Isolated patches of deep-water ('white') corals. 9. Bathyal mud. The upper slope is a transitional area (approximately 135 – 200 m) depending on both the geomorphological features and the input of mud; here the off-shore rocky assemblage is, sometimes, well represented on rocky steps or in emerging knobs of the hard substrate. (After Pérès, 1967, slightly modified).

assemblage (see above). When more fine sediment is deposited, the coastal detritic assemblage disappears altogether in shallower waters and is replaced by one of the 'muddy detritic' or the 'shelf terrigenous mud' group (see also Chapter 8).

The sediment consists of organogenous gravel from existing benthos organisms with an admixture of sand. The percentage of silt is low and muddy particles are generally absent. Many algae and animals of the assemblage show brighter colours (mostly red, yellow, orange) more so than any other assemblage on soft substrates in the circalittoral zone. This striking peculiarity indicates a link with the coralligenous assemblage; in fact, as mentioned above, several species occur in both assemblages, depending on the predominance of either concretioning or destructive processes; some alternation in time may also occur between these two assemblages (Pérès, 1982).

The coastal detritic assemblages have a characteristic species composition (Picard, 1965). Among the algae are the soft rhodophyte *Cryptonemia tunaeformis* and the calcareous species *Lithothamnium calcareum, L. coralloides, L. fruticulosum*; the sponges *Suberites domuncula, Basiectycon pilosus*, Bubaris vermiculata*; the pelecypods *Modiolus phaseolinus, Pecten jacobaeus*, Lima loscombei*, L. elliptica, Laevicardium oblongum, Tellina donacina*, Psammobia faroense**, etc; the gastropods *Turritella triplicata*, Eulima polita, Drillus maravignae*; the decapods *Paguristes oculatus, Anapagurus laevis, Ebalia tuberosa, E. edwardsi*, etc.; the echinoderms *Astropecten irregularis, Aniseropoda placenta, Ophioconis forbesi, Ophiura grubei, Genocidaris maculata, Psammechinus microtuberculatus, Stereoderma kirchbergi*; the ascidians *Molgula oculata, Ctenicella appendiculata, Polycarpa pomaria, P. gracilis*. Apart from these highly characteristic species, some others may occur where presence depends on a certain characteristic of the sediment, e.g. gravel-loving species such as the echinoids *Echinocyamus pusillus* and *Spatangus flavescens*, the decapod *Lambrus massena*, the pelecypods *Astarte fusca* and *Venus fasciata* or — where strong currents often occur — the sand-loving tectibranch gastropod *Philine aperta*. Small scattered concretioned patches may be occupied by the coralligenous assemblage (Picard, 1965).

In the Mediterranean Sea the coastal detritic assemblage exhibits several different facies. Among the most interesting and widespread are those in which calcareous red algae predominate (namely two species of branched Lithothamniae: *Lithothamnium calcareum*, rather thick and purple – pink, and *Lithothamnium corallioides* with a thinner light pink thallus). The very young algae settle on a sediment particle and become rapidly enclosed by the growing thallus. The result is an alga which lives 'free' on the bottom. These 'nullipores' form a constituent of the sediment found on coarse sand with gravel and broken-shell bottoms with strong currents. The weight of the algal fraction may be higher than that of the detritic component of the sediment. Due to their ramified thallus, these Lithothamniae cannot form concretioned structures. However, other rhodophytes (*Jania, Gelidium*) sometimes form a felt-like tissue linking together several Lithothamniae individuals. Fine particles, such as decaying remnants of *Posidonia* leaves, may sediment and slightly increase the organic matter content of such bottoms. In the western basin of the Mediterranean Sea the relatively poor fauna includes mainly the animals marked with an asterisk in the list of characteristic species given above, together with the gravel-loving species *Lambrus massena, Echinocyamus pusillus* and a current-loving species such as *Venus casina, Spatangus purpureus*. This 'nullipore' facies generally occurs at depths between 25 and 40 m sometimes between 60 and 65 m in very clear waters (Pérès, 1982).

The 'burnt almond' (in French 'praline') facies, were first described by Gautier and Picard (1957) on the shelf stretching eastwards from the Hyères Islands, near Toulon; it also occurs in places where either the shelf or isolated banks are washed by offshore waters (Centuri Bank on the north-western coast of Corsica, Hecate Bank between Sicily and Tunisia, and so on). The sediment (fine gravel, and broken shells) is sprinkled with irregular nodules, a few centimetres in diameter, formed by the superposition of layers of a calcareous seaweed belonging to the Melobesiae family; the central part of each nodule generally shows a small gravel or organogenous remnant. Because the alga is living over the whole of the surface of the nodule, it would be expected that these irregular spheres are frequently rolled on the bottom by currents or by motile animals. In this facies the characteristic components of the coastal

Fig. 7.4. The large sea-urchin *Spatangus purpureus* is a bottom, current loving species, which feeds upon micro- and meiofauna and small-sized invertebrates; it is rather common in the coastal detritic assemblage and sometimes also exists in some facies of the shelf-edge assemblage.

detritic assemblage are mixed with other less characteristic species. These include *Venus casina*, a pelecypod favouring the bottoms where there are strong currents, and the large kelp, *Laminaria rodriguezii*, which is epiphytic on the nodules and with its thallus lying on the bottom, in the direction of the current. On both the calcareous seaweed which form the nodules and on the kelp, there are several epibiotic species, such as hydroids and bryozoans.

Another common facies in the Mediterranean Sea is characterized, not by Corallinaceae, but by calcareous rhodophytes of the family Squamariaceae. In these areas the typical coarse sediment of the coastal detritic assemblage is covered by a layer of a very fluid mud which buoys up the algae. *Peyssonnelia rosa-marina* (= *polymorpha*) largely predominates, but some *P. bornetii* (= *harveyana*) are also present. Both algae have a dark purple thallus, consisting of a thin, strongly twisted plate with a more or less spheroidal shape. The thallus consists of a single layer, and 'float' on the fluid mud. They are not immobile and may frequently make see-sawing movements which result from differences in growth rate (caused by displacement of the centre of gravity resulting from different illumination of different parts of the thallus) and from the movements of the animals, mainly *Ophiopsila aranea*, which live inside the anfractuous thallus during the day but move out at night. The assemblage is rather similar to the typical coastal detritic assemblage described above. However, the holes and crevices of the thallus contain varying amounts of muddy sand and are inhabited (in addition to *O. aranea*) by two small pelecypods: *Kellya suborbicularis* and *Mysella bidentata*. On the thallus itself, the bryozoans *Mollia patellaria* and *Chorizopora brongniarti* are common, but not genuinely characteristics, as they occur on every circalittoral rhodophyte with a sufficiently stiff thallus. The current-loving *Venus casina* and *Spatangus purpureus* are also present. The areas supporting the Squamariaceae facies may change in both space and time and its characteristic associated species may sometimes disappear completely, leaving a purely coastal detritic assemblage. This particular facies only develops at the mouth of open bays, (i.e. in areas with alternating periods of storm-swept whirling currents and of sedimentation during smooth sea conditions) (Carpine, 1958; Pérès, 1982).

7.4.2. Muddy-detritic assemblages

According to Picard (1965) it is easy to recognise an assemblage on bottom areas which receive a characteristic sediment of the coastal detritic containing a significant admixture of finer material. The sediment may be a muddy sand, sandy mud or sometimes a relatively firm mud, but gravels, clinkers, broken shells are always present, and the sedimentation rate is sufficiently low to allow the settlement of sessile species. However, the muddy fraction always predominates (Pérès, 1982). In contrast with coralligenous and coastal detritic species, those of the muddy detritic assemblage are not colourful.

Picard (1965) lists 12 characteristic species of the muddy detritic assemblages the sponge, *Raspailia viminalis*; the anthozoans, *Alcyonium palmatum* and *Anemonactis mazeli*; the polychaetes, *Aphrodite aculeata*, *Polyodontes maxillosus*, *Euphanthalis kinbergi*, *Leiocapitella dollfusi* and *Clymene palermitana*; the sipunculid, *Golfingia elongata*; the pelecypod, *Tellina serrata*; the isopod, *Cirolana neglecta*; the holothurioid, *Pseudothyone raphanus*. Mud-loving species such as *Nepthys incisa*, *Pectinaria auricoma* and *Amphiura chiajei*, are also present.

Two species are common in this assemblage. (i) The ophiuroid, with its spiny arms, *Ophiothrix quinquemaculata*, may be very abundant. It clings to the substrate or, more often, to broken shells and dead organogenous concretions with two or three of its arms, the others being lifted up for collecting particles suspended in the bottom water layer 10 – 15 cm above the sediment. (ii) In the *Alcyonium* facies, the rate of fine particles sedimentation is particularly low. This allows a significant development of sessile species, in particular *Alcyonium palmatum*, but also some hydroids and bryozoans and the big compound ascidian *Diazona violacea* and several large solitary ascidians (*Ascidia mentula*, *Phallusia mammillata*, *Microcosmus* spp. and *Polycarpa pomaria*).

7.4.3. Terrigenous mud-shelf assemblage

The marine terrigenous sediment is a relatively mobile (fluid) mud composed of silt and clay. Admixture of sand is rare and always slight. Due to a high sedimentation rate and very soft sediment consistency, hard bodies tend to be buried quickly.

For the Mediterranean Sea the list of the most characteristic species, reviewed by Picard (1965), include the pennatularian *Virgularia mirabilis*; many polychaetes (e.g. *Lepidasthenia maculata*, *Phyllodoce lineata*, *Nereis longissima*, *Nephthys hystricis*, *Goniada maculata*, *Sternaspis scutata* and *Pectinaria belgica*); decapod crustaceans (e.g. *Callianassa truncata* and *Goneplax rhomboides*); the pelecypods *Thyasira croulinensis*, *Mysella bidentata*, *Abra nitida* and *Thracia convexa*; the holothuroid *Oerstergroenia digitata*; the fishes *Caecula imberbis* and *Gobius lesueurei*. Mud-loving species also occur in other assemblages, provided that the sediment contains sufficient mud: the polychaetes *Lumbriconereis fragilis* and *Terebellides stroemi*, together with *Alpheus glaber* and *Amphiura chiajei* are commonly found.

It is possible to distinguish four kinds of such terrigenous mud in the shelf.

(i) On non-viscous muds which have the highest sedimentation rate and are submitted to direct input from rivers, all hard substrates are rapidly buried and sessile species cannot survive. On this kind of bottom the gastropod *Turritella tricarinata* f. *communis* is exceptionally abundant, comprising up to about 95% of all the present individuals.

(ii) In the immediate proximity of an estuary — or delta — the gastropod becomes more scarce and the holothurian *Oerstergroenia digitata* more abundant, especially if the mud is highly reduced.

(iii) On the viscous mud. This is usually grey or greyish in colour, with a sedimentaetion rate that is somewhat lower than in areas (i) and (ii). Characteristically present are the cnidarians *Virgularia mirabilis* and *Pennatula phosphorea*, sometimes in association with *Veretillum cynomorium*.

Fig. 7.5. The sea pen *Pteroides griseum* is relatively common in shelf muddy detritic bottoms; here among broken and dead leaves from the neighbouring *Posidonia* meadows are visible. Depth: 50 m.

Fig. 7.6. This suspension-feeding brittle-star (*Ophiothrix quinquemaculata*) is very common on the muddy detritic assemblage, upon pieces of dead coralligenous concretioned rock or upon agglomerations of large-sized ascidians (e.g. *Microcosmus* sp., *Polycarpa pomaria*, *Styela partita*). Such agglomerations exist also on shelf terrigenous mud assemblages, but without the characteristic brittle-star. Depth: 60 m.

(iv) The last facies corresponds to areas where the sedimentation rate is lower, thus allowing sessile species to settle upon the hard substrates which are scattered on the mud surface. It is predominantly characterized by *Alcyonium palmatum*, the bivalve *Pteria hirundo*, ascidians, e.g. the large compound species *Diazona violacea*, and the large holothurioid *Stichopus regalis*, which requires a rather firm bottom on which to crawl.

7.4.4. Shelf-edge detritic assemblage

In areas where the input of terrigenous material is very low, there is a gradual diminution in the abundance of the species characteristic of the Coastal Detritic assemblage, as the depth increases. The shelf-edge assemblage begins at a depth of 90−95 m and extends down to 120−150 m.

The sediment consists in a mixture of gravel (stream pebbles) together with some sand, silt and many remnants from benthic organisms (dead, subfossil or fossil and all more or less corroded). It is possible to distinguish the remnants from : (i) an assemblage which lived at a depth of about 40 m and which is very

Fig. 7.7. The ascidian *Diazona violacea* is often observed in the terrigenous mud shelf assmblage, especially if the bottom is rather firm, but it exists also in the offshore rocky-bottom assemblage and sometimes in the transitional horizon. One can notice here the arms of the suspension-feeding brittle-star *Ophiopsila* sp., which takes a part of the suspended organic material put in motion by the ascidian.

Fig. 7.8. The sea-lily *Leptometra phalangium* is the most characteristic and abundant species of the shelf-edge detritic assemblage. Also present is the walking sea-urchin *Echinus acutus*. The bottom is studded with burrows of the small and very nimble fish *Gobius lesueuri*. In this frame, the current was flowing from the left, at the top, to right down. Depth: 135 m.

similar to existing coralligenous ones, notably: *Lithothamnium*, some bryozoans (e.g. *Hippodiplosia*, *Myriozoum* and *Sertella*), molluscs (*Cerithium vulgatum*, *Lucina borealis*, *Lima loscombei* and *Pecten jacobaeus*, drilled by the sponge *Cliona*); (ii) some remnants of species formerly living at depths from 50 to 80 m, i.e. the existing pelecypod fauna in circalittoral soft bottoms, for example: the bivalves, *Cardita calyculata*, *Pitaria rudis*, *Venus ovata*, and the gastropod *Turritella triplicata* and *Aporrhais pespelicani*; (iii) dead shells of characteristic existing species, notably: *Dentalium panormum* and *Astarte sulcata*.

The preservation of remnants of calcareous organisms is made more difficult — especially by the activity of borers — in proportion to the rising of the sea level during the transgression which followed the maximum of the Würm regression.

According to Picard (1965), there are only 9 characteristic species: the scaphopod *Dentalium panormum*, the bivalve *Astarte sulcata*, three crustaceans (*Haploops dellavallei*, *Lophogaster typicus* and *Ebalia granulosa*), the ophiuroid *Ophiura carnea*, the holothuroids *Thyone gadeana* and *Neocucumis marioni* and the crinoid *Leptometra phalangium*. The latter is sometimes very common (5 – 15 ind. m^{-2}) and benefits from the strong currents at the shelf edge, flowing mainly perpendicularly to the isobaths. The pelecypod *Chlamys clavata*, also rather common here, is not really typical. The current-loving species, *Venus casina* and *Spatangus purpureus*, are usually present, but neither are characteristic.

Two facies of this assemblage can be recognised: (i) those of large hydroids (*Lytocarpia myriophyllum* and *Nemertesia antennina*), in rich silt and mud; rhizoidal network of these rather firm hydroids strengthen the superficial layer of the sediment in which they reach 20 — 25 mm in length. They support a small microcosm: hydroids (*Lafoea*), the actinian *Gephyra*, the gastropod *Capulus* and a pedunculate cirriped (*Scalpellum*). (ii) When the water circulation is especially strong, the assemblage becomes extremely poor, but the small echinoid *Neolampas rostellata* may still be common and is characteristic of this facies.

7.5. OFFSHORE ROCKY-BOTTOM ASSEMBLAGES

7.5.1. Offshore rocky-bottom assemblage

Observations from Cousteau's diving saucer, between the coralligenous (Laborel *et al.*, 1961) and the 'White-corals' assembalges, led to the recognition of a fairly specific assemblage at the upper part of the slope, containing some very poorly developed calcareous red algae. The rocky substrates here are often covered with a very thin layer of silty sediment. The peduncles or supports of sessile species may penetrate through the sediment layer to attach to the rock. Possibly some benthic oligophotic diatoms

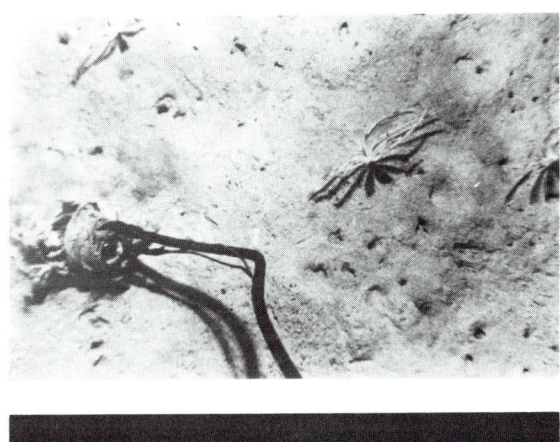

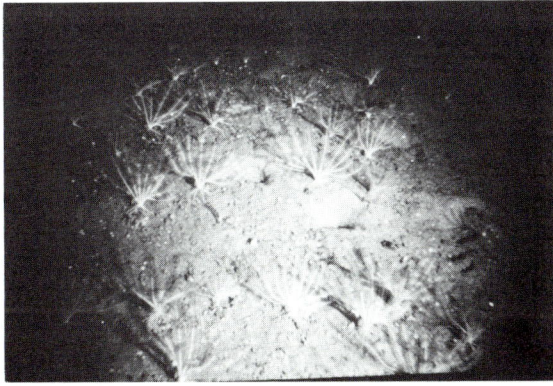

Fig. 7.9. In this close-up picture of *Leptometra phalangium* the orientation is the same as in Fig. 7.8. The plume of this crinoid is not corolla-like (as in *Antedon*) but fan-shaped. Populations may be very dense at depths between 115 — 140 m.

may occur upon rocks and on living substrates, which should be studied, especially from the physiological point of view (Ercegović, 1957).

This offshore rocky-bottom assemblage is dominated by large-sized sponges, for example: *Poecillastra compressa*, *Rhizaxinella pyrifera*, *Phakellia ventilabrum*, *Axinella* spp. and *Petrosia ficiformis*; the yellow scleractinian *Dendrophyllia cornigera* and *Antipathes fragilis*, the alcyonarians *Paralcyonium elegans* and *Corallium rubrum*; many bryozoans (*Hornera* sp., *Sertella*, *Porella cervicornis*), the brachiopods *Terebratulina caput-serpentis* and *Gryphus vitreus*. Some polychaetes are also characteristic, such as *Serpula vermicularis* var. *echinata*, *Placostegus tridentatus* and *Omphalopomopsis fimbriata*. Echinoderms are represented by the ophiuroid *Ophiacantha setosa* (often in the oscula of large sponges) and the opportunist echinoid *Cidaris cidaris* is very common, grazing upon sponges or fragments fallen from the overhanging rock. Decapods are common e.g. the spiny lobster *Palinurus mauritanicus*, *Paromola cuvieri* and *Munida* sp. and the bivalve *Spondylus gussoni*.

The predominant components of the assemblage are suspension-feeders as would be expected from the strong currents and high turbidity. The assemblage is markedly impoverished between 250 – 300 m (the upper limit of the white-corals assemblage).

7.5.2. Bottoms with large brachiopoda

These bottoms were recently investigated by Falconetti (1980), mainly on the coasts in eastern Provence and North Corsica. The substrate is sandy mud, mixed with thanotocoenoses debris and also studded with clumps of living and dead ahermatypic corals. The pilot species is the brachiopod *Gryphus vitreus*, which is sometimes abundant (mainly between 100 – 200 m, but sometimes down to 700 m in depth) especially in straits between large islands or between an island and the continent.

This type of bottoms cannot at the moment be considered as an ecological assemblage because of the difficulty in compiling a list of characteristic species. Nevertheless, they sometimes occupy large areas near the line of inflexion and extend deeper into the bathyal zone. They are thus a transitional 'horizon' between the lower shelf formations, of the circalittoral zone, and the upper slope formations of the bathyal zone (Reyss, 1971, 1973). This transitional horizon seems to result from changes in sea level which occurred during the Pliocene, Pleistocene and Quaternary times and its distribution depends on submarine topography, water temperature, strong currents and granulometric features of the sediment.

The invertebrate fauna is a mixture of species from the Coastal Detritic, Shelf-edge Detritic assemblages, and the bathyal zone (49%, 33% and 18%, respectively). Among the most important zoological groups are, in order of predominance, polychaetes, crustaceans, molluscs and echinoderms.

The closest affinity is with the Coastal Detritic assemblage, especially with two of its facies: that of *Laminaria rodriguezii* and the 'burnt almonds', where encrusting Lithothamniae play an important role, while many species — characteristic or not — are of the coralligenous assemblage. The Shelf-edge Detritic assemblage bears a closer relationship with the *Neolampas rostellata* facies.

The sessile fauna is sometimes rich containing: ramose Scleractinians, such as living *Dendrophyllia cornigera* and *Madrepora oculata* (often dead), some sponges, the zoantharian *Parozanthus marioni*, some bryozoans and serpulids. Among brachiopods, *Gryphus vitreus* and some other species (*Crania anomala*, *Mergerlia truncata*, *Terebratulina caput-serpentis*, *Megathyris subtruncata* and *Platidia anomioides*) are less well represented. Most of these brachiopods were recorded from the concretioned parts of the coralligenous assemblage. A filamentous green alga drills into the shell of *G. vitreus*, a species of the genus *Ostreobium* which is common everywhere on Indo-Pacific coral-reefs, always in the lower parts of the clumps, where light is very dim.

There are also intruders from the offshore rocky-bottom assemblage, notably sponges, cnidarians, serpulids, bryozoans and some others which are truly bathyal. Among the latter is the sea-star

Sclerasterias richardi (a fissiparous species) and the large sea-urchin *Cidaris cidaris*, an opportunistic species, very common on rocky bottoms, sometimes grazing upon sponges, but also existing on flat soft coarse bottom down to 1000 m, and possibly deeper.

7.6. THE BATHYAL ASSEMBLAGES

As in the neighbouring Atlantic Ocean, the assemblages of the bathyal zone (Pérès, 1982) depend on the quality of the bottom, that is, whether hard or soft.

7.6.1. Deep-sea coral assemblage

On the hard substrates of the bathyal zone there is *present* but *not widespread* in the Mediterranean Sea, an assemblage of 'white-corals' generally forming scattered clumps. They are 'ahermatypic' (e.g. devoid of zooxanthellae) and, necessarily, consumers of living organic matter. These deep-reefs were built

Fig. 7.10. In many canyons of the Provence coast, the slope consists of a succession of steps (2 to 5 m in height) separated by more gentle areas where the mud settles. On the steps where the rocky substrate is fissured, the off-shore rocky bottom assemblage develops, and is dominated by the octocoral *Paralcyonium elegans* and many sponges, such as *Poecillastra compressa* f. *calyx*. Depth: 150 m.

Fig. 7.11. A muddy bottom, at 180 m depth, on the north coast of Tunisia: the sea-pen, *Pennatula rubra*, and traces from two sea-stars (possibly *Tethyaster subinermis* or *Luidia sarsi*) which dug for prey in the subsurface layers.

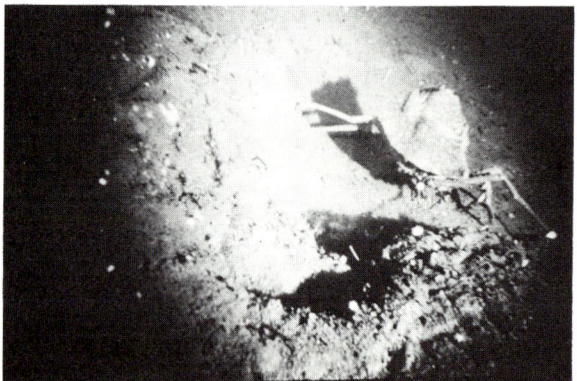

Fig. 7.12. The large crab, *Paromola cuvieri*, prefers sloping bottoms, where bathyal mud and emerging rocky substrates are intermixed, usually deeper than 200 m.

mainly by two large branched species, *Lophelia pertusa* and *Madrepora oculata*, which are associated with solitary species such as *Caryophyllia armata* and *Desmophyllum cristagalli*.

The fauna of the Mediterranean Sea deep sea assemblage is impoverished in comparison with that of the Atlantic slope of western Europe. Sometimes the colonies of the two pilot species are dead; in other places with a large input of fine sediment — possibly accumulated since historic times, due to progressive forest destruction by man — ramified corals become increasingly buried by fine silt, compensating for such silting by increased vertical growth.

Among the associated fauna, sponges are relatively scarce and probably not characteristic; some gorgonians exist, e.g. *Isidiella elongata*, *Primnoa*, *Muricea* and also antipatharians (e.g. *Antipathes fragilis*). Polychaetes are rather common and diversified. Many of them are eurybathic species. The most characteristic is *Eunice floridana* which lives on the coral masses in a parchment-like tube (coated with limestone) which persist after the death of the worm. Most other polychaetes are mainly eurybathic species, e.g. *Harmothoe longisetis*, *Phyllodoce madeirensis* and *Vermiliposis multicristata*. However, besides *E. floridana*, some true bathyal species are present, for example, two serpulids *Omphalopomopsis fimbriata* and *Placostegus tridentatus*, and two species of the Aphroditidae family, which seem to be characteristic and endemic (*Acanthicolepis cousteaui* and *Lagisca drachi*). The chiton *Hanleya hanleyi* seems to be a characteristic species, as well as the bivalves *Arca nodulosa*, *A. obliqua*, *Spondylus gussoni*, which is very common, and *Chlamys bruei*. Among the crustaceans are cirripeds of the genus *Verruca*, the shrimp *Pandalina profunda* and the Anomura *Paromola cuvieri*. The sea-urchin *Cidaris cidaris* is rather common, but not characteristic.

It seems that these white-coral reefs exist only below 300 m, possibly down to 800 – 1000 m, and only when the slope is sufficiently steep to allow some part of the hard substrate to be exposed. The clumps are more or less dispersed on the bottom, according to the topography of the slope, and separated by depressions where the bathyal mud settles.

At present, the only living parts of the corals are the tips of small branches at the summit of larger heads, as high as 50 cm. The rest of the corals are completely buried. The surface of the heads and buried colonies is covered with a blackish crust of limonite and manganese and iron oxides. The 'subfossil' parts which are embedded in the muddy sediment, for example on the Italian coast, near Portofino, at a depth of 450 – 750 m, contain a very rich assemblage, probably of Tyrrhenian age (80,000 – 100,000 B.P.). Among the white corals are found the tubes of the large *Eunice* and many serpulids (*Protula* and *Apomatus*), together with debris of *Spondylus gussoni* and *Cidaris cidaris*, which have large populations. The lowest part of deep-coral clumps is agglomerated with an interstitial greyish cement of microcrystalline limestone and mineral and organogenous material, which contains a thanatocoenosis characterized by the large gastropod *Strombus bubonius* together with some 'senegalian' molluscs (e.g. *Conus testudinarius*, *Spondylus gaederopus*, *Cardita senegalensis* and *Thais haemastoma*) and also some species such as the gastropod *Ranella gigantea* f. *atlantica* and *Pleuromurex lamellosus*, which have now disappeared from the Mediterranean. Their demise probably happened during the regression which corresponded with the maximum peak of the Würm glaciation.

Some of the white-coral reefs of the Algerian coast largely survived (in contrast to similar assemblages in the northern part of the western basin), due to the inflow of nutrient-rich Atlantic waters (Pérès, 1982).

Fig. 7.13. The orange-coloured sea-star *Ceramaster placenta*, is mainly found at depths from 150 to 400 m. Here it is near the 'white' coral *Madrepora oculata*. Depth: 225 m.

Fig. 7.14. In the areas of the slope where mud is deposited, sponges dominate the fauna, especially *Rhizaxinella pyrifera* and *Poecillastra compressa* f. *placentula*. Brittle-star arms rise from the sediment and collect particles transported by the bottom current. Depth: 200–250 m.

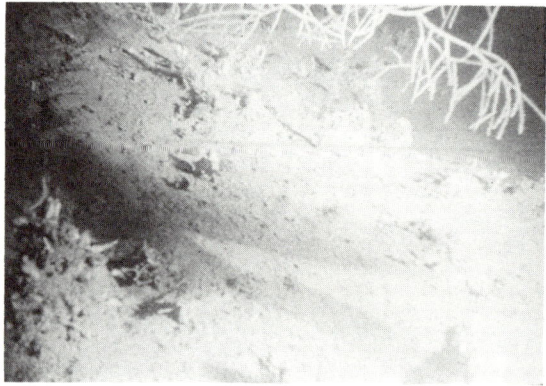

Fig. 7.15. The 'black coral' *Antipathes fragilis* which develops large branched colonies on isolated hard substrates at 200–300 m depth, when the food supply is sufficient.

Fig. 7.16. Live 'white coral' patches (*Madrepora oculata*) on the north-east coast of Corsica. The sea urchin *Cidaris cidaris* is present here, but less abundant than in the upper levels of canyons (150 – 300 m) where steps of hard bottom are more common, as well as the sponges it grazes upon. Depth: about 500 m.

7.6.2. The Bathyal mud assemblage

In the Mediterranean the bathyal soft bottom is a clayish mud, which is usually rather firm, reddish or bluish. In the bathyal zone (from 150 – 250 m down to 2000 – 2500 m) the assemblage is relatively impoverished, both in species and abundance. It is obvious that some Atlantic species cannot at present — or earlier — penetrate into the Mediterranean. They were prevented by the shallowness of the Gibraltar sill (maximum depth 300 m) and the sub-surface current of intermediate water (of high salinity and temperature) flowing towards the Atlantic Ocean. Even those individuals which are, or were, able to overcome these hindrances may not have been able to survive, because of the high and almost constant temperature (about 13°C) below approximately 200 m, or by reason of the food scarcity, due to the relative oligotrophy of the Mediterranean Sea (Pérès, 1982).

Although the bathyal assemblage on soft bottom is poorer (in terms of biomass and species richness) in the Mediterranean than at similar latitudes on the European Atlantic coasts, the bathyal assemblage of the north margin of the western basin is relatively abundant. The neritic waters are generally rather oligotrophic here, but they benefit locally from inflow of rivers (Rhône, Ebro), from local upwelling, for example, along the French coast from Toulon to the Spanish borderline (Millot, 1979), from cascading processes (in winter between the circalittoral and bathyal zone in the Gulf of Lions), from the upwelling in the Alboran Sea and the influence of the Balearic archipelago and the doming of the Ligurian Sea. Moreover, this is the best documented area in the whole Mediterranean, in contrast to the African margin where the important eastward flow of Atlantic surface water is less documented. Such documentation is, of course, crucial for useful comparison of the Atlantic and Mediterranean slope assemblages.

The input of terrigenous material to the depths of the western basin is more important at the European margin than at the African one, the latter having a much steeper slope. In fact, depths over 2500 – 3000 m (exceptional within the western basin and usually considered as the upper limit of the abyssal zone) are all located near the African margin, and must be considered as part of the bathyal zone.

The assemblage of bathyal mud in this area is rather rich in characteristic species, in comparison with many other Mediterranean assemblages.

The following amended list of characteristic invertebrates of the bathyal mud assemblage has been proposed by Pérès (1967), based on the very elaborate data from Picard (1965). The large foraminiferan,

Fig. 7.17. The large sea-pen *Funiculina quadrangularis* is common on the upper subzone of the bathyal mud (mainly between 200 to 400 m, and sometimes deeper) if the food supply is sufficient.

Cyclamina cancellata; the sponges *Asconema setubalense*, *Pheronema grayi* and *Thenea muricata* (with its epibiontic zoanthid *Parazoanthus marioni*), the cnidarians *Funiculina quadrangularis*, *Isidiella elongata*, *Hormathia coronata* and *Actinauge richardi*. Many polychaetes are common, e.g. *Aphrodite pallida*, *Harmothoe johnstoni*, *H. impar*, *Panthalis oerstedi* and *Leocrates atlanticus*. Particularly abundant are benthic crustaceans, (e.g. the cumaceans *Leucon longirostris* and *Diastylis cornuta*) and many decapods the most common being the *Calocaris macandreae*, *Polycheles typhlops* and *Pagurus variabilis*, some species of *Munida*, *Paromola cuvieri*, *Ebalia nux* and *Geryon tridens* and some nektobenthic shrimps, e.g. *Parapenaeus longirostris*, *Aristeus antennatus* and *Aristeomorpha foliacea*. Among the molluscs, three species are common, especially as empty shells: the scaphopods *Dentalium agile* and *Siphonodentalium quinquangulare* and the bivalve *Abra longicallus*. Some other bivalves are present, but rather rare, such as *Modiolus politus*, *Chlamys septemradiata* and *Propeamussium vitreum*, whereas the gastropods, *Platydoris dura*, *Calliostoma suturale*, *Ranella gigantea*, *Sipho torus* and *Xenophora mediterranea*, have been recorded. The cephalopods *Sepietta oweniana*, *Rossia caroli* and *Bathypolypus sponsalis* also exist there. *Siboglinum carpinei* is the first species of the Pogonophora phylum hitherto recorded in the Mediterranean (Carpine, 1970). The most characteristic echinoderms are *Odontaster mediterraneus*, *Ceramaster hystricis*, *Brisingella coronata* (especially common), *Ophiocten abyssorum*, *Mesothuria intestinalis* and *Leptometra celtica*.

Besides the characteristic invertebrate species of the bathyal mud assemblage, there are two categories of species which must be considered, as follows: (i) The mud-loving (pelophilous) species that extend from the circalittoral zone down to the bathyal depths. For example, the cnidarians *Pennatula phosphorea* and *Veretillum cynomorium*, the large polychaete *Aphrodite aculeata*; the echiurids *Echiurus abyssalis* and *Thalassemma gigas*; the gastropods *Triton nodifer*, *Cassis saburon*, *Morio rugosa*, *Scaphander lignarius* and *Doris tuberculata*; the crustaceans are especially numerous, such as the prawns *Pontophilus spinosus*,

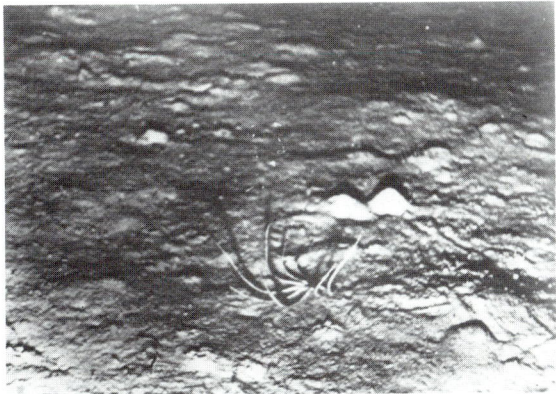

Fig. 7.18. In the straits between Sicily and Tunisia, the bathyal mud is studded with small hills made by burrowing invertebrates. The sea-star *Brisingella coronata* is common here, feeding, by means of its erected and bent arms, on suspended organic particles transported in the bottom current, eastwardly flowing. Depth: 475 m.

Pontocaris cataphracta and *Alpheus glaber*, two species of the genus *Processa*, and, among the Reptantia, *Nephrops norvegicus*, *Goneplax angulata* and *Dorippe lanata*; the enteropneust, *Glandiceps talaboti*; the echinoderms *Amphiura chiajei* (common), *Brissopsis lyrifera*, *Stichopus regalis*, *Molpadia musculus* and *Oerstergrenia digitata*. Large colonies of the ascidian *Diazona violacea* may be occasionally recorded here.

Most eurybathic species in the bathyal mud assemblage belong to the lower part of the shelf, where they benefit from the almost constant temperature below 200 – 250 m. Such species occur from the Coastal Detritic assemblage down to different depths depending on the quality and quantity of the food supply, as the cnidarians *Caryophyllia clavus* and *Lytocarpia myriophyllum*; the crustaceans *Pontocaris lacazei*, *Macropipus depurator*, *Eupagurus prideauxi*, *Dardanus arrosor* and *Anapagurus laevis*; the molluscs *Calliostoma granulata*, *Ovula adriatica* and *Xylophaga dorsalis*; the sea-stars *Ceramaster placenta*, *Astropecten irregularis pentacanthus*, *Anseropoda membranacea* and *Tethyaster subinermis* and the echinoid, *Echinus acutus*; the ascidian *Polycarpa fibrosa*.

Fig. 7.19. The same area as in Fig. 7.18, but deeper (500 – 600 m). *Cidaris cidaris* also exists on the bathyal mud, possibly feeding on foraminiferans. This species is opportunistic and largely omnivorous. In this area (close to that shown in Fig. 7.18) the mud appears rather flat. The middle and lower parts of the slope were literally 'buried' by terrigenous material from the north-eastern part of the Gulf of Genova, which receive a great deal of fine sediments from the Arno river and small coastal streams.

The horizontal and vertical distribution of the diverse components of the mud assemblage are poorly documented. However, it is clear that the bathyal zone of the North African margin (from Gibraltar to Cape Bon) possesses a small number of species which are missing on other regions of the western basin. Among these, are those of Atlantic Ocean origin: the sponge *Asconema setubalense*; the decapods *Plesionika antigai*, *Palinurus mauritanicus*, *Munida iris* ssp. *rutlandi* and *Rochina carpenteri*; the molluscs are represented by the two gastropods, *Calliostoma suturale* and *Sipho torus* and the two bivalves, *Modiolus politus* and *Chlamys septemradiata*. In the same area, the proportion of eurybathyc species seems to be larger and these species go down to greater depths (Pérès, 1967).

Except for the peculiarities of the northern margin of Africa, mentioned above, the bathyal mud assemblage of the western basin are of value from a biocenotic point of view.

Within the zone, three subzones can be distinguished. (i) An upper one which includes a maximum of eurybathic forms, suggesting a transition between the circalittoral and bathyal assemblages on muddy bottom. (ii) The middle subzone, the richest and most diversified, and the best documented. (iii) The lower subzone (poorly documented) shows a marked decrease in diversity and the presence of some species which are missing in the upper and middle subzones, (e.g. the crustaceans, *Nematocarcinus ensifer* and *Polycheles sculptus*, the asteroid, *Plutonaster bifrons*, and some large sized fishes such as *Haloporphyrus* and *Bathypterois mediterraneus* (Carpine, 1970), which were observed on the occasion of bathyscaph dives deeper than 1500 m, and two species of the genus *Chalinura* (*C. mediterranea* and *C. guentheri*) which seem to be the deepest fishes in the Mediterranean (2400 – 2900 m)).

The assemblage of the bathyal mud may show different facies, which are largely correlated with the firmness and compactness of the substrate, in both the upper and middle subzones (Pérès and Picard, 1964; Pérès, 1967).

In the upper subzone local facies vary according to the silt and mud input. For example, very fluid sediment is characterized by the abundance of the irregular echinoid *Brissopsis lyrifera*, mainly in the submarine valleys, while in the middle subzone the big cerianthid, *Branchiocerianthus norvegicus* is common near Arzew (Algeria).

On soft mud on somewhat inclined bottoms, the upper subzone shows large populations of the big sea pen, *Funiculina quandrangularis* together with the shrimp *Parapenaeus longirostris* and *Nephrops norvegicus*. On gentle slope the big sea pen *Kophobelemnon leuckarti* is more common. In the middle subzone the holothurian, *Mesothuria intestinalis* occurs, sometimes associated with the sea-star, *Odontaster mediterraneus*.

On compact soft mud, the species which dominate the assemblage in the upper subzone are the actinian *Actinauge richardi*, the gastropod *Calliostoma suturale*, and the sea-stars *Odontaster mediterraneus* and *Brisingella coronata* (which is a sestonophagous form). In the middle subzone the assemblage is very rich and corresponds to the *Isidiella* bottoms, dominated by the accompanying species listed above: serpulids, *Scalpellum*, *Gephyra*, and the large shrimps, *Aristeus* and *Aristeomorpha* which are preyed upon by cephalopods (*Sepietta*, *Rossia*), and fishes such as *Hymenocephalus italicus*, *Chlorophthalmus agassizi*, *Notacanthus bonaparti*, *Chimera monstrosa* and some sharks (e.g. *Squalus fernandinus* and *Centrophorus uyatus*). However, the bulk of this fish fauna, sometimes rather abundant, displays species which exist in both this subzone and in the lower levels of the circalittoral zone; most of them are nektobenthic forms (e.g. the shark *Etmopterus spinax*, and many teleosts: *Coelorhynchus coelorhynchus*, *Phycis blennoides*, *Gadiculus*, *Merluccius merlucius*, *Macrorhamphosus scolopax*, *Capros aper*, but also some forms which are really benthic such as the two species of *Lophius*, *Sebastes dactylopterus*, *Lepidorhombus boscii*.

Finally, there is an assemblage in the lower subzone on compact soft mud which is very poorly documented.

The sponge, *Thenea muricata*, is common on sandy muds of the upper subzone, rising above the bottom on two or three stalks made of bundled spicules; another sponge, *Rhodiella tissieri*, is also present. In the middle subzone the gastropod, *Apporhais serresianus* is occasionally present. When the

HISTORY OF THE MEDITERRANEAN BIOTA

Fig. 7.20. The large shark, *Centrophorus uyatus*, is sometimes seen in the bathyal mud, especially in the lower levels of canyons, at about 1000 – 1500 m depth. (Near Toulon, 1350 m).

sandy mud is mixed with gravel, and sometimes with boulders or emerging rock, the assemblage is dominated by the brachiopod *Gryphus vitrea* and the sea-urchin, *Cidaris cidaris*, both existing also in the offshore rocky-bottom assemblage.

Finally, mention should be made of a facies which has been observed offshore on the east Corsican coast. In this large area a great deal of decaying organic material goes down and drifts south, accumulating as the depth increases. Presumably a large part of this organic matter comes from the Ligurian Coast, where the neritic assemblages of soft algae are abundant, mainly in the infralittoral and on hard bottoms. A rather strong southwards current exists here in the middle of the slope, which results in a facies characterized by a massive abundance (500 – 800 m) of the hexactinellid sponge *Pheronema grayi*, together with several polychaetes settling upon or living in it. A little more south, at about 1000 m depth, the bottom is apparently almost flat, and *C. cidaris* is sufficiently common to suggest that the food supply is rather abundant there, possibly in relation with organic microparticles or dissolved organics, originated from either the white-coral assemblage or the *Pheronema* bottoms just above mentioned.

7.7. IS THERE AN ABYSSAL ZONE IN THE MEDITERRANEAN SEA?

The abyssal zone begins at 3000 m (Pérès, 1982). The western basin of the Mediterranean Sea does not exceed this depth by more than 100 – 150 m, except in small areas of the Algéro-Provençcal basin. The depth of the Tyrrhenian basin does not exceed 3500 m.

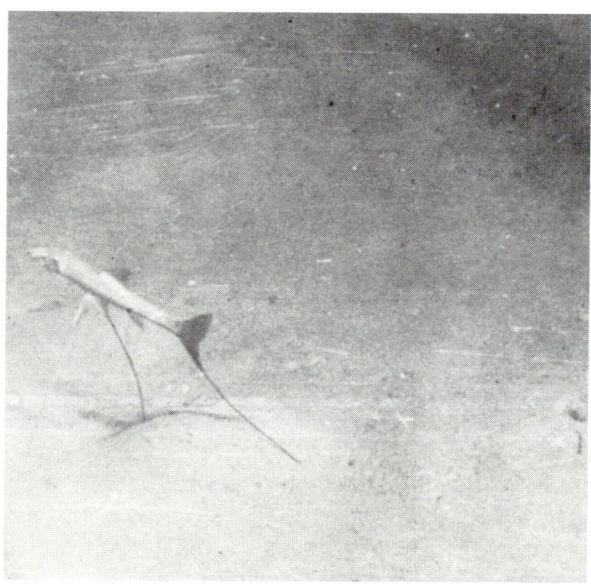

Fig. 7.21. The large fishes, *Haloporphyrus* sp. (a) and *Bathypterois mediterraneus* (b), have been observed several times during French bathyscaph dives, usually deeper than 2000 – 2200 m. (a) is swimming near the bottom (i.e., nektobenthic), while (b) is standing erect on its fins and tail when not disturbed; when startled it jumps to several metres away and again remains impassive.

As already emphasized the assemblages, from deep shelf down to the greatest depths, are markedly impoverished — especially in comparison with those of the Atlantic.

Drastic changes in the water balance during recent times and the current system of the Gibraltar area, were severe hindrances to the establishment of benthic species from outside. Moreover, even those which successfully cross the sill may have not been able to survive because of food scarcity and the high, almost constant temperature.

Obviously, the consequences of both present and past environmental changes, responsible for the composition and density of benthic deep-sea assemblages, must also be taken into account for the abyssal zone. The Mediterranean was formerly believed to be devoid of a true abyssal fauna, but recent French investigations have corrected this view (Chardy et al., 1973; Laubier, 1972 and Reyss, 1971, 1972). Our present knowledge of the Mediterranean Sea abyssal macrofauna is restricted to animals which are retained on a 0.5 mm screen and sampled with the epibenthic sled, complemented with samples obtained with Reineck's corer and trawl and can be summarized as follows.

The number of abyssal pelecypod species in the Mediterranean Sea is estimated to be about 50. However, the French expedition *Polymede I* collected only small numbers of 9 live species, none exclusively abyssal. They were all also recorded from the bathyal zone and may be considered as eurybathic bathyal species, because they also occur in the North Atlantic Ocean bathyal zone. The faunistic diversity of pelecypods is low (i.e. there were only 2–3 different species in each sample, as compared with the 9–10 found in the North Atlantic Ocean abyssal zone). However, some populations may be rather dense: up to more than 200 *Nuculoma tenuis* individuals were recovered in a single sample. None of the pelecypod species hitherto recorded in the Mediterranean deep-sea benthos seems to be endemic.

Polychaetes are much better represented and may be divided into three groups of species, with different bathymetric and geographical distributions. Group 1 lives in the Atlantic Ocean and in the Mediterranean, either in the bathyal and abyssal zones (*Paraonis gracilis, Laonice cerrata, Cirrophorus branchiatus, Glycera rouxi*) or purely abyssal areas (*Fauveliopsis brevis, Tharyx marioni, Prionospio cirrifera*). Group 2 consists of endemic species that live in the bathyal as well as in the abyssal zones (*Aricidea annae, A. monicae, Paraonis lyra*) or only in the abyssal zone (e.g. *Macellicephala annae, M. laubieri, Aricidea aberrans, A. abyssalis, A. trilobata, Aedicira mediterranea*). Group 3 includes only one species of the genus *Macellicephaloides* which to date has been recorded only from the hadal zone.

The cumaceans were studied by Reyss (1972, 1973b,c). Excluding the most eurybathic species (which occur in shelf assemblages and often down to 3000 m) we can recognise two groups among the forms which were never observed above 200 m: (i) a group of 13 species inhabiting depths of 500–2500 m, which is largely composed of what appear to be endemic forms, whose most abundant members seem to be *Diastylis jonesi, Leptostylis bacescoi* and *Procampylaspis bacescoi*; (ii) a group of about 10 species inhabiting both the eastern Atlantic and the Mediterranean Sea, from the shelf edge down to 3000 m. These species are genuine eurybathic forms, of which the most abundant and widespread are *Macrokylindrus longipes, Bathycuma brevirostris, Procampylaspis armata* and *Platysympus typicus* (Reyss, 1973c). The last group is numerically dominant in the Mediterranean, where the composition of the cumacean fauna does not change continuously with increasing depth as in the Atlantic. At depths of below 2000 m, there is the addition of a typical deep Mediterranean fauna to a eurybathic one (from 200–500 to 3000 m and more). Most species of the latter also occur in the Atlantic Ocean.

Quantitatively, the macrofaunal abundance varies considerably in different areas. For example, in the 'abyssal' plain of the western basin (2000–3000 m) the average number of individuals per haul is 750 (min. 475; max. 2131); polychaetes represent 60–70% of the total, cumaceans 10%, pelecypods and isopods less than 10%. For comparison, in the same depth range, the abundance is 1000–2000 individuals in the Bay of Biscay and 5000–10000 on the Gat Head – Bermuda transect. In the latter, abundances as low as those reported from the Western Mediterranean Sea may be observed only between 4000 and 5000 m. In the western basin abyssal plain the specific diversity is fairly low for the four main taxa mentioned above (polychaetes, cumaceans, pelecypods and isopods). In general, two or three species account for more than 70% of the total number of individuals. In comparison with the Atlantic Ocean, the decrease in biomass in the deepest bottoms of the Mediterranean Sea is even more marked than the decrease in numbers of individuals, owing to the smaller average size of Mediterranean species.

It should be emphasized that the bathyal and abyssal soft-bottom assemblages of the Mediterranean Sea tend to mix in some places. Such mixing arises partly from the rather high percentage of eurybathic bathyal species, which are not hindered by decreasing temperatures, and can extend downwards to abyssal depths. However, since true abyssal species closely related to oceanic ones are known to exist on the deepest bottoms of the Mediterranean Sea, it is surprising that some of these may extend upwards to the bathyal zone in some places. Possibly, these formerly cold-loving forms may have become progressively adapted to a temperature around 13°C near the abyssal bottom and are, therefore, no longer prevented from rising by a thermal barrier (Pérès, 1982).

To summarize our present knowledge on the deep sea macrobenthos of the Mediterranean Sea, we conclude that despite the rather important percentage of eurybathic species, a truly abyssal fauna is present. In the western basin where the depth does not generally exceed 3000 m, abyssal forms are always less abundant than eurybathic and bathyal species. In contrast, abyssal species seem to dominate in the Matapan deep. The Mediterranean deep sea macrobenthos is less diversified than in neighbouring areas of the Atlantic Ocean and also less abundant, due to the less favourable trophic conditions prevailing at any depth in the Mediterranean Sea. Most animals collected during the two French cruises *Polymede I* and *II* were small-sized (0.5 – 10 mm) except for such echinoderms as the asteroid *Plutonaster bifrons* and the holothurioid *Pseudostichopus occultatus*. The elasipod holothurioid, *Kolga ludwigi*, although probably belonging to the bathyal zone and formerly found in the Mediterranean, was not collected during these cruises.

Compared with the general scarcity of the Mediterranean deep sea macrobenthos, especially in the abyssal zone, it is surprising that the average abundance of deep sea meiobenthos (i.e. animals smaller than 0.5 mm) amounts to 54 individuals 10 cm^{-2} (min. 33; max. 78) in the western basin, at depths of 2116 – 2855 m. These values are of the same order of magnitude as those in the poorest areas of the Indian and Atlantic Oceans (Dinet and co-authors, 1973; Pérès, 1982).

The depths of the Mediterranean have yet to be fully investigated. However, it is well documented that most of the deep bottoms are covered by silt and mud (corresponding to the very high supply of fine sediments which took place from Early Pliocene up to Quaternary times) while the subsidence movement developed, making our sea really deep.

Perhaps the 'large brachiopod bottoms' investigated by Falconetti (1980), were insufficiently documented for he restricted his research to depths of 100 – 200 m. It was difficult to decide, when writing this chapter, whether to take into account the white corals assemblage. Today this concerns only very small and scattered relicts, isolated amongst muddy substrates, where the branched colonies are often literally buried in the sediment. The list of species recorded from this assemblage in the Mediterranean is very short and it must be emphasized that many of the characteristic species are either missing (e.g. among sponges, cnidarians, ophiurids, crinoids) or replaced by eurybathic opportunistic species. This impoverishment probably should be ascribed to the extreme oligotrophic nature of the surface waters, except in very limited areas.

However, in some parts of the eastern Provence and North Corsica coasts, the distribution of brachiopoda suggests that *Gryphus vitreus* and a few of the accompanying species may be recorded deeper, if the sediment, depending on the bottom current, is very coarse. In fact, *G. vitreus* is often found down to 600 m, and even at 1000 m in the eastern basin.

Pérès and Picard (1964) noticed that the range of the transition between the circalittoral and bathyal assemblages is 150 – 350 m on the Iberian Atlantic coast, while it is restricted to 100 – 200 m (180 near the Gulf of Genova) in the Mediterranean Sea, in a more closed environment. The upper horizon which extends from about 300 down to 900 – 1000 m in the Altantic Ocean, is just restricted between 200 – 500 m in the Mediterranean Sea. Moreover, two different hard-substrate assemblages were described on the south-western slope of the British Isles, which consisted of cirripeds, (*Verruca, Hexalasma*) brachiopods (*Hispanirhychia, Platidia*), ophiurids etc., with different species of these groups from 900 – 1250 and 1500 – 1899 m. *Verruca* and *Platidia* were recorded deep in the Mediterranean, together with some brachiopods, *Cidaris cidaris*, etc., and it is possible that a deeper assemblage exists, the range of which would be 500 – 1000 m and perhaps even a little deeper, if the bottom is sufficiently firm or coarse. It should be located and studied, possibly in passes, for example between Nice and North Corsica and in the Bonifacio Strait.

REFERENCES

Blanc, J. J. (1966) Le Quaternaire marin de la Provence et ses rapports avec la Géologie sous-marine. *B. Mus. Anthrop. prehist. Monaco* 13, 5 – 27.
Blanc, J. J. (1972) *Initiation à la Géologie Marine*, Paris, Doin.
Blanc, J. J., Pérès, J. M. and Picard, J. (1958) Coraux profonds et thanatocoenoses Quaternaires en Méditerranée. *Col. Intern. C.N.R.S. Paris: La topographie et la géologie des profondeurs océaniques*, 185 – 192.
Boudouresque, C. F. and Denizot, M. (1975) Recherches sur le genre *Peyssonnelia*. I. *Peyssonnelia rosea-marina* sp. nov. et *Peyssonnelia bonnetii* nov. sp. *Giorn. Bot. Ital.* 107, 17 – 27.
Carpine, C. (1958) Recherches sur les fonds à *Peyssonnelia polymorpha* (Zan.)Schmitz de la région de Marseille. *Bull. Inst. Océanogr. Monaco* 1125, 50.
Carpine, C. (1970) Ecologie de l'étage bathyal dans la Méditerranée. *Mém. Inst. Océanogr. Monaco* 2, p.146.
Chardy, P., Laubier, L., Reyss, D. and Sibuet, M. (1973) Données préliminaires sur les résultats biologiques de la campagne Polymède I. Dragages profonds. *Rapp. P.-v. Réun. Commn int. Explor. scient. Mer Méditerr.* 21, 621 – 625.
Cita, M. B. (1979) Quand la Méditerranée était asséchée. *La Recherche* 107, 26 – 36.
Di Geronimo, D. and Zibrowius, H. (1983) Le Scléractiniaire *Fungicyathus fragilis* et l'Octocoralliaire Stonolifère *Scyphopodium ingolfi*. *Rapp. P.-v. Réun. Commn int. Explor. scient. Mer Méditerr.* 28 (in press).
Dinet, A., Laubier, L., Soyer, J. and Vitello, P. (1973) Resutats biologiques de la campagne Polymède II. Le méiobenthos abyssal. *Rapp. P-v. Réun. Comm. int. Explor. scient. Mer Méditerr*, 21: 701 – 704
Duclaux, G. and Lafargue, F. (1973) Madréporaires de Méditerranée Occidentale. Recherches des zooxanthelles. Compléments morphologiques et écologiques. *Vie et Milieu* A, 23, 45 – 63.
Ekman, S. (1953) *Zoogeography of the Sea*. Sidgewick and Jackson, London.
Ercegović, A. (1957) Principes et essai d'un classement des étages benthiques. *Recl. Trav. Stn mar. Endoume (Bull. 13)* 22, 17 – 21.
Fairbridge, R. W. (1968) *Encyclopedia of Geomorphology*. Item Quaternary Period. Reinhold Book Corp., New York.
Falconetti, O. (1980) *Bionomie benthique des fonds situés à la limite du plateau continental du Banc du Magaud (Iles d'Hyères) et de la région de Calvi (Corse)*. Thèse Etat, Univ. Nice, 287 pp.
Gautier, Y. and Picard, J. (1957) Bionomie du Banc du Magaud (Est des Iles d'Hyères). *Recl. Trav. Stn mar. Endoume (Bull. 13)* 22, 28 – 40.
Genesseaux, M. and Lefebvre, D. (1980) Le Golfe du Lion et le Paléo-Rhône messinien. In *Les aspects géodynamiques du passage Miocène-Pliocène en Méditerranée* Eds. P. Orssog-Sperber and J. M. Rondry. *Géol. Médit.* 7, 71 – 80.
Holthuis, L. B. and Gottlieb, L. (1958) An annotated list of decapod crustacea of the Mediterranean coast of Israel, with an appendix listing the Decapoda of the eastern mediterranean. *Bull. Res. Conc. Israel* B, 7B (num 1 – 2), p. 126.
Hsü, K. J., Cita, M. B. and Ryan, W. B. F. (1973) The origin of Mediterranean evaporites. *Rep. Deep Sea Drilling Project*, Washington, U.S. Govern. Print. Off. 13, 1203 – 1231.
Laban, A., Pérès, J. M. and Picard, J. (1963) La photographie sous-marine profonde et son exploitation scientifique. *Bull. Inst. océanogr. Monaco* 60(1258), p. 32.
Laborel, J., Pérès, J. M. and Vacelet, J. (1961) Etude directe des fonds des parages de Marseille ole 30 à 300 m avec la soucoupe plongeante Cousteau. *Bull. Inst. Océanogr. Monaco*, 1206, p.16.
Laubier, L. (1972) Découverte du genre abyssal *Fauveliopsis* (Annelide Polychète) en Méditerranée Occidentale. *C.-r. hebd. Séances Acad. Sci., Paris* 274, 697 – 700.
Mars, P. (1963) Les faunes marines et la stratigraphie du Quaternaire Méditerranéen. *Recl. Trav. stn. Mar. Endoume* 28(43), 61 – 97.
Mars, P. and Picard, J. (1958) Notes sur les gisements sous-marins à faune celtique en Méditerranée. *Rapp. P.-v. Réun. Commn int. Explor. Scient. Mer Méditerr.* 15(3), 325 – 330.
Millot, C. (1979) Wind induced upwellings in the Gulf of Lion. *Oceanol. Acta Paris* 2, 261 – 274.
Montadert, L., Letouzey, J. and Maupfret, A. (1978) Messinian events: seismic evidence. *Rep. Deep Sea Drilling Project*, Washington, U.S. Govern. Printing Office, 42 (part 1), 1037 – 1050.
Nesteroff, V. D. (1973) Petrography and Mineralogy of sapropels. *Rep. Deep Sea Drilling Project*, Washington, U.S. Govern. Printing Office 13, 713 – 720.
Pérès, J. M. (1967) Méditerranean Benthos. *Océonogr. Mar. Biol., Ann. Rev.* 5, 449 – 533.
Pérès, J. M. (1982) Major Benthic Assemblages. In *Marine Ecology* Ed. O. Kinne, vol. 5, part 1, pp. 373 – 521, John Wiley and Sons, Chichester.

Pérès, J. M. and Picard, J. (1964) Nouveau Manuel de Bionomie Benthique de la Méditerranée. *Recl. trav. Stn. mar. Endoume* (*Bull.* 31) 47, 5 – 137.

Picard, J. (1965) Recherches qualitatives sur les biocoenoses marines des substrats meubles dragables de la région marseillaise. *Rec. Trav. Stn. mar. Endoume,* 52. 3 – 160

Postel, E. (168) Marine hydrology and buigeography in western Africa. In *West African International Atlas.* Organization for African Unity, IFAN, Dakar, pp. 13 – 16.

Rehault, J. P. (1981) Evolution tectonique et sédimentaire du Bassin Ligure. Paris, Univ. Pierre et Marie Curie, thèse Doc. état, 132 pp.

Reyss, D. (1971) Les canyons sous-marins de la mer catalane: le rech du Cap et le rech Lacaze-Duthiers. 3. Les peuplements de macrofaune benthique. *Vie Milieu,* 22. 529 – 613.

Reyss, D. (1973) Les canyons sous-marins de la mer catalane. Le rech du Cap et le rech Lacaze-Duthiers. 4. Etude synécologique des peuplements de macrofaune benthique. *Vie Milieu,* 23 (1B), 101 – 142.

Ryan, W. B. F., Hsü, K. J., Cita, M. B., Dumitrica, P., Lort, J. M., Maine, M., Nesteroff, V. D., Pautot, G., Stander, H. and Wezel, F. C. (1973) *Report Deep Sea Drilling Project*, Washington, U.S. Govern. Printing Office, 13, 1 – 1447.

Vaissière, R. and Fredj, G. (1965) Etude photographique préliminaire de l'étage bathyal dans la région de Saint-Tropez. *Bull. Inst. océanogr. Monaco* 64(1323), p. 70.

Wright, R. (1978) Benthic foraminiferal repopulation of the Mediterranean after the Mesinian (Late Miocene). *Palaeogeogr., Palaeoclimat. Palaeoecol.* 29, 189 – 214.

Zibrowius, H. (1974) *Oculina patagonica*, Scléractiniaire hermatypique introduit en Méditerranée. *Helgoländer wiss. Meeresunters.* 22, 153 – 173.

Zibrowius, H. (1980) Les Scléractiniaires de la Méditerranée et de l'Atlantique nord-occidental. *Mém. Inst. Océanogr. Monaco* 11, 1 – 227.

Zibrowius, H. (1983) Nouvelles données sur la distribution de quelques Scléractiniaires "méditerranéens" à l'est et à l'ouest du détroit de Gibralter. *Rapp. P.-v. Réun. Commn int. Explor. scient. Mer Méditerr.* 28 (in press).

Zibrowius, H. (1983b) *Spirorbis marioni* et *Pileolaria basheleyana*, Spirorbidae exotiques dans les ports de la Méditerranée. *Rapp. P.-v. Réun. Commn int. Explor. scient. Mer Méditerr.* 28 (in press).

Zibrowius, H. and Rao Omos, A. A. (1983) *Oculina patagonica*, Scléractiniaire exotique en Méditerranée. Nouvelles observations dans le sud-est de l'Espagne. *Rapp. P.-v. Réun. Commn int. Explor. scient. Mer Méditerr.* 28 (in press).

CHAPTER 8

Diving in Blue Water. The Benthos

JOAN DOMÈNEC ROS*, JAVIER ROMERO**, ENRIC BALLESTEROS**, and

JOSEP MARIA GILI**

*Departamento de Ecología, Facultad de Ciencias, Universidad de Murcia, Murcia;
**Departamento de Ecología, Facultad de Biología, Universidad de Barcelona, 08028 Barcelona, Spain

CONTENTS

8.1. The living carpet of the depths	234
8.1.1. The attraction of the depths	234
8.1.2. Zonation	235
8.1.3. Factors	236
8.1.4. A more precise typification of the zones	239
8.2. The supralittoral zone in a sea almost without tides	241
8.2.1. Soft substrates	241
8.2.2. Hard substrates	242
8.2.3. Pools and crevices	244
8.3. The mediolittoral: a zone between 'tides'	244
8.3.1. Soft substrates	244
8.3.2. Rocky substrates	245
8.4. The continuously immersed infralittoral	248
8.4.1. The major communities of algae on hard substrates	248
8.4.2. Diverging adaptations, a characteristic of animal life	253
8.5. The infralittoral soft bottom communities and the sea-grass meadows	257
8.5.1. Infralittoral sands and muds	260
8.5.2. The *Posidonia* biome or formation	260
8.6. The circalittoral hard-bottom communities: the coralligenous	263
8.6.1. Plants of the coralligenous	267
8.6.2. Animals of the coralligenous	272
8.6.3. Facies	273
8.7. The world of submarine caves and tunnels	273
8.8. Biomass, flow of energy and systemic properties of the benthos	281

8.8.1. Succession and regression	281
8.8.2. Ecological strategies	284
8.8.3. Trophic ecology	289

8.1. THE LIVING CARPET OF THE DEPTHS

8.1.1. The attraction of the depths

The *mare nostrum* may be a particularly hospitable sea, for reasons associated with, or derived from, its character of an internal or continental sea: extreme reduction of tides, transparent and blue waters, relatively high salinity and mean water temperature. The different physical, chemical and biological aspects have been presented and discussed in preceding chapters. The attractiveness of the Mediterranean has been enhanced by historical reasons, as ancient cultures blossomed over all its shores. Greeks and Phoenicians, sailors and pirates, and, more recently, pioneers in the craft of scuba diving and the operation of small submarine vehicles, have been very active, especially in the western basin.

The explorations of the Count of Marsigli at the beginning of the 18th century, those of Milne-Edwards a century later, and of the Prince of Monaco at the beginning of the 20th century, can all be considered the forbears of the gigantic momentum of littoral marine biology in the last half century, chiefly on the French coast; although even a century and a half ago, Milne-Edwards dived off Messina in a primitive apparatus 'to pursue marine animals into their hidden retreats'.

Long before, however, the men, mostly Greeks or Italians, who dived for the precious red coral and sponges, had penetrated into the sea, thus becoming the scouts for the later legions of skin and scuba divers. These, first sporadically and then systematically after the thirties, have dived in the clean Mediterranean waters of the Balearic Is., the Côte d'Azur, the Gulf of Naples, Malta or the Costa Brava, practising what has been called submarine or underwater tourism; but also scientifically studying the sea bottom.

What sunken marine landscapes explain the existence of such 'tourism' in waters apparently devoid of the attractive marine life which swarms on coral reefs? It must be said, without sounding chauvinistic, that Mediterranean sea life is second to none in beauty and scientific interest. This chapter aims to provide an elementary description of the fauna, flora and communities of the Western Mediterranean, largely based on the studies of French pioneers in this field, like Feldmann (1937), which have been continued and summarized by Pérès and Picard (1964) and Pérès (1966, 1982a, 1982b, 1982c, 1982d), and in the more recent work done in the Catalan Costa Brava by a team in which the authors are integrated (Romero, 1981; Ballesteros, 1982; Bibiloni *et al.*, 1982; Gili and Ros, 1982; Ros *et al*., 1984; etc.). Other authors have greatly contributed to the knowledge of the Western Mediterranean coasts bionomy, mainly French and Italian; without being exhaustive, the following must be mentioned: Funk (1927), Molinier (1960), Laborel (1961), Picard (1965), Laubier (1966), Ledoyer (1968), Zibrowius (1968), Bellan-Santini (1969), Harmelin (1969), Fiala-Medioni (1970), True (1970), Ben (1971), Boudouresque (1971), Giaccone and Bruni (1972), Boudouresque and Fresi (1976), Coppejans (1980), and many others. References are given in the corresponding sections.

This chapter will be mainly descriptive; the understanding of the dynamics of the benthic communities is in their beginnings and, so far, there is no general agreement concerning the meaning and role of the different species and communities in the functioning of marine littoral benthos. Some general trends, however, can be mentioned, and reference is made to recent comprehensive publications on the subject (Price *et al.*, 1980; Longhurst, 1981; Pérès, 1982e).

8.1.2. Zonation

The environmental changes are generally sharper along the vertical axis than along the horizontal plane. This is particularly evident when the vertical axis cuts important interfaces, as between air and water and between water and bottom. In the first instance, the 'terrestrial', environment changes suddenly into a 'marine' one, and the parameters defining the environments undergo very rapid changes. The environmental gradients are easy to observe; humidity (or humectation) increases from the atmosphere to the water; the battering action of waves is maximal at level zero and decreases both in the shoreward and in the seaward directions; with increasing depth, the temperature of water and light intensity decrease, the pressure increases and diminishes the risk that a storm or a seiche expose a section of the coast. To these quantitative gradients others of qualitative nature are superimposed; the light not only decreases with depth, but its spectrum is changed (Fig. 8.1); the water movement has a more pronounced vertical (wave-generated) component in depth (Fig. 8.2); the grain size of the soft sediment decreases with depth, and the amount of hard, rocky substrates is higher at shallow levels.

The combination at each point of the littoral of the different gradients leads to a macroscopically observable zonation or *étagement* (from the French word for floor, *étage*). It consists of an arrangement of the plants and animals in belts or strips, which are usually parallel to the water surface. The pattern appears on all coasts of the globe; a paradigmatic but by no means exclusive example is the occurrence of intertidal algal belts. With such general zonation, the species that are more tolerant or more adapted (to survive dessication, the battering of waves, intense illumination and extreme temperatures) settle in the upper, more superficial levels.

The different assemblages of plants, animals and microbes that dwell on the sea bottom constitute the benthic community. They form a living, relatively thin, 'film' or 'carpet' consisting of sessile (attached to the substrate) or vagile organisms, which move on or into it, and inhabit the hard (mainly rocky) or soft bottoms (gravel, sand or silt) or the water immediately above them. The benthos is a major component of ocean life and a counterpart of pelagic life. It extends from the shore to the deepest parts of the sea floor in a rather two-dimensional pattern and interacts with the three-dimensional ocean and the littoral terrestrial environment.

Plants need light, and are characteristic of a 'phytal' or 'photic' level, where they play a dominant role in the structure and functioning of the entire community. Primary producers are absent from the deeper 'aphytal' or 'aphotic' level, where the mode of life is restricted to heterotrophy. In the phytal zone a distinction is made between photophilic and sciaphilic species or assemblages, according to their distribution in relation with higher or lower light intensity. The sciaphilic organisms settle at greater depths or if in superficial levels under poorly lit overhangs, crevices or caves. Etymologically, sciaphilic or sciaphilous means one that loves low intensity of light. Yet this is really not the case for algae. More light could be used, but sciaphilic species are able to use nutrients and grow, rather slowly, at low levels of light.

Not only average intensity, but also the variability of environmental factors is important. Only in very stable or constant environments are more conservative strategies advantageous. These include physiological parsimony, a high level of space structuration and a high specific diversity of the communities.

The different zones are macroscopically identifiable near the interface (where the gradients are strongest) but as these become less accentuated, then the zonation and the boundaries become less sharply defined. The gradient on the vertical axis is often accompanied by a patchy heterogeneity in the horizontal plane caused by discontinuities in the type, texture, inclination and exposition of substrate, in its orientation, the existence of predominantly one-directional currents, random perturbations on little-size areas, etc. These factors complicate the simple model, and, in so doing, generate diversity which is still enhanced by the increasing structural role of the organisms themselves.

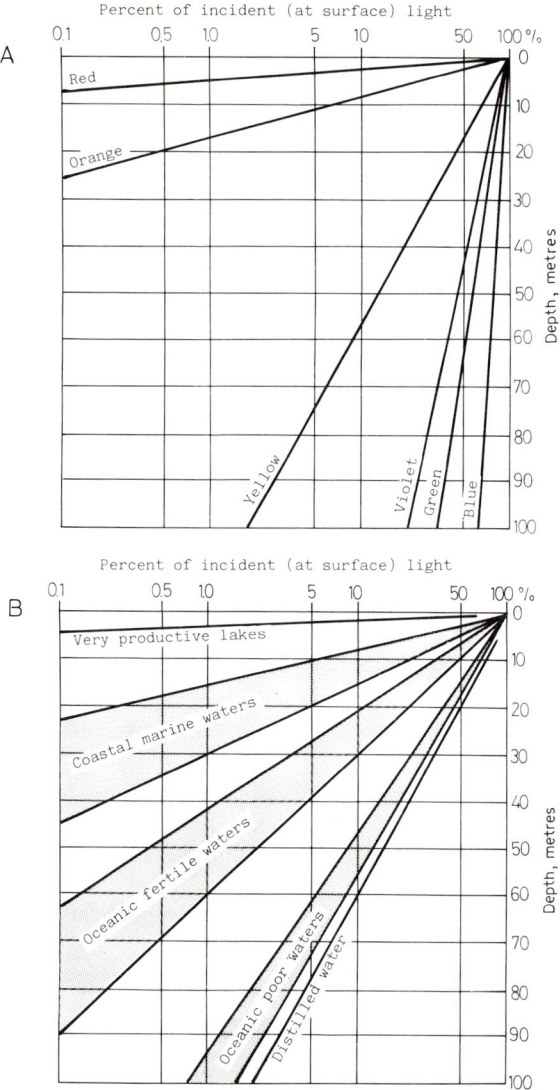

Fig. 8.1. Light extinction in water. A, decrease in the intensity of the colours constitutive of solar light at increasing depth in a lake of optically pure water. B, decrease in the light intensity in natural waters, chiefly marine, of different properties. The pattern of colour extinction depicted in A applies only to poor oceanic and distilled water, being spectrally different in waters of other characteristics. Modified from Clarke (1939).

What appears to the observer's eyes and what the marine biologist usually studies is the result of the prolonged interaction between the different environmental parameters and the species pool. We see the living assemblages resulting from this interaction, but not the values of the parameters themselves.

8.1.3. Factors

The most important of these parameters fall into two categories: the abiotic (climatic and edaphic) and the biotic ones. Climatic factors (Pérès, 1982a) are the humectation, the illumination and the pressure.

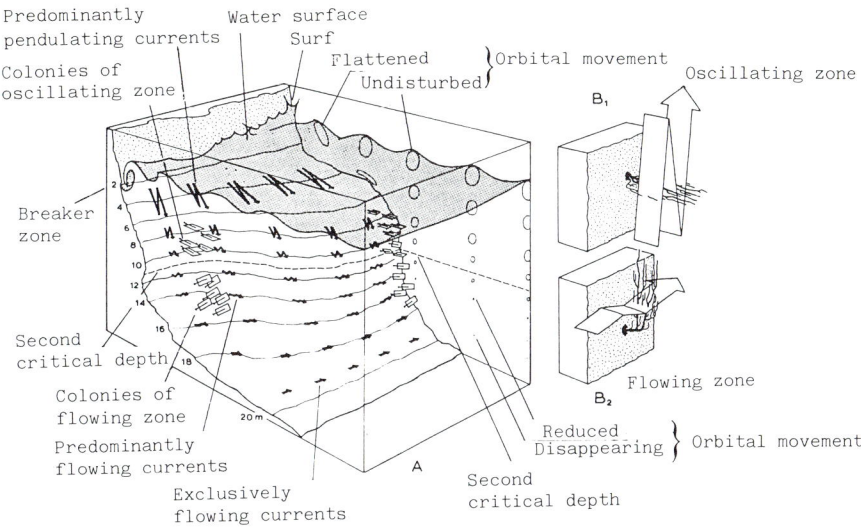

Fig. 8.2. Main types of water movement in a sea-coast without tides. The wave action generate a mainly vertical and shearing movement in the first levels, and in depth this becomes a chiefly horizontal and directional movement. The direct splashing and battering action of waves affects only the emerged littoral. The shape and position of many filter-feeders (such as the gorgonian *Eunicella* in the drawing) reveal the general current pattern. From Riedl (1966; by permission of Paul Parey Verlag).

The humectation is important only in the surface level. It results from the balance between the inputs and outputs of sea-water, due to evaporation or to local, periodical or aperiodical variations of sea level: waves, tides, seiches, etc. The quantitative and qualitative aspects of illumination must be considered, and thermal effects may also be important. The role of hydrostatic pressure on marine organisms is important. Depth affects the solubility of calcium carbonate, and conservation of hard skeletons and shells is harder below the depth of solution or lysocline.

The edaphic factors are so named because they usually act on the substrate and also at a local level, differing in this way from the climatic ones. The following must be mentioned: the physical nature of the substrate (compactness and character of sediments, grain size, sedimentation rate, mobility or instability of the substrate, etc.); the chemical composition of solid materials; physicochemical properties (pH, redox) of interstitial water; the movement of water; its salt content that can be decreased by rainfall and run-off from land and augmented by evaporation; daily and seasonal temperature changes under the influence of local winds or water movement in small basins or littoral pools, etc.; water enrichment in nutrients (eutrophication) or in organic matter (pollution), high rate of sediment deposition and different exploitative and destructive effects by man.

The biotic factors are dependent on organisms and can modify the action of abiotic factors. Examples include: the alteration of the substrate by a community, as the concretion of previously loose sediments or the disgregation of hard substrates; the creation of microclimates as a result of shading by species growing on the top. Biotic factors introduce competition for space, which seems to be one of the more limiting resources in the benthos on rocky bottoms. The result is the selection of shapes which are better adapted for each environment, abundance of colonial forms (ensuring the fast occupation of space by means of asexual reproduction) covering of some species by others, which suffocate and eliminate them or use them as substrate (epibiosis; Fig. 8.3), etc.

The competition, as already mentioned, may be for physical substratum, or for a water 'volume' to be filtered to obtain food, for a territory to be explored for food or to be defended for social or sexual reasons. Certain cyclical changes in the community composition also have biotic origins which are traceable to oscillations of predator-prey relationships, or to seasonal cycles or migrations. More

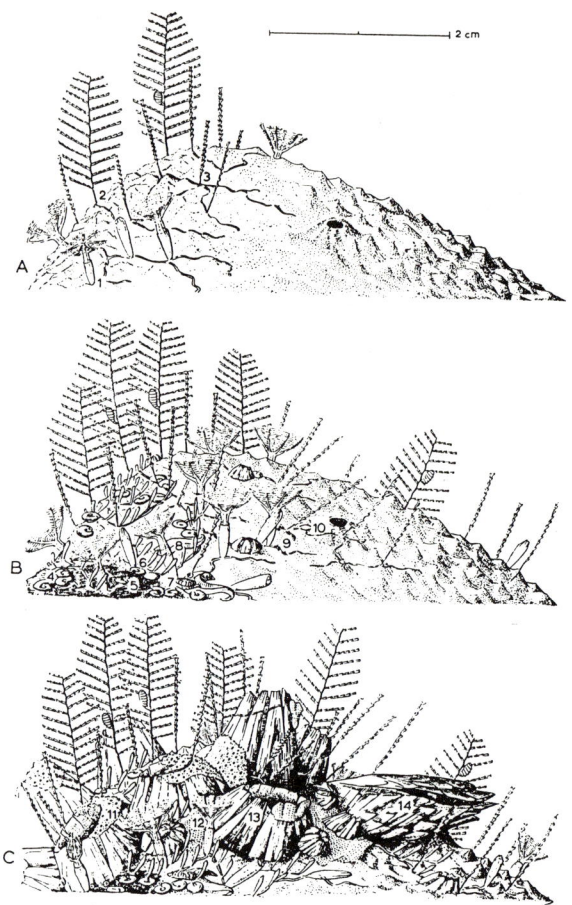

Fig. 8.3. Three successive steps in the colonization of the living surface of the sponge *Ircinia* by sedentary epibiont organisms. 1, *Cornularia*; 2, *Aglaophenia*; 3, *Dynamena*; 4, *Spirorbis*; 5, melobesiae; 6, *Aetea*; 7, *Lichenopora*; 8, *Pomatoceros*; 9, *Miniacina*; 10, rotaliidae foraminiferans; 11, *Cellepora*; 12, *Schizomavella*; 13, *Balanus*; 14, *Ostrea*. From Riedl (1966; by permission of Paul Parey Verlag).

directional changes result from substitution of some species by others, for example, as occurs following the colonization of a virgin substrate or during the slow transformation of a substrate by a community. Evolutionary changes can be mentioned for the sake of completeness.

The nutrient and food supply and the production of organic matter are key factors. Indeed, illumination, water fertilization and many other environmental factors and the interactions between benthic organisms (plants and animals) and between plankton and benthos, are expressed as processes of primary production and transfer of production. Below the level of light penetration, life depends on the input of organic matter. Chemoautotrophy is relevant only in the 'vents' or 'oases' of the recently formed sea floor, and not in the Mediterranean.

In the first, well-lit étages of the phytal system, and 'probably also in the upper levels of the circalittoral zone, the *in situ* primary and paraprimary production meet a high percentage of the food requirements of the...species. In contrast, at deeper levels...multicellular algae...are scarce and their production is low...Thus...the nutrition of [the] species depends in part on exogenous organic matter; in waters below the shelf edge the dependence becomes total...[This] input of organic matter stems from three sources: (i) plankton organisms and their detritus which sink to the bottom from upper water layers; (ii) detritus transport from land or shallow benthos; and (iii) organismic migration' (Pérès, 1982a). The benthos, on the whole, has been considered as a deficitary community, with more consumers than producers, that exploit the planktonic production of the sea through a largely detritic pathway. The primary production of the benthic plants per unit surface can be compared with that of the terrestrial vegetation, being higher than phytoplankton production. But because of the small extension, it amounts to no more than $1-2\%$ of the total marine primary production.

8.1.4. A more precise typification of the zones

As defined in a colloquium on marine biology held at Geneva in 1957, an étage or vertical zone 'is the depth interval of the benthic domain where the ecological conditions related to the main environmental factors...are homogeneous, or exhibit a gradient between two critical levels which corresponds to the boundaries of the zone...[Usually]...the boundary between two adjacent vertical zones...corresponds to a sharp change in [the living] assemblage composition' (Pérès, 1982a).

The existence of clear-cut boundaries has been challenged recently, and (in accordance with present trends in other areas of ecology) it has been suggested that it would be better to speak of a continuum instead of accepting sharp changes in the distribution of living assemblages (Boudouresque, 1971). Yet this will not affect the general account here, and only needs to be borne in mind when considering the mean depths given for a particular zone.

The following zones have been characterized in the benthos for all the seas, and, specifically, in the Mediterranean (Pérès and Picard, 1964; Pérès, 1966, 1982a): supralittoral, mediolittoral, infralittoral and circalittoral (together making up the littoral or phytal system). The aphytal or deep system has three zones: bathyal, abyssal and hadal (Fig. 8.4). In the Mediterranean the two last zones are absent, at least in the western basin. The existence of a true abyssal zone in the deeper parts of the oriental basin (up to 5092 metres of depth in the Hellenic Trough) is under discussion.

Supralittoral

The supralittoral zone is never, or only very rarely, immersed. Yet its habitants require a relatively high degree of humidity, supplied by the spray of the waves. Immersions are irregular due to the small tidal range, and mainly occur during heavy storms or seiches. Even so, storm-generated waves only

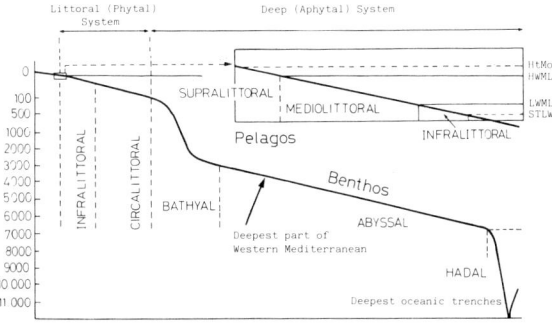

Fig. 8.4. Main divisions of the marine benthos. HtMo, highest level of moistening by waves and sprays (or tides in tidal seas); HWML, high water mean level; LWML, low water mean level; STLW, low water spring tide. The deepest part of Western Mediterranean (3,600 metres in the Tyrrhenian Trough) is indicated. Modified from Pérès (1982; by permission of the author).

rarely cover the supralittoral zone and its upper part is only wetted approximately once per year, although the lower part is almost continually splashed. The supralittoral in exposed places is more extended vertically than in sheltered ones, due to battering by waves. A vertical amplitude of 30 — 50 cm in sheltered areas and of 3 — 4 m in exposed ones are typical. The living assemblages of this zone are remarkably homogenous on a worldwide scale and very simple in composition: the severe environment has selected a few successful strategies. On rocky shores, the main group are unicellular cyanophyceae and chlorophyceae, lichens, littorinid gastropods, cirripeds and mobile isopods, and, on sandy beaches, amphipods and decapods well adapted to a semiaerial life.

Mediolittoral

The mediolittoral zone consists of living assemblages that require or tolerate immersion but cannot survive permanent or semi-permanent immersion: they are adapted to the periodic ebb and flow of the intertidal zone. The Mediterranean tidal range is small (except for some areas such as the Algeciras littoral and the Gulf of Gabès), and the upper limit of this zone corresponds to the highest level of wave inmersion, and its lowest to the level of 'normal' emersion. Occasionally, a combination of high atmospheric pressures, offshore winds and low tides, produce abnormally low levels of exposure, but with no damage to the organisms. It is usual to divide the mediolittoral into two subzones. In the upper one humectation is due primarily to wave action and immersions are rare; in the lower one submersions are more frequent but do not last. On rocky coasts, thoracic cirripeds cover the upper subzone, often closely packed; on the lower zone, less tolerant animals, mussels and gastropods, mainly littorinids and patellids, cling to the wave-battered algal belts (very apparent on tidal coasts) or hide in the crevices of the trottoirs of algae on very exposed shores. The algae are rich and varied. The living assemblages on soft substrates consist chiefly of typical burrowers: polychaetes, bivalves, isopod and amphipod crustaceans, and also very mobile walking crabs.

Infralittoral

The upper limit of the infralittoral zone is defined by species which cannot endure emergence, and its lower one is marked by the disappearance of marine phanerogams (sea-grasses) and photophilic algae. The limits are variable, the upper depending on the exposure of the shore (*see* above) and the lower one on light penetration, which in turn is dependent on water turbidity. In the Western Mediterranean this limit is around 35 m depth, but varies greatly according to geographical situation, for example: 15 — 20 m in Medes Islands and other Costa Brava localities, and 35 — 40 m in Majorca. It is possible to subdivide this zone physically according to the general pattern of water movement (Fig. 8.2). It is important, among other things, in bringing nutrients to plants and food to filter-feeding animals. According to Riedl (1966), water movements are multidirectional from the surface down to about 3 metres deep. This is the breaker zone, whose incidence on the two previous ones has already been mentioned. From 3 m to about 11 m, the water moves in a plane, oscillating alternatively in both directions; and from 11 to 25, or more, metres the wave movement is a unidirectional flow, depending on deep- or wave-generated currents. The hydrodynamic regimes of each zone affect the bionomy of the living assemblages. The deepest zone is a transition towards the circalittoral with an increasingly high number of dark microhabitats. On rocky substrates the photophilic algae predominate with associated animal communities; in tropical seas the corresponding zone is the domain of the coral reefs. Soft substrates are characterized by burrowing endofauna (bivalves, polychaetes, echinoderms, gastropods, etc.), meadows of marine phanerogams (in the Mediterranean mainly *Posidonia*, but also *Cymodocea* and others) and some algae. The main plant assemblages are taken as a reference for this zone, because its biomass is predominating, despite a very rich and varied associated fauna.

Circalittoral

The circalittoral zone extends from the lower level of the photophilic algae and sea-grasses to the maximal depth consistent with the life of multicellular algae which can thrive in relatively dim light, below 5% of surface radiation and down to 100 m and more. This zone coincides approximately with the lower continental shelf and its edge. Silty substrates do not support algae, but it is always possible to find criteria to place these unvegetated bottoms in the frame of the plant-defined zones. This may be helped, for example, by a rocky outcrop which behaves as an 'island' amid a 'sea' of mud, containing characteristic sessile, mainly crustaceous (coralline) algae and animals. The loss of sharpness in the gradients and the coincidence with the shelf edge (with different hydrodynamic and sedimentation regimes) blurs the boundary between this zone and the bathyal. The seasonal development of a thermocline at the boundary between the infra- and circalittoral zones may confuse or complicate their mutual relationship. Fixed and colonial animals (sponges, cnidarians, bryozoans, tunicates) and red calcareous algae predominate on hard bottoms. The algae growing on coarse gravelly substrates coalesce the calcareous sediments to form a continuous, organogenous substrate on which the coralligenous community develops; this assemblage is the most structured and species-rich in the Mediterranean benthos. In dark overhangings and caves, an impoverished coralligenous community, with very interesting endemics and dark-adapted species appears. The absence of plants in the open muddy bottoms and in other situations leads inevitably to detritus-based food chains. These are typical of the aphytal system. Food chains are rich and fragile (caves) or very productive and exploited by man (trawl fishing bottoms).

The bathyal zone extends from the lower boundary of the circalittoral to the deepest, muddy bottoms of the Western Mediterranean. Characteristically there is an impoverished fauna of commercially exploitable shrimps and colonial cnidarians. This zone has been described in detail in Chapter 7.

8.2. THE SUPRALITTORAL ZONE IN A SEA ALMOST WITHOUT TIDES

The supralittoral zone starts where the terrestrial halophilic animals and higher plants (which characterize the so-called adlittoral zone) end. Some authors (Fletcher, 1980; Ballesteros, 1982) consider the zone dominated by terrestrial, but typically littoral lichens, included in the supralittoral zone. According to these workers, lichen communities with *Lecanora helicopis*, *Xanthoria parietina*, *X. aureola*, *Dirina repanda*, *Lecidea sulphurea*, *Roccella phycopsis*, *Anaptychia fusca*, *Ramalina* spp., *Caloplaca* spp. and so on belong to the supralittoral. This is not the point of view of other authors: Molinier (1960), Pérès and Picard (1964) and Boudouresque (1971).

8.2.1. Soft substrates

On soft substrates (usually gravel, sand or silty sand, in beaches or sandbanks) two communities are found which are related to the humectation conditions. When the grain size is large, and there is little or no organic debris on the beach, dessication is fast and the community is characterized by two crustaceans which feed on small detritic particles in the sand: the amphipod *Talitrus saltator* and the isopod *Tylos europaeus*; both remain buried in the sand during the day and emerge from their burrows by night. Less frequent associates are Coleoptera (*Bledius*) and Diptera, mainly found in the wetter part of the beaches.

Small-grained sands are more resistant to dessication and are more-or-less covered by organic debris (e.g. leaves of *Posidonia* and other sea-grasses and mixed algal remains) and here therefore a number of detritus feeders and predators are usually found. Amphipods of the genus *Orchestia*: *O. mediterranea*, *O. montagui*, and others, are common on *Posidonia* balls. Further components of the fauna are *Talorchestia*; *Tylos sardous* and *Halophiloscia couchii*, gastropods (*Alexia*), chilopods, coleoptera, pseudoscorpionidae, myriapods, dermapterans and dipterans. If there are boulders around, the faunal assemblage resembles that on rocky substrates, with littorinids and idoteids, the small pulmonate *Truncatella subcylindrica*, *Alexia firmini*, etc. The sand or muddy sand between boulders gives shelter to the same *Orchestia* assemblage reported above.

On the silt of mud-flats, under the growth of glassworts (*Salicornia* and *Arthrocnemum*), marine blue-green algae develop. Among the animals, burrowing insects are abundant: beetles (*Bledius*) and dipterans, as well as the isopods (*Halophiloscia*) and gastropods (*Alexia*) already mentioned. Because of the continued humectation of this type of substrate, its assemblages could be classified as mediolittoral.

8.2.2. Hard substrates

On rocky or hard substrates (Fig. 8.5) the plant and animal assemblages of the supralittoral zone are remarkably homogeneous on all marine shores, except the polar ones, perhaps because not many adaptive solutions could be found to the environmental stresses associated with so harsh a habitat. In the Western Mediterranean coasts the floristic and faunistic assemblages have the following components:
(a) Blue-green and green algae, normally epilithic but also with endolithic blue greens on calcareous substrates. The plastic morphology of these unicellular plants adapts to the abrupt environmental gradients of the supralittoral producing a clear zonation of species and ecophenes (*sensu* Drouet; Le Campion-Alsumard, 1979); the most common genera are *Calothrix*, *Entophysalis* and *Plectonema*, which can compete with the lichens for light and space. In this chapter we have followed the traditional taxonomy and nomenclature of the Cyanophyceae, but we suspect that many of the reported species may only be ecophenes, morphological variations induced by environmental factors (Drouet and Daily, 1956; Drouet, 1968, 1973, 1977).
(b) Lichens, especially *Verrucaria symbalana*, black on the more paler rocks, rocks that can be cyanophyceae-pitted or carpeted with cirripeds. The nomenclature for lichens follows that of Hawksworth *et al.* (1980).
(c) The gastropod *Melaraphe* (=*Littorina*) *neritoides* and other vicariant species from all over the world (such as *M. punctata*, which overlaps with *M. neritoides* in parts of the Spanish and North African coasts); they are ecologically more akin to pulmonates than to prosobranchs. This littorinid is almost the only herbivore in the supralittoral (if one excludes the nocturnal incursions of the mediolittoral limpets). It is very abundant and characterizes, together with *Verrucaria*, the living assemblage of the supralittoral rocky zone.
(d) The isopod *Ligia italica*, a detritivore, is also the sole occupant of its niche, having many related species and genera in the shores of other seas. Unlike *Melaraphe*, this isopod is very active, but both species go down to the mediolittoral level, especially at night, in search of food.
(e) The cirriped *Chthamalus depressus*, in the lower part of the zone and in crevices and sheltered areas, shows also a very unusual adaptation. It is a filter-feeder but it inhabits places which are only irregularly sprayed with water. The richness of organic particulate matter in the surface layer of the water no doubt accounts for the adequacy of such a feeding.

Fig. 8.5. Typical aspect of the supralittoral and mediolittoral zones of a Western Mediterranean sea-coast (based on that of Medes Islands, Girona, NE Spain). Foreground, from top to bottom: supralittoral rock, with *Chthamalus depressus*, *Verrucaria symbalana*, *Melaraphe neritoides*, *Ligia italica*, cyanophyceae and an intruder from the mediolittoral, the crab *Pachygrapsus marmoratus*. In the wave-eroded base of the limestone the upper mediolittoral zone appears, with *Ralfsia verrucosa*, *Patella rustica*, *Chthamalus stellatus*, *Monodonta turbinata*. The lower third of the illustration is occupied by the *trottoir* (lower mediolittoral rock); its upper, well-lit surface is formed by the encrusting growth of *Lithophyllum tortuosum*; other algae present are *Corallina elongata*, *Ulva rigida*, *Laurencia pinnatifida*, *Bryopsis muscosa*, *Chaetomorpha capillaris* and, over the limpets' tests, *Gelidium pusillum*, *Polysiphonia sertularioides* and more *Bryopsis*. The animals depicted are *Patella coerulea*, *Acanthtochiton fascicularis*, *Mytilus galloprovincialis*, *Actinia equina*, *Eriphia spinifrons*. The lower surface of the *trottoir* entablature is a sciaphilic assemblage belonging to the infralittoral rock; the species represented are the algae *Schottera nicaeensis*, *Plocamium cartilagineum*, *Lomentaria articulata* and the animals *Balanus perforatus*, *Aglaophenia kirchenpaueri*, *Spongionella pulchella*, *Paracentrotus lividus* and *Blennius cannevae*, besides more mussels not covered by *Lithophyllum* like its mediolittoral counterparts. In the background, a general landscape of a wave-battered coast is depicted; the bird is a young *Larus cachinnans michahellis*, a herring gull, and the algal belt between the supra and the mediolittoral zone is made of *Enteromorpha compressa*, a nitrophilic species thriving where abundant sea bird's faeces enrich the water with extra nutrients. Note the importance of the *trottoir*, which in fact is an overgrowth of the hard littoral substrate. For the sake of clarity, the sea surface has been depicted below its mean level, as if the water ebbed all along the shore depicted. (Drawing by M. Zabala in *Els sistemes naturals de les illes Medes*, Ros et al., 1984, by kind permission of the Institut d'Estudis Catalans, Barcelona).

8.2.3. Pools and crevices

Besides the main communities, two further habitats deserve mention: the littoral pools (with inconstant temperature and salinity and usually short lived) and the rock crevices (more humid and receiving less radiation than the surrounding rock surfaces). Both can be considered as extensions (less and more stable, respectively) of the mediolittoral fringe.

In the hypo- to hyperhaline pools unicellular flagellate algae of the genera *Brachiomonas*, *Chlamydomonas*, *Tetraselmis*, *Pyramimonas* abound, together with cyanophyceae (*Oscillatoria*, *Microcoleus*, *Schizothrix*) and multicellular chlorophyceae of the genera *Enteromorpha*, *Rhizoclonium* and *Cladophora*. Three animals can be considered characteristic in the Costa Brava rock pools: the harpacticoid copepod *Tigriopus brevicornis*, the beetle *Ochthebius quadricollis* and the larvae of the mosquito *Aëdes mariae*. Other equally euryoic species can be found as vicariants in other places of the Western Mediterranean coasts.

The crevice habitat in the Mediterranean (Kensler, 1965) provides shelter for marine species that in deeper levels of the littoral are normally found under the rocks, and also for terrestrial species well adapted to resist in these small, dark and relatively stable spaces. The more peculiar and curious of these non-marine inhabitants are acarina (*Hydrogramasus*, *Halotydeus*, *Bdella*), pseudoscorpions (*Garypus*) and chilopods (*Henia*). When the size of these habitats increase to reach the dimension of caves, the terrestrial species disappear, and the marine supralittoral species can be accompanied by species that normally are found in medio- and infralittoral open ground: that is the case with *Hildenbrandia rubra* (nomenclature for the algae follows Ballesteros and Romero, 1982) and *Phymatolithon lenormandii*, red algae which encrust the supralittoral humected walls of dark caves. The humidity, more than the lack of illumination, seems to be the main factor responsible for the 'ascension' of species in the supralittoral and mediolittoral caves and overhangings (*see* section 8.7 of this chapter).

8.3. THE MEDIOLITTORAL: A ZONE BETWEEN 'TIDES'

The boundaries of the mediolittoral zone are not clear-cut on soft bottom, where the degree of humectation not only depends on wave action and changes in sea level, but also on the grain size of the sediment — an important factor that defines the capacity for retaining water. The behaviour of many burrowing species, which migrate in the sand according to the degree of wetness, adds further difficulties to the delimitation of this usually narrow fringe.

8.3.1. Soft substrates

Two communities have been described from the mediolittoral pebble, gravel and sand bottoms and a third assemblage comes from muddy bottoms, which are generally brackish. The first, the community of the mediolittoral coarse detritic bottoms, has as its main components the amphipod *Gammarus olivi* and the isopod *Sphaeroma serratum*, together with the ubiquitous runner crab (*Pachygrapsus marmoratus*) and the amphipod *Allorchestes aquilinus*, more common in the infralittoral. These crustaceans, with the polychaete *Perinereis cultrifera* and some oligochaetes, live between the pebbles, eating detritus and moving between, below or on boulders. When the beach is made of gravel or sand, the community includes the polychaetes *Nerine cirratulus*, more abundant in fine sand, and *Ophelia bicornis*, which prefer coarser grains, the isopod *Eurydice affinis* and the bivalve *Mesodesma corneum*; they occupy different levels of the beach according to their preferences for humidity.

On muddy bottoms, generally associated with brackish waters, cyanophyceae can cover the sediment, which is burrowed by nereids and other polychaetes; the emerging stratum, usually of phanerogams, shades, stabilizes and adds food to the proper aquatic environments.

8.3.2. Rocky substrates

Upper mediolittoral rock

Two main subzones can be recognized on the basis of a number of factors that in the order of decreasing importance are humectation, light intensity, nutrient availability and type of substrate. The presence of unusual combinations of environmental factors can lead to a confusion of these two subzones, or else permit the recognition of only one, normally the upper subzone as in the Spanish coast near Gibraltar. The community of the upper mediolittoral rock is characterized by the barnacles *Chthamalus stellatus* and *C. montagui*, which can cover the rocks, especially in exposed places, giving them a characteristic ochraceous hue, and by cyanophyceae, which establish in more sheltered areas. These blue-green algae are epi- or endolithic, and can perforate the tests of the barnacles as well as the limestones. The more common epilithic species are *Entophysalis granulosa*, *Brachytrichia quojii*, *Gloeocapsa crepidinum*, *Lyngbya confervoides* and *Calothrix* and *Rivularia* species; among the endolithic ones, *Plectonema terebrans*, *Mastigocoleus testarum*, *Hyella* species, etc.

The lichen *Arthopyrenia halodytes* is present both on the rock and on the tests of barnacles, while the brown alga *Mesospora macrocarpa*, common only on granitic substrates, also covers the shells of limpets. Of these, *Patella rustica* (= *P. lusitanica*) is very abundant, and like other species (such as *P. ferruginea*, restricted to the insular shores), feeds on the attached algae. *Melaraphe neritoides* and *Ligia italica* can be present in this level, coming from the supralittoral in calm sea, and the crab *Pachygrapsus marmoratus* walks over this and the lower levels, as well as the gastropod *Monodonta turbinata*, which is more abundant in the infralittoral.

The algae form belts. The most commonly present is that of *Rissoella verruculosa*, which is also the lower one and the only perennial, although solely the basal part of the thalli persist in autumn; this belt is absent or inconspicuous on limestones, where *Ralfsia verrucosa* can take its place; *Nemalion helminthoides* can accompany *Rissoella*. Higher in the rock and restricted to the winter months, *Porphyra leucosticta*, *P. umbilicalis* and others can appear, both in exposed and semi-exposed shores. When a rich nutrient supply is present, for example through the depositions of coastal marine birds, *Ulothrix*, *Rivularia mesenterica* and others can cover the regularly splashed rocks. At the very level of the sea in calm places, *Ralfsia verrucosa*, *Nemoderma tingitanum*, *Scytosiphon lomentaria*, *Enteromorpha compressa* and, if there is pollution by organic matter, *Blidingia minima* are common, as is *Bangia atropurpurea* in unstable hard substrates. All of them are browsed up by the above-mentioned gastropods.

The lower mediolittoral rock

The occurrence of facies, that is, the local abundance of one or few species, as a response to some specially favourable environmental factor, complicates the systematization of the living assemblages of the lower mediolittoral rock. This is an anticipation of the progressively complex plant and animal communities that will appear along any vertical transection across and down any typical Western Mediterranean shore.

The communities of the lower mediolittoral rock are characterized by the presence of all, or at least of some, of the following species: three melobesiae, red encrusting calcareous algae: *Lithophyllum tortuosum*, which forms little cushion-like masses; *L. papillosum* (= *Goniolithon papillosum*), with a papillate thallus; and *Neogoniolithon notarisii*; some cyanophyceae; *Nemalion helminthoides*, a red algae growing on exposed rocks, and two molluscs, the chiton *Lepidochitona corrugata*, on the melobesiae or in shallow burrows in the limestone, and the limpet *Patella aspera*, whose shell is often covered by calcareous algae of the same species that this gastropod eats.

Other characteristic species are generally restricted to the next facies we shall consider. Yet there is a set of common algae and animals which also appear in lower levels and, notwithstanding, are associated with the common species in this zone. They are, among the seaweeds: the green algae *Bryopsis muscosa* and *Chaetomorpha capillaris*, the red algae *Ceramium rubrum*, *Gastroclonium clavatum*, *Polysiphonia sertularioides*, *Laurencia pinnatifida*, *L. papillosa*, *Callithamnion granulatum* and others. The animals living in the holes or crevices, among the algae and on other mediolittoral substrates are very numerous, as a response to a less harsh environment and a more diversified food supply, consisting chiefly on the plants of this zone. Mention can be made of a colonial foraminiferan (*Miniacina miniacea*), sponges (*Halichondria* spp.), cnidarians (the sea anemone *Actinia cari*, the hydrarian *Sertularella ellisi* and the stoloniferan *Cornularia cornucopiae*), molluscs (*Achanthochiton fascicularis* and *Lepidochitona corrugata*; the bivalves *Mytilaster minimums*, *Musculus costulatus*, *Modiolus barbatus*, *Cardita calyculata*, *Venerupis irus*, *Striarca lactea* and young mussels), many nematodes and polychaetes (mainly syllids), bryozoans (*Schismopora armata*), sipunculans (*Phascolosoma granulatum*), crustaceans (the amphipod genera *Allorchestes*, *Amphithoe*, *Hyale*; the isopods *Ischyromene lacazei* and *Dinamene bidentata*; and the decapod *Eriphia spinifrons*), tunicates (*Diplosoma gelatinosum*) and some others. All these species have means of fixation to the substratum or of fastening to the algae, as required by the strong movement of the water at this level.

The more characteristic communities or facies of the lower mediolittoral rocks are as follows:

(a) The belt of the black phaeophycean *Ralfsia verrucosa*. This is usually found on sheltered coasts extending over all the mediolittoral rocks if there is no competition from other algae, chiefly melobesiae, or is restricted to the upper stretches of the lower mediolittoral rock when such competition is effective.

(b) The aspect of cyanophyceae, which are sometimes very abundant in the absence of other algae; the rock, when calcareous, is heavily eroded by the blue-greens and by *Lepidochitona corrugata*.

(c) The community of *Nemoderma tingitanum* in exposed rocks near the infralittoral level.

(d) The community of *Neogoniolithon notarisii*, an encrusting calcareous alga which can eliminate all other species except the two belonging to the genus *Lithophyllum*, and which cements small rocks and shell fragments in a coherent structure; it appears also in the infralittoral (Fig. 8.10).

(e) The community of harbours and heavily polluted waters, with species that grow in eutrophic environments, like *Enteromorpha intestinalis*, *E. compressa*, *Scytosiphon lomentaria* and other opportunistic or pioneering species.

(f) The facies of *Pollicipes cornucopiae*, a pedunculate cirriped common in exposed Atlantic shores, but which can also appear in very turbulent, splashed areas of the Algerian coast.

The trottoir

A place of special distinction must be reserved for:

(g) the characteristic community of *Lithophyllum tortuosum*, the so-called *trottoir* from a French word meaning sidewalk or *corniche* (entablature). According to the position of the viewer it looks like a sidewalk from above and a corniche from below. It is a very common community in the northern part of the Western Mediterranean, especially in exposed coasts, where it can cover all the lower mediolittoral rocks. *L. tortuosum* grows on vertical or subvertical rock surfaces, and may attain up to 2 m in breadth and 1 m in thickness; a section of this 'sidewalk' shows its structural complexity (Fig. 8.6), comparable to other deeper organogenous formations in the infralittoral and, especially, in the circalittoral (coralligenous). The exterior surface is covered by a layer less than 1 cm thick of living *L. tortuosum*. Below it is an intermediate, amorphous layer consisting of the uncompacted skeletons of the dead algae. The innermost zone is very hard and white and is composed of almost pure calcite. This results from the clogging and consolidation of the skeletons of calcareous algae and some accompanying plant and animal species. A fourth, lower zone is perforated by algae, sponges, molluscs, fungi, etc. and harbours a very specialized fauna and flora (*see* below). Finally, a fifth zone, not exactly belonging to the *trottoir* can be recognised in the assemblage of infra- and even circalittoral algae and animals, which seek the shadow and protection that the organic building provides; it has been termed precoralligenous console and will be dealt with in the appropriate section. A very abundant fauna, including many juvenile forms, occupies the fissures, crevices and holes of the *trottoir*, especially in the first and fourth zones. The following are characteristic mediolittoral species: the nemertean *Nemertopsis peronea*, the molluscs *Gadinia garnoti, Onchidella celtica, Fossarus ambiguus* and *Lasaea adansoni*, the isopod *Campecopea hirsuta* and the spider *Desidiopis racovitzai*.

Besides the characteristic algae mentioned above, *Dermatolithon confinis* encrust the dark crevices of the *trottoir*. The typical structure just described cannot develop on calm coasts and on not vertical substrates, and in such places only small cushions of *L. tortuosum* occur (Fig. 8.10). Conversely, in very exposed places the *trottoir* has its maximum development. However, the wave impact during seasonal storms may break off large fragments of the *trottoir*, which thus is continuously eroded and rebuilt.

Mediolittoral rock pools

The mediolittoral rock pools have a more regular supply of sea-water than the supralittoral ones and are less fluctuating. The occupation of many of these pools by opportunistic species of the genera *Cladophora* or *Enteromorpha*, among others, is typical; many tolerant photophilic species of the

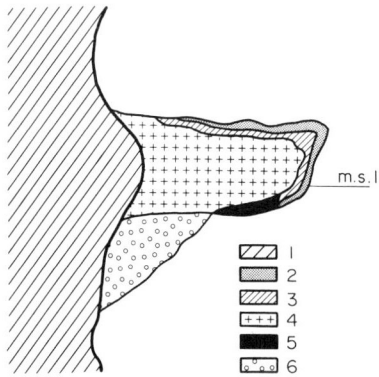

Fig. 8.6. Diagrammatic section of a *trottoir* of *Lithophyllum tortuosum*. 1, rocky substrate; 2, living *Lithophyllum tortuosum*; 3, dead *L. tortuosum* whose interstices are not clogged; 4, biogenous, solid rock made from dead *L. tortuosum* with clogged interstices; 5, amorphous layer with infralittoral assemblage; 6, sciaphilic (precoralligenous) enclave. msl, mean sea level. (Redrawn from Pérès and Picard, 1964; by permission of the authors).

infralittoral (such as *Padina pavonica* and *Ulva rigida*) can also invade these well-lit microhabitats. The community of the mediolittoral holes and caves is not very different from that of the supralittoral ones, with *Hildenbrandia rubra* as the main characteristic species.

8.4. THE CONTINUOUSLY IMMERSED INFRALITTORAL

8.4.1. The major communities of algae on hard substrates

The communities of relatively large seaweeds can be very complex and diverse (Fig. 8.7). They begin a few centimetres below the mean sea level, where the typically mediolittoral species (i.e., those which cannot endure a continued immersion, such as *Lithophyllum tortuosum* and *Nemoderma tingitanum*)

Fig. 8.7. Typical aspect of the photophilic algae community of a Western Mediterranean coast (based on that of Medes Islands, Girona, NW Spain). Foreground, from top to bottom; left: algae (*Padina pavonica, Codium vermilara*) and animals (*Paracentrotus lividus, Verongia aerophoba, Halichondria panicea, Blennius gattorugine, Aiptasia diaphana*); right: algae (*Asparagopsis armata, Dictyota dichotoma, Amphiroa rigida, Plocamium cartilagineum, Lithophyllum incrustans, Corallina elongata*); animals (mullets, *Flabellina affinis* on *Eudendrium* sp., *Clavellina lepadiformis, Trypterigion tripteronotus*, a pagurid in a gastropod shell, *Chiton olivaceus*, and *Pomatoceros triqueter* tubes). Background: infralittoral algal belts (left) and (centre): *Ircinia fasciculata, V. aerophoba, Arbacia lixula, Hymeniacidon sanguinea, Anemonia sulcata, Octopus vulgaris* and *Codium bursa* amid the same species identified above. (Drawing by M. Zabala in *Els sistemes naturals de les illes Medes*, Ros et al., 1984, by kind permission of the Institut d'Estudis Catalans, Barcelona).

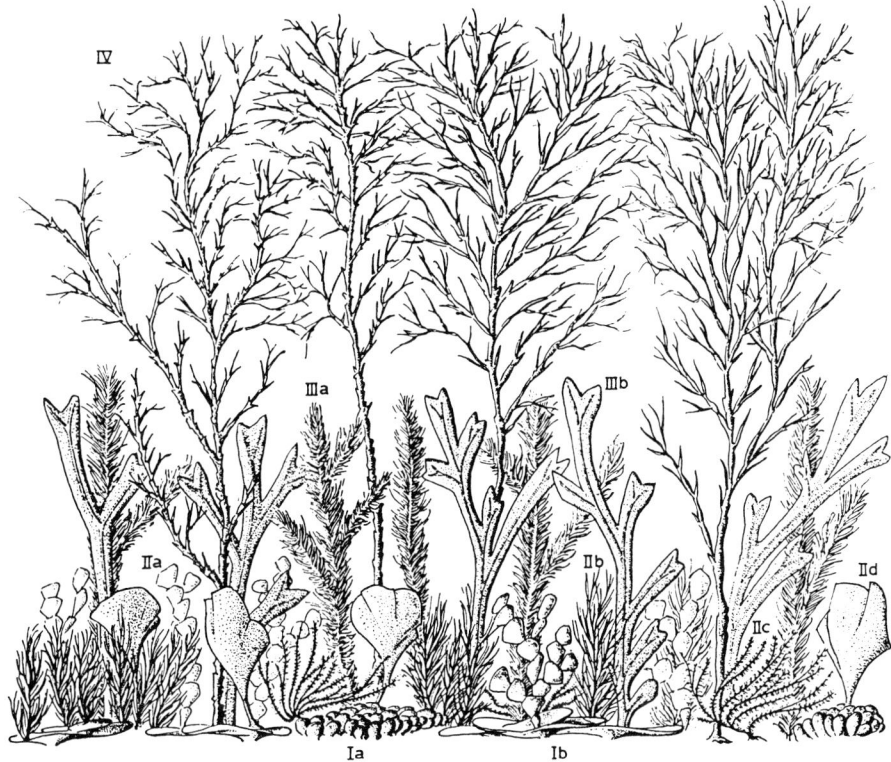

Fig. 8.8. Diagrammatic multi-strata disposition of the algal vegetation of a *Cystoseira* assemblage of the infralittoral étage as example of stratification in the photophilic algae community. There are four different strata: I, encrusting and cushion-like forms (*Valonia, Peyssonnelia, Mesophyllum*, etc.); II, turf-like forms (*Halimeda, Cladophora*, etc.); III, low erect ('arbustive') forms (*Digenea, Dictyopteris*, etc.); IV, high erect ('arboreal') forms (*Cystoseira*). The composition of each stratum can vary depending on multiple factors, and some strata can be absent. Usually, the alga or algae of the higher stratum gives its name to the facies or community (besides *Cystoseira* spp., other common components of this stratum are *Halopteris scoparia, Jania rubens, Corallina elongata, Polysiphonia* spp., and others). (From Ernst, in Riedl, 1966; by permission of Paul Parey Verlag).

disappear. At the same level, species that are unable to resist a long emersion (such as *Cystoseira* species, *Halopteris scoparia*, etc.) make their appearance. In many cases an assemblage of *Ceramium ciliatum* and *Gelidium pusillum* forms a fringe between the medio- and the infralittoral zones. This infralittoral zone ends with the disappearance (from lack of light) of photophilic species such as the algae *Padina pavonica*, *Cladostephus hirsutus* and others, and the sea-grass *Posidonia oceanica*.

A well-developed community of photophilic algae is like a minute terrestrial forest: a one-foot-high carpet with up to four different strata, which sway with the movement of the waves (Fig. 8.8). There is an encrusting, basal stratum consisting of calcareous algae as well as skeletal material of polychaete worms, bryozoans and gastropods. Above this a caespitose, intermediate layer can appear, composed by small calcareous or soft algae. These species and the ones building crusts and cushions in the lower stratum are sciaphilic species when the two remaining superior strata are dense and reduce the available light. The third stratum consist of low ('arbustive') erect forms, and the fourth and upper one is made of large phaeophyceans or rhodophyceans, which in turn can have an important epiphytic covering (equivalent to the 'arboreal' stratum, to continue the forest analogy). This layered pattern contributes to the success of the photophilic algae and to the high diversity of their communities. Depending on the environmental conditions, or on the biogeographic restrictions, each stratum is formed by different species, and the possibilities of combination are very high. This complexity adds to the relatively broad

TABLE 8.1. Infralitoral communities with dominance of algae on hard ground. The group to which the animals belong is indicated: hydr., hydrarians; bryo., bryozoans; gast., gastropods; cirr., cirripeds; biv., bivalves; anth., anthozoans; spon., sponges; ech., echinoderms. High light intensity; photophilic communities.

		High stratum	Lower strates (including epiphytes)	Animals
With dominance of *Cystoseira* species. Between 0 and 0.5 m deep.	In exposed substrate	*Cystoseira stricta* *Cystoseira mediterranea*	*Polysiphonia deludens* *Ceramium rubrum* *Jania rubens* *Feldmannia caespitula* *Lithophyllum incrustans*	*Coryne muscoides* (hydr.) *Sertularella ellisi* f. *lagenoides* (hydr.) *Schismopora armata* (bryo.) *Vermetus triqueter* f. *gregarius* (gast.) *Mytilus galloprovincialis* (biv., upper levels) *Balanus perforatus* (cirr.)
	In sheltered or moderately exposed waters	*Cystoseira crinita* *Cystoseira compressa* *Cystoseira caespitosa* *Cystoseira eregovicii* *Cystoseira elegans* *Cystoseira balearica*	*Padina pavonica* *Halopteris scoparia* *Sargassum vulgare* (With notable seasonal changes)	Rich and varied fauna, with many sedimentivores and detritivores.
Without *Cystoseira* species	In more exposed waters	Absent or reduced	*Asparagopsis armata* (more frequent in the tetrasporic phase of *Falkenbergia rufolanosa*) *Corallina elongata* *Laurencia obtusa* *Jania rubens*	Facies (less algae): **Mytilus galloprovincialis* and other mytilids **Anemonia sulcata* and associated crabs (*Inachus dorsettensis*, *Macropodia longirostris*) and other decapods (*Scyllarus arctus*) **Hydrarians (*Eudendrium capillare*, *E. racemosum*, *Bougainvillia ramosa*, *Sertularella ellisi*, *Halecium* ssp.), accompanied by other cnidarians (*Clavularia ochracea*), tunicates (*Clavellina lepadiformis*) and sponges (*Hymeniacidon sanguinea*)
	In calm waters		*Padina pavonica* *Cladostephus hirsutus* *Halopteris scoparia* *Dilophus fasciola*	*Amphiroa rigida* *Lithophyllum incrustans* *Acetabularia acetabulum* *Dasycladus vermicularis* Many deposit-feeders, molluscs, polychaetes, crustaceans, echinoderms

Local perturbations	Grazing by sea urchins, sedimentation, sand abrasion	None	Lithophyllum incrustans	Arbacia lixula, Paracentrotus lividus (ech.) Anemonia sulcata (anth.)
	Pollution	None	Lithophyllum incrustans Gelidium pusillum Corallina elongata Ulva rigida Enteromorpha spp.	
Solid, massive structures (see Fig. 8.10)		None		*Vermetus (Spiroglyphus) cristatus *Serpulids: Serpula, Pomatostegus, Protula
Low light intensity: sciaphilic communities				
	Exposed places, under the overhangs of the trottoir of Lithophyllum tortuosum	Schottera nicaeensis Cladophora pellucida Pterocladia capillacea	Gelidium melanoideum Rhodophyllis divaricata Gymnothamnion elegans Valonia utricularis (a) Plus cold-water species Lomentaria articulata Plocamium cartilagineum Callithamnion tetragonum (North of Western Mediterranean) (b) More warm-water species Botryocladia botryoides Polyphysa parvula (Tyrrhenian, etc.)	Actinia equina (anth.) Coryne muscoides (hydr.) Halichondria panicea (spon.)
	Lower level of the infralittoral; calm waters. Transition to the coralligenous	Cystoseira spinosa Halopteris filicina Codium vermilara Sphaerococcus coronopifolius Codium bursa Halimeda tuna Udotea petiolata	Rhodymenia ardissonei Peyssonnelia rubra Peyssonnelia squamaria Codium effusum Callithamnion tripinnatum Acrosorium uncinatum	Facies (less algae): *Alcyonium acaule *Sponges: Ircinia fasciculata, Hymeniacidon sanguinea, Petrosia ficiformis, Hamigera hamigera

distribution of this zone and maintains a rich faunal composition which, although with not as large a biomass as the algae, nevertheless shows a high species diversity characteristic of the richest communities or facies of the zone.

The most conspicuous or important ones of these are shown in Table 8.1. The most characteristic photophilic algae communities have *Cystoseira* species as the main algal component. These phaeophyceans represent most of the biomass of these communities and their structural role has already been emphasized. In exposed areas, *Cystoseira mediterranea* and *C. stricta* are vicariants. The first has been found in the French and Spanish Catalan coasts, Almería, Naples, etc., and the second appears in some parts of the French and Italian coasts and in Corsica. The specific composition of these vicariant assemblages are, for the most, identical. Two filter-feeders often accompany this community: *Mytilus galloprovincialis* (in the upper, boundary levels) and *Balanus perforatus* (in the middle ones). The former species can also form a facies of its own (*see* Table 8.1).

Communities in sheltered waters dominated by *Cystoseira* species have a well marked seasonal pattern depending on their geographical distribution. In the Costa Brava, for example, this assemblage appears in the spring and gives way in the summer to the local development of such opportunistic species as *Ulva*, *Enteromorpha* and *Cladophora*. The fauna is rich, and the sedimentivores and detritivores abound, exploiting the somewhat higher sedimentation linked to the decreased hydrodynamism.

Photophilic algal communities without *Cystoseira* species are usually dominated by other phaeophyceans. A community of *Padina pavonica* and *Cladostephus hirsutus* is found in calm and warm waters, with relatively high deposition rates which favour the presence of an understratum of molluscs, polychaetes, crustaceans and echinoderms and other algae, including some with pantropical affinities (*see* Table 8.1).

Assemblages on more exposed places can be regarded as transitional ecotones between the exposed *Cystoseira* communities (near the surface) and the photophilic or sciaphilic ones, calmer due to depth.

Altogether, these assemblages, with the *Cystoseira* communities constitute the typical underwater landscapes of the infralittoral, interspersed with more sciaphilic enclaves and homogeneous facies (*see* below). They also abound in warm rock pools.

In certain places the upper stratum of soft algae is absent or strongly impoverished, due to local factors (e.g. high rate of grazing by sea-urchins, sedimentation, sand abrasion and pollution). In such cases the encrusting algae or sessile calcareous animals form a basal stratum which can constitute large, unispecific assemblages. These appear as solid, massive structures resembling the mediolittoral *trottoir*, but with a lesser degree of spatial organisation (Fig. 8.10). The associated fauna is usually poor, lacking the protection and spatial complexity of the multistrata structure mentioned earlier. However, when crevices and easily burrowable substrate exists (e.g. as provided by the shells of *Vermetus* or the chalky tubes of the serpulids) the accompanying fauna can be rich and show remarkable tropical affinities (no doubt related to the environmental conditions of these shallow 'reefs' or *trottoirs*).

Mytilus galloprovincialis can appear in very dense assemblages, forming an almost continuous carpet of shells, on exposed rocks in clean water or in sheltered places with a fresh water supply, or even in slightly polluted waters. The accompanying animals, hiding in the interstices, and the epibionts make up an abundant but monotonous fauna, especially in polluted locations (*see* below). *Anemonia sulcata* forms a band at the foot of rocky sections, in open or enclosed littorals and also at the boundary with soft, sandy bottoms.

In polluted regions, algae such as *Ulva rigida* or *Corallina* species dominate (possibly there has been some confusion in the determination of *C. elongata* and *C. officinalis*; the facies of polluted places traditionally attributed to *C. officinalis* could be formed by an ecotype of *C. elongata*). The *Cystoseira* species are dependent on environmental stability; when pollution is high, or if there is a strong stress (such as sedimentation or freshwater input), the *Cystoseira* communities disappear and are replaced by a community dominated by opportunistic species (*see* Table 8.1).

Sciaphilic algal communities dominated by *Schottera nicaeensis* occur mainly on exposed places, usually located under the overhangs of the *trottoir* of *Lithophyllum tortuosum* (thus in very exposed, surface waters; Figs. 8.6 and 8.10) or in dark infralittoral crevices. Two communities have been distinguished in the Western Mediterranean (Boudouresque and Cinelli, 1976); the first, with many species with cool-water affinities, is restricted to the Costa Brava and Côte des Albères, while the second, with some warm-water species, appears in the Tyrrhenian and other areas.

In the lower levels of the infralittoral (or in the shade of overhangs and in crevices, but almost always in calm waters) the so-called precoralligenous community appears. This has been described as an impoverished component of the coralligenous, but can be identified as a true lower infralittoral assemblage of sciaphilic algae. The main constituent species are very variable and a large number of facies can be distinguished (Fig. 8.11). The faunistic composition is always very rich and represents a transition community between those of photophilic algae and the coralligenous.

Newly emerged natural or man-made substrates (large piers, rocks, boat and ship hulls) are readily covered by the progressive stages of marine benthic succession (*see* section 8.8). The persistence of these initial stages (through the successive cleaning of the ship hulls, for example, in the same way as grass is kept growing by mowing) led some authors to speak of a 'fouling' community, characterized by, among others: sponges (*Sycon*), hydroids (*Tubularia, Campanularia, Obelia*), bryozoans (*Zoobothryon, Bugula*), polychaetes, ascidians and other equally opportunistic and pioneering algal species belonging to the genera *Ulva, Cladophora* and *Enteromorpha*. Such a community is, however, only a pioneer stage of some other, more mature assemblages (*see* section 8.8).

8.4.2. Diverging adaptations, a characteristic of animal life

A number of animal species are conspicuous in such communities by being of good size, notably sea-urchins, or, because they gather in large numbers such as *Mytilus* (mussels), or build extensive concretions with their dwellings and tubes, such as *Vermetus cristatus*.

The quantitative and qualitative dominance of the plants in the infralittoral level allows recall of forest analogy used to describe algal stratification. The benthic animals are present in the rocky infralittoral communities as are the terrestrial animals in a forest. There they find food, hiding places and appropriate microenvironments, but their biomass is only a small fraction of the total biomass. Almost all can easily withstand relatively violent water movements and high light intensities. They are mainly herbivores, browsers on animal colonies and filter-feeders, with a remarkable representation of species in symbiotic relationship with algae. The multiplicity of microhabitats makes it easy to group the animal species of the rocky infralittoral in at least seven major categories of ecological strategies.

(1) Encrusting species, that cover the substrate and, like the calcareous algae, can modify the surface structure of the underlying rock and thus affecting the eventual settlement by other species. Many sponges (*Hymeniacidon sanguinea*), serpulid polychaetes, the cirriped *Balanus*, bryozoans and colonial ascidians, among others, act in this way.

(2) Sedentary species, which are strongly attached to the substrate. These are chiefly molluscs, as *Acanthochiton fascicularis* and *Mytilus galloprovincialis*, but crustaceans and members of other groups also occur.

(3) Species that spend part of their life in the rock crevices or in the interstices left or formed by other species. These can move around but usually their activity is limited. They also occur in the mediolittoral and other zones. Examples include many polychaetes (*Nereis zonata, Lepidonotus clava, Spirobranchus polytrema,* etc.), crustaceans (amphipods and isopods), molluscs (*Jujubinus gravinae, Muricopsis cristatus* and others), the brittle-stars *Ophiothrix fragilis* and *Amphipholis squamata* and, especially, species which

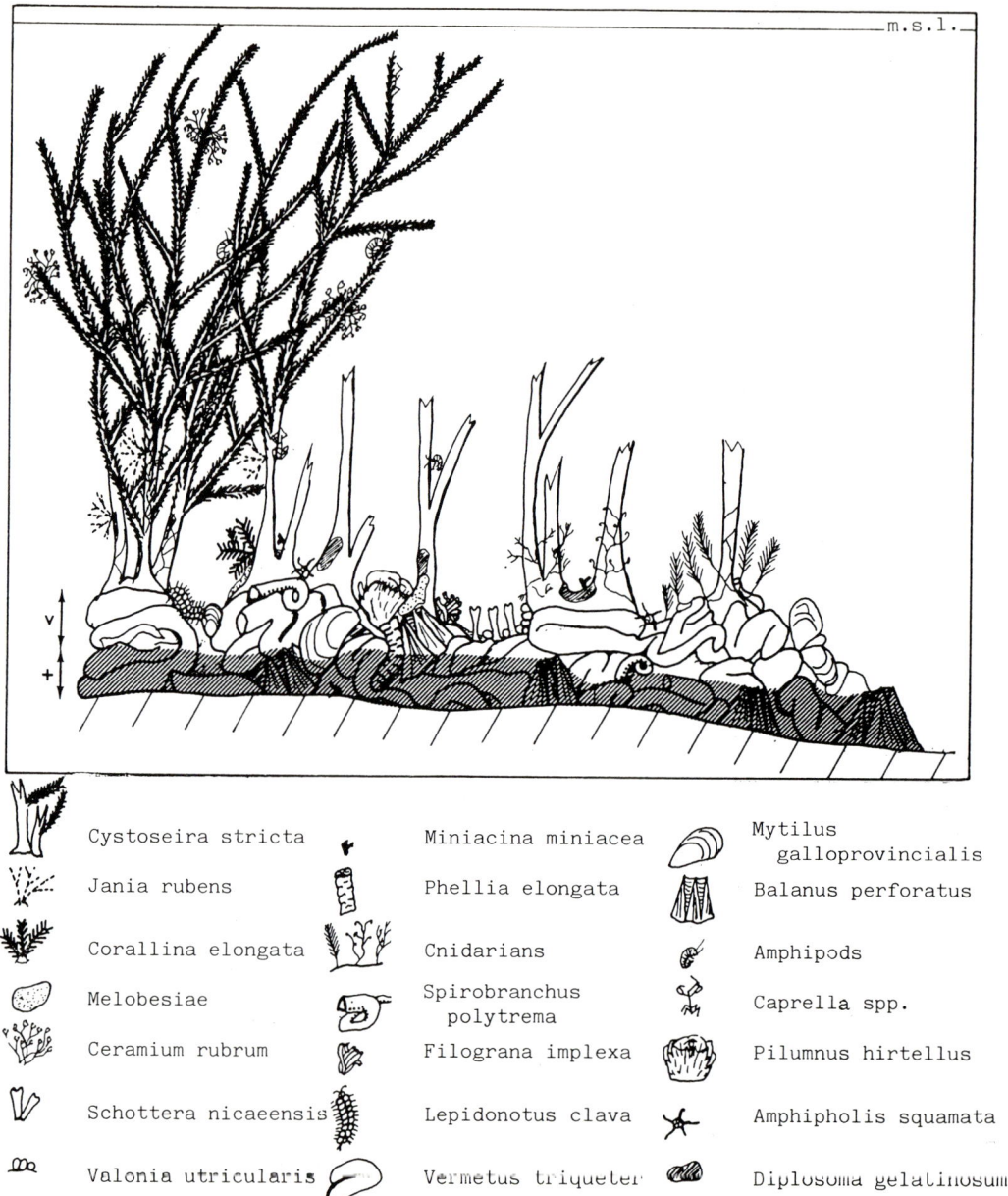

Fig. 8.9. Semi-diagrammatic representation of the *Cystoseira stricta* community; v, living fraction of the basal encrusting concretion; +, dead fraction of the same; msl, mean sea level. (From Bellan-Santini, 1969; by permission of the author).

characteristically cling to the underface of boulders (such as *Coscinasterias tenuispina*, *Asterina gibbosa*, *Haliotis tuberculata*, and *Dendrodoris limbata*). The endobionts of sponges and ascidians also belong to this group, although only from a structural point of view, since ecologically the relationship between a living 'substrate' and its endobionts is more complex, including immune and defence reactions, parasitism or commensalism.

(4) Species of varying size which move from one algal tuft to another, adhering to the fronds or to the animals of the next group; browsers belong to this group. There are many amphipods (e.g. *Dexamine spiniventris*, *Hyale* spp., *Maera inaequipes*); *Cymodoce truncata* and other isopods, together with pantopods, molluscs, chiefly prosobranch and opisthobranch gastropods, and free-living polychaetes, such as *Platynereis dumerilii* and *Lysidice ninetta*. The fauna accompanying the facies of hydrarians (Table 8.1)

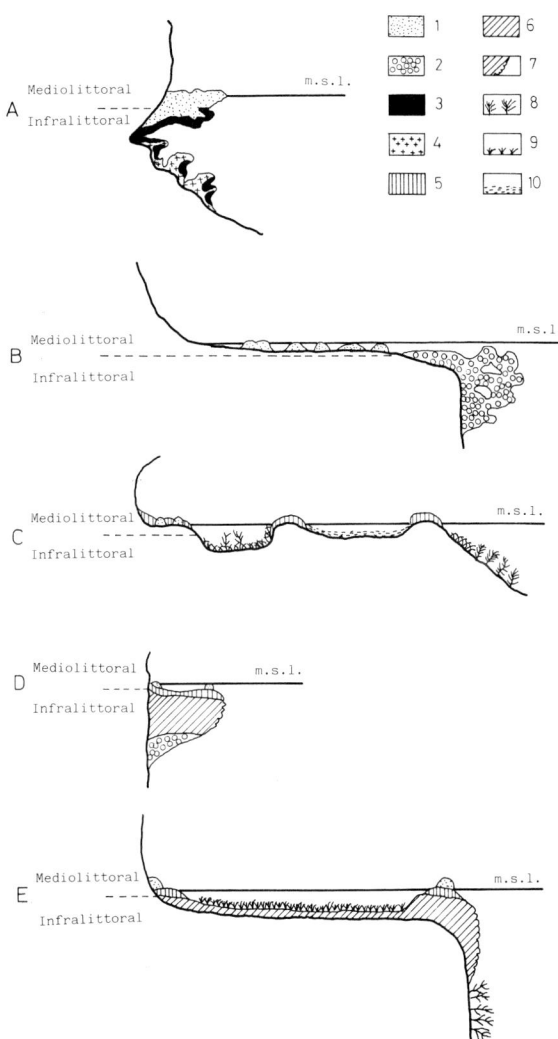

Fig. 8.10. Different types of '*trottoirs*' in the littoral of the Western Mediterranean. A, typical mediolittoral *trottoir* of *Lithophyllum tortuosum* (see also Fig. 8.6). B, infralittoral '*trottoir*' of *Lithophyllum incrustans*. C, false mediolittoral '*trottoir*' of *Neogoniolithon notarisii* (in fact, these are thin coverings of the underlying rock). D, infralittoral '*trottoir*' of vermetids (usually *Vermetus (Spiroglyphus) cristatus*). E, infralittoral platform of vermetids. 1, entablature or little cushions of *Lithophyllum tortuosum*; 2, entablature of melobesiae, mainly *L. incrustans*; 3, sciaphilic assemblages growing as enclaves below the infra- and mediolittoral entablatures; 4, pads of *Corallina* sp. and *Phymatolithon lenormandii*; 5, thin covering of *Neogoniolithon notarisii*; 6, organogenous formation of *Vermetus (Spiroglyphus) cristatus*; 7, living (seaward) vermetids; 8, *Cystoseira* and other exposed infralittoral assemblages; 9, *Laurencia* and other calm infralittoral assemblages; 10, mediolittoral pool assemblages. (Redrawn with modifications from Pérès and Picard, 1964; by permission of the authors).

Fig. 8.11. Typical aspect of the sciaphilic algae community ('precoralligenous') of a Western Mediterranean coast (based on that of Medes Islands, Girona, NW Spain). At left, from top to bottom: *Alcyonium acaule*, *Hymeniacidon sanguinea* on *Spondylus gaederopus*, *Cystodites dellechiajei*, *Myriapora truncata*, *Microcosmus sabatieri*, *Agelas oroides*, *Hemimycale columella*, *Sertularella ellisi*, *Ophiothrix fragilis*, amid *Halimeda tuna*, of which a close up is shown at bottom left (with, on it, *Dermatolithon* sp., *Halecium halecinum*, *Campunularia* sp., *Aetea truncata*, an unidentified ectoproct and, feeding on it, *Polycera quadrilineata* with spawn mass below). At centre and right, from top to bottom and in addition to the above-mentioned species: *Eunicella singularis*, *Codium bursa*, *C. vermilara*, *Cliona viridis*, *Pentapora fascialis*, *Salmacina dysteri*, *Scorpaena porcus*, *Sabella* sp., *Parazoanthus axinellae*, *Peyssonnelia rubra*, *Oscarella lobularis*, *Ircinia dendroides*, *Caryophyllia* sp., *Palaemon serratus*, *Conger conger*, *Botryllus schlosseri*, *Hymeniacidon sanguinea* and *Scyaena umbra*, all amid *Udotea petiolata*. (Drawing by M. Zabala in *Els sistemes naturals de les illes Medes*, Ros et al., 1984, by kind permission of the Institut d'Estudis Catalans, Barcelona).

includes many species of *Caprella*, *Ammothella* and other amphipods, and the aeolidiacean nudibranchs *Flabellina affinis*, *Hervia costai* and *Coryphella pedata*, among others.

(5) Sessile and large species, forming the intermediate stratum (Figs. 8.8 and 8.9) together with algae, with which they compete for space. This group includes sponges (such as *Ircinia fascicullata*, *Hamigera hamigera*), cnidaria (such as *Aiptasia diaphana*, *Anemonia sulcata* and *Balanophyllia regia*), many hydrarian species, some bryozoa (*Crisia occidentalis*, *Pentapora ottomulleriana* and *Turbicellepora magnicostata*) as well as molluscs (*Ostrea edulis*, *Spondylus gaederopus* and *Arca noae*), tunicates (*Microcosmus sabatieri*, *Pyura dura* and *Polycarpa pomaria*) and sedentary, tubicolous polychaetes. While the animals of the first group are chiefly substrate-occupiers which compete strongly for space, these are mainly substrate-suppliers which contribute to the spatial structuring of infralittoral communities, and which are grown over by species of group 7 (*see* Fig. 8.3).

(6) Vagile and large species which are highly mobile and linked in varying ways to the photophilic algal communities. Most of them are predators and include many decapods (such as *Pilumnus hirtellus*, *Alpheus dentipes* and *Calcinus ornatus*), molluscs (including many gastropods and *Octopus vulgaris*), echinoderms (*Paracentrotus lividus*, *Arbacia lixula*, *Genocidaris maculata* and *Echinaster sepositus*) and many fishes, especially the most characteristically benthic representatives of the families Gobiidae and Blenniidae.

(7) Epibiont species, chiefly on algae, but also on sponges, ascidians, molluscs, etc. These include many hydrarians, bryozoans, foraminiferans, polychaetes, tunicates (Fig. 8.3) which settle and live on the surface of other organisms.

This account of the animal species present in the infralittoral zone is as incomplete as the previous lists of algae. Furthermore, many of the species are not exclusive to this zone. However, our purpose is only to summarize and provide a quite complete panorama of the major components of the infralittoral flora and fauna.

8.5. THE INFRALITTORAL SOFT BOTTOM COMMUNITIES AND THE SEA-GRASS MEADOWS

Soft bottoms, especially infralittoral ones, are intrinsically more unstable than hard ones. Thus, the communities established on them support a higher energy flow and show a less spatial organization. These characteristics account for two basic features of these bottoms: reduced species richness (despite a greater environmental diversity) and increased community diversification. It is also possible that perhaps, however, the large number of different communities from soft bottoms results from the more fluctuating nature of the substrates which produces communities almost as variable as the planktonic ones. The classification of these communities is laborious, and their boundaries, even of those that have been well described, are blurred for there are as many transitional stades as there are well-defined communities. Only one of these, the meadow of *Posidonia oceanica*, which is the more mature and rich in species, seems to maintain its identity or to vary only slowly, along what has been described as an ecological succession (*see* Fig. 8.13). This community will be described in more detail than the other, simpler, ones.

The nomenclature used, as in other parts of this account, is that of Pérès and Picard (1964) and Pérès (1967, 1982d). This classification of soft bottoms is mainly based upon physical parameters (such as grain size, exposure to waves or currents and salinity) and corresponds, more-or-less, with the composition of the living assemblages summarised in Table 8.2.

TABLE 8.2. Some infralittoral communities of soft bottoms. The nomenclature adopted is from Pérès and Picard (1964) and Pérès (1982d), from which the specific composition has been summarized.

Communities on coarse soft bottoms. Upper infralittoral	Infralittoral pebbles, in coves in rocky coasts Size: 64 – 256 mm	Amphipods: *Melita hergensis*, *Allorchestes aquilinus* Decapods: *Xantho poressa*, *Porcellana bluteli* Fishes: *Gouania wildenowi*, *Lepadogaster gouani* Turbellarians, Nemertines, etc.
	Coarse sands and fine gravels stirred up by waves. Size: 1 – 4 mm	*Saccocirrus papillocercus* (archiannelid) *Lineus lacteus* (nemertine)
Community on sandy or muddy shallow bottoms. Grain size of sediments, between 0.02 and 2 mm.	Sands protected against breaking waves	Bivalves: *Kellia corbuloides*, *Loripes lacteus*, *Divaricella divaricata* Decapods: *Callianassa tyrrhena*
	Muddy sands in sheltered areas	Plants: *Cymodocea nodosa*, *Caulerpa prolifera*, *Zostera noltii*, *Zostera marina* (rare). Endofauna: *Aricia foetida*, *Heteromastus filiformis*, *Paraonia lyra* (polychaetes); *Loripes lacteus*, *Tapes decussatus* (bivalves); *Upogebbia pusilla* (decapod) Vagile epifauna: *Holothuria poli*, *Holothuria tubulosa*, (echinoderms); *Cerithium* spp. (gastropod); *Clibanarius misanthropus*, *Carcinus mediterraneus* (decapods) Epifauna of plants (when these exist): *Eudendrium rameum*, *Aglaophenia harpago* (hydroids); *Bunodeopsis strumosa* (anthozoan); *Electra pilosa* (bryozoan)
	Fine sands in shallow water of the surf zone	Bivalves: *Donax trunculus*, *Donax multistriatus*, *Macoma tenuis*, *Lentidium mediterraneum* Gastropods: *Cyclonassa donovani* Polychaetes: *Nerinides cantabra* Isopods: *Iphinoe inermis*, *Idothea baltica* Echinoderms: *Echinocardium mediterraneum*

Communities under a certain degree of stress	Brackish waters	Plants: *Zostera noltii*, *Zostera marina* (rare); less salinity, *Ruppia cirrhosa*; less stable; *Ruppia maritima*, charophyta, seasonal growth of filamentous chlorophyceae Bivalves: *Cardium lamarcki*, *Abra ovata*, *Scrobicularia plana* Crustaceans: *Sphaeroma hookeri*, *Idothea viridis*, *Gammarus locusta*, *Microdeutopus gryllotalpa* Bryozoa: *Conopeum seurati* Polychaetes: *Phicopomatus enigmaticus*
	Polluted sediments	Polychaetes: *Capitella capitata*, *Magelona papillicornis*, *Scolelepis ciliata* Many protozoa
Communities of sandy deep bottoms Depth: 3 to 20 – 30 m.	Fine, well-sorted sands. Size of materials: 0.02 – 1 mm	Hydrarians: *Hydractinia echinata* Polychaetes: *Sigalion mathildae*, *Onuphis eremita*, *Exogone haebes*, *Diopatra neapolitana* Bivalves: *Cardium tuberculatum*, *Mactra corallina*, *Tellina fabuloides*, *Tellina nitida*, *Tellina pulchella*, *Donax venustus* Gastropods: *Acteon tornatilis*, *Nassa mutabilis*, *Nassa pygmaea*, *Neverita josephinia* Crustaceans: *Idothea linearis*, *Eocuma ferox*, *Macropipus barbatus* Echinoderms: *Astropecten* spp. Fishes: *Callionymus belenus*, *Gobius microps*
	Coarse sands and fine gravels under bottom currents *Amphoxus* sand, in patches around and between the *Posidonia* meadows Grain size: 1 – 4 mm	Algae: *Lithophyllum racemus* Molluscs: *Dentalium vulgare*, *Diplodonta apicalis*, *Venus casina*, *Venus fasciata*, *Dosinia exoleta*, *Tapes rhomboides*, *Tellina pusilla*, *Tellina crassa* Polychaetes: *Polygordius lacteus*, *Sigalion squamatum*, *Euthalenessa dendrolepis*, *Glycera lapidum*, *Armanda polyophthalma* Crustaceans: *Cirolana gallica*, *Thia polita*, *Macropipus pusillus*, *Anapagurus breviauculeatus* Echinoderms: *Ophiopsila annulosa*, *Astropecten aurantiacus*, *Echinocardium fenauxi*, *Sphaerechinus granularis* Prochordata: *Branchiostoma lanceolatum* Fishes: *Gymnammodytes cicerellus*

Posidonia oceanica meadows (see text)

8.5.1. Infralittoral sands and muds (excluding the *Posidonia oceanica* meadows)

Only a few macrophytes are adapted to survive on sandy or muddy bottoms. The main problem is the necessity for suitable attachment organs, which virtually only phanerogams possess although a few algal species have developed rhizoidal systems (such as the chlorophycean *Caulerpa prolifera*). These macrophytes are usually found in calm conditions, often in hypo- and sometimes hypersaline waters, where they can form dense meadows. In relatively stable environments, the phanerogam *Cymodocea nodosa* appears and sustains a rich epifauna. In more unstable habitats and reducing sediments, it is substituted by *Caulerpa prolifera*.

In brackish waters other phanerogams can be dominant, such as *Zostera noltii* (and, more rarely, *Z. marina*) and *Ruppia*. Mixed populations with *Zostera noltii*, *Cymodocea nodosa* and *Caulerpa prolifera* are not uncommon. Faunal inhabitants of brackish water communities are bryozoans, swimming crustaceans, bivalves and rissoid gastropods. The tubiculous polychaete *Phicopomatus enigmaticus* (= *Mercierella enigmatica*) deserves special mention. It forms small, friable reefs which cover any suitable substrate, in warm and shallow lagoons. In these environments, the degree of stability of environmental parameters (such as salinity and temperature) seems to be the main factor on which the specific composition of the living assemblages depends.

In bottoms which are devoid of vegetal cover, decreasing water movement (resulting from wave protection or increasing depth) usually leads to a richer fauna. Two faunal assemblages can be found: the sediment endofauna and the vagile epifauna. A third one should be added in the case of a macrophyte stratum (the epifauna of plants). There is also a fourth assemblage, the meiofauna, which is very important from qualitative, quantitative and economic points of view (by virtue of its role in secondary production processes). However, it is irrelevant for the physionomical description of the underwater seascapes because the very minute and modified animals in this category live in the interstices between the sand grains (nematodes, copepods and other crustaceans, turbellarians, opisthobranchs, rotiferans and solenogastres).

8.5.2. The *Posidonia* biome or formation

Marine phanerogams (sea-grasses) represent the end of a long evolutionary journey, from chlorophyceans to monocotyledons. Their relatively recent return to the sea is linked to the acquisition of new physiological and ecological capabilities which have enabled the 50 or so species throughout the world to compete successfully with the algae, which have been adapted to the sea for much longer. The sea-grasses live preferentially on soft (sandy) sediment into which their rhizomes penetrate. *Posidonia oceanica* is an endemic sea-grass forming luxuriant submarine 'prairies' or meadows, which can be considered as the archetype of the Mediterranean sea-grasses formations (Fig. 8.12). It is a photophilic species of high primary production distributed in function of the available light and resisting a limited agitation of water. *Posidonia* offers continuous substrates (leaves, rhizomes) for settlement and colonization by many sessile organisms, stabilizes the sediment and strongly changes it. The leaves act as sediment traps, which determine the grain size of the settled materials. The rhizomes grow vertically, raising and 'hardening' the bottom. The plant also influences the chemical and physicochemical properties of the water and sediment. From the ecological point of view the *Posidonia* meadows can be considered as mature ecosystems; both biomass and production are higher than on other sandy bottoms, as also are the residence time of the elements and the species diversity.

The annual cycle of the plant is well-known. It consists of a season of foliar growth (autumn-summer), which permits the colonization of leaves by a very diverse and well-adapted epiphytic

Fig. 8.12. Typical aspect of the *Posidonia oceanica* meadows community of a Western Mediterranean coast (based on that of Medes Islands, Girona, NE Spain). Amid the *Posidonia* plants, in the meadow proper or in the coarse sand and fine gravel under bottom currents community, from top to bottom and from left to right: *Coris julis* (male and female), *Spirographis spallanzani* (with *Halecium* on its tube), *Holothuria forskali*, *Codium bursa*, *Pinna nobilis*, *Maena maena* (courting), *Echinaster sepositus*, *Paracentrotus lividus*, *Sphaerechinus granularis*, shells of *Lima lima*. On the *Posidonia* leaves, *Sertularella perpusilla*. On the close up, from top to bottom: *Obelia dichotoma*, *Electra posidoniae*, *Lichenopora radiata*, *Disporella hyspida*, *Castagnea irregularis*, *Giffordia mitchelliae*, *Spirorbis* sp., *Fosliella lejolisii*, *Fenestrulina joannae*. On the *Posidonia* bases: *Schizobrachiella sanguinea*, *Platonea stoechas*. (Drawing by M. Zabala in *Els sistemes naturals de les illes Medes*, Ros et al., 1984, by kind permission of the Institut d'Estudis Catalans, Barcelona).

community (see below). This colonization peaks in the summer months. The leaves decay noticeably by September and the autumn storms uncover the next foliar generation. The energy transfer from the *Posidonia* leaves to the surrounding community seems to follow two different pathways: (1) the detritic pathway, which, as we have seen, is very important in bringing organic matter to the littoral communities (the movement of water may break up the leaves while amphipods and isopods contribute to their fragmentation) and (2) the herbivore pathway. The latter is less important than might be expected, because of the limited palatability for marine animals of this secondarily marine terrestrial plant. Nevertheless, some plant-eaters, chiefly sea-urchins — *Paracentrotus lividus* — , fishes — *Sarpa salpa* — , isopods, some molluscs and polychaetes graze the leaves. The rhizomes remain untouched and their breakdown is very slow.

Today, some of the *Posidonia oceanica* meadows in the Western Mediterranean seem to be in regression. There is no general agreement on the causes and a number have been suggested (i.e. chemical or organic pollution, suspension of sediment in water and mechanical degradation by fishing trawls). Nevertheless, the disappearance of the *Posidonia* communities from the neighbourhood of harbours and estuaries indicates the high sensitivity of this sea-grass to environmental disturbances.

The associated flora and fauna is varied and has been extensively studied (*see* revision in Pérès 1967, and Ballesteros *et al.*, 1984). The organisms living in the *Posidonia* meadows have been grouped in many categories. For the present purpose, three main groups are distinguished:

(a) Inhabitants of the leaves which are characterized by small size, short life cycles and high growth rate (all as a result of being adapted to the cycle of growth and decay of the *Posidonia* leaves). The unusual nature of this habitat, which includes the motility of leaves in the water, is responsible for a great deal of specificity. Organisms found here are rare in other environments. Hydrozoans, bryozoans and algae are the dominant groups. Hydrozoans grow through stolons and this increases their survival on a continuously growing substrate. They are represented by, among others: *Sertularia perpusilla*, *Campanularia assymetrica* and *Plumularia obliqua* f. *posidoniae*. The most frequent bryozoans are *Electra posidoniae*, *Fenestrulina joannae*, *Chorizopora brongniartii* and *Haplopoma impressum*. The algae attain a considerable growth in the older, apical regions of leaves, and several species can segregate themselves and divide the space at a reduced scale. The basal stratum is formed by encrusting forms: *Fosliella lejolisii*, *F. farinosa* and *Myrionema magnusii*. On them, erect or pulvinular species can settle, such as: *Giraudia sphacelarioides*, *Castagnea irregularis*, *C. cylindrica* and *Myriactula gracilis*. Many of these are characteristic of the *Posidonia* meadows. Foraminiferans and polychaetes have also epibiont representatives.

(b) Inhabitants of the rhizomes. They make up an assemblage of species which is scarcely representative, but shows affinities both with the photophilic and with the sciaphilic communities, depending on the degree of cover of the submarine prairie. When it is dense and the foliar canopy is well-constituted, the organisms settling on the rhizomes can be partly similar to those found in the coralligenous community. Examples are: algae of the genera *Peyssonnelia*, *Mesophyllum*, *Rhodymenia*; sponges such as *Sycon ciliatum*, *Dysidea fragilis* and *D. avara*; bryozoans such as *Schizobrachiella sanguinea*, ascidians such as *Halocynthia papillosa* and *Microcosmus sulcatus*. With sparse cover, the species do not differ from those of the 'precoralligenous' or photophilic algae communities. The sediment trapped between the rhizomes is also occupied by a soft-bottom community that includes copepods, nematodes, bivalves, polychaetes and decapods.

(c) The vagile associated fauna. This assemblage is more heterogeneous than the other two, because it includes different feeding types and ecological strategies. Five main groups can be mentioned. The molluscs are represented chiefly by gastropods, such as *Tricolia speciosa*, *Turbona cimex*, *Alvania lineata* and other rissoids. Among the crustaceans are isopods and amphipods which feed on debris of leaves (*Dexamine spiniventris*, *Maera inaequipes* and *Idothea baltica*) and some carnivorous decapods such as *Alpheus dentipes*, *Pilumnus hirtellus* and others. The relation of fishes with the *Posidonia* meadow is rather ill-defined, since only a few species, such as *Sarpa salpa*, feed on leaves while others, such as *Maena*

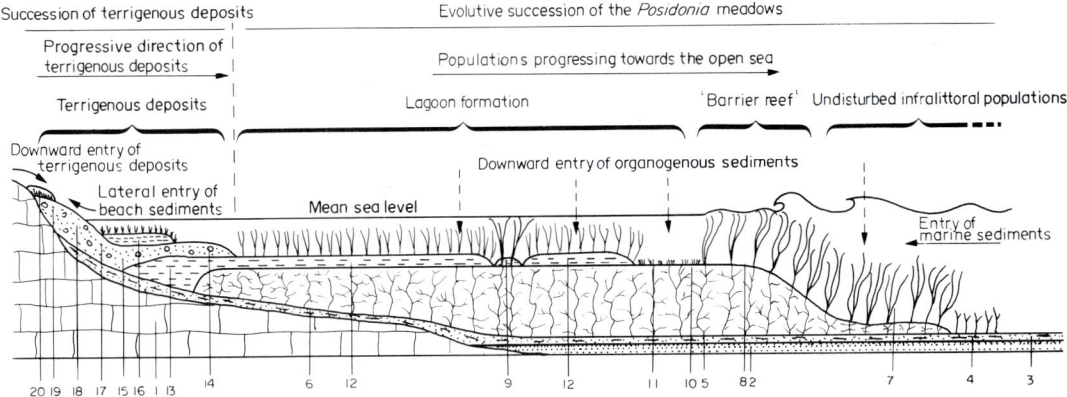

Fig. 8.13. Diagrammatic section through a *Posidonia oceanica* 'barrier reef' and its lagoon, illustrating also the suggested sense of the ecological succession leading to a well-constituted *Posidonia* meadow (at right) and to an impoverished lagoon of photophilic algae (at left). 1, original rocky substrate; 2, original sandy substrate formerly inhabited by the well-sorted fine-sand assemblage; 3, sediment inhabited by the same community or by the coarse sand and fine gravel under bottom currents assemblage; 4, sediment with humus content increased by retention of organic matter by *Cymodocea* lawn; 5, same, covered by *Posidonia* terraces; 6, sediment fastened on rocks by the photophilic algae assemblages and humus enhanced by *Cymodocea* facies; 7, *Posidonia* meadow (very reduced in the illustration horizontal sense); 8, barrier reef of *Posidonia* due to terrace rising; 9, vestigial patch of *Posidonia*; 10, *Padina pavonica* facies of the photophilic algae community on (11); 11, dead *Posidonia* rhizomes; 12, area of silt sedimentation with *Cymodocea nodosa* lawn on dead *Posidonia* matte; 13, silty sediment covered by (14); 14, coarse detritic but silty and muddy sediment with *Upogebbia* facies of the superficial muddy sands in sheltered areas; 15, same with *Upogebbia* burrows filled by later sediment; 16, silty sediment with *Zostera noltii*; 17, same covered by terrigenous deposits; 18, mediolittoral beach; 19, supralittoral beach; 20, mass of dead *Posidonia* leaves with supralittoral community. (From Pérès and Picard, 1964; by permission of the authors).

chryselis, use the meadow as a courting and spawning ground. The only group that seems to maintain a relatively constant relationship with this community are the Syngnathidae: *Syngnathus acus*, *S. typhle*, *Hippocampus hippocampus*. The more important group which influences the general dynamics of the system is without doubt the echinoderms. The sea-urchin *Paracentrotus lividus* browses on the leaves at night, while resting on the bottom by day. It can reach remarkable densities and is suspected of controlling the dynamics of the *Posidonia* meadow. *Sphaerechinus granularis* is another common species in the deeper parts, and preys on the rhizomes inhabitants. Holothurians (*Holothuria* spp.) are sediment-eaters, while among sea-stars, *Asterina gibbosa* and *Echinaster sepositus* feed on molluscs, sponges and compound ascidians. The crinoid *Antedon mediterranea* is a plankton and detritus filter-feeder, which settles on leaves.

The relationships between the *Posidonia* meadows and other soft- and hard-bottom communities have been studied, from a successional point of view, by Molinier and Picard (1952) (Fig. 8.13), and the dynamics of the prairie itself by several workers (*see* Ott, 1980).

8.6. THE CIRCALITTORAL HARD-BOTTOM COMMUNITIES: THE CORALLIGENOUS

Beyond the limit of the *Posidonia* meadows or of the photophilic algae assemblages, and following the transect which started at the wave-battered littoral, there is the circalittoral zone. This domain is semi-dark, the seascape is lit with blue light. The animals, however, show the brightest colours when artificially illuminated underwater or brought to the surface. Currents are steady, and may be strong,

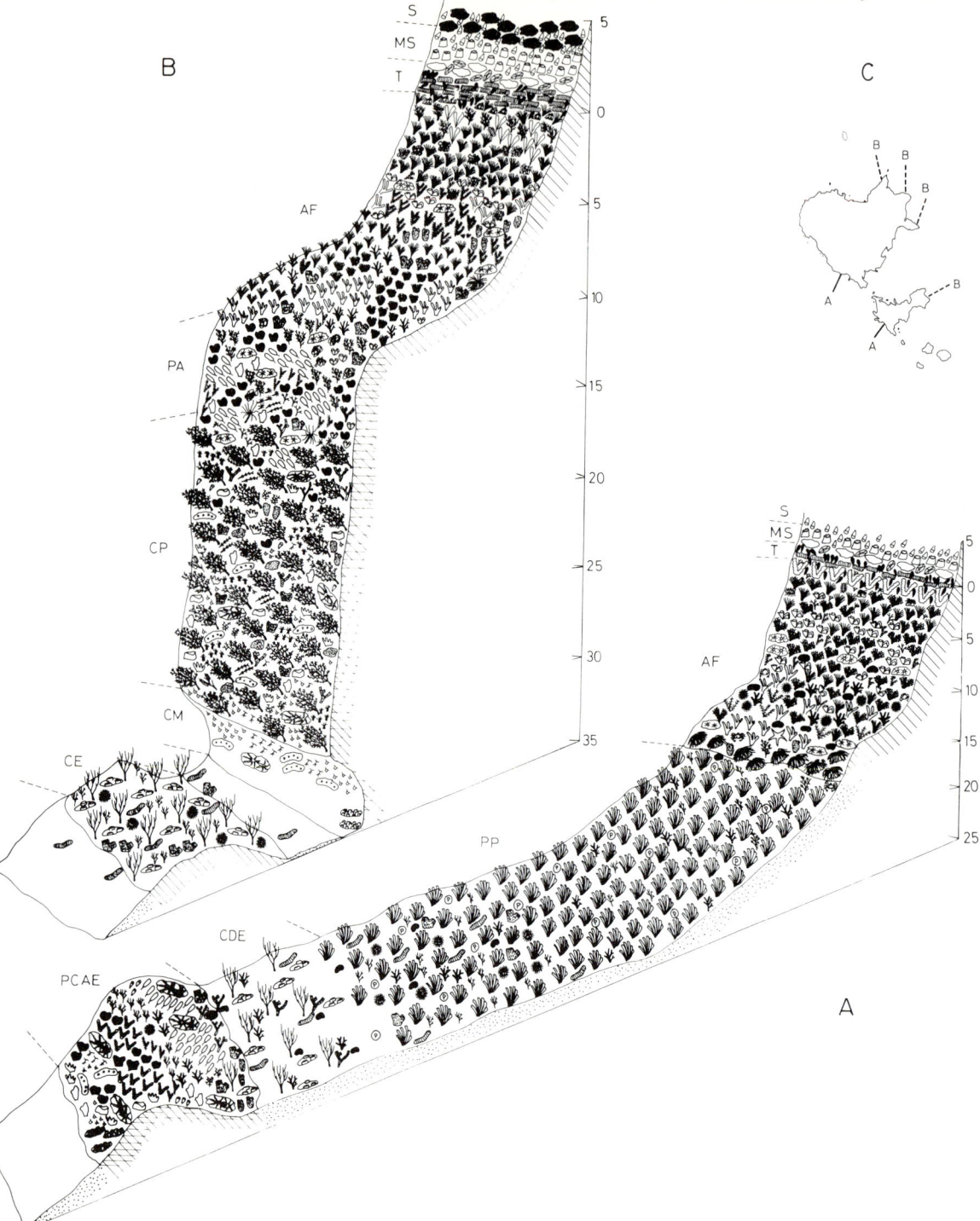

Fig. 8.14. Diagrammatic view of: A, a transect between 0 and 25 metres deep in the relatively calm and slowly slanting south coast of Medes Islands (Girona, NE Spain), and B, a transect between 0 and 35 m deep in the exposed and almost vertical underwater cliffs of the northern littoral of the same Western Mediterranean archipelago. C, location of transects. D, explanation of the species-diagrams used in figures A and B, and meaning of the abbreviations of the communities or facies depicted. (Modified from Gili and Ros, 1982; by permission of the authors).

DIVING IN BLUE WATER. THE BENTHOS

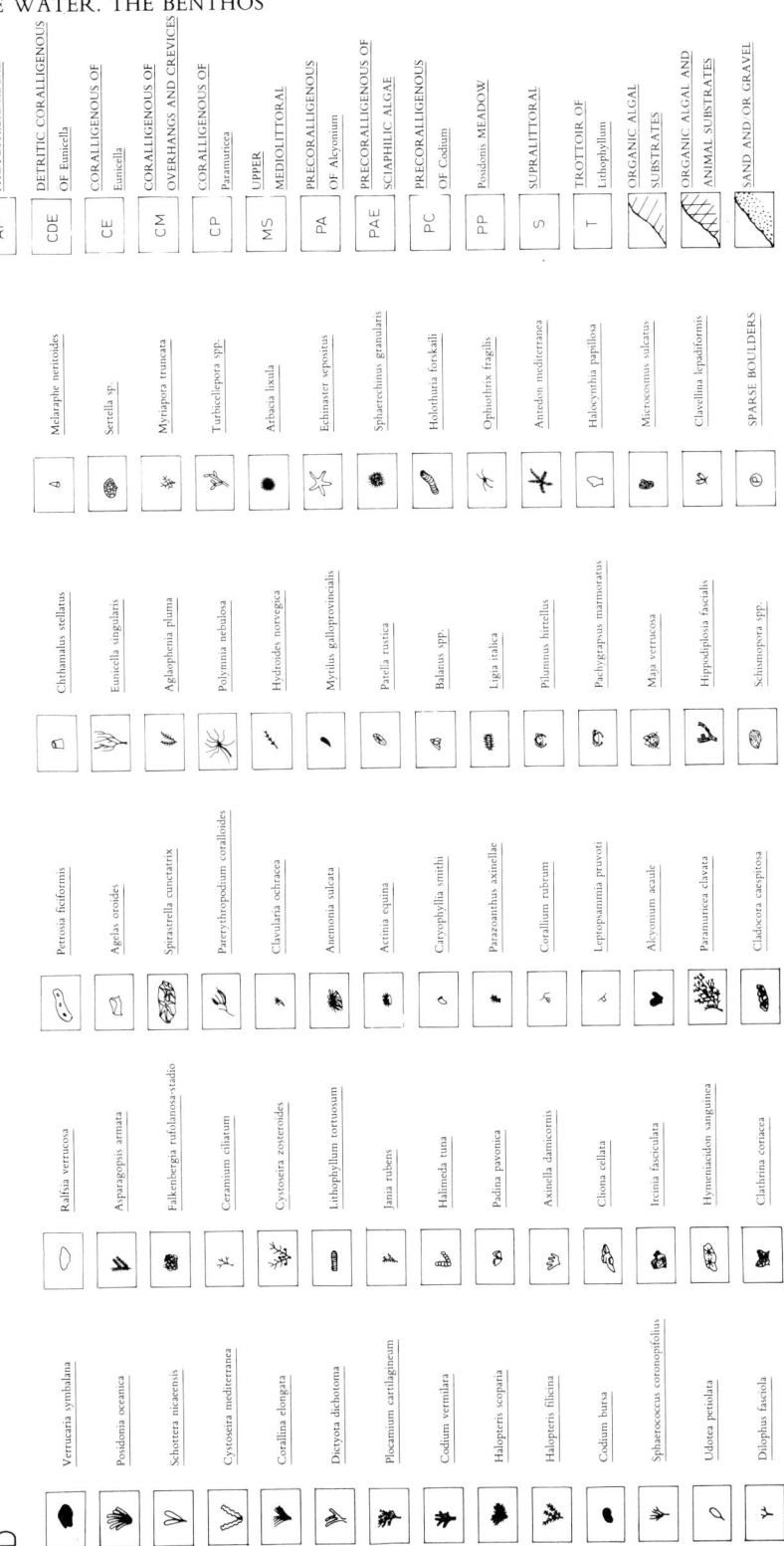

Fig. 8.14. cont.

and water is constantly cool. The plants are no longer the main contributors to the biomass, although they can be dominant in some communities and its role is extremely important as builders of the main circalittoral supracommunity, the coralligenous.

The main variables acting in the delimitation of communities and facies are the light and climate changes due to the microreliefs in hard substrates, the grain size in the sediment, the slope of the substratum and alteration between hard and soft bottoms. The coralligenous, whose diversity and riches in animal species has been compared to those of the tropical coral reef, deserves a special mention. The cave environment will be dealt with in another section, and the soft circalittoral bottoms in Chapter 7.

The term 'coralligenous' originated from a misunderstanding. It was applied to circalittoral bottoms with coarse gravel which were considered 'generators of coral' because precious coral, *Corallium rubrum*, appeared in trawling hauls together with the debris of the many calcareous organisms forming such bottoms. This has been proved false, for, in fact, *C. rubrum* belongs to the semi-dark caves community, and not to the coralligenous proper. However, the name has been retained. This may not be totally inappropriate, because the community to which the name is applied is formed by the accretive action of many calcareous algae, chiefly corallinaceae, but also peyssonneliaceae, among other organisms, as will be seen later. In the general description given below we will follow mainly Laborel (1961), Laubier (1966) and Gili and Ros (1984).

Two main types of coralligenous are recognised: one which appears covering more-or-less thickly the circalittoral rocks or those lying in the shadow in the infralittoral, and the one which originates on coarse, soft, relatively deep bottoms, the so-called bank or platform coralligenous (Fig. 8.15). The platform coralligenous appears as irregular, craggy and many-perforated masses (Figs. 8.16 and 8.19) resulting from the aggregation of gravels and sands of organic origin with broken shells and other calcareous skeletal fragments. The coalescence results from the overgrowth of corallinaceae (*Lithophyllum expansum*, *Mesophyllum lichenoides* and *Neogoniolithon mamillosum*) and peyssonneliaceae (*Peyssonnelia polymorpha*, *P. rubra*, *P. rosa-marina* and others). Between the foliose thalli of these rhodophytes, the calcareous debris of a number of circalittoral animals accumulate. This material is compacted and aggregated by algae and animals. These are responsible for no more than 20% of the total concretion, chiefly by bryozoans, molluscs and sponges. Madreporarians and serpulid polychaetes play a secondary role. The result is masses with a fragile, living surface layer, covering a hard core of cemented organic rock, as full of holes as Gruyère cheese.

The shape and consistence of the coralligenous banks depend on the algal more active building species and on differential or heterogenous growth. In some places, algae grow faster than in others, become enclosed, deprived of light or of the nutrients the algae themselves need to continue growing. The result is the formation of small canals, crevices and irregularities which give the coralligenous mass its characteristic appearance (Fig. 8.19). Associated with the construction processes there are destructive ones, first by physical erosion (due to water movement and overturning of the blocks), secondly by burial in sedimentation, and thirdly by biotic destruction, mainly brought about by sponges of the genus *Cliona* but also by other animal species, such as the polychaete *Polydora*, several bivalves and the sea-urchin *Sphaerechinus granularis* (Fig. 8.16).

The coralligenous on hard rock can appear almost at any level, if the shading suffices (Fig. 8.15), but the concretion is not usually as thick as in the platform type. The complexity of the organogenic mass in the deep formation is replaced here by a more important vertical stratification (Fig. 8.17), which contributes significantly to the richness and diversity.

The following strata can be recognised. There is an upper stratum, usually composed of large gorgonians and sponges, which in turn serve as substrate for a host of epibiont animals. In silted bottoms only this upper stratum remains. There is an intermediate stratum consisting of large bryozoan colonies and sponges, in addition to some polychaetes and ascidians. Here the large calcareous algae are not completely covered by epiphytic fauna, although the epibionts may be abundant. A third, lower, stratum is composed of several small animal species, living in dim light. Finally, there is an understratum, with endobionts, burrowing or simply hiding microfauna, which live under or between the rest of the biomass (Fig. 8.17).

The environmental conditions allowing the formation and the full development of a coralligenous multi-community complex change along the Western Mediterranean coasts, and it is not surprising to find that whereas in the Catalan coasts the coralligenous bank can appear at 18 – 40 metres deep (because light extinction is higher in the relatively rich waters) in the Marseilles region it sinks to 30 – 50 m, and in the exceptionally clear waters of Corsica and Majorca it does not appear above 60 – 80 m deep. Thus, in the latter places, and still more in the eastern basin, there is a very broad gap between the rock and the platform coralligenous, which almost overlap on the Spanish and French continental coasts. The situation is essentially similar with the so-called precoralligenous (Fig. 8.14), which in the NW Mediterranean coasts precedes the coralligenous (in the bathymetric, but not in the successional sense), with some overlapping at the boundary, whereas in the SW littoral the two communities are fully separated. Light is an important factor, and in some places the coralligenous platforms receive less light than the others. Similarly the lower infralittoral algal assemblages, the rhizomes of *Posidonia*, the crevices and overhangs are poorly lit (Fig. 8.18).

8.6.1. Plants of the coralligenous

The algae of the hard, coralligenous substrates of the circalittoral zones are, as we have seen, mainly melobesiae. The major species are *Mesophyllum lichenoides* and *Lithophyllum expansum*, often accompanied by other building corallinaceae of the genera *Lithothamnium*, *Phymatolithon*, *Neogoniolithon* and *Lithophyllum*. Unencrusted algae are also common, either foliose (as with many *Peyssonnelia* and

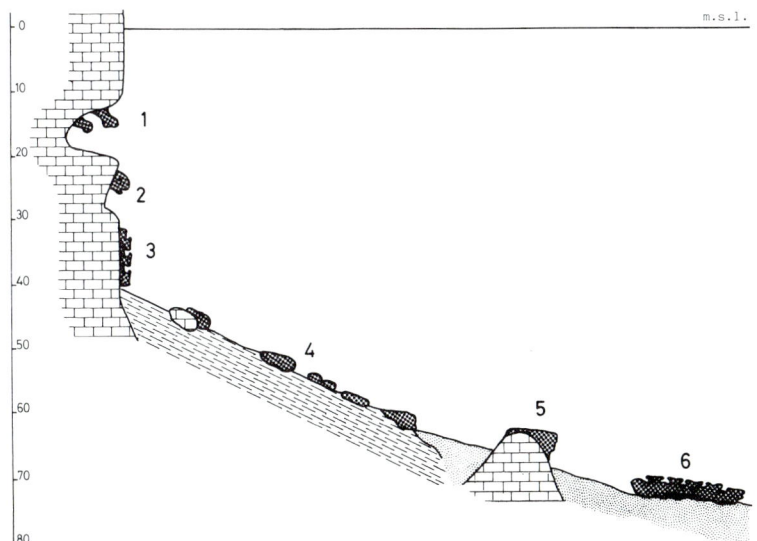

Fig. 8.15. Different kinds of coralligenous concretioning, according to their position on the littoral and type of substrate. 1, cave entablatures; 2, overhang entablature; 3, cover and entablature on vertical walls; 4, crumbling blocks from vertical walls, with coralligenous surface; 5, cover over rock socle (blocks; these five types belong to the so-called coralligenous of the littoral rock); 6, platform or bank coralligenous (coralligenous block on soft substrate). (Redrawn from Laborel, 1961).

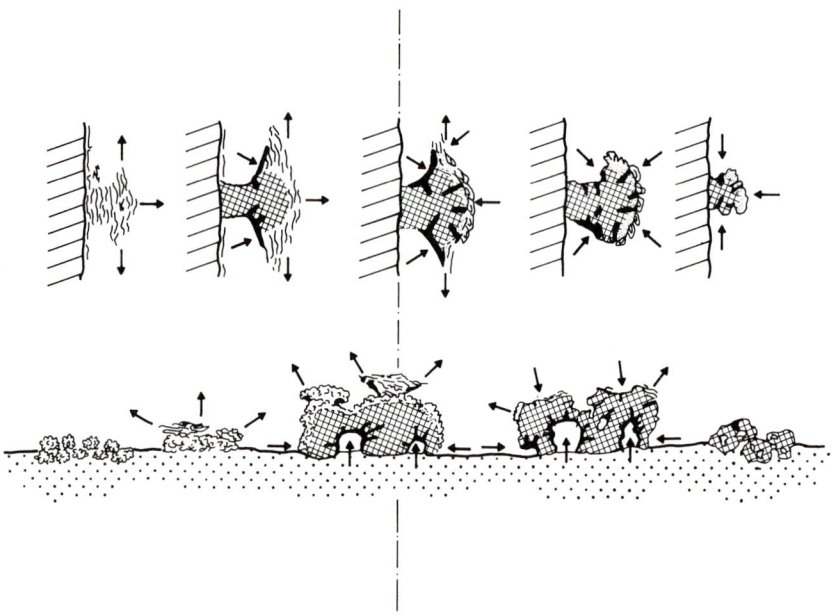

Fig. 8.16. Diagrams showing the growth (at left) and destruction (at right) patterns of the two types of coralligenous concretioning, the coralligenous of the littoral rock (above) and the coralligenous platform (below), the first on a hard (in this case vertical) substrate and the second on a soft one. Growing zone, thin drawing; destruction by borers, black; cemented zone, squared; the arrows mean intensity of growth and of biotic erosion. (Redrawn from Laborel, 1961).

Kallymenia) or arborescent (such as *Bonnemaisonia asparagoides, Dudresnaya verticillata, Gulsonia nodulosa, Phyllophora crispa* and *Rodriguezella* species). New *Cystoseira* species (such as *C. zosteroides* and *C. spinosa*), besides the infralittoral ones, are added at this level. Less common phaeophyceans are *Sargassum hornschuchii, Phyllaria reniformis* and *Laminaria rodriguezii*.

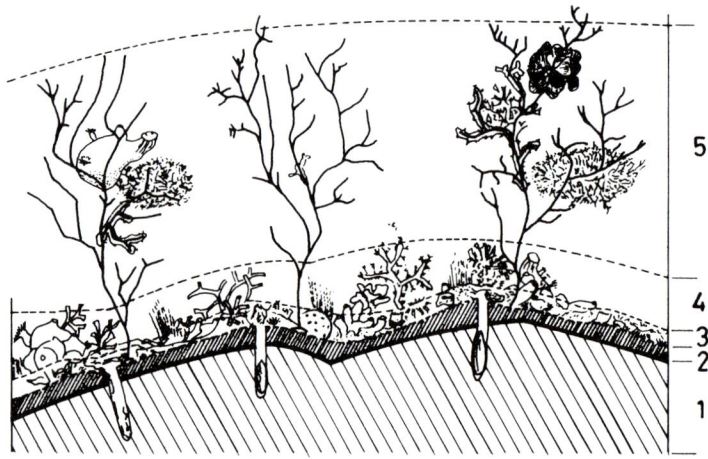

Fig. 8.17. Diagrammatic section across the living assemblage of the coralligenous of the littoral rock on a vertical or subvertical wall and representative genera. 1, calcareous rock with endoliths (*Lithophaga*); 2, dead encrusting stratum (coralligenous concretion); 3, living encrusting stratum, with encrusting algae, sponges, cnidarians, molluscs, bryozoans, etc.; 4, intermediate stratum, with erect cnidarians, sponges, briozoans and ascidians; 5, upper stratum, with: *Paramuricea* and (epibionts) *Salmacina, Sertella, Parerythropodium*, tunicates, etc. (Redrawn with modifications from True, 1970).

Fig. 8.18. Idealized representation of a typical coralligenous community, based on that of Medes Islands (Girona, NE Spain). Left, from top to bottom: *Paramuricea clavata* (and, on it, *Halecium halecinum, Pteria hirundo*), *Aglaophenia septifera, Cliona viridis, Alcyonium acaule, Acanthella acuta, Lithophyllum expansum, Agelas oroides, Palinurus elephas, Parazoanthus axinellae, Spirastrella cunctatrix, Chondrosia reniformis, Petrosia ficiformis* (and, on it, *Porella cervicornis* and *Peltodoris atromaculata*), *Serpula* sp., *Caryophyllia inornata, Halocynthia papillosa, Clathrina coriacea, Corallium rubrum, Chromis chromis*. Right, from top to bottom (excluding the above-mentioned species): *Anthias anthias, Eunicella singularis, Diplodus sargus, Epinephelus guaza, Galathea strigosa, Aglaophenia septifera, Synthecium evansi, Dysidea tupha*. (Drawing by M. Zabala in Els sistemes naturals de les illes Medes, Ros et al., 1984, by kind permission of the Institut d'Estudis Catalans, Barcelona).

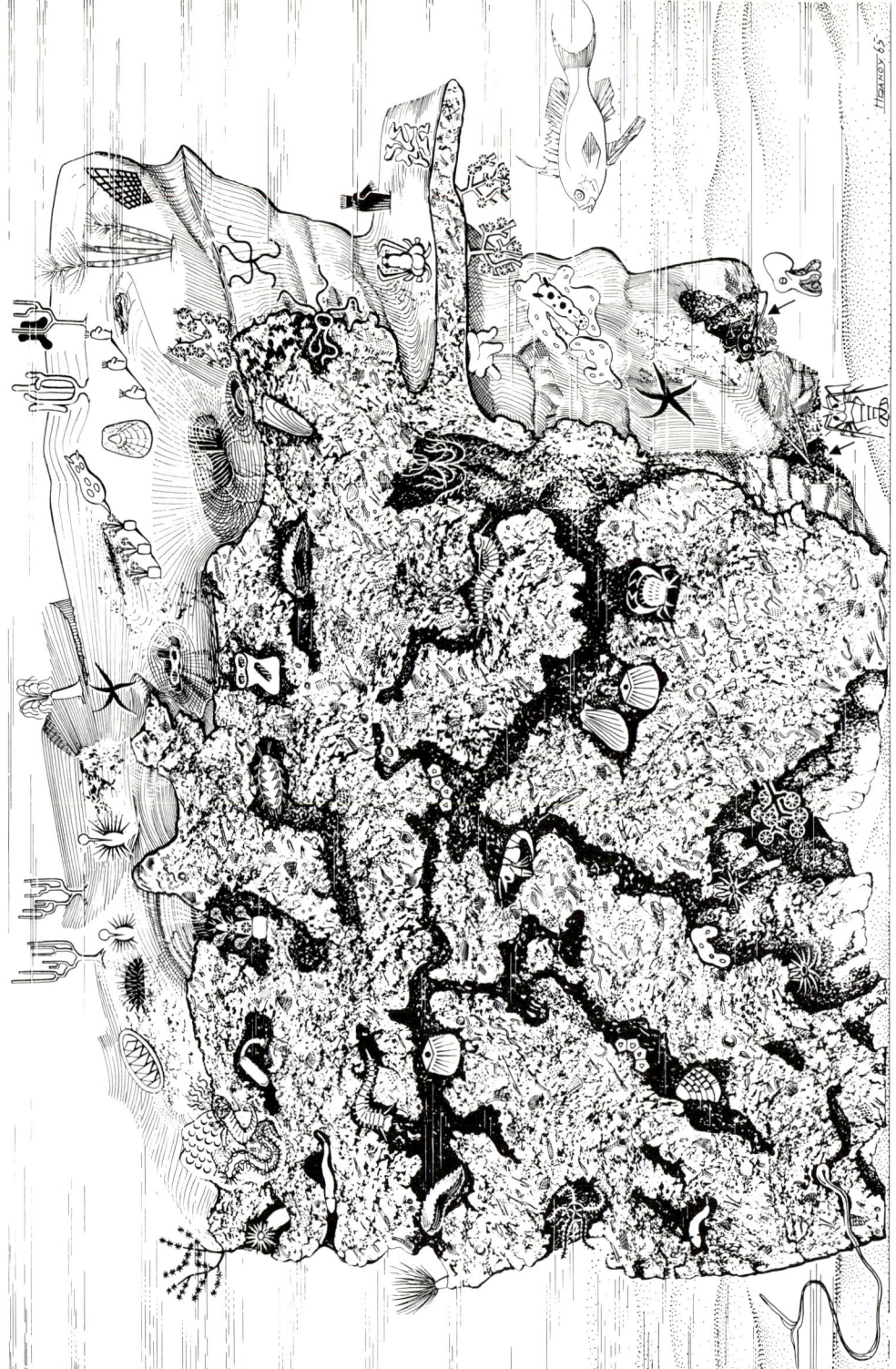

Fig. 8.19. Diagrammatic section of a coralligenous bank, presenting the micro-habitats and the location of fifty motile, epilithic and endolithic animal species of the coralligenous community. For explanation see Fig. 8.20. (From Laubier, 1966; by permission of the author).

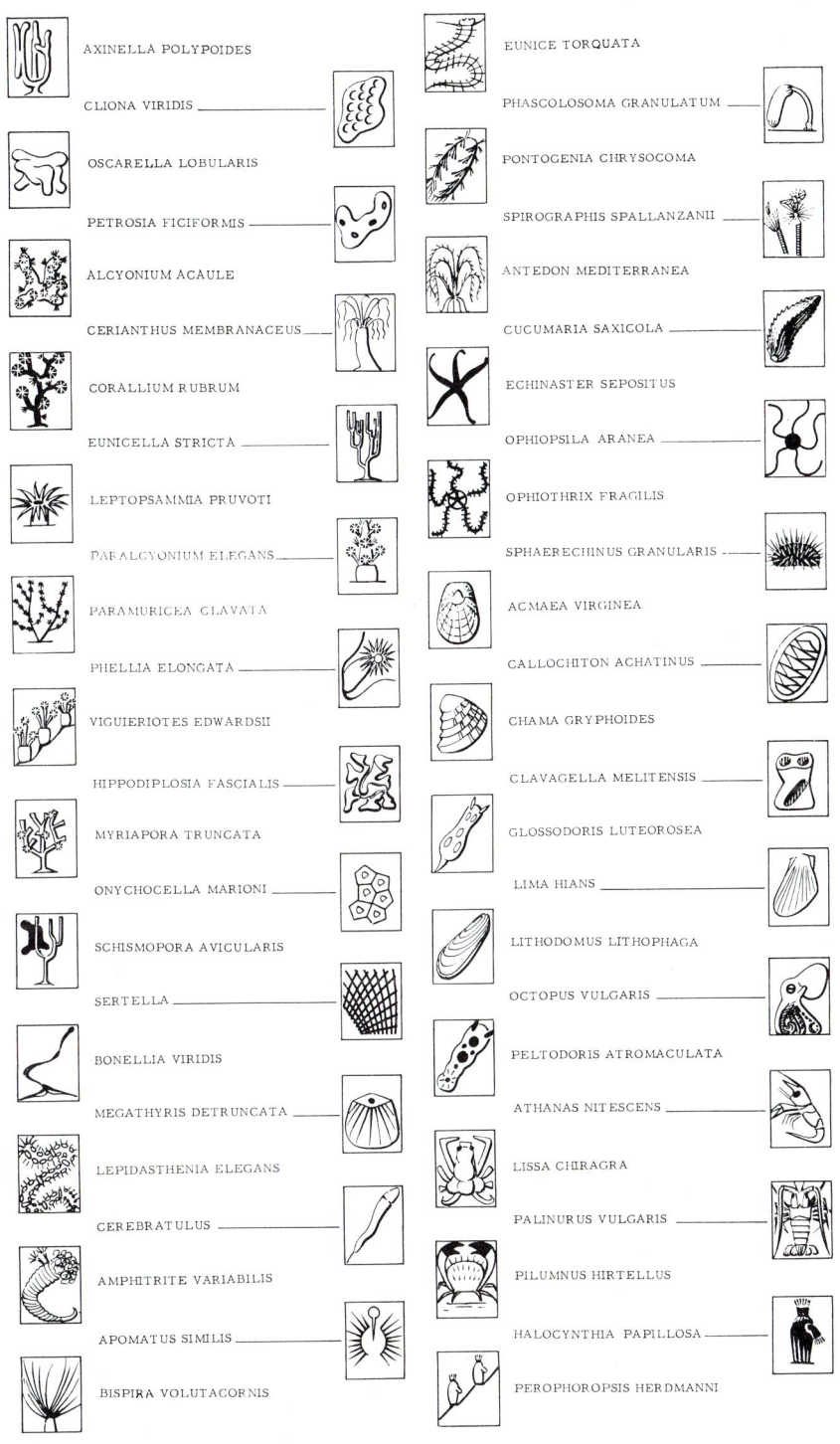

Fig. 8.20. Identification of the species sketched in Fig. 8.20. (From Laubier, 1966; by permission of the author).

8.6.2. Animals of the coralligenous

Many sponges intermix with algae in a reciprocal encrusting process, and often only the oscula of the animals can be seen. Among these are *Halisarca dujardini*, *Suberites carnosus* f. *incrustans*, *Batzella inops* and *Reniera* and *Haliclona* species. Other sponges, such as *Cliona celata* and *C. viridis*, are involved in the previously mentioned destructive processes. Many sponges appear abundantly in the lower stratum of the coralligenous: *Anchinoe fictitius*, *Clathrina coriacea*, *C. clathrus* and *Spirastrella cunctatrix*, among the laminar and, among the massive forms, *Chondrosia reniformis*, *Axinella verrucosa*, *Acanthella acuta*, *Ircinia fasciculata*, *I. dendroides*, *Spongionella pulchella*. *Axinella polypodes* and *Ircinia muscarum* form part of the higher stratum.

The cnidarians are also well-represented by encrusting, erect and stolonial species. Some of these are typical of the community, and are regarded as particular facies. The crawling species compete for space with algae, sponges and bryozoans, and produce the patchwork pattern of the more colourful coralligenous assemblages. Among the cnidarians, there are the massive *Cladocora caespitosa* and the carpeting *Leptopsammia pruvoti*, *Astroides calycularis*, *Caryophyllia smithi* and *Hoplangia durothrix*; *Corynactis viridis* and *Parazoanthus axinellae* can also live as epibionts on sponges and other substrates, while *Parerythropodium coralloides* overgrows the large gorgonians *Eunicella singularis*, *E. cavolinii* and *Paramuricea clavata*. *Sertularella crassicaulis* belongs, like these gorgonians, to the higher stratum, but other hydrarians are also epibionts or grow attached to the substrate: *Nemertesia antennina*, *Aglaophenia kirchenpaueri* and *Synthecium evansi*, among others.

Bryozoans are major components of the coralligenous. When living, they cover a large fraction of the substrate and the fragments of dead colonies help to build the organogenous concretions. The commonest species are *Pentapora fascialis*, *Myriapora truncata* (on horizontal and lit substrates) *Turbicellepora avicularis* (on *Eunicella* spp.), 'Porella' cervicornis, *Hippodiplosia foliacea*, *Beania magellanica*, *Adeonella calveti*, *Cellaria salicornioides* (on vertical, shaded walls) and *Prenantia inerma*, *Cribrilaria radiata* and others (in the dark). The polychaetes occupy the interstices. The most conspicuous species are the encrusting (*Serpula vermicularis*) or colonial ones (*Salmacina dysteri*), but many others are found inside the banks or crawl over them, such as *Vermiliopsis infundibulum*, *Chrysopetalum caecum*, *Lysidice ninetta*, *Syllis hyalina*, *Ceratonereis costae*, *Glycera tesselata*, *Eunice vittata*, *E. siciliensis*, *Nereis zonata* and *Lumbrineris funchalensis*. The molluscs are as widespread as the polychaetes, but they exploit still more numerous habitats. The more characteristic or common species are the gastropods *Fusinus rostratus pulchellus*, *Raphitoma linearis*, *Triphora perversa* and *Muricopsis cristatus*, and the bivalves *Chlamys multistriatus*, *C. pesfelis*, *Pteria hirundo*, *Anomia ephippium* and *Kellia suborbicularis*. It would be easy to construct a longer list: some species burrow in the rock or the organogenous formation, such as *Lithophaga lithophaga*. Others are attached as epibionts (*Pteria hirundo*) or on the substrate (*Arca noae*, *Spondylus gaederopus*). There are many endobionts, *Diodora mamillata* and *Lima lima* among them, and endoliths such as *Alvania cimex*, *Rissoina bruguieri*, *Ammonicera fischeriana* and *Chama gryphoides*. The vagile species are many and specialized. As an example in only one group, the opisthobranchs, and in only one feeding strategy, that of browsers of animal tissues, there are many sponge-eaters (including *Tylodina perversa*, *Peltodoris atromaculata*, *Chromodoris*, *Hypselodoris* and *Dendrodoris* species), bryozoan-eaters (*Diaphorodoris papillata*, *Limacia clavigera* and *Platydoris argo*), hydrarian-eaters (*Flabellina affinis*, *Godiva banyulensis* and *Trinchesia coerulea*), and so on.

The same can be said of the crustaceans. A large spectrum of life styles exists from encrusting species, as *Balanus perforatus* (which participates in the concretioning) through small-sized forms (such as harpacticoids and the amphipods *Lilljeborgia brevicornis* and *Maera inaequipes*) to the decapods, such as the large lobsters *Palinurus elephas* and *Homarus gammarus*, and medium-sized, very numerous crab and shrimp species, including: *Lissa chiragra*, *Pandalina brevirostris*, *Athanas nitescens*, *Processa macrophthalma*, *Balssia gasti* and *Periclimenes scriptus*.

The echinoderms are very frequent components of the vagile macrofauna, but other species also creep in holes and crevices, such as *Ophiothrix fragilis*, *Amphipholis squamata*, *Ophiopsila aranea* and *Cucumaria saxicola*. Typical coralligenous species are *Sphaerechinus granularis*, *Genocidaris maculata*, *Echinaster sepositus*, *Ophidiaster ophidianus*, *Marthasterias glacialis*, *Holothuria tubulosa* and *H. forskali*. In deeper coralligenous platforms, *Echinus acutus*, *E. melo*, *Stylocidaris affinis* and *Astrospartus mediterraneus* are common. Other common circalittoral species are the brachiopod *Megathyris detruncata* and the sipunculan *Phascolosoma granulatum*.

The tunicates are well-represented in the coralligenous. Some species belong to the sessile macrofauna, such as *Halocynthia papillosa* and *Microcosmus sabatieri*, and others live more or less sunk in the organogenic concretions, notably: *Polycarpa pomaria*, *Pyura dura*, *P. microcosmus*, *Rhodosoma verecundum*, *Diazona violacea* and *Cystodites dellechiajei*. Many fish species hide in the maze inside the coralligenous blocks or rest on the coralligenous carpet. These include the conger (*Conger conger*), the moray eel (*Muraena helena*), the grouper (*Epinephelus guaza*) and many scorpionfishes (*Scorpaena*).

8.6.3. Facies

Several communities and facies have been described for the coralligenous. Some of these are as follows (*see also* Fig. 8.13):

(a) The community of *Cystoseira zosteroides*. This is found in well lit and horizontal substrates influenced by water currents. The accompanying algal assemblage is rich and many encrusting organisms grow on these phaeophyceans.

(b) The community of *Eunicella singularis* occurs on horizontal substrates which are well-washed by relatively strong currents, and at deeper levels than the preceding facies.

(c) The community of *Paramuricea clavata*. This is found on vertical, dark substrates with strong currents. The lower strata are very dense and rich.

(d) The community of *Eunicella cavolinii*. This occurs on deeper areas with slight deposition of sediment and moderate currents.

(e) Facies of *Astroides calycularis* with *Ophidiaster ophidianus*, on quite warm waters.

(f) Facies of large sponges, such as *Chondrosia reniformis*, *Spongionella pulchella* and *Agelas oroides*, with bryozoans and madreporarians.

(g) The community of the large blocks in the coralligenous banks. This is very rich and grows on soft stable bottoms, as has already been described (Fig. 8.19).

(h) Facies of transition to the semi-dark cave community. This includes *Parazoanthus axinellae*, *Corallium rubrum* and other sciaphilic species. It is found in boundary zones with a rapidly declining light intensity such as cave entrances, overhangs and vertical cliffs.

(i) Facies of transition towards the detritic soft bottoms. The impoverishment of this community is due mainly to sedimentation and the destruction of the coralligenous cornice, with large sponges, gorgonians and echinoderms (Fig. 8.14).

8.7. THE WORLD OF SUBMARINE CAVES AND TUNNELS

The yawning mouths of submarine caves appear sometimes interspersed among other open-bottom communities. The attraction these black patches convey for the scuba diver is matched by the scientific interest of their rare, isolated and occasionally blind inhabitants (Fig. 8.23). Although there are submarine caves at every level of the littoral, the special environmental conditions provided by these dark and cool habitats are essentially circalittoral. Nevertheless, the fauna has clear connections not only with the coralligenous, but also with communities of deeper waters. The total absence of primary producers is a fundamental similarity of these singular biotopes with the communities of the aphytal zone. There are a variety of transitions between the caves which are close to the surface and open to the air and the deep grottos with several channels and galleries, where there are no effects of the waves or of daily or seasonal changes in the properties of the water (whether abiotic or biotic).

The extreme reduction or the total absence of light is, of course, the most important environmental feature of submarine caves. From the entrance to the innermost region of the cave there is a gradation from light to completely dark zones. Plants disappear somewhere along this gradient. A similar gradient is seen in the distribution of the sessile fauna in the horizontal plane, which (although more compressed in space) parallels the vertical distributions in open habitats. Species from deep open water appear at shallower depths when caves are available to them (Fig. 8.21).

Water movement plays a key role in determining the rate at which the water mass can restore its oxygen and food supply, eliminate wastes and disperse the cave inhabitants' offspring. In long and narrow caves the exchange may be solely by diffusion. The existence of more than one opening in a cave provide currents, which brings in a steady supply of food, in the form of plankton and detritus. Most caves are formed in karstic, calcareous and permeable rock, with freshwater leakage from above. A limited circulation of the water is seen by some authors as the reason for the rather heterogeneous distribution of food, oxygen, temperature and salinity. This creates deep, isolated stretches which are virtually devoid of eukaryotic life. But recent accurate measurements of hydrographical parameters within a cave in the Catalan coast, however, suggest that the main factor hindering the development of a normally constituted assemblage is the reduced hydrodynamism and not the absence of water circulation; even in rather closed caves, there is always enough water exchange to ensure the homogeneous distribution of temperature, oxygen and other parameters (see Gili *et al.*, in press). Between these abiotic sections and the dimly-lit and well-mixed water of the entrance, an intermediate zone exists of considerable environmental constancy and stability which favours the settlement of specialized animal species. The food chains appear to change (not necessarily becoming simpler, as with the communities on open ground) and each cave has insular properties for colonization and evolution. For these and other reasons, the fauna of the caves is interesting but poorly understood.

A major, initial faunistic component is the circalittoral or coralligenous fauna which has managed to pass through the severe ecological filter of the environmental conditions of a submarine cave. There are no herbivores, due to the lack of primary producers, and the predators are limited. Food is not available in enough quantity, or its supply is too irregular for the normal pattern of successive trophic levels. Detritivores are common, and filter-feeders abound especially when plankton or detritus are supplied by water currents. The vagile fauna, chiefly its more active species (fishes and crustaceans), do not usually reflect these limitations, for they use the cave environment as a hiding place, from which they periodically emerge to forage on the open bottoms. This strengthens the horizontal transport required to ensure energy inputs to the cave. There is, besides, a closed cycling of energy and matter inside the cave, which in some complex cave systems can lead to abnormal trophic relationships.

Two communities of dark circalittoral habitats have been described: semi-obscure caves and totally dark caves and tunnels.

The community of the semi-obscure caves occurs in overhangs, vertical walls, crevices and entrances of caves and tunnels throughout all the circalittoral étage and in many infralittoral isolates. Many of the species of the coralligenous formations belong to this community, chiefly the ones limited to holes, crevices and interstices of the coralligenous platforms. The community of totally dark caves covers the main walls or the vertical corridors or chimneys, the crevices and holes of deeper caves and the intermediate, darker stretches of submarine tunnels. The following faunistic description is based mainly in the inventories of samples taken in the caves and tunnels of the Medes Islands (Gili *et al.*, 1982;

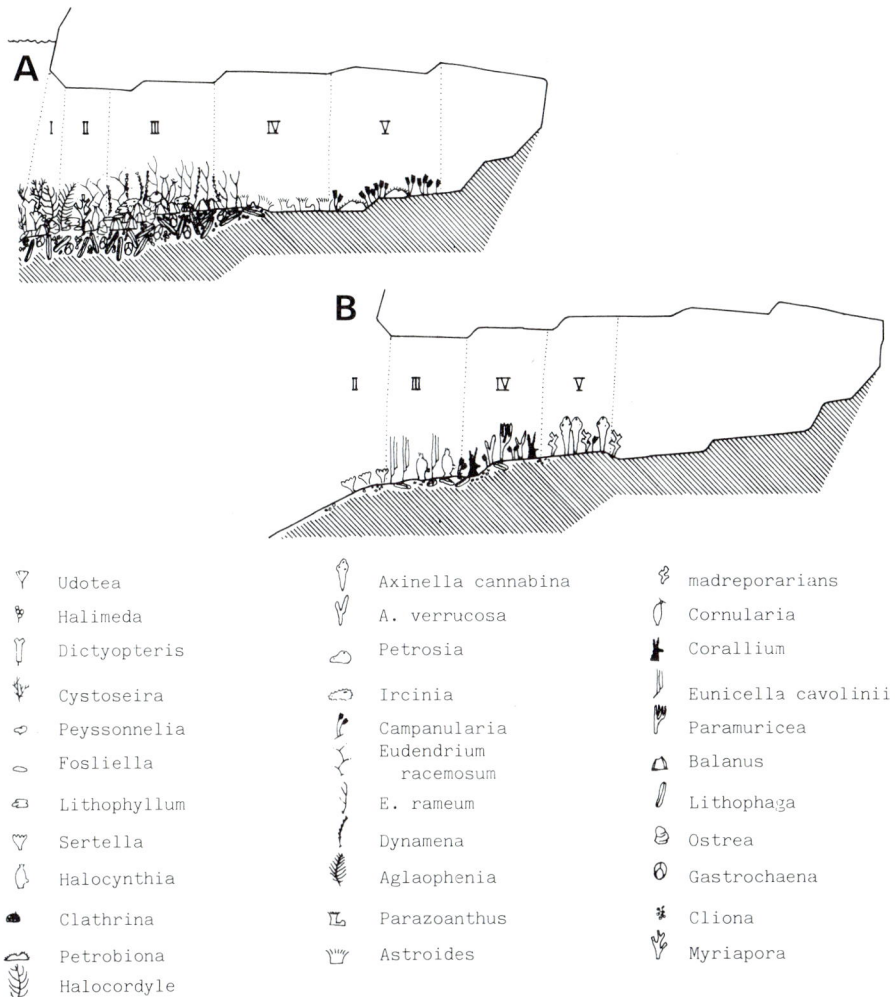

Fig. 8.21 Zonation of sedentary or attached species along submarine caves. Only the floor assemblages of two typical caves, a superficial (1-5 m deep) one (A) and a deep (15-40 m) one (B) have been represented. The identified zones are: I, phytal (photophilic and sciaphilic algae); II, cave entrance; III, outer cave assemblages; IV, intermediate cave assemblages; V, inner cave assemblages (followed by a void innermost zone). (Modified from Riedl, 1966; by permission of Paul Parey Verlag).

Bibiloni et al., 1984), Majorca (Bibiloni and Gili, 1982), and those of the French (Laborel and Vacelet, 1959; Zibrowius, 1978, etc.) and other coasts (Riedl, 1966). For a thorough study the account of Riedl (1966) provides an authoritative summary of the submarine cave environment.

The community of semi-obscure caves has a vertical distribution in strata or layers similar to that of the coralligenous (Fig. 8.22). However, the concretion stratum is reduced due to the absence of algae, while the elevated stratum is usually absent. The hard, rocky substrate is unsuitable for burrowing animals and sustains only some molluscs. The intermediate stratum is better developed, and contains the majority of the species which characterize the community.

The sponges, typical filter-feeders, are an important group which can cover other animals of the same or of other groups. The more common species are the brightly-coloured *Spirastrella cunctatrix*, *Oscarella*

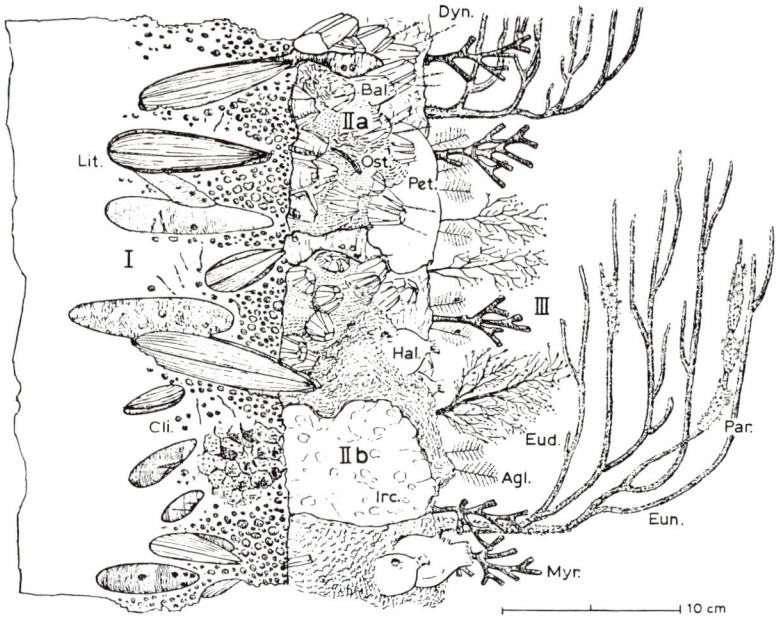

Fig. 8.22 Diagrammatic section through the wall of a marine cave to show the typical three-strata disposition. I, endolithic species (*Lithophaga* and *Cliona*); II, encrusting or intermediate stratum (with *Balanus, Ostrea, Halocynthia, Petrosia, Ircinia*); III, upper stratum and epibionts (with *Dynamena, Eudendrium, Aglaophenia, Myriapora, Eunicella, Parerythropodium*). (From Riedl, 1966; by permission of Paul Parey Verlag).

lobularis, Agelas oroides, Spongionella pulchella, Ircinia dendroides, I. spinulosa, Chondrosia reniformis, Suberites carnosus, Acanthella acuta, Aplysilla sulphurea, Anchinoe fictitius and *Spongia virgultosa*. Anthozoans are the dominant cnidarians: *Corallium rubrum, Leptopsammia pruvoti, Rolandia rosea, Hoplangia durothrix, Parazoanthus axinellae* and *Caryophillia inornata*. The hydrarians, mainly epibiont on sponges, are less well suited for living in dark places, but nevertheless, *Eudendrium racemosum, Campanularia bicuspidata* and *Halecium beani* are common in caves.

Small bryozoans are filter-feeders that colonize every suitable substrate. Medium-sized colonies characterize special facies. *Sertella septentrionalis, Crassimarginatella maderensis, Cribrilaria radiata, Prenantia inerma, Adeonella calveti, Costazzia caminata* and *Smittoidea marmorea* are the commonest. Other sessile filter-feeders include polychaetes and ascidians. The polychaetes are found under, between or on sponges and other sessile organisms, sometimes only the filtering fans and the chalky tube openings being visible. *Serpula vermicularis, Pomatoceros triqueter* and *Spirobranchus polytrema* are very common. The tunicates include *Pyura vittata, Cystodites dellechiajei, Didemnum maculosum* and *Botryllus* species. A few sessile molluscs characterize the same community: *Lithophaga lithophaga* boring into the substrate and *Acar pulchella, Barbatia barbata, Rocellaria dubia* and *Aequipecten opercularis* settling on to it. There is also the ubiquitous colonial foraminiferan, *Miniacina miniacea*.

Among the vagile species, usually detritivores or carnivores, are the polychaetes *Ceratonereis costae, Glycera tessellata, Lumbrinereis coccinea, Lysidice ninetta*; the molluscs *Calliostoma zizyphinum, Luria lurida* and, especially, the nudibranchs *Peltodoris atromaculata, Hypselodoris elegans, H. fontandraui, Phyllidia pulitzeri* and others, that feed on sponges, cnidarians and ascidians. The crustacea include medium to large-sized representatives of the vagile macrofauna: *Homola barbata, Scyllarus arctus, Maja verrucosa, Dromia vulgaris, Galathea strigosa* and some hermit crabs (Paguridae). Small size species live into or among the large sponges and other cave organisms: *Elasmopus pocillimanus*, *Lembos websteri*, *Amphithoe vaillanti* and *Dexamine spiniventris*.

The commonest echinoderms are the sea-stars *Marthasterias glacialis* and *Coscinasterias tenuispina*, but there are also brittle-stars (*Ophiocomina nigra*, *Ophioderma longicauda*) and sea cucumbers (*Holothuria tubulosa*). Some fishes, such as conger eel and large Serranidae, hide in caves while others live there permanently. These include the following species in semi-obscure caves: *Anthias anthias* and *Apogon imberbis*, swimming in the semidark entrance, and the more sedentary *Scorpaena porcus*, *S. notata* and *Gobius niger* resting on the bottom or in wall ledges.

Sessile species can entirely carpet the substrate in the same caves. There are marked substrate preferences; for example, in a typical underwater tunnel the sessile organisms which line the lateral walls are, in order of abundance: sponges, cnidarians and the remainder, mainly bryozoans. On the tunnel roof, however, the cnidarians are the more abundant filter-feeders, followed by sponges and others (Fig. 8.24). The covering of the normally silt-loaded floor is poor in filter-feeders but detritivores can abound. Cnidarians and anthozoans are numerous on walls and madreporarians on the roof. Hydrarians are commoner near the tunnel entrance and the other cnidarian groups along the intermediate sections (Fig. 8.24; Gili et al., 1982).

There are some very characteristic facies in this community: (a) that of the yellow *Parazoanthus axinellae* in the entrance, lighted zones; (b) of the bright red precious coral, *Corallium rubrum*, in the entrance roof and in dark overhangs and crevices; (c) of *Leptopsammia pruvoti* with *Agelas oroides* and *Spirastrella cunctatrix*, in overhangs and cave entrances; (d) of the madreporarians *Polycyathus muellerae*, *Caryophyllia inornata* and *Hoplangia durothrix*, whose small polyps appear in the darker parts; (e) of big bryozoan colonies, such as *Sertella septentrionalis*, in the cave entrances; (f) of hydrarians, considered as an impoverishment or stressed facies, with *Sertularella*, *Eudendrium* and others.

The dark cave community is poorer and covers less than 60 – 50% of the floor and walls. The majority of species are similar to the preceding community, but their relative abundances change and very rare, relict forms, appear. Among the inhabitants of these totally dark, terminal caverns are the sponges: *Penares helleri*, *Rhabderemia minutula*, *Verongia cavernicola*, *Reniera valliculata*, *R. sarai*, *Diplastrella bistellata*, *Petrosia ficiformis*, *Anchinoe tenacior* and *Haliclona* species. Nearly all of them are laminar and unpigmented, or else grow abnormally, building erect and impossibly fragile structures. When they grow in the open, under light, as *P. ficiformis*, their morphology is quite different. There the rare species *Discodermia polydiscus* (bathyal) and *Petrobiona massiliana* and *Pectoroninia hindei* (pharethronid sponges) are found.

Among cnidarians, *Guynia annulata*, *Conotrochus magnanii*, *Polycyathus muellerae* and *Madracis pharensis* can form characteristic facies. Their polyps are small, as are the colonies of the bryozoans *Celleporina lucida*, *Crassimarginatella crassimarginata*, *Setocellina cavernicola* and *Coronellina fagei*. Food scarcity is undoubtedly responsible for this small size. Other sessile, facies-forming species are the polychaetes *Omphalopoma aculeata*, *Vermiliopsis monodiscus*, *V. infundibulum* and other serpulids; the brachiopod *Megathyris decollata* is also an inhabitant of the dark caves.

Among the rare vagile species, are the crustaceans *Aristias tumidus*, *Herbstia condyliata*, *Stenopus spinosus*, *Palaemon serratus* and many free-swimming mysids forming a cave zooplankton characteristic of enclosed biotopes; the molluscs *Alvania reticulata*, *Leptothyra sanguinea*, *Bouvieria aurantiaca*, *Susania testudinaria*, *Peltodoris atromaculata* and *Discodoris cavernae*, and the echinoderms *Ophiopsila aranea* and *Genocidaris maculata*. The blind fish, *Grammonus ater* is confined to these dark habitats, as are other rare and even unnamed species.

The assemblages which constitute the coralligenous and the cave communities are not the only ones on hard circalittoral substrates, since a fairly specific assemblage (that of the offshore rocky bottom) develops in deep waters whose feeble illumination still support some poorly-developed calcareous rhodophyceae. The faunistic component is formed mainly by sessile pedunculate species whose basal part adheres to the hard substrate, notably: sponges, cnidarians, bryozoans, brachiopods and large decapods and echinoderms. Filter-feeders predominate, exploiting the rain of suspended matter brought by the quite strong currents of these depths.

Fig. 8.23 Idealized representation of a typical cave community, based on that of Medes Islands cave system (Girona, NE Spain). On the cave roof, right to left, and then on the cave wall, top to bottom: *Petrosia ficiformis*, *Eudendrium racemosum*, *Oscarella lobularis* (with *Dromia vulgaris* on it), *Verongia cavernicola*, *Caryophyllia inornata*, *Polycyathus muellerae*, *Anchinoe tenacior*, *Chondrosia reniformis*, *Hoplangia durothrix*, *Haliclona cratera*, *Leptopsammia pruvoti*, *Spirastrella cunctatrix*, *Reniera sarai*, *Clathrina coriacea*, *Dysidea avara*, *Spongia virgultosa*, *Petrosia ficiformis*, *Sertella septentrionalis*, *Gobius niger*, serpulid worms. Swimming freely: *Apogon imberbis*, mysids and other cave zooplankters (alight by the searchlight) and *Phycis phycis*. (Drawing by M. Zabala in *Els sistemes naturals de les illes Medes*, Ros et al., 1984, by kind permission of the Institut d'Estudis Catalans, Barcelona).

But the circalittoral zone is chiefly the domain of the gentle slanting, soft shelf bottoms abutting to the continental slope and to the bathyal zone. Between the shelf-edge and the limit between the infra- and the circalittoral zones there is an almost continuous organogenous or 'detritic' bottom, with a more or less important fraction of mud, and a vegetal component made chiefly by free-living corallinaceae.

Beyond the shelf-edge, the absence of multicellular plants is total and we enter the bathyal zone, in the aphytal or deep system. This is the domain of soft, muddy bottoms. Many Mediterranean bathyal assemblages include species which, in other areas, are truly abyssal. The presence of a true abyssal zone in the Western Mediterranean has been questioned (see Introduction) and no doubt the mixing with the bathyal assemblages makes clear-cut identification of such a deep zone even more difficult. A more accurate description of these bottoms has been provided in Chapter 7.

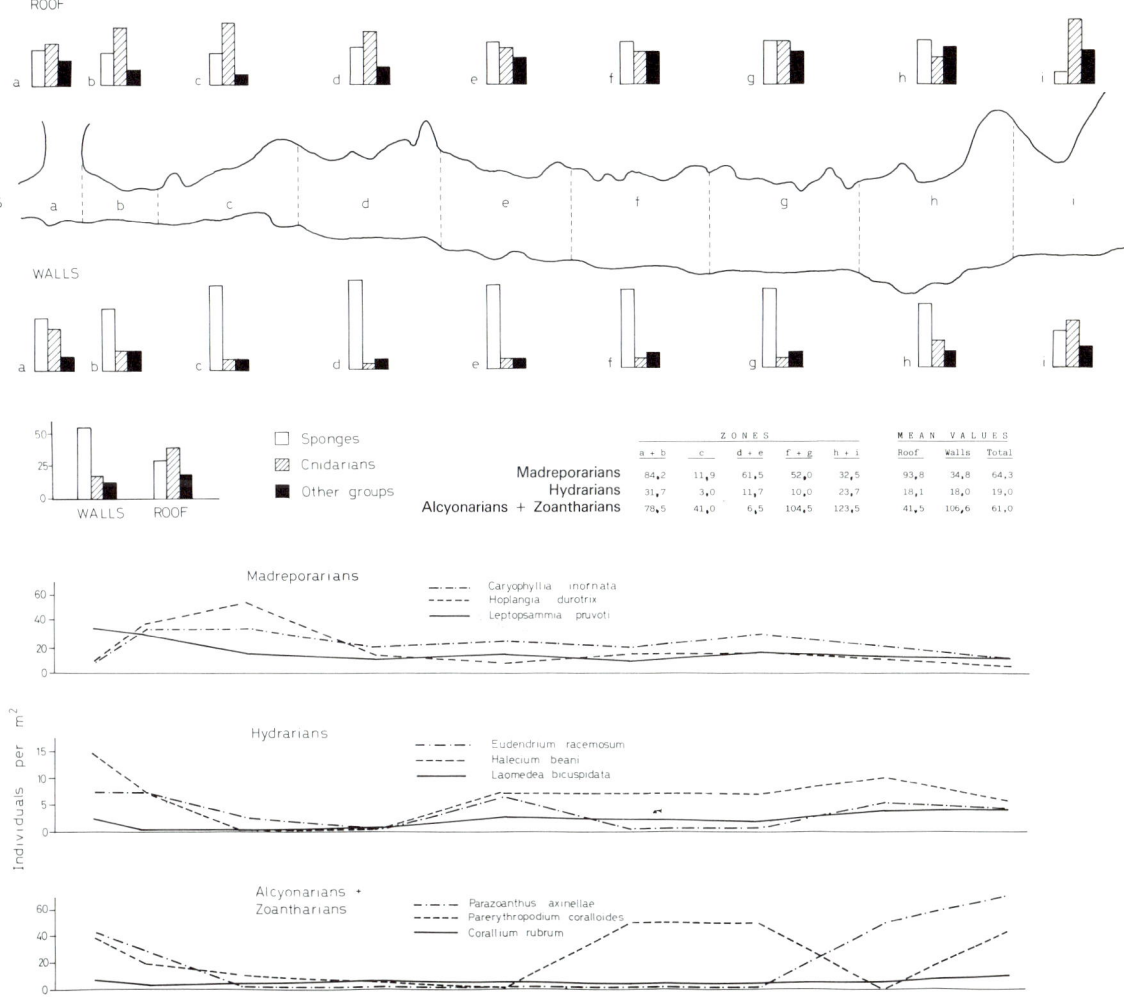

Fig. 8.24 Longitudinal section of the Túnel Llarg de la Meda Petita (Medes Is., Girona, NE Spain), an underwater tunnel 87 metres long and between 0 and 24 m deep. Top, percent cover on walls and roof of sponges, cnidarians and other animal groups. In the graph at left, mean values; visual inventories. Bottom, density (in individuals per square metre) of the nine more common cnidarian species along the tunnel walls and roof. Middle centre and right, densities of the five zones in which the tunnel can be divided according to light penetration (two outermost, dimly lit; one central, dark; two intermediate, semi dark zones) and mean values; visual inventories. (Modified from Gili et al., 1982; by permission of the authors).

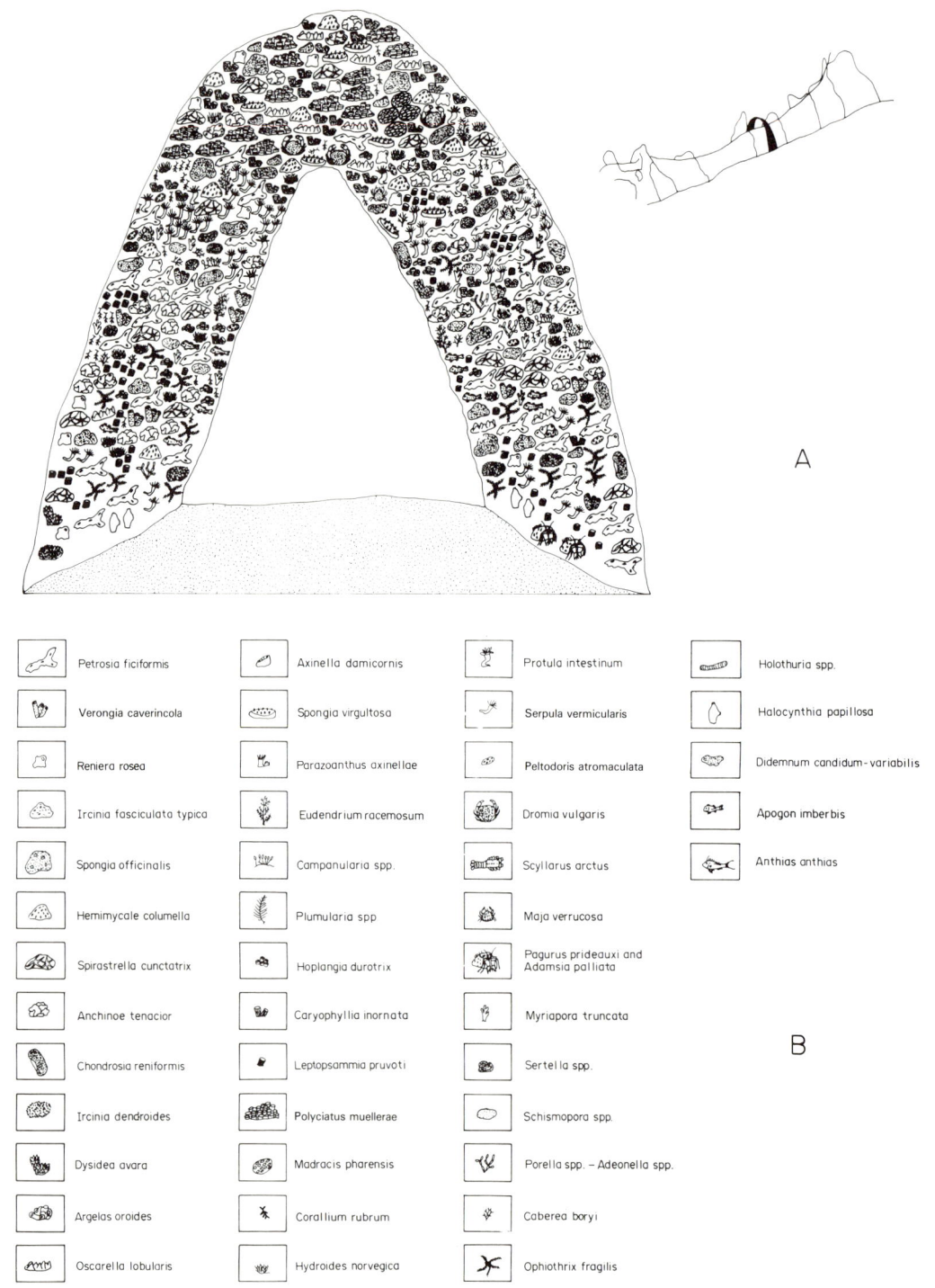

Fig. 8.25 Diagrammatic representation (A) of a transect in the intermediate zone of the Túnel Llarg de la Meda Petita (inset), and symbol equivalence (B). (Modified from Gili and Ros, 1982; by permission of the authors).

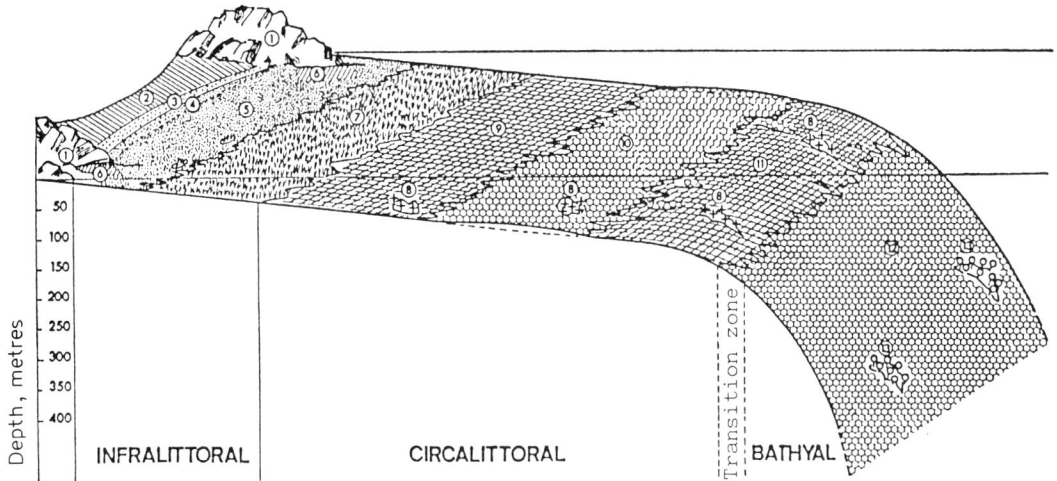

Fig. 8.26 Diagrammatic representation of the distribution of the main benthic assemblages on the shelf and upper part of the slope in the Mediterranean. 1, rocky points; 2, alluvial area; 3, upper and middle beach (supra- and mediolittoral étages and upper part of the infralittoral); 4, upper clean-sand assemblages; 5, well-sorted fine-sand assemblage; 6, photophilic algae assemblages on rocky substrate; 7, *Posidonia* meadows; 8, coralligenous assemblage; 9, coastal detritic assemblage; 10, terrigenous and mud shelf assemblage; 11, shelf-edge detritic assemblage; 12, bathyal mud assemblage; 13, deep-sea corals assemblage. (After Pérès, 1967; by permission of the author).

8.8. BIOMASS, FLOW OF ENERGY AND SYSTEMIC PROPERTIES OF THE BENTHOS

So far more attention has been paid to the description of the Western Mediterranean benthic communities than to their function. On the human scale benthos is seen more as structure and plankton as a process, and interest has developed accordingly. Furthermore, the understanding of the dynamics of benthos lags far behind. The description of the Western Mediterranean benthos will be supplemented by a brief summary of benthic marine succession, bionomic strategies, trophic ecology and other functional aspects of benthic communities.

8.8.1. Succession and regression

Consider a rocky surface, devoid of any organisms and located in the infralittoral zone; the causes of being bare are unimportant to the following description of the process of colonization and development of the living assemblages. In the change from a bare area to an area indistinguishable from the neighbouring, undisturbed ones, diverse populations settle, persist for varying periods, and are substituted by other assemblages, in an ecological succession. The general features have been well-studied (*see* Margalef, 1974 for a summary and Fig. 8.27).

At the beginning, the empty surface is readily covered by bacteria and a minor number of fungi from open waters. These profit from the relative increase in nutrient concentration due to the physical effect of a solid surface, and grow as a thin living film. Organic coatings produced by the very organisms and not only the organisms themselves are substantial for their properties. From the second day onwards there is a progressive development of diatoms (unicellular algae of high turnover rate) which give the substrate a brownish colour. After one or two weeks the minute thalli of ephemerophycean algae appear; they become fully dominant after a month. This colonization phase is dominated by cyanophyceae and opportunistic multicellular algae. The species vary according to place, environment and season, but are primarily of the families Ulvaceae, Cladophoraceae, Ectocarpaceae, Scytosiphonaceae, Bangiaceae and Ceramiaceae. An expansion and alternating development of one or a few plant species is accompanied by the appearance of a collection of filter-feeding animals (including cirripeds, mussels, bryozoans, hydrarians, tubicolous polychaetes and ascidians) which readily settle on the substrate. They share a relative indifference to environmental factors and a capacity for rapid development.

The pioneer phases of any succession are remarkable for spatial homogeneity and, therefore, for low structural complexity. At between three to twelve months, in the Western Mediterranean, the assemblage changes radically with the settlement of larvae and young stages of plants and animals from surrounding communities which progressively displace the opportunistic, pioneer species. At this stage of the colonization (predestination phase) the assemblages are becoming spatially heterogeneous and structurally more complex, thus progressively ressembling the normally built communities.

The succession process continues with dynamic phases punctuated by transition or interphase, relatively less dynamic, periods. According to Niell (1981), these transitional periods are characterized by discontinuities, that is, sudden increases in energy flow within the system, as an exaggerated increase in the production/biomass (P/B) ratio may show (Fig. 8.28). These increases are more important in the early than in the later interphases, and are followed by a progressive decrease in the relative energy flow, which levels-off at the commencement of the next phase. Stabilizations may occur in the diatom and

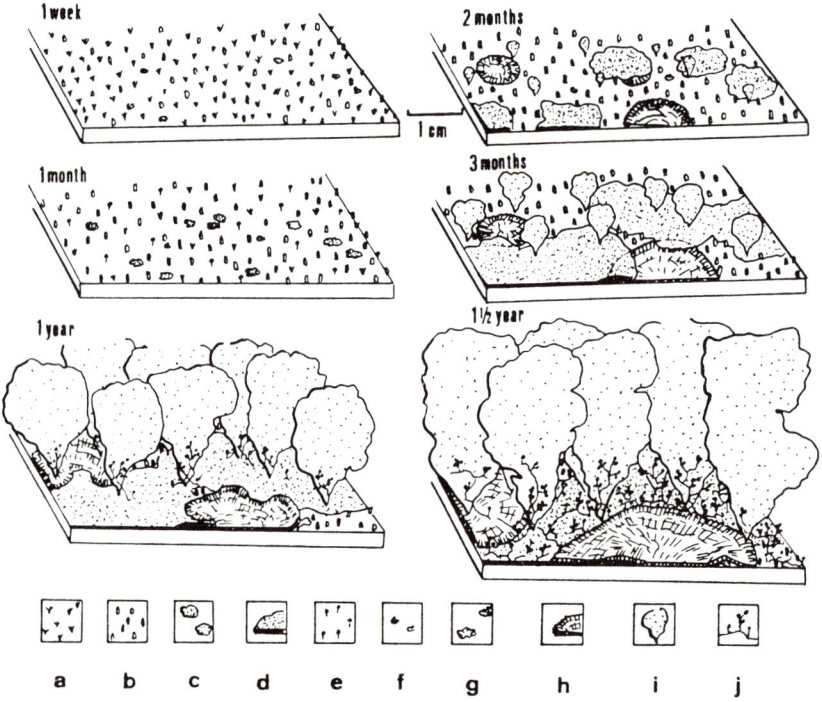

Fig. 8.27 Successive stages in the colonization of an artificial immersed substrate by sessile plant and animal species in the infralittoral zone (0 – 6 m). The time intervals are given only as a general indication, as the rate of colonization can vary as a function of place, season and so on; nor the final stage corresponds to a 'mature', well-developed community, but to the ulvacean or opportunistic algae phase mentioned in the text. a, colonial diatoms; b, small-sized multicellular algae; c, encrusting calcareous rhodophyta recently settled; d, the same, later; hydroids recently settled; f, *Spirorbis* and other sessile polychaetes; g, bryozoans recently settled; h, the same, later; i, *Ulva rigida*; j, hydroids, later. (From Huvé, 1953; by permission of the author).

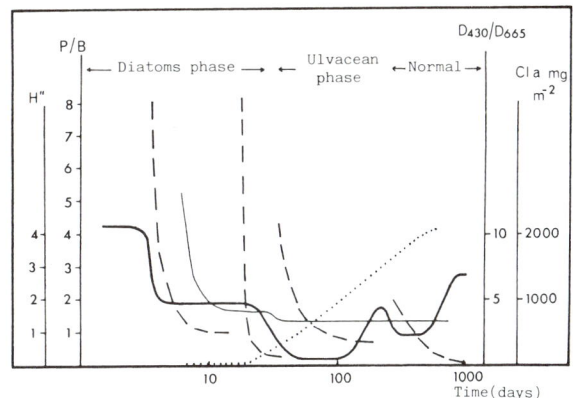

Fig. 8.28 Variation along the benthic succession of four parameters of ecological signification: specific diversity (H', continuous coarse line), P/B ratio (broken line), D_{430}/D_{665} ratio (Margalef index, continuous fine line) and an estimate of the chlorophyll a amount (dotted line). (From Niell, 1981; by permission of the author).

predominance phases, as well as at the end of the predestination phase (when the structural complexity peculiar to the assemblages of the area is achieved). These stabilizations at the end of each development phase can be considered as stable points (*sensu* Sutherland, 1974). The more advanced the succession the longer is the period of stabilization, while conversely the P/B ratio at the end of each phase declines. Often no evidence would be found to prefer this dialectical view of succession to a more continuous interpretation of the events.

Succession begins with a scrambling for niches (niche preemption model). At this stage the diversity (H') is high but decreases rapidly as the succession proceeds, reaching a minimum in the predominance phase. From there onwards it increases, peaking in the channelling phase. Spatial structuring is highest at the end of this phase, then remains more-or-less constant, with the usual ups and downs of the annual cycle of the normal assemblages. The development of a complex patchiness can be seen in the diversity spectra, in the sense that a universe made of areas all uniformly mixed and diverse evolves into a pattern of increasing spatial segregation. An increase in biomass, in production and of chlorophyll a per unit surface area, follow along the succession.

Although the succession process never actually ends, the last phase, that of maturation, is characterized by stabilization of the essential features of the assemblage. If these do vary, then the variation is cyclical and seasonal and related to the internal dynamics of the system. In reality a stable, unchanging situation virtually never exists and if cyclical variations appear, they are by no means identical. A particular fragment of community is thus *unique* in space and time and is never duplicated. Surveys of communities with time confirm the qualitative and quantitative variation in species composition and the persistence of structural complexity.

Several patterns of benthic succession have been described, both in Western Mediterranean and elsewhere, and the processes in general agree well with the above model. Castric (1977), for example, distinguishes the following phases in the colonization of artificial substrates submerged in Mediterranean waters:

— Recruitment, which last two months and corresponds to the settlement of the first larvae and young forms.

— Settlement (between the second and the eleventh month), when the species settle and grow and its numbers stabilize, depending on the season of the year and on other factors.

— Convergence (from the eleventh month onwards), when the assemblages on the different artificial substrates tend to converge and to resemble one another;

— Predictability (from the 16th to the 26th month), when the assemblages approach a qualitative climax, which is nevertheless very different from the surrounding communities. Total similarity with adjacent, undisturbed communities, can take many years to achieve.

The ecological succession does not always follow the above pathway. The path, which can be termed 'progressive', can be truncated, to yield what some authors speak of as a fouling community (see section 8.4.1). The path can go in the opposite direction as well, and this is called regression. In marine benthos a regression can be instantaneous, due, for example, to a storm action or to a strong seiche. Very often disrupted communities can heal with a successional process that do not differ essentially from that described above. But a regression can also extend in time, as a result, for example, of an excessive and lasting exploitation by some of the system components. If their population density is high, the littoral sea urchins are, for example, able to strongly modify photophilic algae communities — see page 250 — and even *Posidonia oceanica* meadows. Very often, human action is responsible for regressions. Organic and chemical pollution, the alteration of the sedimentary inputs and of the coastal circulation due to the building of piers, the dumping of fine sediments on rocky coasts, the exploitation of benthic populations all result in abnormal stresses on the benthic communities, that appear to return to earlier phases of the succession.

During regression the communities become qualitatively and structurally simpler, spatial heterogeneity is destroyed, biomass decreases, the P/B rises, the r-strategists overcome the K-strategists and, in general, the affected grounds lose their riches and beauty. With extreme regression the opportunistic species cannot survive, and benthic assemblages are forever restricted to the first successional phases (i.e., to heterotrophic fungi and, especially, bacteria, and, eventually, some autotrophs: cyanophyceae, *Derbesia tenuissima*, etc.).

The microstructure of a benthic community is often regarded as a series of patches and of patches of patches, as a result both of concurrence, affinity, interspecific epi- or endobioses, and of dynamic and structural characteristics more deeply rooted in the community. Two examples can illustrate this diversity.

A community of *Schottera nicaeensis*, *Plocamium cartilagineum* and *Lomentaria articulata* (see page 252) develops in turbulent, poorly-lit areas of cool water in the Western Mediterranean; it is clearly patchy (Boudouresque, 1973), due to the simultaneous occurrence of different regeneration stages of variable age. This mosaic results from wave destruction of the structurally complex and older parts of the community. These are renewed and in places re-start a new successional progress towards maximum overall complexity.

Another example is that of the dynamics of coralligenous communities (see pages 263 and 267 and Fig. 8.16). In them, the building of complex structures (consisting of corallinaceae algae, sponges, cnidarians, bryozoans and other organisms) parallel their destruction by other organisms, subsidence and by erosion, and the simultaneous replenishment of the holes and crevices of the coralligenous mass by sediment and organic debris. The result is the growth of the coralligenous socle in which the patchiness of the formation is obvious.

When spatial heterogeneity is greatest, other parameters also characterize the community, notably: slope diversity spectra or high β-diversity, low turnover rate, relative dominance of K-strategist species, etc. Thus, as in other living assemblages, there is a strong relation between structure and dynamics in the benthic assemblages and its patchiness is a useful measure of the dynamics of a benthic community.

8.8.2. Ecological strategies

Structural and dynamic complexity enables the presence of many organisms with different biological strategies. It is assumed that the more complex a system is, the more ecological niches there will be. Irrespective of its precise role in the community (as producers, filter-feeders, omnivores, detritivores, etc.), species fall into a bionomic or ecological continuum between extreme results of the operation of natural selection, the so-called *r* and *K* strategies (Pianka, 1970; Margalef, 1974; see Ros, 1982, for a summary).

Hiscock and Mitchell (1980) have ranged these organisms into four large groups:
— long-lived species, which stabilize and characterize the system;
— seasonal species that appear only at certain times of the year;
— opportunistic species, with a high reproductive potential;
— species which respond to small alterations in environmental conditions, with usually known and temporally variable structural signification.

The *r* strategy (named after the conventional symbol for reproductive rate) is employed by fugitive, pioneer, opportunistic, or generalist species. These exploit the lack of organisation and environmental stability, with a high production of offspring in which the great majority inevitably disappear. They are highly adaptable and characterize very unstable or stressed environments. At the other end of the spectrum are the *K*-strategists (*K* is a symbol for the asymptotic limit to the population growth, the carrying capacity of the environment). Such species, which are described as specialists or strategists, can

Fig. 8.29 An example of zonation: the area occupied by the barnacle *Chthamalus stellatus* leaves way to a narrow band of the red alga *Rissoella verruculosa*, below which poorly-developed individuals of *Litophyllum tortuosum* (a calcareous red alga), not forming a typical *trottoir*, appear. The picture, taken in the moment of the wave ebb, shows also the underlying community of the brown alga *Cystoseira mediterranea*. Horizontal scale, ca. 2 m. Cap de Creus, NE Spain. Photo by E. Ballesteros.

Fig. 8.30 The community of the phaeophycean *Cystoseira mediterranea*, appearing in battered places, supports one of the greatest plant biomass and has the higher primary production in the Western Mediterranean marine benthos. A sound explanation can be found in the input of auxiliary energy in the form of wave action, which brings to the plant the renewal of water and, consequently, a steady nutrient input. Horizontal scale, ca. 3 m. Tossa de Mar, NE Spain. Photo by E. Ballesteros.

stabilize their populations in a given environment and produce few, very protected offspring which are well adapted to ecologically stable microenvironments.

These strategies fit in the succession process as we have stressed above: there is a general trend towards the replacement of *r* for *K*-strategists. The mechanisms for the substitution include all the adaptive features which tend to favour longevity and size, and lower reproductive effort. The reduced expenditure is used for other functions, such as movement, defence and homeostasis in general.

The benthos can be regarded as being more advanced than the plankton, both ecologically and evolutionarily, and that on hard substrates as more advanced than that on soft bottoms. The reason can be related to physical stability. A muddy bottom is more stable than the water above it. Increasing particle size (from silt to pebble through sand, and from pebble to solid and massive rock) also increases

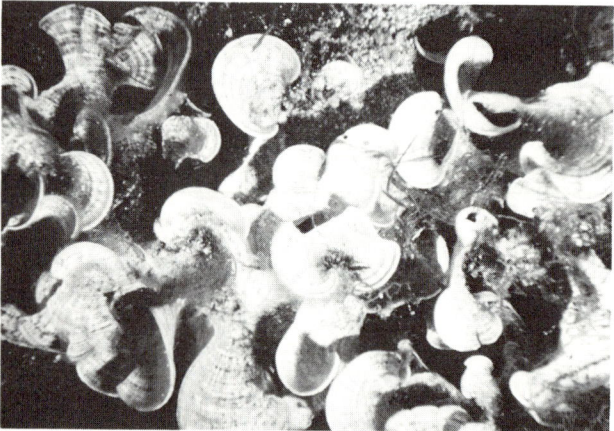

Fig. 8.31 In well-lit, shallow areas only poorly wave-agitated, communities whose species show some tropical affinities often develop. This is one of them, *Padina pavonica*, a brown alga whose paper cone-like fronds are loosely encrusted with lime. Horizontal scale, ca. 25 cm. Medes Islands, NE Spain, 5 m deep. Photo by A. García.

Fig. 8.32 *Caulerpa prolifera* is a green alga (chlorophycean) showing some convergence with sea-grasses, one of which (*Cymodocea nodosa*) appears also in this picture: buried stolons, rhizoidal hairs, etc. It lives in bottoms of silt or fine sand, in sheltered and shallow places, forming extensive meadows with a notoriously poor epifauna, a phenomenon apparently caused by the secretion of a substance, caulerpine, which inhibits epiphytic fixation. Horizontal scale, ca. 30 cm. Els Alfacs, Ebro Delta, NE Spain, 3 m deep. Photo by J. Romero.

stability. Similar relationships exist with other environmental factors, such as water movement, chemical and physical properties of water and food supply. The stability of the water mass increases with depth, but that for the substrate decreases from a given, say circalittoral, point in both directions. Towards the surface, because of wave and storm and towards the depths, because of the progressive importance of the soft bottoms over the hard ones. Light, the paramount factor, also decreases with depth.

Thus the situation can be summarized as follows. At upper, littoral levels, the benthic strategies favoured are those which tend to counteract the mechanical stresses associated with such surface habitats. The balance between physical hindrances and high illumination rate results in productive, multi-layered plant communities which support a relatively rich fauna. Although production is high, the degree of organisation is intermediate due to the above-mentioned restraints. The most successful ecological strategies are therefore the *r*-selected ones: high reproduction rates, fast turnover and high production.

At intermediate, for example, circalittoral levels, the stress is not mechanical but refers to the availability of light. The supply of food is adequate (since the animals profit of the excedents of the upper levels) and competition for space is maximal. The degree of organisation is high. Many species are builders and the construction processes seem to fit in with a pattern of ecological succession which is well exemplified by the coralligenous community-complex. Production is high, but the investment in spatial structuration is also large. Thus, the most successful ecological strategies are *K*-selected: slow growth rates and slow turnover.

At deep levels, or in dark littoral caves, the stresses are greater, due to the light reduction or absence and high sedimentation; there is a scarce food supply. Such environments can harbour only heterotrophs, or animal assemblages, formed by a majority of *K*-strategists, which evolution has often been marked by the acquisition of far-reaching specializations.

The peculiar ecological features of the supralittoral étage, which we have purposely left to the end, can be explained by the harshness of this zone. Its inhabitants, therefore, share a high degree of opportunistic characters with specializations akin to those of *K*-strategists.

Recently, among other workers, Zabala (1982) has reviewed the ecological strategies of benthic filter-feeders, and Ros (1981, 1982) has discussed those of opisthobranch molluscs. Zabala concludes, on the basis of studies of the Medes Islands in the Western Mediterranean, that there is a convergent trend among the different groups towards a few adaptive options, which are themselves diverse according to the particular communities. The most suitable strategy for spatial competition in a filter-feeder includes a colonial, sessile organization, with a skeleton, unlimited growth, a large capacity for covering ground, long life and efficient capabilities for filtering and defence. Instances of epibioses and simbioses are frequent. This is clearly a *K*-selected strategy.

At the other extreme are the *r*-strategists: individual organisms, vagile and of limited growth, with short life cycles, high reproductive capacities, with planktonic, long-lived larvae and less capable of defending themselves. They are minor competitors, profit from the unstable surface levels, opportunistically covering any area which is left open to them or, when occupying more stable substrates, show strong spatial and temporal fluctuations.

Something similar can be said of the ecological strategies of the opisthobranch molluscs (Ros, 1981, 1982; Todd, 1981). In opposition to the feeding uniformity of the many taxonomic groups reviewed by Zabala (1982), many food regimes exist in the opisthobranchs. The browsers of plant and animal tissues

Fig. 8.33 The long, blade-like leaves of the sea-grass *Posidonia oceanica*, covering great expanses of the infralittoral zone in the so-called *Posidonia* meadows, and its perpetual swinging following the wave movement, make up a unique underwater landscape. In the picture there is also a big individual of the tubiculous, filter-feeding polychaete *Spirographis spallanzani*, whose spectacular branchial tuft makes it one of the underwater photographer's favourite subjects; *Asparagopsis armata*, a red alga, is also present. Horizontal scale, ca. 1 m. Medes Islands, NE Spain, 10 m deep. Photo by J. M. Gili.

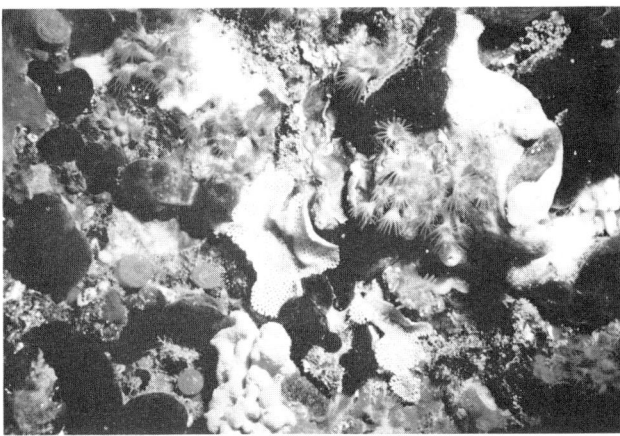

Fig. 8.34 Living assemblage on the wall of a cave entrance. Here, as in other sciaphilic, animal-dominated communities, the competition between organisms in an environment where the more limiting factor seems to be the space leads to a complex and variegated community, with a high specific and pattern diversity. In the picture: sponges (*Axinella damicornis*, *Anchinoe tenacior*, *Spirastrella cunctatrix*), algae (*Lithophyllum expansum*), cnidarians (*Parazoanthus axinellae*, *Leptopsammia pruvoti*, *Rolandia rosea*), bryozoans (*Sertella septentrionalis*), tunicates (*Cystodites dellechiajei*, *Diazona violacea*), etc. Horizontal scale, ca. 25 cm. Medes Islands, NE Spain, La Vaca cave entrance, 17 m deep. Photo by J. M. Gili.

show, perhaps, the greatest diversity and a high predator-prey specificity. The general absence of skeleton is counterbalanced by an extraordinary array of defence mechanisms, of which the homotypy and the complex mimicries are the more dramatic. In the most specialized forms there is a high incidence of direct (type III) development (without planktonic larval stages) or they possess planktonic larvae with large yolk reserves (type II development). On the other hand there are forms which are planktonic, with long-lived, plankton-feeding larvae (type I development). These are the *r*-strategists. The occurrence of these and other characters (e.g. symbioses and lengthening of life cycles) are in good agreement with the degree of stability, or organization, of the habitats of these gastropods.

Many general reviews on benthos biology and strategies are available. Jackson (1977) for example, has discussed the advantages of colonial vs. solitary strategies. Other workers have provided interpretation in terms of selection of the strategies found in meiobenthos or in reef communities.

8.8.3. Trophic ecology

Primary producers

The primary producers in the marine benthos can be grouped into three broad categories: macroscopical algae, sea-grasses and the microphytobenthos. The prevalence of one or another group is mainly a function of the substrate. On hard bottoms, the macroscopical algae (Chlorophyceae, Phaeophyceae and Rhodophyceae) dominate, while the sea-grass meadows settle on sandy bottoms and the microbenthos, made up chiefly of diatoms, can be dominant on soft substrates (such as sand and silt). The representatives of the Cyanophyceae are often abundant under extreme conditions, for example, the supralittoral zone and polluted areas.

What conditions limit and define the benthic primary production? There are four key factors in addition to the nature of the substrate and different stress conditions.

(a) The nutrient concentration (chiefly compounds of nitrogen and phosphorus). As they can be stored in the organisms, benthic plants do not show as clearly as the planktonic ones the effects of the

Fig. 8.35 One of the more common aspects of the coralligenous of the basis of the littoral rock is the one appearing in this picture, where *Paramuricea clavata* makes up the higher stratum. This gorgonian, of vivid and variable hues, lives very often in vertical walls or overhangs, but appears also on top of big coralligenous boulders, as the one depicted. Horizontal scale, ca. 1 m. Medes Islands, NE Spain, 23 m deep. Photo by A. García.

scarcity of such nutrients. Some macrophytes have the ability to photosynthetise in the absence of these nutrients, by storing them in the periods they are available and later using them. Although this is not a common strategy, it is a useful example of the trend which organisms show towards freeing themselves from their environment.

(b) The intensity and quality of light. It is easy to understand the decrease of primary production with depth.

(c) Temperature effects on photosynthesis and respiration. For equivalent temperature increases (especially under conditions of light saturation) respiration increases at a higher rate than the photosynthesis. Thus at lower temperatures a compensation point can be reached at lower light intensities. This could explain the sciaphilic character of some species in relatively cool areas (such as *Halimeda tuna*, see page 255).

(d) Water agitation. This is an external energy input that enhances the renewal of nutrients and breaks up the microgradients of nutrients and inorganic carbon sources which form around plants by absorption.

Fig. 8.36 The red or precious coral, *Corallium rubrum*, has been a commercially exploited species since ancient times, and accordingly its present distribution is seriously restricted. In the picture a conger eel (*Conger conger*) shows its head through one of the multiple crevices characterizing the coralligenous of semi-dark environments in which the red coral and other cnidarians, sponges, etc. (see Fig. 8.34) thrive. Horizontal scale, ca. 40 cm. Medes Islands, NE Spain, 30 m deep. Photo by J.M. Llenas.

All these factors can vary seasonally, and are affected by the physical and hydrographical features of the coast. The following zones can be recognized, according to their primary production (*see also* Figures 8.14 and 8.26).

(i) A supralittoral zone in which the physical conditions scarcely permit primary producers. There are no reliable estimates of primary production at this level, but a suitable figure could be about 5 g of dry weight per square metre per year (5 g(dw)/m^2year). Using organic carbon weight as a reference, this equals 2 gC/m^2year. An average figure for the biomass of the supralittoral fringe would be around 5 gC/m^2.

(ii) The mediolittoral, a zone under a certain stress, but in which the immersion-emersion alternation enhances nutrient absorption. The energy flow (P/B) is high and the species have a high turnover rate, with a production of 700 g(dw)/m^2year (some 300 gC/m^2year), and a biomass around 250 gC/m^2. (In the *trottoir*, due to the lime skeleton, the figure rises to 3500 g(dw)/m^2 year.)

(c) The upper infralittoral zone (*see* page 250) in which the immersion is constant and the environmental conditions are less fluctuating, but in which there is still enough mechanical energy (wave movement) to favour production. Maximal values of biomass and production are found, chiefly in the communities of *Cystoseira mediterranea*. The primary production can reach 4000 g(dw)/m^2year (900 gC/m^2year) and the biomass up to 1000 gC/m^2.

(d) The lower infralittoral zone where dominated by 'soft' or non-calcified algae. Both production and the P/B ratio decrease with depth. This is effectively a deceleration of energy flow in which two differing strategies of production and of space occupation coexist. The algae grow rapidly (one or more generations per year), using available illumination but do not build large and permanent structures. The sea-grasses, chiefly *Posidonia*, turn over slowly, with appreciable accumulation and preservation of biomass. The biomass of photophilic algae ranges between 100 and 800 gC/m^2, and its primary production between 500 and 1800 g(dw)/m²year (200 – 700 gC/m²year), with a daily estimate of between 1 and 5 gC/m². The production of *Posidonia* meadows amounts to 1500 g(dw)/m²year (600 gC) in addition to 300 g(dw)/m²year (50 gC) of the epiphytes. The biomass of *Posidonia oceanica*, not including the rhizomes, is estimated at between 300 and 800 gC/m².

(e) A zone dominated by 'hard' (calcareous) algae: a very efficient strategy in maintaining the occupation of space. It is perhaps an inevitable consequence of the slow growth in dim light. There are few estimates available for the production of the plant communities of the circalittoral zone, but a figure of between 50 and 500 g(dw)/m²year (about 20 – 80 gC), depending on the degree of calcification can be given.

From the data on primary production, and assuming a production of 200 g(dw)/m²year for the plankton (*see* Chapter 5), it is possible to compute that the primary production of the Western Mediterranean benthos approaches the primary production of plankton occupying a coastal band of only about 1500 metres width. This emphasizes two important factors. First that the benthic contribution to the total marine primary production is small, but secondly, that its contribution to the littoral system economy must, nevertheless, be considered.

Animals

If primary production is inadequate to supply the benthic animals, where does the trophic input to the benthic fauna arise?

The benthos as a whole is a deficitary community (more consumers than producers) which sustains itself mainly by exploiting other marine, excedentary communities, for example, planktonic ones. There are only a few animals which feed directly on the primary producers of the benthos, and a great part of the primary production of the first benthic level (photophilic algae, sea-grass meadows, sciaphilic algae) enter the planktonic system as detritus. Only about 1% of the Mediterranean bottom is in the photic zone, and thus there is a functional coupling between plankton and benthos. The food supply to deep bottoms populations is through sinking and depletion processes. The large carcasses of fishes and other animals, plant detritus, crustacean moults and faecal pellets and other small particulate detritus of plant or animal origin, and the dissolved organic matter are all potential food sources for the deep sea organisms. The vertical transport is mediated by physical sinking, by the zooplankton trophic ladder, migration of bottom fishes and, especially in the Western Mediterranean, by turbidity currents and slumps along submarine canyons (*see* Chapter 2).

The secondary producers of the benthos posseses diverse food-gathering mechanisms for limited food supplies have led to the evolution of sophisticated feeding strategies. Filter-feeders and the sedimentivores are most abundant (in relation to the availability of particulate organic material) but macrophages or predators (including the browsers on animal tissues) show a much larger array of specializations than the microphages. These include advanced co-evolution in prey-predator systems which is so intense and specific that it can lead to extremely specialized and even symbiotic relations.

Extensive reviews of the feeding mechanisms and trophic groups in marine benthos will be found in Crisp (1964), Jørgensen (1966), Purchon (1979), Pandian (1975) and Pérès (1982c).

A fraction of the energy used as food by the benthic animals is returned to the environment as faeces and moults, and is then used by other animals and by decomposers. However, two-thirds or more of the assimilated food supports the metabolism of the animal itself, and the remainder is used for growth and reproduction. The allocation of energy to each of these three main processes differs according to the ecological strategy of particular species (*see* page 285). Roughly speaking, more energy is invested in reproduction in the *r*- than in the *K*-strategists, and more in growth and space structuration in the *K*-

than in the *r*-strategists. Opportunistic species usually assimilate a smaller fraction of the ingested food, while *K*-strategists usually use a larger percentage of the assimilated food in respiration (as a result of their higher activity).

The relation between plankton and benthos cannot always be seen as an exploitation by the benthos. Planktonic larvae of benthic animals (which are preyed upon or are dead during their stay in free water, and/or yield organic matter) frequently contribute to the plankton in excess of the biomass of the few larvae that return safely to settle in the benthos community. From the usual understanding of profit in ecology, this sounds like a paradox.

Nevertheless, for each individual species, there are the potential advantages of dispersal and the possibility of colonizing fresh habitats or perhaps accelerated evolution along a life stade of high mortality or even to an adoption in adult life of an *r*-strategy by *K*-strategists. There are obvious examples of evolutionary schizophrenia found in all species in which different age classes are subjected to widely different, and even opposing selective pressures (for example, insects, especially those with aquatic larvae, and frogs). They emphasise complex, and not yet clearly understood relationships, that link species in symbiosis, mutualism, defence mechanisms and other related phenomena, which, although not especially mentioned in this account, are very common and numerous in marine benthos.

As a result, the cycling of organic matter becomes very complicated. The simple path leading from producer to consumer to decomposer and to producer again, is thus diverted many times and bypassed, delayed and further complicated — leading to a pattern of multiple connections between species and, probably, to a high ecological stability. The *Posidonia* meadow (on soft bottoms) and the coralligenous (on hard ones), the most organised and richest communities in the Western Mediterranean, are good examples. They are, also, the most beautiful and captivating ones. If they deserve our admiration, they also require effort to understand more fully their structure and functioning, and the will to ensure their continued existence by urgent and serious protection.

REFERENCES

Ballesteros, E. (1982) Primer intento de tipificación de la vegetación marina y litoral sobre sustrato rocoso de la Costa Brava. *Oecologia aquatica* 6, 163 – 173.

Ballesteros, E. (1984) Els estatges supralitoral i mediolitoral de les illes Medes in *Els sistemes naturals de les illes Medes*, Eds. J. D. Ros, I. Olivella & J. M. Gili. *Arxius de la Secció de Ciències* 73, 645 – 659. Institut d'Estudis Catalans, Barcelona.

Ballesteros, E. and Romero, J. (1982) Catálogo de las algas bentónicas (con exclusión de las diatomeas) de la costa catalana. *Collectanea Botanica* 13(2), 723 – 765.

Ballesteros, E., García, A., Lobo, A. and Romero, J. (1984a) L'herbei de *Posidonia oceanica* de les illes Medes in *Els sistemes naturals de les illes Medes*, Eds. J. D. Ros, I. Olivella & J. M. Gili. *Arxius de la Secció de Ciències* 73, 739 – 759. Institut d'Estudis Catalans, Barcelona.

Ballesteros, E., Romero, J., Gili, J. M. and Ros, J. D. (1984b) L'estatge infralitoral de les illes Medes: les algues fotòfiles in *Els sistemes naturals de les illes Medes*, Eds. J. D. Ros, I. Olivella & J. M. Gili. *Arxius de la Secció de Ciències* 73, 661 – 675. Institut d'Estudis Catalans, Barcelona.

Ben, D. van der (1971) Les épiphytes des feuilles de *Posidonia oceanica* Delile sur les côtes françaises de la Méditerranée. *Mém. Inst. Roy. Sci. Nat. Belgique* 168, 1-101.

Bellan-Santini, D. (1969) Contribution à l'étude des peuplements infralittoraux sur substrat rocheux. *Rec. Trav. Sta. Mar. Endoume* 63(47), 1 – 293.

Bibiloni, M. A. and Gili, J. M. (1982) Primera aportación al conocimiento de las cuevas submarinas de la isla de Mallorca. *Oecologia aquatica* 6, 227 – 234.

Bibiloni, M. A., Cornet, C. and Ros, J. D. (1982) Estudio bionómico del litoral de Blanes (Girona) entre Punta de Santa Anna y Cala Sant Francesc. *Oecologia aquatica* 6, 185 – 198.

Bibiloni, M. A., Gili, J. M. and Ros, J. D. (1984) Les coves submarines de les illes Medes in *Els sistemes naturals de les illes Medes*, Eds. J. D. Ros, I. Olivella & J. M. Gili. *Arxius de la Secció de Ciències* 73, 707 – 737. Institut d'Estudis Catalans, Barcelona.

Boudouresque, C. F. (1971) Contribution à l'étude phytosociologique des peuplements algaux des côtes varoises. *Vegetatio* 22(1-3), 83 – 104.

Boudouresque, C. F. (1973) Étude in situ de la réinstallation d'un peuplement sciaphile de mode battu après sa destruction expérimentale, en Méditerranée. *Helgol. wiss. Meeresunters* 24, 202 – 218.

Boudouresque, C. F. and Cinelli, F. (1976) Le peuplement algal des biotopes sciaphiles superficiels de mode battu en Méditerranée occidentale. *Publ. Staz. Zool. Napoli* 40, 433 – 459.

Boudouresque, C. F. and Fresi, E. (1976) Modelli di zonazione del benthos fitale in Mediterraneo. *Boll. Pesca Idrobiol.* 31, 129–143.
Castric, A. (1977) Recrutement et succession du benthos rocheux sublittoral in *Biology of Benthic Organisms*, Eds. B. F. Keegan, P. O. Ceidigh & P. J. S. Boaden, pp. 147-154, Pergamon Press, Oxford.
Clarke, G. L. (1939) The utilization of solar energy by aquatic organisms. Problems in lake biology. *Publ. Am. Assoc. Advance. Sci.* 10, 27–38.
Coppejans, E. (1980) Phytosociological studies on Mediterranean Algal vegetation: rocky surfaces in the photophilic infralittoral zone. In *The Shore Environment*, Eds. J. H. Price, D. E. G. Irvine & W. F. Farnham, pp. 371–393, Academic Press, London.
Crisp, D. J. (ed.) (1964) *Grazing in Terrestrial and Marine Environments*. Blackwell, Oxford.
Drouet, F. (1968) *Revision of the Classification of the Oscillatoriaceae*. Acad. Nat. Sci. Philadelphia Monographs.
Drouet, F. (1973) *Revision of the Nostocaceae with Cylindrical Trichomes (formerly Stygonemataceae and Rivulariaceae)*. Haffner, New York.
Drouet, F. (1977) *Revision of the Nostocaceae with Constricted Trichomes*. Gantner Verlag, K.G.
Drouet, F. and Daily, N. A. (1956) *Revision of the Coccoid Myxophyceae*. Butler University Botanical Studies, Haffner, New York.
Feldmann, J. (1937) Recherches sur le végétation marine de la Méditerranée. *Rev. Algol.* 10, 73–254.
Fletcher, A. (1980) Marine and maritime lichens of rocky shores: their ecology, physiology and biological interactions. In *The Shore Environment*, Eds. J. H. Price, D. E. G. Irvine & W. F. Farnham, pp. 789–842, Academic Press, London.
Fiala-Medioni, A. (1970) Les peuplements sessiles des fonds rocheux de la région de Banyuls-sur-mer. *Vie et Milieu* 21(3B), 591–656.
Funk, G. (1927) Die Algenvegetation des Golfs von Neapel nach neuren ökologischen Untersuchungen. *Pubbl. Sta. Zool. Napoli* 7, supl., 1–507.
Giaccone, G. and Bruni, A. (1972) Le cistoseire e la vegetazione sommersa del Mediterraneo. *Atti Ist. Ven. sci. Lett. Art.* CXXXI, 59–103.
Gili, J. M. (1984) El detrític costaner de les illes Medes. In *Els sistemes naturals de les illes Medes*, Eds. J. D. Ros, I. Olivella & J. M. Gili. *Arxius de la Secció de Ciències* 73, 761–766. Institut d'Estudis Catalans, Barcelona.
Gili, J. M. and Ros, J. D. (1982) Bionomía de los fondos de sustrato duro de las islas Medes (Girona). *Oecologia aquatica* 6, 199–226.
Gili, J. M. and Ros, J. D. (1984) L'estatge circalitoral de les illes Medes: el coral·ligen. In *Els Sistemes Naturals de les Illes Medes*, Eds. J. D. Ros, I. Olivella & J. M. Gili, *Arxius de la Secció de Ciències* 73, 677–705. Institut d'Estudis Catalans, Barcelona.
Gili, J. M., Olivella, I., Zabala, M. and Ros, J. D. (1982) Primera contribución al concomiento del poblamiento bentónico de las cuevas submarinas del litoral catalán. *Actas Ier Simp. Ibér. Est. Bentos marino II*, 813–835.
Gili, J. M., Riera, T. and Zabala, M. (in press). Physical and biological gradients in a submarine cave on the Western Mediterranean coast (NE Spain). *Marine Biology*.
Harmelin, J. (1969) Briozoaires des grottes sous-marines obscures de la région Marseillaise. Faunistique et écologie. *Téthys* 1(3), 77–98.
Hawksworth, D. L., James, P. W. and Coppins, B. J. (1980) Checklist of British lichen-forming, lichenicolous and allied fungi. *Lichenologist* 12, 1–115.
Hiscock, K. and Mitchell, R. (1980) The description and classification of sublittoral epibenthic ecosystems. In *The Shore Environment, 2: Ecosystems*, Eds. J. H. Price, D. E. G. Irvine & W. F. Farnham, pp. 323-370, Academic Press, London.
Jackson, J. B. C. (1977) Competition on marine hard substrata: the adaptative significance of solitary and colonial strategies. *Amer. Nat.* 3, 743–767.
Jørgensen, C. B. (1966) *Biology of Suspension Feeding*. Pergamon Press, Oxford.
Kensler, C. B. (1965) The Mediterranean crevice habitat. *Vie et Milieu* 15(4), 947–977.
Laborel, J. (1961) Contribution à l'étude directe des peuplements benthiques sciaphiles sur substrat rocheux en Méditerranée. *Rec. Trav. Sta. Mar. Endoume* 33(20), 117–173.
Laborel, J. and Vacelet, J. (1959) Etude des peuplements d'une grotte sous-marine du golfe de Marseille. *Bull. Inst. Océonogr. Monaco* 1120, 1–20.
Laubier, L. (1966) Le coralligène des Albères. Monographie biocénotique. *Ann. Inst. Océanogr. Paris* 43(2), 139–316.
Le Campion-Alsumard, T. (1979) Les Cyanophycées endolithes marines. Systématique, ultrastructure, écologie et biodestruction. *Oceanol. Acta* 2(2), 143–156.
Ledoyer, M. (1968) Ecologie de la faune vagile des biotopes Méditerranéens accessibles en scaphandre autonome. IV. Synthèse de l'étude écologique. *Rec. Trav. Sta. Mar. Endoume* 44(60), 125–296.
Longhurst, A. R. (Ed.) (1981) *Analysis of Marine Ecosystems*. Academic Press, London.
Margalef, R. (1974) *Ecología*. Omega, Barcelona.
Molinier, R. (1960) Étude des biocoenoses marines du Cap Corse. *Vegetatio* 9, 120–192.
Molinier, R. and Picard, J. (1952) Recherches sur les herbiers de phanérogames marines du littoral méditerranéen Français. *Ann. Inst. Océanogr.* 17(3).
Niell, F. X. (1981) Estructuras disipativas en la organización de los sistemas bentónicos. *Oecologia aquatica* 5, 239–245.
Ott, J. A. (1980) Growth and production in *Posidonia oceanica* (L.) Delile. *P.S.Z.N.I., Marine Ecology* 1, 47–64.
Pandian, T. J. (1975) Mechanisms of heterotrophy. In *Marine Ecology, II(1)*, Ed. O. Kinne, pp. 61–249, John Wiley, Chichester, London.

Pérès, J. M. (1966) Benthonic zonation. In *The Encyclopedia of Oceanography*, Ed. R. W. Fairbridge, pp. 128 – 132, Reinhold, New York.
Pérès, J. M. (1967) The mediterranean benthos. *Oceanogr. mar. Biol. ann. Rev.* 5, 449 – 533.
Pérès, J. M. (1982a) Zonations. In *Marine Ecology, V(1)*, Ed. O. Kinne, pp. 9 – 46, John Wiley, Chichester, London.
Pérès, J. M. (1982b) General features of organismic assemblages in pelagial and benthal. In *Marine Ecology, V(1)*, Ed. O. Kinne, pp. 47 – 66, John Wiley, Chichester, London.
Pérès, J. M. (1982c) Structure and dynamics of assemblages in the benthal. In *Marine Ecology, V(1)*, Ed. O. Kinne, pp. 119 – 186, John Wiley, Chichester, London.
Pérès, J. M. (1982d) Major benthic assemblages. In *Marine Ecology, V(1)*, Ed. O. Kinne, pp. 373 – 522, John Wiley, Chichester, London.
Pérès, J. M. (1982e) Specific benthic assemblages. In *Marine Ecology, V(1)*, Ed. O. Kinne, pp. 523 – 582, John Wiley, Chichester, London.
Pérès, J. M. and Picard, J. (1964) Nouveau Manuel de Bionomie Benthique de la mer Méditerranée. *Rec. Trav. Sta. Mar. Endoume* 31(47), 5 – 137.
Pianka, E. R. (1970) On r- and K-selection. *Amer. Nat.* 104(940), 592 – 597.
Picard, J. (1965) Recherches qualitatives sur les biocoenoses marines des substrats meubles dragables de la région marseillaise. *Rec. Trav. St. mar. Endoume* 35(52), 3 – 160.
Price, J. H., Irvine, D. E. G. and Farnham, W. F. (Eds.) (1980) *The Shore Environment. 1, Methods. 2, Ecosystems.* Academic Press, London.
Purchon, R. D. (1979) *The Biology of the Mollusca*. Pergamon Press, Oxford.
Riedl, R. (1966) *Biologie der Meereshöhlen*. Paul Parey, Hamburg.
Romero, J. (1981) Biomasa de comunidades de algas bentónicas de las islas Medes (Girona). *Oecologia aquatica* 5, 87 – 93.
Ros, J. D. (1981) Desarrollo y estrategias bionómicas en los Opistobranquios. *Oecologia aquatica* 5, 147 – 183.
Ros, J. D. (1982) Tipos biológicos en los Opistobranquios. *Actas Ier Simp. Ibér. Est. bentos marino*, I, 413 – 440.
Ros, J. D., Olivella, I. and Gili, J. M. (Eds.) (1983) *Els Sistemes Naturals de les Illes Medes. Arxius de la Secció de Ciències* 73. Institut d'Estudis Catalans, Barcelona, 828 pp.
Sutherland, J. P. (1974) Multiple stable points in natural communities. *Am. Nat.* 108, 859 – 873.
Todd, C. D. (1981) The ecology of Opisthobranch molluscs. *Oceanogr. Mar. Biol. Ann. rev.* 19, 141 – 234.
True, A. M. (1970) Étude quantitative de quatre peuplements sciaphiles sur substrat rocheux dans la région marseillaise. *Bull. Ins. océanogr. Monaco* 69(1401), 1 – 64.
Zabala, M. (1982) Algunas consideraciones sobre estrategias en los organismos bentónicos filtradores. *Actas Ier Simp. Ibér. Est. Bentos Marino*, II, 451 – 497.
Zibrowius, H. (1968) Étude morphologique, systématique et écologique des Serpulidés (Annelida Polychaeta) de la région de Marseille (provenant essentiellement des grottes sous-marines). *Rec. trav. St. mar. Endoume* 59(43), 81 – 252.
Zibrowius, H. (1978) Les Scléractinaires des grottes sous-marines en Méditerranée et dans l'Atlantique nord orientale (Portugal, Madère, Canaries, Açores). *Pubbl. Staz. Zool. Napoli* 40, 516 – 545.

CHAPTER 9

Fishes and Fishermen. The Exploitable Trophic Levels

CARLES BAS, ENRIQUE MACPHERSON and FRANCESC SARDA

Instituto de Investigaciones Pesqueras, Paseo Nacional, s/n, 08003 Barcelona, Spain

CONTENTS

9.1. The fishing tradition in the Western Mediterranean	296
9.2. Fishing vessels and gear	298
9.3. The state of the fishery in the Western Mediterranean. Past and present	299
9.4. Pelagic resources	301
9.5. Other vertebrates	303
9.5.1. Marine mammals	303
9.5.2. Turtles and seabirds	304
9.6. Demersal resources	305
9.7. Resources other than fishes	306
9.8. Management and policy	307
9.9. The future	310
9.10 Prospectives in aquaculture	313

9.1. THE FISHING TRADITION IN THE WESTERN MEDITERRANEAN

Evidence of fishing activity going back to ancient times can be found all around the Mediterranean Sea and numerous centres of population along the coast can trace their origins to this activity. Progress has been rapid in recent decades, with new technology being particularly applied to navigation and changes in some of the most commonly used fishing techniques. Formerly it was possible to divide fishing into two main categories: that carried out from shore, using nets and fish traps, and that requiring the use of one or more fishing boats. Offshore, a wide variety of gillnets, pots and longlines were used, together with primitive forms of trawl gear, including beam trawls, pair trawls, and otter or bottom trawls. The latter has undergone the greatest development, with constant technical advances both in the gears themselves and in the trawl vessels. In the pelagic fisheries — primarily those directed at pilchard and anchovy — the use of special gillnets for pilchard and the like has given way to purse seines, in which the bottom of the net is drawn shut to prevent the surrounded shoal from escaping.

Progress has not been smooth and occasionally has produced great tension, as, for instance, when gillnets were replaced by purse seines in the pilchard fishery. Other changes have resulted from alterations in the behaviour of the shoals, as, for example, with the disappearance of practically all tuna traps as the migrating tunas moved further and further offshore. Generally speaking, the increased economic value of fish products has been the major incentive underlying the development of modern fisheries in the Western Mediterranean.

Another important factor in fisheries development in the region is the great seafaring spirit of the peoples dwelling along these shores, a trait seen not only in fishing activities, but also in sailing in general. There were many notable shipyards for building crafts of all kinds, and the Catalonian and Genoese fleets, to name but two, were famous not only for their ships but for their sailors as well. This seafaring spirit also bore fruit in the field of fishing and explains the abilities of these peoples as entrepreneurs; for they ranged far and wide, founding important fishing and fish processing centres. There was a marked tendency for coastal fishermen, from the south-western Mediterranean, Italy and Spain, to move north, especially during this century, so that to a great extent it can be said that technological innovation has been concommitant with this migratory process. Thus as they settled in ports further and further north, the southern fishermen have supported or motivated technological progress. Even today, whenever concentrations of fish are detected, they are caught by large contingents of fishermen arriving from southern ports.

However, the basic origin of fishing as an artisan activity has not been abandoned, not even in recent, highly developed forms of exploitation, for the industrial side of fishing operations has penetrated the fishing strategy of these shores only with great difficulty. A certain spirit of adventure and a catch-oriented ethos, rather than the concept of industrial production, have been the driving force behind development. In other words, the industrial concepts of markets and production were totally foreign, with the result that, while fishermen progressed slowly and had to seek personal motivation in family tradition, in many cases centuries-old, the buyers and the markets rapidly improved their economic capabilities, manipulating the product under the stimulus of its economic value, becoming the true arbiters of fishing activity. The considerable technological advances in fishing methods and increasing contacts among fishermen from different ports have accelerated the transformation of artisan fishing into an industrial process: an activity whose objectives include not only the catches but also mass production and cost-effective performance, with proper pricing and marketing of the product.

The Mediterranean Sea as a whole, and the western basin in particular, has a relatively low productivity rate, with high mean temperature and salinity. It is, therefore, understandable that abundance should be reduced as far as exploitable fish concentrations are concerned. Nevertheless, using one innovation after another, Mediterranean fishermen are moving resolutely forward — as though the Mediterranean was really a sea rich in resources. This is, perhaps, one of the most distinctive features of the fishing tradition in this region.

It is important to analyse the recent history of developments in the field of fishing. The most striking aspect is undoubtedly the gradual disappearance of artisan gear (trammel nets, bogue gillnets, entangling gear in general, longlines, hand lines, and so forth) whereas certain others, such as pots, still remain in use. This apparent decline does not reflect a lack of interest in fishing, even in the face of higher paying jobs in the expanding building and tourist sectors. Rather it emphasizes a fundamental change in fishing methods towards greater effectiveness and selectivity. Consequently, although the number of small boats using artisan gear has decreased tremendously, and many of their crews have abandoned fishing altogether, other fishermen with a true love of their work have, in contrast, attempted with renewed vigour to develop the new trawling and seining techniques which are currently so extraordinarily successful.

The tendency to expand fishing activities towards offshore waters is confined to the ports of southern and south-eastern Spain. Elsewhere the general tendency is to exploit waters in the vicinity of the home port; greater mobility being apparent only in the purse seine fishery.

9.2. FISHING VESSELS AND GEAR

Despite important differences between regions and countries, the development of the inshore fisheries in the Western Mediterranean has, in recent years, exhibited a strong tendency towards uniformity in both the gear and vessels employed. Two important factors have contributed to this: the uniform topography of the sea bed and the fact that the species exploited are practically the same. Yet this does not mean that there are no differences in the relative proportions of these.

Generally speaking, bottom trawling and purse seining in surface waters are the two main fishing techniques, for artisan fishing methods — gillnets, longlines, different kinds of fish traps, etc. — have declined considerably. This has led to much uniformity between vessels, gear, and technological advances at the different ports.

There is a growing tendency to replace wood with steel in the building of ships' hulls, especially with trawlers, and some experimentation is under way with hulls built of glass reinforced plastic. Although technological development started with the trawlers, it is now the purse seiners that are undergoing rapid modernization, using vessels which are more like trawlers. Similarly, there is an extensive, but still not totally widespread, use of variable-pitch propellers, multi-cylinder high rpm engines, two-way radio communication, significant improvements in navigation and position-finding systems and the use of echo sounders to detect bottom features and fish schools, and many other innovations, especially on medium and larger vessels. Small boats engaging in different kinds of artisan fishing have been converted from older, smaller, low-power trawlers. Some of these boats are of recent construction, normally of wood with rather rudimentary support technology. Such boats carry auxiliary equipment to facilitate the handling of the most commonly used gears.

The average length of trawlers is around 25 m, corresponding to a mean GRT of some 75 tonnes. They are powered by engines averaging around 400 h.p. (the range of engine power running from 250 to 1,000 h.p.). The hold capacity for storing the catch is usually small, as fishing is normally carried out on a daily basis. On trawlers the crew size is around 10 and purse seiners about 14 men although these are decreasing as technology improves. For example, the crews on these boats have shrunk from 20-22 men since the introduction of power blocks. The catch is preserved on ice, which is loaded on board prior to sailing, although refrigeration plants are beginning to appear on some boats. On trawlers, the winch equipment has undergone the most intensive mechanical development, with considerable increase in power as the drive systems have been modified. Formerly, winches were driven by belt drives from the main engine, but they now have their own individual, normally hydraulic, drive systems. Purse seiners usually carry two small purse boats equipped with lights with luminous intensities of between 12,000 and 24,000 candlepower. Finally, relatively simple winches or hauling blocks have been installed on most artisanal boats to facilitate the setting and retrieval of the different kinds of gear.

Mediterranean trawl gear is considerably larger than that employed in other fishing areas. There are a number of different designs, nearly all aimed at achieving a wider mouth opening and to increase the fishing power of the net. Trawl net openings of up to 20 m are now common. A striking feature of net design is the length of the wings, which makes it possible to sweep a broad section of the bottom. The trawl otter boards are in most cases rectangular in shape (few trials have been carried out using oval or polyvalent boards). Long sweeplines (usually 200 m from trawl board to wing tip) are necessary to separate the wings and thereby increase the area swept by the net. Pelagic trawling gear is not used, although similar designs are towed along the bottom. Polyamide and polypropylene fibres are used in the manufacture of this equipment. The codend mesh size is usually between 36 and 40 mm (measured diagonally with the mesh stretched).

Seine gear, mostly purse seines, vary in size according to the home port (ranging between 400 – 700 m in length by 100 – 240 m in depth, stretched mesh size). The commonest mesh sizes vary only slightly, around 18 mm (again measured diagonally with the mesh stretched). Much smaller gear is used in Tunisia (170 – 280 m in length by 40 – 50 m in depth, with a slightly larger mesh size of around 20 mm). All this gear is used to catch surface pelagic fish — chiefly pilchard and anchovy — that tend to form rather dense shoals, which can be kept together with powerful lights carried in the purse boats. Formerly, purse seines used to take different species of tunas of varying, generally small, sizes. These nets were generally much larger (ranging from 1,200 to 1,500 m in length and from 150 to 240 m in depth). The mesh size was also much larger than that used for pilchard and anchovy, between 150 and 180 mm, stretched.

A comprehensive list of all the different types and variants of artisan gear could go on indefinitely: gillnets, longlines, hand lines, jigs, trammel nets, pots, dredges, different kinds of traps normally used at river mouths, and so on. However, special mention should be made of tuna traps, which, because of their large size, required very complicated setting techniques. The scarcity of migrating tuna in close inshore has, in turn, brought about the almost complete disappearance of these devices, especially along the coast of Spain. Although artisanal gears have practically disappeared at many ports, they are still extensively used at others, notably along the Mediterranean coast of France, Corsica, ports near the mouth of the Ebro River in Spain, the coast of Tunisia, and some parts of Italy. The so-called beach gear — shore hauled nets — are still used. These take not only a limited number of species commonly found in the coastal habitat — stargazers, scorpionfish, weevers, and sandeels — but also, and most importantly, large quantities of larvae and juveniles of deeper-water species, which are caught before they have had time to be normally recruited to their fishery.

9.3. THE STATE OF THE FISHERY IN THE WESTERN MEDITERRANEAN. PAST AND PRESENT

The fisheries in the Western Mediterranean have made steady technological progress, leading to the development of certain fisheries and to the almost total decline of many others. Artisan fisheries, using gillnets, trammel nets, longlines and pots, have practically vanished from many ports, although their use continues at others and are even staging a comeback in places from which they had all but disappeared. In contrast, the technology employed in the trawl and purse seine fisheries has made great advances, and this progress is continuing. The substantial increase in fishing power brought about by these technological innovations has resulted in a considerable decline in the available biological resources. This can be seen in the steady decrease in the catch per unit of effort (cpue), especially with species from specific habitats. The quantity of pelagic species taken by purse seines also fluctuates, although this does not seem to be directly related to technological development. Nevertheless, the pelagic fishery has undergone a profound transformation during the present century. Thus, although pilchard and anchovy used to be primarily caught with special drift gillnets, these species are now caught with purse seines which enable large quantities of fish to be taken at each haul. This change in fishing power is very important and, in addition, has converted a selective fishery into a mass fishery.

The pelagic fisheries exhibit another important trend: a continuing tendency for the replacement of the pilchard by the anchovy. This can be seen not only off the coast around Castellón de la Plana (southeastern Spain), but also off northern Morocco. This change may be encouraged by the higher economic value of anchovy, though it is likely that ecological factors are also involved, for similar trends are apparent in other important fisheries.

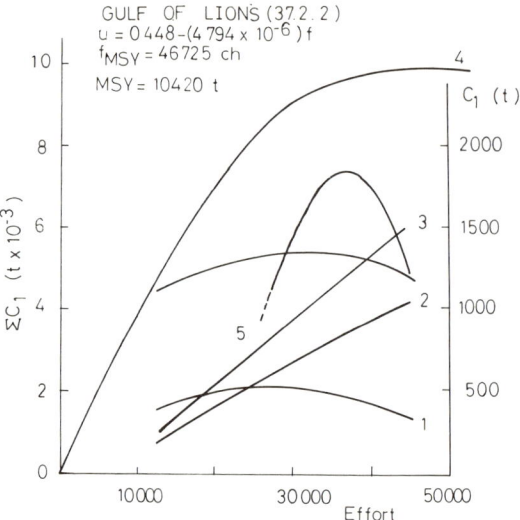

Fig. 9.1 Total catch and effort in the Gulf of Lions and particular relationship for selected species (1: horse mackerel; 2: foor cod; 3: hake; 4: mackerel and 5: cephalopods) MSY: maximum sustainable yield. (From FAO – GFMC)

The development of pelagic fisheries has not been unduly affected by these changes, for catches have increased by 40% off the coast of Spain between 1964 and 1976. The estimated potential of the resources around the Balearic Islands seem to be able to sustain such higher levels of exploitation. Off Morocco, the total pelagic catches of pilchard are estimated to be some 6,000 tonnes, whereas the maximum sustainable yield for the Sea of Alborán as a whole is calculated to be around 22,000 tonnes. Off southern France the exploitation rate of pilchard and other pelagic species is moderate, although insufficient information is available for Italy, the three large islands of Corsica, Sardinia, and Sicily. The pelagic fishery of Algeria is more important in the western sector.

The most prominent pelagic species are the pilchard, anchovy, horse mackerel, and mackerel. In some areas the latter two species probably support higher levels of fishing.

The importance of the different species of tuna, a major pelagic resource, has been decreasing with the decline of trap fisheries, but increasing offshore fishing occurs with large purse seines.

Trawl fisheries have also undergone significant technological change. These innovations involve the introduction of faster, keeled vessels, replacement of sails by motors (at first low r.p.m. one-cylinder engines and now high r.p.m. multicylinder engines), the use of radar and radio communications and modern navigational equipment, and, most recently, continuing improvement of net materials, design and handling. This technical innovation, with increasing power and tonnage of vessels, is still continuing with consequent high levels of overfishing. Overall landings show very little variation, despite the constant increase in fishing power, although slight fluctuations have been recorded in different areas. Trawl catches in the Gulf of Lions totalled 10,420 tonnes in 1979: only slightly more than that of 8,667 tonnes taken off south-eastern Spain (Alicante). A decline in blue whiting has been observed in this area, while hake appear to be on the increase. Fishing on the continental slope is important off Morocco and Algeria, the pink shrimp (*Aristeus antennatus*) being the main species caught, together with the norway lobster (*Nephrops norvegicus*), the mainstay of the slope fisheries throughout the Western Mediterranean. Nevertheless, a sharp drop in the total catch of pink shrimp has occurred during the past 2 years, especially off the coast of Spain.

The trawl fishery on the continental shelf is essentially similar in all regions. The principal species caught are the red mullet (*Mullus barbatus*), young hake (*Merluccius merluccius*), and the curled octopus

(*Eledone cirrosa*). A wide variety of other species are taken as well; while none of these are particularly outstanding, taken as a whole, they represent an important part of the fishery.

The first two species are really important only in regions where the continental shelf is wider — for example in the Gulf of Lions and off south-eastern Spain. Over-fishing has been so intense in these areas that rigorous management measures have become necessary (for example around Castellón, in south-eastern Spain, where positive results have been achieved).

In areas where the shelf is narrower and rockier, the red mullet *Mullus surmuletus* is abundant and is taken in with rough bottom species, such as scorpionfish (*Scorpaena*) and gurnards (*Trigla*). The proximity of the slope to the coast facilitates exploitation, especially of blue whiting (*Micromesistius poutassou*), which is particularly important in the northern part of the Western Mediterranean, greater forkbeard (*Phycis blennioides*), and wreckfish (*Polyprion americanus*), together with the scarce, but highly-prized, specimens of full-grown hake.

The steady increase in fishing capacity has produced competition between vessels for the scant resources available. This competition is motivated by the high economic value of the catches, which more than compensate for the high costs involved.

Artisan fisheries vary greatly in extent. They are still important in some parts of the Gulf of Lions and on the islands of Corsica and Sardinia, but they have lost their former importance along the Spanish coast. However, it is precisely in Spain where this type of fishing is now making some recovery. This is seen in the improved fishing technology used on the numerous small boats still engaged in these fisheries. The catches include a variety of general coastal species, primarily sea breams, and large hake which are taken by longlining or by handlining.

Finally, there are the coastal lagoon fisheries — for example, in the Bassin de Thau, the Mar Menor and the Ebro delta. These support important fisheries for the prawn, *Penaeus kerathurus*, the common sole, *Solea solea*, as well as mullets (*Mugilidae*) and the bass, *Dicentrarchus labrax*. Management measures to rationalize these fisheries could yield substantial benefits for these coastal fisheries.

9.4. PELAGIC RESOURCES

Pelagic fishing is quantitatively more important than the demersal fishery, the mean catch for the period 1968 – 1978 being estimated at some 200,000 tonnes a year (Levi and Troadec, 1974; Massutti, 1981).

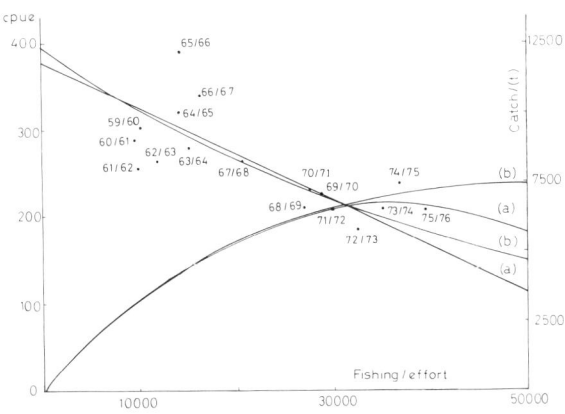

Fig. 9.2 Trawl fishery in the Gulf of Lions. Relationship between cpue and fishing effort and resulting yield curve. (a) linear fitting, (b) exponential fitting. Numbers refer to years. (From FAO – GFMC)

The pelagic fauna in the Western Mediterranean is rather similar to that in the subtropical Atlantic, the most important species belonging to the families Clupeidae, Engraulidae, Scombridae, and Carangidae.

There are two primary commercial species which comprise two-thirds of the total catch: the pilchard (*Sardina pilchardus*) and the anchovy (*Engraulis encrasicholus*). Both are taken all along the coast, especially in inshore areas.

The main catches are made off Catalonia and Castellón. Within this area, the main spawning area for pilchard is located near the mouth of the Ebro River. As the fish grow, one stock moves north and another south, towards the Columbretes Islands, when they reach 2 – 3 years of age. Exploitation of pilchard stocks begins at 7 – 8 cm, age groups 0 and 1 being the main contributors to the catches.

Pilchards are usually found nearest the coast in the winter spawning season, and then move seaward during the warmer seasons. The main catches are made during these warmer months, in part due to the higher commercial value of the fish (the quality of the product being enhanced by increased fat content). The thermocline plays an important role in the distribution of this clupeid, which aggregates in the thermocline.

Anchovy are normally found further offshore, catches being taken mainly at depths of from 50 to 200 m all year round. Age groups 0 and 1 are again the main contributors to the catches.

The depth distribution of anchovy follows the same pattern as in the case of pilchard, with younger individuals being found closest inshore. This behaviour is very well defined in regions like Castellón (Suau, 1979), but the pattern seems to change further north.

Catches of these two species have fluctuated over the years. Around Castellón pilchard underwent a sharp decline in 1965, and was followed by a sharp increase in anchovy. These changes seem to be less the result of fishing pressure and more the consequence of ecological factors, caused by variations in the composition of the phytoplankton on which the larvae feed, which affect larval mortality and subsequent recruitment (Larrañeta, 1981).

Certain carangid species are commercially important, particularly the horse mackerel (*Trachurus trachurus*, and to a lesser extent *T. picturatus*), some 18,000 – 23,000 tonnes of which, on average, were taken yearly over the period 1968-78. There are also less important species such as *Caranx rhonchus* and *Seriola dumerili*. These are generally migratory, for example: *S. dumerili*, greater amberjack, which usually moves inshore during the summer and forms aggregations before moving offshore during the winter.

The Scombridae, Thunnidae, and Scomberomoridae are extremely important in the region, not only because of their relative abundance but also for their high commercial value. These fishes include the bonito (*Sarda sarda*), mackerel (*Scomber scombrus*), bluefin tuna (*Thunnus thunnus thynnus*), tuna (*Euthynnus euthynnus quadripunctatus*), and frigate mackerel (*Auxis rochei*).

These species support an important trap fishery in the Strait of Gibraltar near Ceuta, where large quantities of juvenile tuna (younger than 1 year old) are taken at the end of summer and beginning of autumn. Tuna born the previous spring aggregate in an inshore area near Castellón during the autumn months.

The longline fishery is directed primarily at swordfish (*Xiphias gladius*), and to a lesser extent at tuna, the catches ranging between 100 and more than 500 tonnes a year.

Most of these species make long migrations. Adult bluefin tuna are normally found near the surface during the summertime and deeper in the wintertime, whereas young fish concentrate near the surface in the autumn and early winter. Spawning takes place inshore at the end of winter, one spawning area being located near the Balearic Islands. First-year fish then move south and are caught near the Strait of Gibraltar in the autumn. This seems to confirm the hypothesis that they migrate to the Atlantic. Other species such as bonito and tuna also migrate along the coast to the Atlantic (Rey and Cort, 1981). The main catches of bonito are made in autumn close inshore using gillnets and purse seines.

Mackerel is another commercially important resource with a yield of between 5,000 and 10,000 tonnes yearly, taken chiefly in the spring.

There is also a series of secondary species of small yield, but relatively high commercial value, when compared to that for the previously mentioned ones. The most important of these are the Mugilidae (*Mugil cephalus* and *M. capito*), mullet, which inhabit shallow waters all along the coast, and yield between 3,000 and 5,000 tonnes annually. The bogue (*Boops boops*), though not a highly esteemed species, is rather abundant, with between 13,000 and 15,000 tonnes caught each year, some with bottom gear. Other less important species are the pilot fish (*Naucrates ductor*), which usually forms small shoals a particular distance offshore and which is taken chiefly around the Balearic Islands, *Argentina sphyraena*, found primarily in autumn and winter, and a mixture of species, taken in varying amounts using bottom gears (*Spicara maena*, *S. chrysellis*, etc.).

Small quantities of shad, *Alosa* spp., are taken near the mouths of rivers, as are small clupeids of the general *Sardinella* (gilt sardines) and *Sprattus* (sprats), but catches of these have rarely exceeded 2,000 tonnes in recent years.

Some of the smaller boats often catch transparent goby (*Aphia minuta*) from November to March. Because of its economic value this species constitutes a resource of some importance, especially along the southern coast of Spain. However, major components of the catches of this gobiid are in fact larval stages of other species.

The dolphin (*Coryphaena hippurus*) is an epipelagic migratory species which is caught mainly in waters close to the Balearic Islands and Sicilia.

It is difficult to assess the state of the pelagic resources in this part of the Mediterranean, not only because of inadequate statistical information, but also because of a series of biological and environmental factors which have not been taken into account, when applying stock assessment models. An example of this is the effect of changes in plankton composition and the impact of such changes on recruitment, which is of particular importance in the case of short-lived species like pilchard and anchovy.

Most available information refers to anchovy and pilchard. Both of these species appear to be underexploited over almost the entire area, especially off the coast of Africa, where maximum sustainable yield (MSY) for pilchard in this region is estimated at some 75,000 to 85,000 tonnes, mostly from Algeria. The maximum sustainable yield is lower off the coasts of Spain, France and Italy.

9.5. OTHER VERTEBRATES

9.5.1. Marine mammals

A total of eighteen different marine mammals are found in the Western Mediterranean, five Mysticeti species, twelve Odontoceti species, and one Pinnipedia species (Casinos and Vericad, 1976; Deguy et al., 1983).

The Mysticeti group is poorly represented, and only the fin whale can be regarded as common. Its abundance increases in the summer, chiefly in the region between Corsica and the French Côte d'Azur. At the end of summer it migrates towards the south-western Mediterranean. The other species, (the minke whale (*Balaenoptera acutorostrata*), Sei Whale (*B. borealis*), humpback whale (*Megaptera novaeangliae*), and Biscayan right whale (*Baleana glacialis glacialis*) are seldom encountered and their presence is usually accidental. The latter species was only reported during the 19th century, and its disappearance from the Mediterranean is considered certain.

Of the Odontoceti, the family Delphinidae accounts for the most abundant and the largest number of species in the region. The striped dolphin (*Stenella coeruleoalba*) and the common dolphin (*Delphinus delphis*) are found along the entire coastline and form large schools, although they differ slightly in their geographical distribution: the former is more common along the coasts of Spain and France, while the latter is more common off the coast of Africa. These species appear to have declined in abundance in recent years.

The remaining dolphin species, the bottlenosed dolphin (*Tursiops truncatus*) and the rough-toothed dolphin (*Steno bredanensis*) are rarer, though the former is relatively abundant in the waters off North Africa and Spain (especially in the Ebro delta). In contrast to the striped dolphin and the common dolphin, *Tursiops truncatus* lives in small groups.

The long-finned whale (*Globicephala melaena*) is common all along the Western Mediterranean coastline, particularly off North Africa and southern Spain and in the Ligurian Sea, where large numbers congregate in July, then scattering in August. This concentration, like that of the fine whale indicated above, also seems to be due to increased food abundance.

The other species of the family Delphinidae, the false killer whale (*Pseudorca crassidens*), killer whale (*Orcinus orca*), Risso's dolphin (*Grampus griseus*), and harbour porpoise (*Phocoena phocoena*) are rare. The latter species was common last century and enters the Mediterranean following the cold Atlantic current.

Other Odontoceti species, the sperm whale (*Physeter catadon*), Cuvier's beaked whale (*Ziphius cavirostris*), and Blainville's whale (*Mesoplodon densirostris*), are relatively common, except for the latter, which has only been reported once.

The only Pinnipedia species found in the Mediterranean is the monk seal (*Monachus monachus*), which is occasionally seen in the western part of the sea and, though in danger of extinction, is somewhat more common in the Adriatic and Eastern Mediterranean.

These marine mammals are not presently an important commercial resource, although some individuals, mainly dolphins, are taken incidentally or to avoid damage in the purse seine fishery. The resource was, however, fairly important during the last century and early in the present century, and there was a factory in Algeciras.

The impact of these species on fish and cephalopod stocks (their usual food) may be considerable, although unfortunately the parameters for assessing this impact have not yet been established.

9.5.2. Turtles and seabirds

Two species of turtles are relatively common in the Mediterranean Sea: the loggerhead turtle (*Caretta caretta*), which is found offshore and the green turtle (*Chelonia mydas*) which lives near the coast. Several years ago their eggs were highly esteemed as food and they were found in the beaches around the Mediterranean. At present, they have no economic importance because of the scarcity of both species and only the carapace is used for decorative purposes.

A third species of turtle, the leather turtle (*Dermochelys coriacea*) is occasionally seen in the Mediterranean, following the Atlantic current.

The most common seabirds belong to the families Laridae, Procellaridae and Phalacrocoracidae.

The species of Laridae accounts for the most abundant in the region. The herring gull (*Larus argentatus*) is found living in groups in cliffs, beaches and estuaries. The parent population is estimated in 40,000 pairs (60% in France, north-east of Spain and Corsica). The black-headed gull (*Larus ridibundus*) is also common along the western coastline, particularly in France, and the breeding population is evaluated in 9,000 pairs. Furthermore, the hivernal stock, coming from Central Europe, is being estimated, for France and north-east of Spain at some 20,000 to 250,000 pairs. The Audouin's gull (*Larus audouinii*) is a very interesting species, because the most important breeding population in the world (around 4000 pairs) is found in the Western Mediterranean. Other species, as the common tern

(*Sterna hirundo*) is estimated in 3500 – 4000 pairs (breeding population) inhabit mainly in France and Ebro delta.

The family Procellaridae is important in the region, in particular the Cory's shearwater (*Procelaria diomedea*: 10,000 pairs) usually living in islands. The shag (*Phalacrocorax aristotelis*; family Phalacrocoracidae) shows a similar behaviour, living in islands and the breeding population is estimated in 1500 pairs.

The seabirds are not presently an important commercial resource. However, several years ago, their eggs were used as food. On the other hand, their incidence in the pelagic fishes, which are a very important prey in the diet of the seabirds, could be important, though this effect is poorly studied.

9.6. DEMERSAL RESOURCES

Due to the variety of the features which characterize the fishing grounds in the Western Mediterranean basin, demersal resources must be classified according to their different biocenoses: those located on the continental slope, comprising deep, narrow submarine canyons often ringed by rocky barriers (as in the case of Italy, Corsica and Sardinia, Sicily, Catalonia, the Balearic Islands, or Algeria); those located on a broad, gently sloping continental shelf (such as the south-eastern coast of the Iberian Peninsula and the Gulf of Lions); and those located over shallow rocky or sandy bottoms running along the shoreline.

Certain other special habitats are worthy of note, such as large river deltas like those of the Ebro and the Rhône, and the Alborán Sea, which contains a number of Atlantic species as a result of the currents flowing through the Strait of Gibraltar. There is, thus, a wide variety of benthic species in the Western Mediterranean, distributed according to the different types of continental shelf. The demersal fishery in this area is, for this reason, a multispecies fishery, with no single species predominating. The principal demersal species on the slope are the hake (*Merluccius merluccius*), blue whiting (*Micromesistius poutassou*), greater forkbeard (*Phycis blennioides*), and pouting (*Trisopterus minutus capelanus*), among others, which make up the greater part of the biomass.

The first species, hake, is an important resource in the Mediterranean area. It is widely distributed at depths ranging from 40 to 400 m, depth distribution depending on size, which runs from 9 to 60 cm, with adults dwelling at the greatest depths.

Nevertheless, the hake yield can be considered to be relatively low compared with that taken in the Atlantic Ocean, although the species' high reproduction rate causes temporary population spurts composed of small individuals that contribute to an incidental but highly profitable fishery. The effect of this is to limit the adult stock, resulting in the low yields obtained in this fishery.

This is not the case for blue whiting, the species with possibly the highest biomass in the Mediterranean, and which certainly gives the highest yield, with catches still rising in recent years. The distribution of this species is also regular (the size of individuals ranging between 16 and 24 cm) and, again, the larger individuals are found at greater depths. The largest concentrations are located at depths of around 400 m, with substantial aggregations forming during the spawning season. Spawning takes place in winter, large quantities of juveniles subsequently gathering at about 80 m. However, in contrast to hake, the mushy consistency of its flesh and its low commercial value means that this fishery is not directed at the smaller length groups, which thus sustain a large adult stock.

In addition, other similar species such as greater forkbeard and pouting are important in terms of their biomass within the community. Anglerfish (*Lophius*) should not be overlooked in view of its economic importance, though larger fish are uncommon in the Mediterranean fishery.

Norwegian lobsters (*Nephrops norvegicus*) typically inhabit and form shoals at depths of between 300 and 400 m. It is the most abundant and highly prized of the lobsters, though stocks have been depleted of late through overfishing. Nevertheless, because this species is largely nocturnal in its habits (dawn and

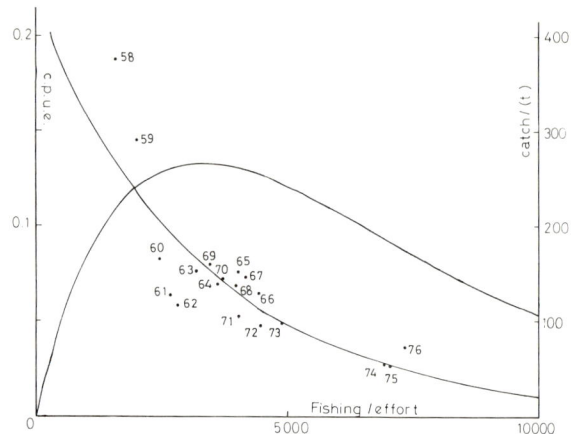

Fig. 9.3 Fishery of shrimp in Balearic Islands. Exponential fitting. Numbers refer to years. From FAO – GFMC.

dusk), and because it dwells in burrows which it digs on the bottom, exploitation entails certain difficulties, which in part helps to sustain stocks.

Below these depths, and particularly around 600 m, the slope supports several species of shrimp, some common to the Mediterranean and nearby Atlantic waters, such as the rose shrimp (*Parapenaeus longirostris*), the pink shrimp (*Aristeus antennatus*), and the red shrimp (*Aristeomorpha foliacea*), all highly popular food items.

Together with hake, blue whiting, greater forkbeard, and pouting, other species are typically found on the slope at depths greater than 100 m, such as pandora (*Pagellus*), sargo bream (*Diplodus*), and, closer inshore, dentex (*Dentex*), Couch's sea bream (*Pagrus*), gilthead (*Sparus*), gurnards (*Trigla*), and others. The red mullet (*Mullus barbatus* and *M. surmuletus*), also inshore species, stand out due to their importance. The former dwells over sandy and muddy bottoms and the latter in more or less rocky areas; juveniles form concentrations near the coast during the spawning season, later scattering. There is also a relatively large variety of flatfishes (Pleuronectidae), most notably the common sole (*Solea solea*).

Among these shallow-water species the prawn (*Penaeus*) occurs near river mouths, where it is presently in decline.

Large crustaceans like the European lobster (*Homarus*) are occasionally found. The spiny lobster (*Palinurus*) is more common and is taken with creels in rocky areas. Spider crabs (*Maja*) are encountered, less frequently (either in shallow water or below 300 m) and a similar genus (*Paromola*), which is large but not considered particularly good eating, is also taken.

Cephalopod species, which are widely distributed along the coast, constitute another important resource, commonly exploited genera being cuttlefish (*Sepia*) and squids (*Loligo* and *Illex*), as well as octopus (*Octopus* and *Eledone*), which have a wider distribution range.

The above species can all be considered common throughout the entire Western Mediterranean region, their diversity and distribution varying (as has already been emphasized) only according to special local features and characteristics of the continental shelf and slope.

9.7. RESOURCES OTHER THAN FISHES

Due to the variety of geographical and physical features there are a large number of specialized fisheries, mostly artisanal, for a wide variety of species; although these fisheries do not attain high levels of production, they do generate substantial economic value, their output being utilized as select speciality food items.

Bivalve molluscs are the most commonly exploited, and are mainly raked up from shallow, sandy bottoms. Among the most productive species, especially in France, is the cupped oyster (*Crasostrea angulata*). The common oyster (*Ostrea edulis*), a species with a very high economic value, is harvested in rocky areas in most of the coastal countries around the Mediterranean, and it is also cultured in areas especially set aside for this purpose. The blue mussel (*Mytilus edulis*) is produced in greater quantities and is harvested both from naturally-occurring and cultured populations. Italy is the largest producer of the Mediterranean mussel (*Mytilus galloprovincialis*).

Spain harvests small bivalves like the common cockle (*Cardium edule*), striped venus (*Venus gallina*), grooved carpet shell (*Venerupis decussatus*), carpet shell (*Tapes pullastra*), razor clams (*Solen* spp.), and others, primarily in the Ebro river delta and along the adjacent shores.

Some gastropods are eaten as appetizers, for example, the spinous and apple murex (*Murex brandaris* and *M. trunculus*).

Turning to crustaceans, all the species subject to mass exploitation have been considered; however, there are substantial stocks of crabs, in particular the general *Macropipus* and *Carcinus*, which are commonly taken and marketed and among the stomatopods, the mantis shrimp (*Squilla mantis*) is popular in Spain and particularly in Italy.

Lastly, some echinoderm species are worthy of mention, as they have attained some importance, at least in Spain. Among these are the sea cucumber (*Stichopus regalis*), which, though not particularly abundant, is paid better than other species for the white muscular strands which are appreciated as a delicacy.

Eggs of the sea-urchin (*Strongylocentrotus lividus*) are edible and are eaten mainly in France. In relation to France, mention should also be made of the sea squirt (*Microcosmus* spp.), which inhabits large areas of the Gulf of Lions at depths between 100 and 200 m.

Certain species which were highly exploited in the past, like the red coral (*Corallium rubrum*) still exist. However, today it has been depleted in many areas (especially in waters shallower than 50 m) a victim of mass exploitation and its slow growth rate. The coasts of Italy and France, where this species has been harvested down to depths of 140 m, have been particularly affected. Large populations are still present off Algeria and Morocco, where they are protected by stringent conservation measures.

Sponges have also been considerably exploited in the Mediterranean, *Spongia officinalis*, *Spongia zimicca*, and *Hippospongia communis* being the most important commercial species. The most important grounds are presently located off Algeria and Tunisia, though production is low as a result of over-exploitation.

Some attempts at aquaculture have been made based on the high regeneration rate of certain species, but these trials have not been very successful due to the slow growth rate and to disease.

The commercial importance of most of these species is relative, inasmuch as their popularity varies from country to country, and even from region to region within individual countries. Moreover, as they are, in the main, harvested with small-scale artisan gear, there is no reliable statistical control.

9.8. MANAGEMENT AND POLICY

Adequate management of fish resources in the Western Mediterranean is extremely difficult for a variety of reasons, primarily for the tendency of each small region or country to establish its own particular management measures, whereas generally speaking environmental conditions, including the topography of the sea bed, and biotic characteristics are quite similar throughout the entire area. In this regard the work carried out by the General Fisheries Council for the Mediterranean (GFCM), which operates under the auspices of FAO, is vitally important, with its emphasis on close coordination of the management of existing resources on both a national and region-wide basis as the most appropriate fisheries policy for the region and, in certain special cases, even for areas more restricted in scope.

Fisheries management must be based on one basic concept: that a clear state of over-fishing exists in the case of the demersal resources. With respect to pelagic resources as a whole, strict monitoring is called for, especially in view of the high instability of these resources and the possibility of replacement between some of the principal species, namely pilchard and anchovy. The first priority of a truly comprehensive fisheries policy in scope must be to achieve the optimum fishing effort/catch relationship (i.e., maximum sustainable yield).

Another very important difficulty stems from the very structure of the particular demersal biocenosis exploited. It is characterized by a large number of species, all of which are subject to some degree of exploitation, although certain species are more highly sought than others. From this it follows that the basic biological behaviour of the different species — growth rate, reproduction, age at recruitment, etc. — will differ. Furthermore, conservation measures to be applied in each case also differ, both with respect to management as well as to the policies necessary to achieve optimum protection. There are numerous examples which clearly illustrate these discrepancies. A particular case is the reactions of hake, *Merluccius merluccius*, and the red mullet, *Mullus barbatus*, to the use of codend mesh of trawl gear. The appropriate mesh size, to ensure adequate protection for mullet, is 40 mm (measured diagonally, stretched mesh). However, this size is insufficient to afford proper protection to hake, which require larger mesh. Furthermore there is no spatial and temporal segregation of these species and separation on the basis of size difference and their economic value is also not feasible. It was decided, therefore, to give priority to conservation of mullet off the south-eastern coast of Spain (because of its greater abundance and its far from negligible economic value) despite the fact that such a measure would have a detrimental effect on the hake stocks, by catching excessive numbers of immature hake by the small mesh. If the opposite measure were taken (i.e., if priority protection were afforded to hake), the result would be the loss of substantial amounts not only of striped mullet and red mullet but also of many other species which, due to their generally reduced size, would escape in large numbers through the mesh. The recommended codend mesh size may vary according to the popularity of one or another of the numerous species caught in each region. However, as previously mentioned, the high degree of similarity existing among the different biotic and environmental aspects in the Western Mediterranean basin has conferred a certain uniform character on the most common management measures all through the area.

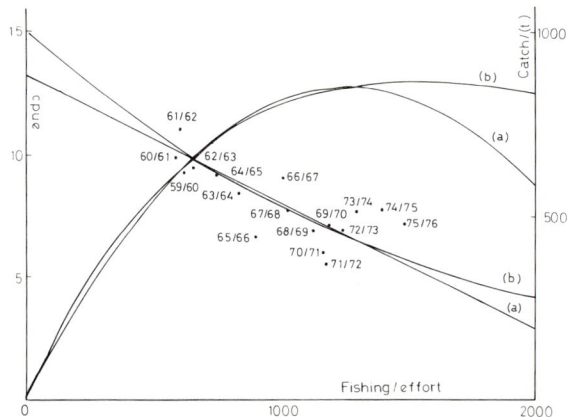

Fig. 9.4 Bottom fisheries in Corsica: cpue/effort and resulting yield curve. (a) linear fitting, (b) exponential fitting. Numbers refer to the years 1960 – 76. (From FAO – GFMC).

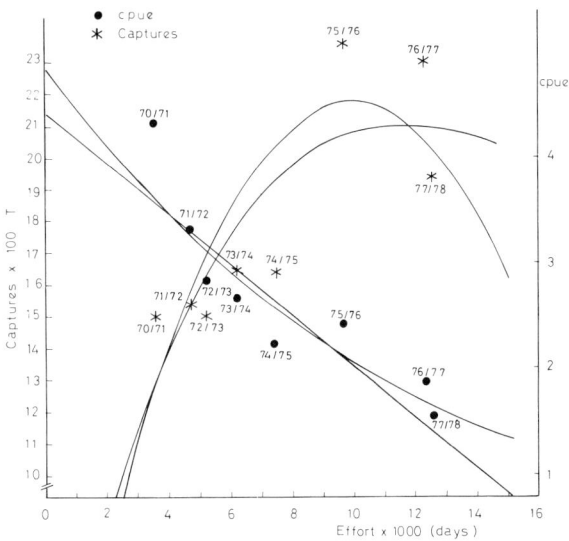

Fig. 9.5 Pilchard fisheries in Alboran Sea. Relationship between cpue and fishing effort and yield curve. Numbers refer to years. (From FAO – GFMC).
(a) linear fitting $u_i = 3.82 - 1.68 - 10^{-4}$
(b) exponential $u_i = 4.92 \times e^{-0.98} \times 10^{-4}$

A summary of the main management measures would have to include the following: in the trawl fishery, a codend mesh size of 40 mm, stretched mesh; in the purse seine fishery, a mesh size of around 16 mm, stretched; limitation of time at sea in the daily trawl fishery; limits on the light intensity used to attract pelagic fish shoals; and finally, regulatory measures establishing closed areas and seasons. There is also a large body of regulations relating to artisan gear, as extensive as the many different types of gear. Last of all, minimum legal size regulations, though their degree of effectiveness is a very controversial point, are extremely varied, depending on the species and area concerned.

The purpose of a true management measure is none other than to ensure continuity of exploitation in optimal conditions or, if conditions have deteriorated, to try to guide the situation back to the desired point. This cannot be achieved unless there is a policy based on the biological, ecological, and economic characteristics of the fishery. In any case fisheries policy in the Western Mediterranean has to bring about a sweeping reduction in fishing power or, better still, in the amount of fishing effort expended. While, on the one hand, for socio-economic reasons, it is difficult to hold back technological development leading to a steady increase in fishing power, the necessary steps must be taken to limit effort — fishing power times operating time — which can only be achieved by drastically reducing fishing time and the area where it occurs. There are two types of time limitations: shortening the fishing day or week, or prohibition of fishing for extended periods of time, which in most cases also tends to ensure good recruitment. Closing certain areas to fishing is advisable in order to improve the chances of survival of juveniles prior to recruitment or in some cases to protect spawning adults, especially when the size of the shoals of spawning fish is a critical factor, and even to protect certain kinds of artisan fisheries from competition from purse seine and trawl gear.

An extremely urgent measure, particularly in the trawl fishery, is the limitation of engine power, which is dangerously excessive at the present time.

In the coastal countries as a group, these measures are not applied in a uniform manner, and in fact only the 40 mm mesh size in the trawl fisheries and 16 mm in the purse seine fisheries have been widely adopted.

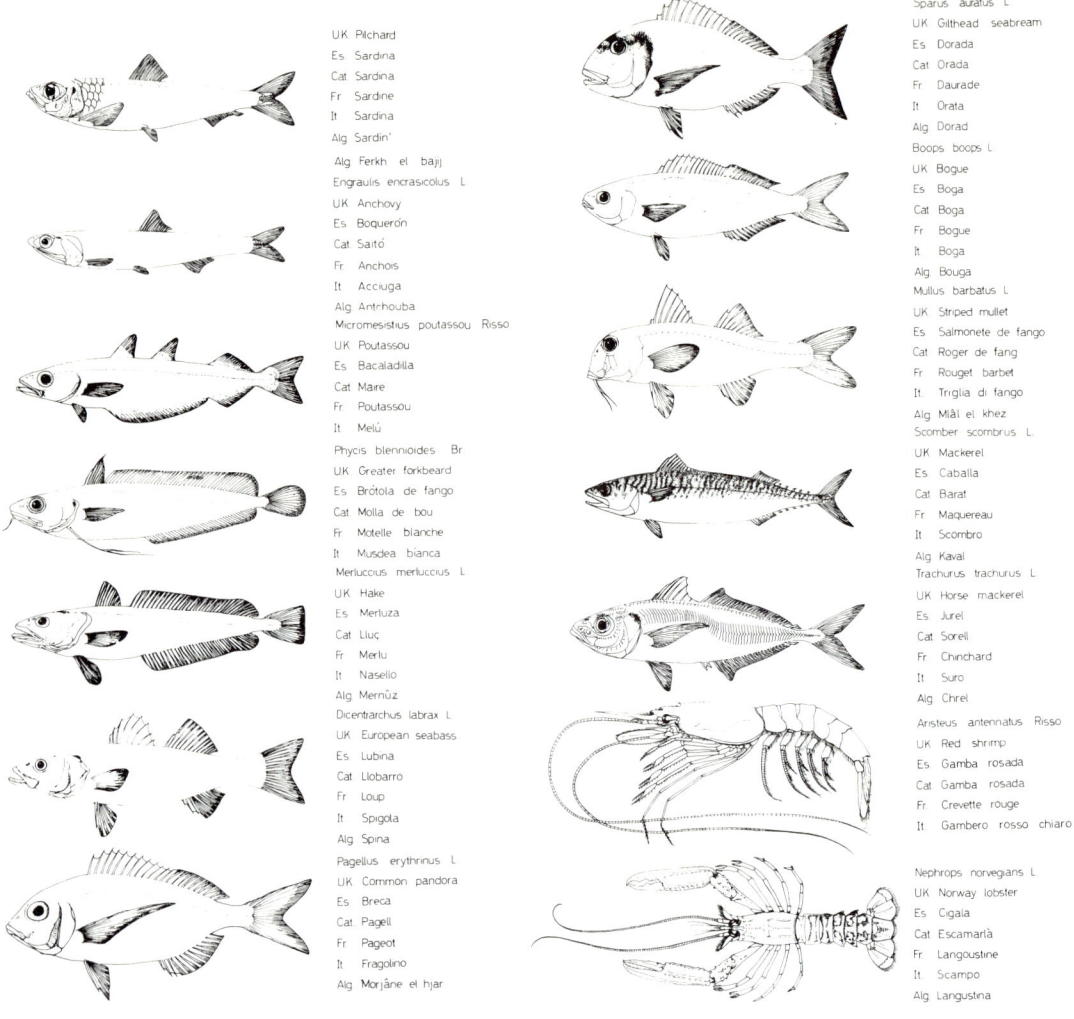

Fig. 9.6/7 Animals used as food.

9.9. THE FUTURE

The low level of resource availability in the Mediterranean in general and in the Western Basin in particular is primarily due to the low productivity of these waters. The production rate in carbon per m^2 is considerably lower than in the Atlantic, and the immediate result of this is the existence of a highly diversified ecosystem with a low renewal rate. A comparison of the biotic composition of the Western Mediterranean with that of any of the ocean regions characterized by a high primary production rate shows that there is a wide variety of species in the Mediterranean but that none of these clearly predominates over the others. As important as they are, neither the hake nor the mullet (among the demersal fishes) are pre-eminent as are, for example, hake in Namibia or cod in the north-west Atlantic.

TABLE 9.1. Annual catches by major species groups in West Mediterranean (From FAO – GFMC)

Species	Total Annual Catches (Metric Tonnes)										
	1970	1971	1972	1973	1974	1975	1976	1977	1978	1979	1980
Anguilla spp.	3917	5003	2766	3018	2807	1872	2532	2040	2602	1970	1842
Alosa spp.		77			24	34	9	57	988	22	5
Psetta ssp.	85	61	82	90	87	94	104	92	81	92	92
Pleuronectiformes	2679	2904	3045	2537	2618	2444	2593	2407	2347	2074	2325
Gadiformes	19076	23184	23134	14464	18363	17386	18529	21465	20159	28651	27298
Sparidae	14687	15740	16946	12429	16790	19065	19795	22843	20478	21784	22026
Maena spp.	4593	4696	4920	1289	1539	1989	2138	1637	1006	1290	1211
Mullidae	8193	9080	9292	6962	6905	6651	8745	8247	6825	6855	6620
Gobiidae	929	1093	1095	529	495	345	467	396	461	392	508
Demersal percomorphs	7130	10549	13095	11902	17118	9002	10782	14192	11937	16431	8112
Mugil spp.	2738	2602	2213	3032	4345	4470	4261	4960	4309	4945	4302
Carangidae	18137	17503	17352	13249	12700	13889	19786	20642	15015	17079	15216
Clupeidae	101959	106907	109592	85859	101547	111746	105397	113440	103205	101573	95609
Engraulis encrasicholus	45540	32311	38981	51409	46250	43134	60001	51147	51267	60840	54114
Pomatomus saltatrix	1	4	4		1	36	24		46	39	18
Scomber spp	18049	12392	8564	6938	7564	8126	3371	6513	6464	6778	7432
Pelagic percomorphs	719	1298	1429	1135	850	339	331	1210	1802	379	328
Sharks, Rays, Chimaeras	3660	3896	4154	2089	2943	2403	3031	3147	2592	2836	4066
Crabs	16	84	115	12	264	468	531	531	364	13	588
Lobsters	1613	1881	2094	1684	1462	1269	1953	1678	1422	1496	1360
Shrimps & prawns	7607	7145	7839	4995	5458	4624	5005	4752	5018	4962	
Marine crustacea	1384	1585	1733	1599	4743	3255	3479	2728	2257	2113	1639
Pelecypods/Gastropods	13171	17267	20773	14293	18242	13954	18143	17920	11464	10472	13717
Cephalopoda	22056	23949	24821	16179	16791	18783	20751	17162	16512	14907	17270
Marine molluscs	1546	1376	1369	1705	2614	2326	3164	2399	1105	2136	1083
Sponges									0	4	1
Marine Fishes, others	23226	19293	21211	12014	22132	40160	36738	29169	35030	28842	34487
TOTAL	322712	321879	336620	269411	314651	327865	351246	351028	324494	339027	326231

Not even the pelagic community as a whole (which is closer to the primary production level) exhibits high abundance levels. There exists a dynamic equilibrium among the components of the ecosystem, and maintaining this equilibrium is of the utmost importance in any attempt to regulate exploitation.

In addition, the system possesses some important defence mechanisms, particularly in certain areas and in connection with certain species. The blue whiting, *Micromesistius poutassou*, undergoes an extended juvenile stage in which it lives scattered throughout the pelagic layer, thus making capture impossible and ensuring good recruitment. The decapod, *Aristeus antennatus* (the pink shrimp), which is very abundant on the continental slope throughout the Western Mediterranean, inhabits areas distant from the usual fishing grounds, which, as with the blue whiting, favours recruitment. Some species which normally dwell on the continental shelf commence their benthic existence in relatively shallow water. The red mullet (*Mullus barbatus*), for example, moves to deeper water as it grows older (because the shelf is narrow along much of the coast), but the habitat of the juvenile stages of this species is located so close to the shoreline that it is impossible to take them with trawl gear. Lastly, the many rocky areas of the sea bed existing inshore furnish a source of adequate shelter for many species, enabling them to survive in spite of intensive fishing.

These circumstances must be considered when evaluating the future of the fisheries along these coasts, since, on the one hand, they help to protect certain species, while, on the other, they are necessary for the proper management and administration of the renewable resources.

The future of the fishery in the Western Mediterranean basin depends upon the management measures and policies adopted, as well as upon adequate co-ordination among all the coastal countries. The exploitation of the resources which are primarily pelagic in nature will be discussed first, followed by a consideration of the demersal resources.

The main pelagic species, the pilchard and anchovy, are currently in a critical state in that the latter species is replacing the former. At least in the coming years, the outlook is for a preponderance of anchovy which, if correctly managed — minimum mesh size and minimum legal size — will surely support large catches. Increasing catches of juveniles are being made in the springtime, especially off the coast of Catalonia (north-western Mediterranean). The presence of large aggregations of juveniles clearly indicates that shoals of adults could become very abundant merely by avoiding the catching of this species until it attains a minimum size to be established in accordance with the findings of fishery biology studies. As long as conditions remain minimally favourable, the rapid growth rate of this species combined with its high reproduction rate promise sustained high catches at the present level of fishing effort. In contrast, the pilchard, with a longer than average life span, is more directly dependent on years of good recruitment, which produce generations that are progressively fished to depletion. Pilchard shoals are normally composed of 2- to 3-year-old individuals, but 4-year-olds are also relatively numerous. The difference in the turnover rates between anchovy and pilchard are likely to favour replacement of the latter by the former.

Horse mackerel and mackerel are two other species of interest. Both species alternate between pelagic and demersal habitats, which results in higher pressure on shoals of these species. Horse mackerel, *Trachurus trachurus trachurus*, *Trachurus trachurus mediterraneus*, and *Trachurus picturatus*, especially the first two subspecies, is abundant in the commonly exploited fishing areas. *Trachurus picturatus* is not particularly abundant and inhabits outlying areas. Mackerel, *Scomber scombrus*, undergoes considerable fluctuations in abundance, and its future is unclear. The chub mackerel, *S. japonicus*, similar to the mackerel, is seldom encountered. Other pelagic species with a certain local importance are the dolphinfish *Coryphaena hippurus*, picarels and smares, *Spicara* spp., and different species of tuna, which

TABLE 9.2. Annual catches by major species groups in the Balearic Zone (From FAO – GFMC)

Species	Annual Catches (Metric Tonnes)										
	1970	1971	1972	1973	1974	1975	1976	1977	1978	1979	1980
Anguilla spp.					700		65	66	55	113	
Alosa spp.				24	23	8	56	82	22	4	
Psetta ssp.										5	3
Pleuronectiformes	483	799	793	894	994	905	1179	1002	1106	1054	1106
Gadiformes	9512	12786	11319	8491	11253	8802	8269	10367	12350	20817	18708
Sparidae	3716	4116	5388	4545	8835	11034	11314	14568	13693	15148	15795
Mullidae	3010	3234	3309	3244	3477	3620	4344	4015	3511	3854	3850
Demersal percomorphs	841	4176	5598	5827	11090	3041	4699	10120	4787	9142	3852
Mugil spp.	211	13	76	119	902	858	1019	788	934	992	561
Carangidae	11845	11065	10529	9984	10030	11065	16899	17656	12192	13659	12046
Clupeidae	53603	60012	68541	55273	70967	77026	78829	89116	78219	72041	69259
Engraulis encrasicholus	16735	11158	12269	25769	20122	23626	33020	30388	37077	48267	41656
Pomatomus saltatrix	1	3	1				24		3		
Scomber spp	7856	6378	3646	3343	4594	4836	770	3758	3779	3293	3158
Pelagic percomorphs	222	678	637	754	420			545	1222	52	59
Sharks, Rays, Chimaeras	237	213	237	274	1009	238	276	515	696	700	1590
Crabs					200	389	403	393	358		588
Lobsters	397	517	606	609	585	307	863	748	657	641	598
Shrimps & prawns	3040	2453	2150	2674	3340	2477	2825	3278	3651	3494	3450
Marine crustacea	1	20	24	21	3193	1545	1613	957	857	627	47
Pelecypods/Gastropods	1221	1065	1305	160	4135	2952	3138	4307	3947	1866	953
Cephalopoda	8947	9802	8767	7059	6999	7170	8087	7656	8141	6371	8610
Marine mollucs		16			497	77	1113	697		750	12
Marine Fishes, others	2568	119	278	872	10969	30531	25636	16254	23430	16729	22625
TOTAL	124446	128623	135473	129912	174335	190522	204392	217251	210747	219637	208530

TABLE 9.3. Annual catches by major species groups in Lions Zone (From FAO – GFMC)

Species	Annual Catches (Metric Tonnes)										
	1970	1971	1972	1973	1974	1975	1976	1977	1978	1979	1980
Anguilla spp.	3071	3866	1696	1878	1166	987	1659	1295	1762	1176	908
Alosa spp.		77							901		
Psetta ssp.	85	61	82	90	87	94	104	92	81	87	89
Pleuronectiformes	511	526	533	573	467	389	360	566	491	278	356
Gadiformes	1180	1057	1008	1385	2103	2960	2745	2713	2205	2770	3685
Sparidae	301	438	295	1076	1124	1284	1357	1551	1631	1410	1452
Maena spp.				225	211	340	370	269	225	218	174
Mullidae				264	199	212	258	426	214	165	180
Gobiidae									38		24
Demersal percomorphs	2207	1971	1967	3116	3123	3069	3043	1622	5267	5292	2016
Mugil spp.				470	496	604	559	842	1291	1950	1681
Carangidae	412	452	450	334	388	462	410	482	552	463	810
Clupeidae	22090	17528	18341	13727	11315	15485	11308	12389	11413	14514	15371
Engraulis encrasicholus	2300	1547	1041	1905	3843	3259	2857	1293	2122	2462	2441
Scomber spp	1482	1601	1413	1136	1201	1415	1082	1142	928	1464	2299
Pelagic percomorphs								64	8		8
Sharks, Rays, Chimaeras	581	592	553	517	485	403	358	342	268	386	542
Crabs	6	66	98			47	96	138	6		
Lobsters	39	36	40	69	71	59	53	37	34	33	19
Shrimps & prawns	95	212	172	134	46	38	30	34	34	31	42
Marine crustacea	71	125	110	132	56	63	46		4	86	8
Pelecypods/Gastropods	8984	9736	13319	9903	12409	8896	12405	11735	5368	5739	8998
Cephalopoda	197	932	1291	2438	2830	4256	4146	1778	1604	1672	1684
Marine molluscs	857	579	491	660	802	504	443	324	179	343	178
Sponges									0	2	
Marine Fishes, others	4220	3386	2943	707	1310	1070	1112		828	978	680
TOTAL	48689	44788	45843	40739	43732	45896	44801	39135	37455	41517	43645

sometimes form large shoals. Looking to the future, it should be noted that the tunas have been abandoning their traditional inshore migratory routes, which has led to the disappearance of tuna traps from much of the Western Mediterranean shoreline. No prediction can be made with regard to shoals which migrate further offshore, as assessment of these is practically impossible at the present time.

Many demersal species are clearly undergoing a downward trend, if not in terms of the resource as a whole, which has even shown signs of increasing slightly, at least in terms of catch per unit of effort (yield). As long as the present overfishing persists, the prospects can only be for a minimum yield. Total depletion will thus be avoided (for the reasons mentioned above) by the special protective mechanisms which exist for certain species, with small fluctuations related to the general primary production cycles (i.e. 7 to 11 years). Because of the steady increase in fishing capacity brought about, firstly, by the high economic value of fish products and, secondly, by the small cyclic variations already referred to, the future situation can improve only if measures aimed at reducing effort to the maximum sustainable yield level are applied. If such measures are not adopted, the prediction would be for the persistence of an extremely reduced parent stock which will maintain a more or less abundant first generation depending on environmental conditions.

9.10 PROSPECTIVES IN AQUACULTURE

Aquaculture in the Western Mediterranean is a promising prospect. There is generally a mild water temperature, with a number of areas which are particularly well-suited for exploitation (coastal lagoons, river deltas, abandoned salt workings and the like). There is also a long-standing practice of semi-culture of molluscs, such as the raft culture of mussels, or the retaining of certain species in lagoons by preventing their escape back to sea so that they can grow under favourable conditions. Despite the interest aroused by this activity, aquaculture is still in the early stages of research into the many problems affecting production. In most countries aquaculture is, in fact, only just getting under way, with preliminary research being directed to scientific study of such problems as feeding, reproduction, and disease. Unfortunately, basic physiological and behavioural studies, without which it will be difficult to lay a solid groundwork for the industrial development of aquaculture, are still rare.

In the Western Mediterranean basin, research facilities, and pilot operations for preliminary trials of industrial aquaculture production, exist in France, Spain, and Italy. In Italy aquaculture is more developed in the Adriatic than in the Ligurian and Tyrrhenian Seas, probably because of its ancient practice in the 'Valli' along the coast around Venice.

Extensive aquaculture is also being carried out along the coasts of France and Corsica (over an area of some 2,500 ha) to produce around 100 tonnes of various species such as mullets, eels, basses, breams, and soles. The techniques employed are similar to those used in the 'valli' of the upper Adriatic. There is also intensive culture of certain freshwater species (such as the Coho salmon *Oncorhynchus kisutch* and the rainbow trout *Salmo gairdnerii*) which are raised in sea-water in special pens, as is done on a large scale in other countries like Japan. However, the high water temperature in the summer months prevents year-round culture of these species, so that this kind of fish culture is restricted to the colder months. Cages and pens are also being used for marine species such as the bass, *Dicentrarchus labrax*. As already mentioned, most of this work is at the research stage, primarily under the auspices of ISTPM and CNEXO, sometimes in cooperation with industry. As with fish, the problems encountered in the

TABLE 9.4. Annual catches by major species groups in the Sardinia Zone (From FAO – GFMC)

Species	Annual Catches (Metric Tonnes)										
	1970	1971	1972	1973	1974	1975	1976	1977	1978	1979	1980
Anguilla spp.	846	1137	1070	1140	941	885	808	680	785	681	934
Alosa spp.						11	1	1	5		1
Pleuronectiformes	1685	1579	1719	1070	1157	1150	1054	839	750	742	863
Gadiformes	8384	9341	10807	4588	5007	5624	7515	8384	5604	5064	4905
Sparidae	10670	11186	11263	6808	6831	6748	7124	6723	5153	5226	4779
Maena spp.	4593	4696	4920	1064	1328	1649	1768	1368	782	1072	1037
Mullidae	5183	5846	5983	3454	3229	2819	4143	3806	3100	2836	2590
Gobiidae	929	1093	1095	529	495	345	467	396	423	392	484
Demersal percomorphs	4082	4402	5530	2959	2905	2892	3040	2450	1883	1997	2244
Mugil spp.	2527	2589	2137	2443	2947	3088	2683	3331	2085	2003	2060
Carangidae	5880	5986	6373	2931	2282	2362	2477	2504	2271	2957	2360
Clupeidae	26266	29367	22710	16859	19264	19235	15260	11935	13573	15018	10979
Engraulis encrasicholus	26505	19606	25671	23735	22285	16249	24124	19466	12068	10111	10017
Pomatomus saltatrix	0	1	3		1	36			43	39	18
Scomber spp	8711	4413	3505	2459	1769	1875	1519	1613	1757	2021	1975
Pelagic percomorphs	497	620	792	381	430	339	331	601	572	327	261
Sharks, Rays, Chimaeras	2842	3091	3364	1298	1449	1762	2397	2290	1628	1750	1934
Crabs	10	18	17	12	64	32	32			13	
Lobsters	1177	1328	1448	1006	806	903	1037	893	731	822	743
Shrimps & prawns	4472	4480	5517	2187	2072	2109	1737	1693	1067	1493	1470
Marine crustacea	1312	1440	1599	1446	1494	1647	1820	1771	1396	1400	1584
Pelecypods/Gastropods	2966	6466	6149	4230	1698	2106	2600	1878	2149	2867	3766
Cephalopoda	12912	13215	14763	6682	6963	7357	8518	7728	6767	6864	6976
Marine molluscs	689	781	878	1045	1315	1745	1608	1378	926	1043	893
Sponges									2		1
Marine Fishes, others	16438	15788	17990	10435	9853	8559	9990	12915	10772	11135	11182
TOTAL	149577	148468	155304	98760	96584	91447	102053	94643	76292	77873	74056

TABLE 9.5. Annual catches of *Sardina pilchardus* in the Bilearic Zone (From FAO – GFMC)

Species	Contries	Annual Catches (Metric Tonnes)										
		1970	1971	1972	1973	1974	1975	1976	1977	1978	1979	1980
Sardina pilchardus	Algeria	15497	15442	19048	15918	20197	25665	21069	27753	19707	21479	21479
	Morocco	6349	10022	13455	11249	14325	10260	15651	19142	15272	13930	9403
	Spain	31657	34382	35926	27638	35032	40542	41335	40247	41460	34783	34640
	TOTAL	53503	59846	68429	54805	69554	76467	78055	87142	76439	70192	65522

culture and fattening of the prawns, *Penaeus japonicus* and *P. kerathurus*, are being investigated, although the high costs involved in the culture of these species makes successful commercial operation appear unlikely. In the main current research centers on feeding during the early stages of the life cycle prior to the growth and attainment of marketable size.

The situation along the coast of Italy is similar, both with respect to the species being studied and to the areas of research. Again, effort is mainly concerned with nutrition and lowering mortality rates. Very interesting preliminary work is being done on bluefin tuna, *Thunnus thynnus*, in close cooperation with Japan, for the purpose of aiding the recovery of shoals of this species off the coast of Sicily. Research is in progress on reproduction and feeding problems in peneid crustaceans and some molluscs, such as *Tapes decussatus*.

On the coast of Spain, two institutes, the Instituto Español de Oceanografía (Spanish Institute of Oceanography, at its laboratory on the Mar Menor) and the Instituto de Acuicultura (Aquaculture Institute, at Torre de la Sal in Castellón), have been conducting several lines of research into the culture of the prawns *Penaeus kerathurus* and *Palaemon serratus*, especially on post-larval feeding. High survival rates have frequently been achieved during the first few days, but there are still serious difficulties associated with feeding in the post-larval stage. The culture of *Artemia salina* is important for its use as food, and Spanish research workers feel that this approach is very promising and that it will also be important for industrial production. Research on fish species has concentrated on bass, the gilt-head bream, *Sparus auratus*, the common sole, *Solea solea*, and mullets (*Mugil* spp.). Survival rates of 20 – 30% have been attained for bass through the first 60 – 67 days of life. Experimental studies on the common sole *Solea solea* are under way at the Instituto de Acuicultura in Torre de la Sal (Castellón), where promising growth rates have been achieved.

As in the other countries bordering the Western Mediterranean basin, the problems associated with feeding have received the most attention, not only for fish, but also for crustaceans and molluscs. The main and most pressing problems relate to feeding of the juvenile stages.

As has been emphasized, aquaculture in this part of the Mediterranean is still in its preliminary and largely experimental stages. The initial research has been satisfactorily concluded (and that with only partial success) in the case of bass, which has been reared from eggs to marketable size in small-scale projects. The bream seems to hold out better prospects than bass in many areas, although current successes are less numerous. On the whole more attention is paid to fish culture than to that of crustaceans, and even less to molluscs. Experiments have mostly been limited to those problems that seem to have a direct bearing on the industrial development of aquaculture. With respect to feeding and

TABLE 9.6. Annual catches of Shrimps and Prawns in Spain in the Balearic Zone (From FAO – GFMC)

Species	Annual Catches (Metric Tonnes)										
	1970	1971	1972	1973	1974	1975	1976	1977	1978	1979	1980
Shrimps and prawns	956	829	535	646	1200	982	1514	1777	1930	1915	1845

the production of suitable granular feeds for these fish, indications of reduced fertility rates found for certain species may be related to improper diet. The main objectives of research in the immediate future will undoubtedly be to improve conversion rates while maximizing survival rates.

Research into the pathology of animals in confinement is rapidly growing in importance in response to the many diseases that have been detected. It appears that marine species are less resistant to attack by different disease organisms and are also more subject to stress.

The physical features of the coast and the water temperature would seem to be ideal for the development of aquaculture in the Western Mediterranean. However, thriving industries of this nature will require highly diversified research, with trials at pilot stations of varying types (pools offering controlled conditions, breeding and culture in estuaries and coastal lagoons, as well as mariculture in special cages and pens). Lastly, more effective and purposeful support from both the public and the private sectors is needed to bring about widespread industrial development.

REFERENCES

Bas, C. (1957) La geographie du fond et l'etat actual de la pêche des especes d'intérêt industriel. *C.G.P.M. Deb. et Doc. Techn.* N° 4 Doc. Techn. 31, 235 – 241.
Bas, C. (1960) Variación en la pesca de crustáceos de fondo. *IV Reunión sobre Prod. y Pesq. Instituto de Inv. Pesqueras*, 91 – 93.
Bas, C. (1964) Fluctuations de la pêche de *Merlangus poutassou* et quelques considerations sur son controle. *C.G.P.M. Deb. et Doc. Techn.* N° 7 Doc. Techn. 41, 417 – 420.
Bas, C. and Morales, E. (1951) Nota sobre la talla y la evolución sexual de las sardinas de la Costa Brava (Septiembre de 1949 a Septiembre de 1950). *Publicaciones del Inst. de Biol. Aplicada* VIII, 161 – 181.
Bas, C., Morales, E. and Rubio, M. (1955) La pesca en España. I. Cataluña. *Ed. Inst. Inv. Pesq.* Barcelona, 468 pp.
Casinos, A. and Vericad, J. R. (1976) The Cetaceans of the Spanish coasts. *Mammalia* 40(2), 267 – 289.
Duguy, R., Casinos, A., Di Natale, A., Filella, S., Ktari-Chakroun, F., Lloze, R. and Marchessaux, D. (1983) Repartition et frequence des mammiferes marins en Méditerranée. *Rapp. Comm. int. Mer Medit.*, 28(5): 223 – 230.
Comité des Vertebrés marins et Cephalopodes, 9 pp.
FAO (1977) Anuario estadístico de Pesca. Capturas y Desembarques vol. 44 (FAO), 328 pp.
C. G. P. M. (1970) Les ressources vivantes des eaux profundes de la Méditerranée occidentales et leur exploitation. *Etudes et Revues* N° 44 (FAO), 38 pp.
C. G. P. M. (1979) L'Evaluation des stocks dans les division statistiques Baléares et Golfe du Lion. C.G.P.M., FAO Rapport sur les Pêches n° 227, (FAO) Palma de Mallorca, 155 pp.
Larrañeta, M. G. (1981) La pesquería de sardina, *Sardina pilchardus* (Walb.) de Castellón. *Inv. Pesq.* 45(1), 47 – 91.
Levi, D. and Troadec, J. P. (1974) The fish resources of the Méditerranean and the Black Sea. *Stud. Rev.* GFCM 54, 29 – 52.
Margalef, R. (1963) El ecosistema pelágico de un área costera del Mediterráneo Occidental. *Mem. R. Acad. Ciencias y Artes de Barcelona* 35(1), 48 pp.
Massuti, M. (1981) Les ressources et la production de petites espèces pélagiques dans la región. In *GFCM/CGPM, 1981. Report of the Technical Consultation of the Utilization of Small Pelagic Species in the Mediterranean Area*, pp. 33 – 51, FAO, *Fish. Rep.* (252), 159 pp.
Rey, J. C. and Cort, J. L. (1981) Contribution à la connaissance de la migration des Escombridae. *Rapp. Comm. Int. Mer. Médit.* 27(5), 97 – 98.
Samir Zaky, Rafail (1974) Study of fish population by capture data and the value of tagging experiments. *Etudes et Revues.* N° 54, FAO, 53 pp.
Suau, P. (1979) Biología del Boqueron (*Engraulis encrasicholus* L.) de las costas de Castellón (E. de España). *Inv. Pesq.* 43(3), 601 – 610.

CHAPTER 10

The Footprints of Life and of Man

JOAN ALBAIGÉS*, M. AUBERT** and J. AUBERT**

*Environmental Chemistry Unit, Institute of Bio-Organic Chemistry (C.S.I.C.), Barcelona, Spain;
**C.E.R.B.O.M., Nice, France

CONTENTS

10.1. Biological markers in the Mediterranean Sea	317
10.1.1. The organic matter in the marine environment	318
10.1.2. The organic matter in the Western Mediterranean	321
10.2. A few preliminary words about pollution	334
10.2.1. The sources of chemical pollution in the Mediterranean	335
10.2.2. Microbiological pollution	339
10.2.3. Thermal pollution	341
10.2.4. The consequences of pollution	342
10.2.5. Conclusion	347

The perfect crime does not exist. Any activity, any stress leaves a trace. With people, the trace of their activities is obvious, and may prove to be unbearable in the long run. But, pollution is just one of the many everyday events in the economy of nature; only scaled-up to the present day capacity of mankind and in particular to his mobility.

To be part of a wider game does not excuse man, even though he is not the sole polluter. Most activities of living organisms leave some traces, and we are happy if they can be understood. We do not intend to minimise the importance of pollution, only to place it in realistic perspective, from which it will perhaps be easier to find eventual solutions.

10.1. BIOLOGICAL MARKERS IN THE MEDITERRANEAN SEA*

Half of the organic matter synthetized on Earth is produced in the ocean. This organic matter, together with its transformations along metabolic pathways and trophic chains, includes a very wide range of molecules. Some of these are the everyday currency of life, are quickly synthesized and decomposed and leave behind few tell-tale signs. Other molecules are much more diversified and persistent. When we learn their ways we shall understand much of the transport of chemicals in the sea and of the biogeochemical cycles involved.

---

* J. Albaigés is indebted especially to his colleagues J. Grimalt, J. M. Bayona and J. Algaba for much of the information reported in this part. J. Albaigés is responsible for part 10.1; M. Aubert and J. Aubert have contributed part 10.2.

TABLE 10.1. Inputs and Reservoirs of Organic Carbon in the Sea (According to Mopper and Gens, 1979)

	gC
Annual inputs	
Net primary productivity (assuming 100 g C/m² year)	3.6×10^{16}
River (assuming 5 mg C/l)	1.8×10^{14}
Rain (assuming 1 mg C/l)	2.2×10^{14}
Reservoirs	
Dissolved Organic Carbon (assuming 700 µ C/l)	1×10^{18}
Surface (0 – 200 m)	0.9×10^{17}
Deep Sea (>200 m)	8.6×10^{17}
Particulate Organic Carbon (assuming 20 µg C/l)	3×10^{16}
Plankton	3×10^{15}
Bacteria	0.2×10^{15}
Organic Carbon in marine sediments	1.1×10^{22}

10.1.1. The organic matter in the marine environment

Sources

The organic matter in the marine environment originates from two sources. Autochthonous compounds are derived from organic inputs within this ecosystem, including those derived from microscopic and macroscopic vegetation (phytoplankton and macrophytes), benthic and pelagic fauna (principally zooplankton) and heterotrophic microorganisms (bacteria). Autochthonous organic matter consists of compounds synthetized *de novo* by organisms and compounds derived from these during the microbiological, chemical and geochemical alterations generated within the water column and the sediments.

Allochthonous compounds, on the other hand, are of non-marine origin and reach the marine environment by water transport (mainly riverine) and eolian fall-out. Input sources include residues and extracts of terrestrial vegetation and soils and the products of forest fires. In addition to these naturally occurring inputs, there are anthropogenic sources such as urban and industrial waste discharges and the surface run-off from cultivated areas.

The relative contribution of these sources to the marine organic matter is variable, both spatially and temporally. In this respect, the results of stable carbon isotope analyses (Gearing *et al.*, 1977) and direct chemical analysis of specific continental markers (Stuermer and Harvey, 1974; Mopper *et al.*, 1977) indicate that, at the present time, autochthonous sources are dominant in the sea. In fact, the major input of organic carbon in the ocean, at least in surface waters, appears to be primary productivity. Ryther (1969) calculated primary production rates of 50, 100 and 300 gC/m²year for open ocean, coastal and upwelling areas, respectively, which yield an estimate for the total primary oceanic production of 3.6×10^{16} gC per annum (*see* Table 10.1).

In contrast, the contribution of allochthonous organic matter appears to be significant only in continental slopes, where the influence of land inputs are evident. The organic content of rivers can be quite high, but the quantitative significance of this source in the sea is low (*see* Table 10.1), because of the considerable organic flocculation in deltas and estuaries (Kranck, 1973; Grimalt, 1983).

Atmospheric sources, such as dry fall-out and rainfall, are of minimum significance, on a global scale, when compared with primary productivity (Duce and Duursma, 1977). However, it is important to consider not only the quantitative values but also the qualitative composition of these sources. In fact, fluvial and atmospheric compartments are dominated by anthropogenic inputs. These usually contain compounds of sufficient biological activity and geochemical stability to threaten marine environmental processes. The study of the organic matter in areas of mixing waters and in the sea-air interface is therefore crucial for the understanding of the transfer processes from those compartments to the sea.

Distribution

Many of the earlier investigations on the distribution of marine organic matter were largely concerned with simple measurements, such as the total organic carbon determined by oxydative methods, which yielded no qualitative information. However, the involvement of the organic matter in several marine physicochemical, biological and geological processes, has, in the last decade, stimulated interest in the qualitative composition of these sources, as well as in their distribution among the different marine compartments.

The accepted conventional separation between the dissolved and particulate phases is based on filtration of the sea-water through glass fibre filter of 0.45 μm pores. According to this definition living organisms, organic detritus (dead plankton cells, faecal pellets, etc.) and suspended sediments are included in the particulate fraction, while some colloidal material (besides the more soluble excretory and decomposition products of organisms) is included in the dissolved fraction.

At present, little is known about the mechanisms and processes by which dissolved (DOM) and particulate organic matter (POM) are produced, transported or accumulated, although they are, apparently, related. In addition, some exchange has been observed at the boundaries between the water and both the atmosphere and the sediment (Daumas, 1980).

In the water column, the DOM is generally found to exceed the particulate forms by a factor of 10 or 20 (*see* Table 10.1). It is mainly stored in the deep sea, although the concentration is higher in the surface layers, namely in the euphotic zone (Cauwet, 1978). POM on the other hand, accumulates at the air-sea water interface where concentrations of up to 200 – 300 μg C/l are currently found, with DOM – POM ratios of less than 10 (Wangersky, 1976; Daumas, 1976).

In the deeper layers, the concentration of DOM (in the order of 500 – 700 μg C/l) remains constant or decreases slightly with depth. In the interstitial waters of the superficial sediments it increases again to about ten times that of the overlying waters.

The distribution of the organic matter between the different marine compartments exhibits not only quantitative, but qualitative differences as well. These have been inferred from the studies carried out in different marine regions (Boehm, 1980; Goutx and Saliot, 1980; Boussuge *et al.*, 1980; Grimalt *et al.*, 1983), which indicate an uncoupling of the organic matter with respect to the source, in the dissolved, particulate, surface microlayer and sediment interstitial phases.

Proteins, for example, are generally found only in the particulates, while carbohydrates predominate in the dissolved phase. The major accumulation of lipids occurs in the air-sea water and sea water-sediment interfaces where enrichment factors of 4 and 50, respectively, with respect to surrounding waters, have been reported (Jullien, 1982; Boussuge *et al.*, 1980).

Separate consideration of these reservoirs will clarify the origins, transport paths and eventual fates of the organic matter in the marine environment, and, moreover, their interactions. Thus, for example, water column-sediment coupling appears to be centred on the particulate organic matter, whereas the principal uptake pathway of organic compounds into the biota appears to be through the dissolved phase. These interfaces are, therefore, vital for physical and diagenetic processes involving marine organic matter.

Cycling

The nature of the organic matter changes through the water column. Generally, the major proportion of the autochthonous biomass in the upper layers of the ocean (i.e. in the euphotic zone) consists of phytoplankton (Horne, 1969), while deeper (in the aphotic zone), heterotrophic organisms dominate. The heterotrophic activity is always linked to primary production and is mainly responsible for the recycling of the labile organic compounds which are produced or released by primary producers.

The more hydrolyzable compounds, such as proteins and nucleic acids are rapidly lost. The residence time of free amino acids in surface waters is very short, due to the rapid turnover by organisms and the concentrations in deeper waters are almost negligible. Carbohydrates react with other compounds, including refractory amino acids, to form humic materials, and lipids appear to be fairly stable components of primary organic matter (Gagosian, 1980).

Although it is difficult to calculate the flux of organic matter leaving the euphotic zone, sediment-trap studies have yielded estimates of the order of $4.1 - 10.0$ g C/m^2year (Mopper and Degens, 1979). Assuming an average surface productivity of 100 g C/m^2year, this represents a recycling efficiency of $90-95\%$.

The extent of this recycling in the water column and the subsequent transfer of the refractory carbon to the deep-sea pools depend, however, upon a variety of physicochemical parameters, primarily the oxic/anoxic conditions (Didyk *et al.*, 1978).

In oxygenated environments, such as most of the near-shore waters, the downward flux of autochthonous organic debris from the euphotic zone to the underlying waters and sediments represents only about 2% of the organic carbon. In anoxic sedimentary environments, however, this downward flux includes up to 20% of the organic matter generated in the euphotic zone.

In fact, biological metabolism and sedimentation appear to be the crucial processes which determine the ultimate fate of organic matter in the sea.

The total annual losses of organic carbon by sedimentation are higher in the pelagic than in the near-shore zones (9.2×10^{13} gC and 2.7×10^{12} gC, respectively; Mopper and Degens, 1979). Although not of primary importance in the total annual inputs, this loss from the biochemical cycle permits the organic carbon to last longer and converts the sediments into the major reservoir of organic carbon in the sea (*see* Table 10.1).

Organic markers

The chemical composition of marine organic matter (including DOM, POM and sedimentary organic matter) is largely unknown. Only 10 to 20% of it has been rigorously identified and found to consist of amino acids, carbohydrates, lipids and phenols (Duursma and Dawson, 1981). The rest includes high-molecular weight complexes such as humic-like materials, melanoidin condensates and, in sediments, the solvent-inextractable kerogen (Tissot and Welte, 1978).

Although the identified component constitutes only a minor proportion of marine organic matter, the structural specificity of the constituent compounds are of particular interest. Such compounds can be of direct biological origin or derived from those which have undergone some structural modification that has not obscured their diagnostic features. The compilation of inventories of these components for precise geographical areas (of known physical, chemical and biological parameters) will permit to relate each of them to definite precursors, the assessment of their cycling and fate in the sea and, hence, the understanding of certain aspects of the depositional environment.

ERRATUM

On page 321 (para. 7, lines 1 and 2)
for figures 10.1 and 10.2 *read* maps 10.1 and 10.2.

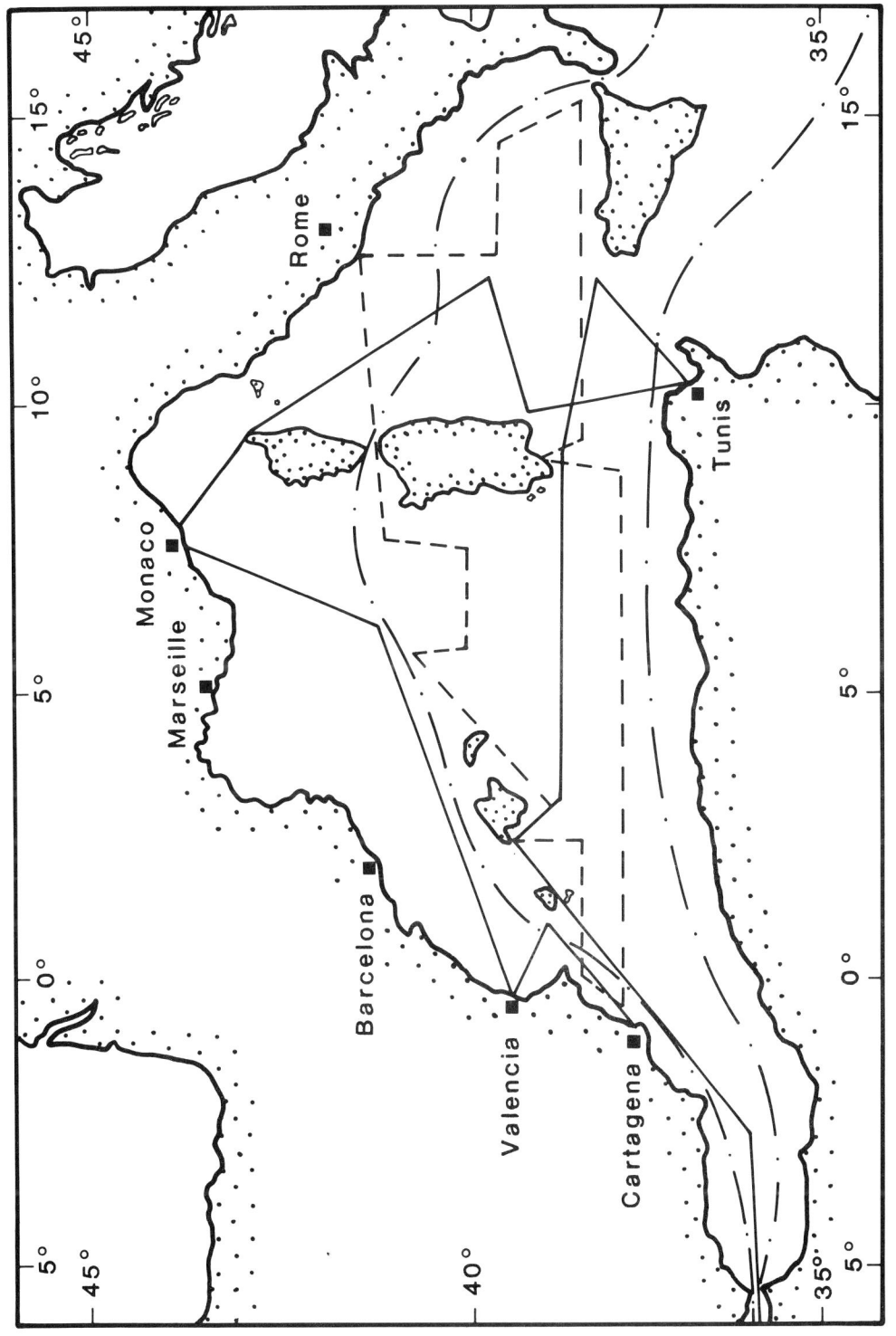

Map 10.1 The Western Mediterranean showing cruises of R/V Westward (1974–1975) and R/V Cornide de Saavedra (——— 1975 and – – – 1977) for monitoring pelagic tar balls and dissolved hydrocarbons.

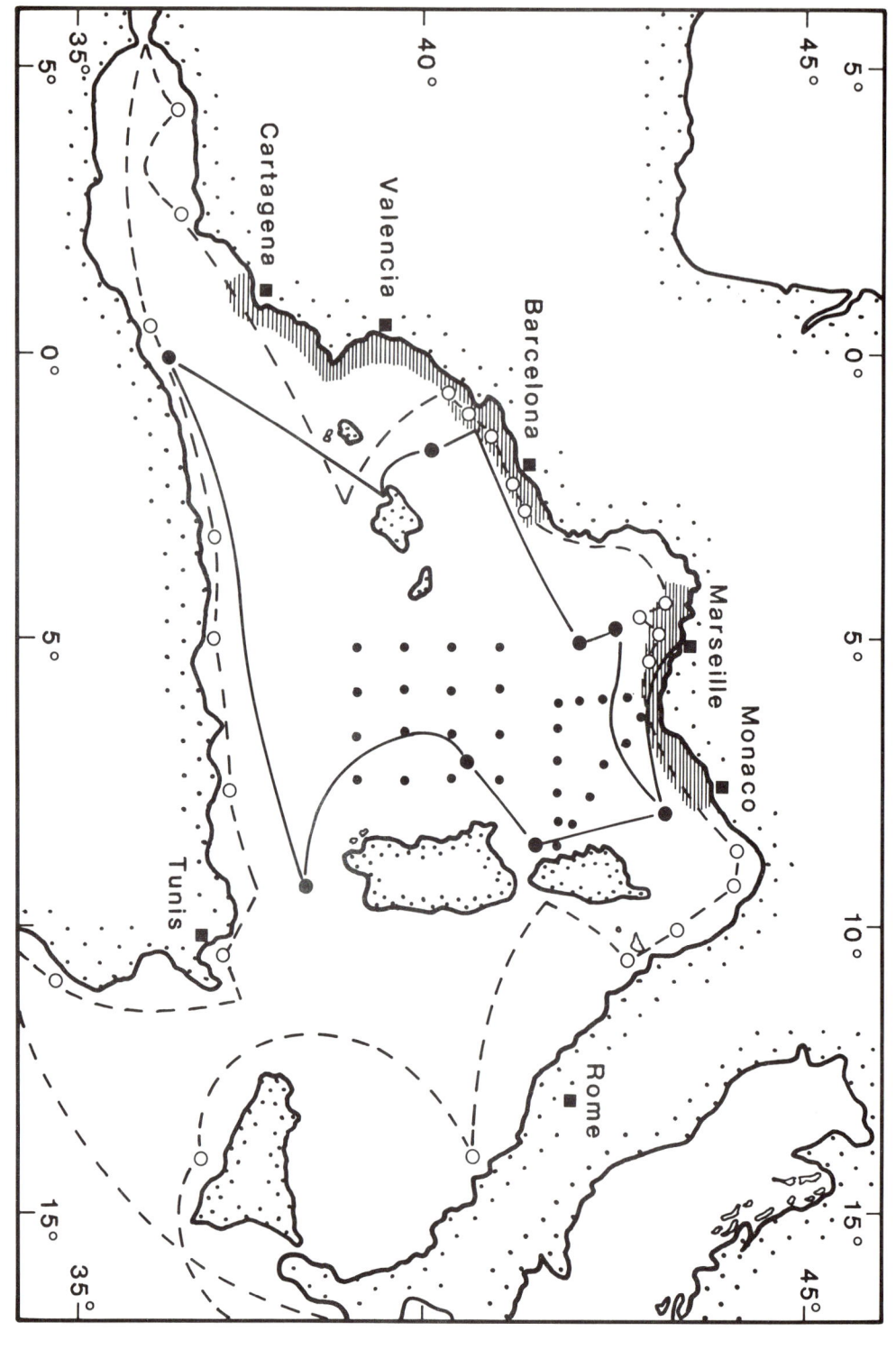

Map 10.2 The Western Mediterranean showing CIESM (R/V Calypso, 1977, – – –) and PHYCEMED (R/V Le Suroit, 1981, —) cruises. The grid of dark points corresponds to stations sampled in cruises BIOMED I (R/V Capricorne, 1976) and BIOMED II (R/V Noroit, 1981). The shaded zones represent coastal areas where monitoring activities are currently undertaken.

The number of compounds of obvious biological origin has increased dramatically in recent years, in marine regions and generally in the geosphere as a consequence of the development of highly-sophisticated analytical techniques, including computerized gas chromatography-mass spectrometry (GCMS) (Blumer, 1975). However, at present, the most convenient part of the organic matter for use as organic marker is the lipid fraction.

Lipid molecules (hydrocarbons, carboxylic acids, sterols, etc.) do, in fact, exhibit great specificity as indicators of direct contributions from marine algae, terrestrial higher plants and bacterial sources (Brassell et al., 1980). They are also remarkably stable in water and sediments, thus permitting the tracing of transport routes or investigation of historical records through the study of sediment cores.

The full molecular characterization of these molecules is essential to the proper description of organic markers. Parameters of special value are: stereochemical data, homologous series, relative abundances and carbon isotopic ratios (Brassell and Eglinton, 1980). Nevertheless, the critical point in devising a successful survey strategy for the marine environment, using organic markers is, as already mentioned, the adequate sampling of the water column, the air-sea and sea-sediment interfaces and the sediment itself.

The following section illustrates the application of some of these concepts to the specific case of the Western Mediterranean.

10.1.2. The organic matter in the Western Mediterranean

An overview of research activities

Due to the widespread public concern much of the research and monitoring activities have been concerned with the apparently increasing pollution problems in the Mediterranean. An important effort has been made since 1976 (with the Mediterranean Action Plan, sponsored by the United Nations Environment Programm: UNEP, 1977) for the estimation of organic pollutant levels in the sea, as well as for the evaluation of the threat they may represent to the ecosystem.

Obviously the Western Mediterranean has been extensively studied, because of the major concentration at its boundaries of urban populations, industrial activities, river discharges, which are likely to be the principal contributors to the total pollution load of the Mediterranean (Helmer, 1977).

Several coastal and open sea areas have been investigated in the Western Mediterranean (Figs. 10.1 and 10.2). The first large surveys were devoted to the evaluation of pelagic tars. After the cruise by the R/V *Atlantis II* in 1969 (Horn et al., 1970), the area was surveyed in 1975 by R/V *Westward* (Morris et al., 1975) and in 1975 and 1977 by R/V *Cornide de Saavedra* (Faraco and Ros, 1979).

In addition to pelagic tar balls, the dissolved particulate hydrocarbons were also measured (Zsolnay, 1979; Ros and Faraco, 1979). An analytical method for the source identification of these petroleum residues was developed (based on fingerprinting of specific series of organic markers such as steranes (I) and triterpanes (II), by computerized GC-MS and applied for the first time to samples collected in the latter cruise (Albaigès et al., 1979).

Monitoring of other organic compounds, such as organochlorinated hydrocarbons (PCB's and DDT's), specific polycyclic aromatic hydrocarbons and phthalates, was carried out during BIOMED I and II cruises (1976 and 1981), to study deep-sea sediments from the Balearic basin (Arnoux et al., 1981 and 1983) (Fig. 10.2).

In December 1977, the R/V *Calypso* gathered samples of water and sediments at more than 60 stations distributed along the coast for the assessment of base line levels of PCB's and DDT's (Cousteau,

1979) (Fig. 10.2). Since then, monitoring programmes on the Spanish and French coasts have been established for the detection of such compounds as hydrocarbons, organochlorinated compounds and detergents (Fig. 10.2) (ICSEM/IOC/UNEP, 1979, 1981, 1983).

A great deal of information is already available, as a result of these activities (see ICSEM/IOC/UNEP, 1983) although a major drawback is the interpretation of the large quantity of data generated (due to the lack of precise information about the sources of the organic inputs, the paths they travel in particular areas and their environmental fate).

This is why, more recently, attention has been focused on the exploration of biogeochemical processes which affect the distribution, transport and ultimate fate of the organic matter in the sea. Thus in 1981 and 1983, the PHYCEMED cruises (Fig. 10.2) were concerned with the evaluation of the atmospheric budget of organic and inorganic substances in the Western Mediterranean and to the investigation of mechanisms of exchange of these materials across the air/sea water interface (Arnold et al., 1983; Hô et al., 1982 and 1983).

In addition, semi-permanent sediment traps have been set up by the Laboratory of Marine Radioactivity in Monaco (at 100 m depth in a 250 m water column approximately 2 km off the Monaco coast) to obtain information on the downward flux of anthropogenic substances in the Ligurian Sea and on the dominant processes controlling the transport to and retention of organic compounds in the sediment reservoir (Fowler et al., 1979). These include biological uptake and concomitant faeces production, biological/chemical degradation and physical/chemical partitioning between marine compartments (Burns and Villeneuve, 1983).

However, the use of organic markers, as outlined in the preceding section, has proved to be the most successful monitor of marine pollution. For this reason we have undertaken in our Unit the qualitative and quantitative analysis of organic extracts from water, particulates and sediments along the Spanish continental shelf, with special emphasis on the characterization of riverine inputs to the marine environment (González et al., 1983; Albaigés et al., 1983; Grimalt et al., 1983; Bayona et al., 1983). Amino acids, carboxylic acids, wax esters, ketones, aldehydes, alcohols, alkanes + alkenes and aromatic hydrocarbons have been identified and used as indicators of terrestrial, aquatic (non-bacterial) and bacterial contributions of organic matter. Evidence has been obtained of several diagenetic degradation pathways of organic matter due to microbial activity in sediments and to depositional conditions. These studies illustrate the value of this approach and the need for major organic geochemical studies in the Western Mediterranean. Some of our results will be presented here.

Organic markers in the Western Mediterranean

The global budget of the organic matter in the Mediterranean is far from complete. Preliminary data have recently been obtained by Jullien (1982) and Hô (1982) for waters along the French coast and by Grimalt (1983) for sediments along the Spanish coast.

The results, shown in Tables 10.2 and 10.3, provide insight for the further selection of organic markers. In the water column, organic matter is usually concentrated in the top hundred metres, primarily in the dissolved phase. The surface and subsurface waters exhibit enrichment factors of 7 and 15 respectively, in agreement with previous findings in other marine areas (Cauwet, 1978; Williams, 1967; Daumas, 1976).

In these two compartments, 65 to 90% of the organic carbon is in complex material which is difficult to characterize. However, in the dissolved phase, carbohydrates appear to be the dominant form of organic matter, while proteins are slightly more abundant in the particulate fraction, according to their different physicochemical properties.

THE FOOTPRINTS OF LIFE AND OF MAN

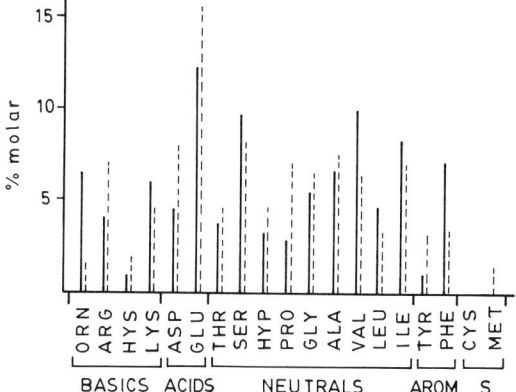

Fig. 10.1 Histogram of the amino acid distribution in a Spanish coastal (Mediterranean) sediment. The distribution in suspended matter is indicated with dotted lines, for comparison. (From Gonzalez et al., 1983).

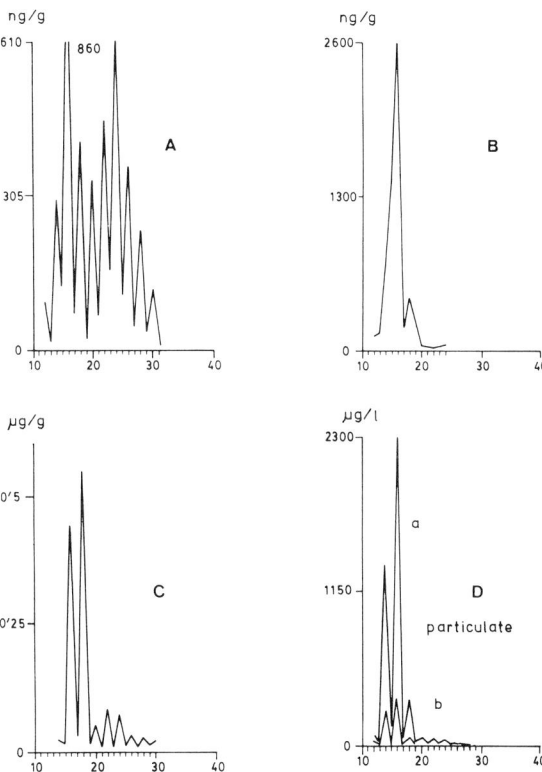

Fig. 10.2 Carbon number distribution of n-alkanoic acids in samples from the Western Mediterranean (A. inshore sediments; B. estuarine sediment; C. deep-water sediment; D. suspended particulate, (a) esterified, (b) free acids).

TABLE 10.2. Mean Concentrations (in μg C/l) of the Different Constituents of Marine Organic Matter off the French Coast (Adapted from Jullien, 1982 and Hô, 1982). (M: Surface Microlayer; W: Subsurface Water; e.f.: Enrichment Factor)

		Organic carbon	Complex organic matter	Proteins	Carboxy-drates	Lipids	Distribution of organic carbon with water depth (m)				
							surface	100	1000	2500	bottom
Particulate	M	368	241	49	17	61	360				
	W	146	101	23	6	15	270	35	25	30	20
	e.f.	2,5	2,4	2,1	2,8	4,0					
Dissolved	M	2583	2094	—	426	61	1450				
	W	2317	2067	—	218	26	3600	500	500	—	200
	e.f.	1,1	1,0	—	2,0	2,3					

TABLE 10.3. Composition of Representative Sediments off the Spanish Coast (Adapted from Grimalt, 1983)

	% sands (>66 μm)	%C org.	%C inorg.	organic matter (mg/g of dry sediment)		
				aminoacids	carbohydrate	lipids
Estuarine	44	7.8	4.1	8.7	13.6	0.17
River mouth (Ebro)	47	3.3	3.6	2.7	2.2	0.02
	26	4.0	3.5	1.7	1.7	0.05
Coastal	94	0.9	3.5	1.0	0.5	0.02
Shelf	34	2.2	4.3	2.0	1.8	0.01

The lipid components are more evenly distributed between the dissolved and particulate phases, thus providing a higher enrichment factor (e.f.) in the surface microlayer. This enrichment is especially noticeable for the particulate fraction (e.g. 4.0) and is even more significant in polluted areas (e.f. 22.5) (Jullien, 1982), demonstrating the importance of the exchange processes which occur at the air/sea water interface.

An important process affecting the distribution of marine organic matter is the sedimentation of particulate material. Data on vertical flux of total carbon in the Mediterranean are very scarce because of the lack of adequate installations (sediment-traps) for measuring it. However, at 50 m water depth, off the French coast, it has been estimated to be 96 mg POC/m^2 day (Staresinic et al., 1981). That is about an order of magnitude lower than those measured along coastal upwelling regions (e.g. Peru), but still very high for primary production in the Mediterranean.

Organic carbon is then incorporated into the sediments and, as can be seen in Table 10.3, the final content is closely related to the sediment lithology as well as to the deposition conditions of the environment. Thus, sediments from eutrophic estuaries (where some reductive environmental conditions may occur) exhibit organic carbon contents of 5-7%, or even higher, whereas at the shore (where the sediments are well-aerated and generally consist of coarse sands) the organic carbon rarely exceeds 1%, unless there are local inputs such as urban effluents. Similar values have also been reported for the French coast (Cauwet, 1975; Cenciarini et al., 1981; Arnoux et al., 1981).

On the continental shelf, the organic carbon contents are moderate (1−20%), from the primary productivity of the Mediterranean, which decreases strongly with depth.

The data of Vigneaux et al. (1980), on the distribution of organic carbon in other ocean sedimentary environments, also reveals a general decrease in organic carbon, from coastal to deep-sea sediments. This gradient contributes to the degradation of the organic matter in the marine environment and to the origin of the deep sea pools. However, quantitative analyses are difficult to interpret and specific molecular indicators are better organic markers.

The qualitative analyses in Table 10.3 show how the depositional conditions affect the composition of the organic matter. In confined areas the relative concentration of carbohydrates (the least stable components) increases. The total amounts of carbohydrates are relatively small, and together with amino acids and lipids accounts only for 5 – 15% of the total organic carbon. Nevertheless *carbohydrates* and *amino acids* are valuable indicators of the biological or biochemical events for they have relatively short residence times in marine environments (Michaelis et al., 1980). A characteristic distribution of amino acids in Mediterranean coastal sediments is shown in Fig. 10.1.

A general feature of the collected samples was the virtual absence of free amino acids. These were released only after drastic acid hydrolysis of the sediment. This implies that at the very early stages of deposition, they were present as intact proteins or, more probably, some form of association with the sediments (e.g. as part of the more chemically resistant humic fraction) (Pelet and Debyser, 1977).

The absolute concentrations of amino acids decrease significantly with water depth (Siezen and Magne, 1978). This may be due either to declining organic inputs, or to the particular environmental characteristics of the pool. Degens (1970) suggested that as absolute concentrations are higher near shores water column depth plays a decisive role. The heterotrophic turnover of amino acids and carbohydrates near the surface of the Western Mediterranean amounts up to 30 – 50% per day (Morris and Culkin, 1975). Yet the distribution closely resembles those reported for other marine areas (Whelan, 1977) and also for plankton (Fig. 10.1), with two slight differences: the disappearance of the unstable sulphur-containing compounds, and the formation of ornithine (probably from arginine released by degenerating plankton material).

Glutamic acid predominates over aspartic acid because it is the major amino acid in plankton (Morris, 1975), marine bacteria (Henrichs and Farrington, 1980) and oceanic suspended particulate matter (Siezen and Magne, 1978). It is difficult to account for the high concentration of phenylalanine, although Nissenbaum et al. (1972) working with sediments from the Dead Sea have related predominance of this amino acid to the oxidising conditions.

The above data indicate a general contribution of autochthonous organic matter to continental shelf sediments, which can, nevertheless, be modified by prevailing depositional conditions, notably by the oxidising conditions.

The lipid fraction

Lipids are useful indicators both for the characterization of input sources and the degradation pathways of organic matter.

This fraction rarely forms more than 5% of the total organic matter in the sea (*see* Tables 10.2 and 10.3). It is, however, the most clearly understood and useful indicator in molecular organic geochemistry (*see* McKenzie et al., 1982; Brassell, 1980 for references). The proposed lipid indicators of direct contributions to the marine and environment from autochthonous organisms (algae and zooplankton) and higher terrestrial plants and bacterial sources are given in Table 10.4. These indicate not only the presence or absence of a specific compound, but also the characteristics of homologous and pseudohomologous series (i.e. their carbon range and carbon preference index). Stereochemical data are also valuable in some cases. Examples listed are not exhaustive, but are illustrative of the compounds that are of widespread occurrence in the Western Mediterranean.

Three classes of compound (alkanoic acids, alkanes and sterols) are characteristic of lipids of different origins and are therefore potentially useful organic markers. An overall investigation of organic markers in the Western Mediterranean is under way (Grimalt, 1983; Bayona, 1984; Algaba, 1983).

Chemical structures cited

TABLE 10.4. Proposed Lipid Indicators of Organic Inputs to the Marine Environment (Adapted from Brassell, 1980)

Lipids of marine origin		
Planktonic inputs		
Straight chain alkanes	C_{17} dominant	III
alkenes	Heneicosahexane (C_{21}:6)	IV
alkanols	$C_{14} - C_{22}$ (even-odd predominance)	V
alkanoic acids	$C_{14} - C_{18}$ (even-odd predominance)	VI
Acyclic isoprenoid alkenes	Phytenes ($C_{20:1}$)	VII
alkenols	Phytol (C_{20})	VIII
Sterols	5α(H)-stanols	IX
	23,24-dimethyl sterols (C_{29},C_{30})	X
	22,23-methylenesterols (C_{29}-C_{31})	XII
Bacterial inputs		
Straight chain alkanes	$C_{14} - C_{28}$ (low carbon predominance)	III
Branched chain alkanoic acids	$C_{10} - C_{22}$ (iso and anteiso)	VI
alkenoic acids		XIII
Acyclic isoprenoid alkenes	Squalene (C_{30}:6)	XIV
Triterpenoids	Hopanoids ($C_{31} - C_{35}$)	XV
	Hopenes (C_{30})	XVI
	Fernenes (C_{30})	XVII
Lipids of terrestrial origin		
Higher plants		
Straight chain alkanes	$C_{23} - C_{33}$ (odd-even predominance)	III
alkan-2-ones		XVIII
alkanols	$C_{23} - C_{33}$ (even-odd predominance)	V
alkanoic acids		VI
Diterpenoids	Fichtelite	XIX
	Abietic acid (C_{20})	XX
Triterpenoids	Friedelan-3-one (C_{30})	XXII
	α-and β-Amyrine (C_{30})	XXIII
Sterols	24R-Ethylsterols (C_{29})	XXIV
Bacterial reworking	derived from 3-oxytriterpenoids	XXV – XXIX
	Retene	XXX

Alkanoic acids

Predominant components of the lipid fraction are *carboxylic acids*, both in water and in sediment compartments. Among them, n-alkanoic acids represent 50 – 95% of the total acid fraction. In general, the concentrations are an order of magnitude higher inshore and in estuaries, than in the open sea (Table 10.5). The particulate forms tend to be concentrated in the surface microlayer and in coastal areas, clearly showing its relationship with primary productivity or with eutrophication in estuaries.

Qualitatively, the distributions of carboxylic acids in the marine environments are often comparable to the populations of organisms, reflecting either marine or terrestrial inputs. Fig. 10.2 shows typical profiles of different marine samples. The acids of lower molecular weight (around C_{16}) are of autochthonous origin, whereas maxima at around C_{24} indicates allochthonous sources (Cranwell, 1978). Although the profiles generally exhibit a marked even-odd carbon number periodicity, in some cases the distribution is more uniform, as in Fig. 10.2B. Such distribution has been found in sediments, associated with microbial activity, from certain coastal zones.

These features are useful for the characterization of depositional environments in terms of organic inputs. As an example, Fig. 10.3 shows, at the mouth of the Ebro River, a zone of high concentration of $C_{20} - C_{32}$ n-alkanoic acids. The thick dashed zone provided samples with concentrations in the range of 13 – 16 µg/g dry wt, while in the outer zone concentrations are only 3 – 5 µg/g.

TABLE 10.5. Mean concentrations (in µg/l or µg/g) of some organic markers in the Western Mediterranean (M: Surface Microlayer; W: Subsurface Water) (Adapted from Goutz and Saliot, 1980; Hô, 1982; Grimalt, 1983)

		n-Acids	n-Alkanes
Water (coastal)			
Dissolved	(W)	100 – 300	1 – 20
Particulate	(W)	500 – 2000	0.2 – 15
Water (open sea)			
Dissolved	M	10 – 40	0.2 – 2
	W	1 – 15	0.1 – 2
Particulate	M	80 – 90	0.2 – 5
	W	2 – 10	0.1 – 3
Sediments			
Estuarine		40 – 90	1 – 5
Coastal		2 – 10	0.6 – 3
Deep Sea		1 – 2	0.2 – 1

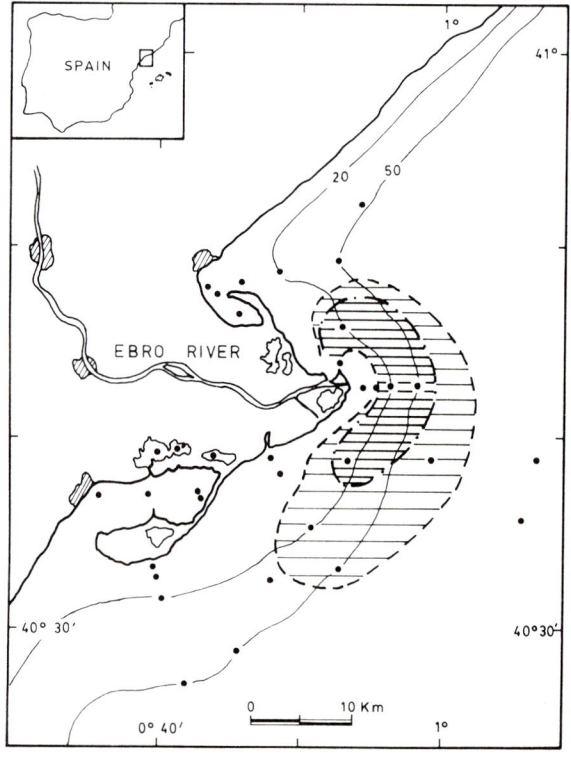

Fig. 10.3 Map of the Ebro Delta, showing the grid of sampled stations. Dark and light dashed zones indicate, respectively, distribution of allochthonous n-alkanoic acids and hydrocarbons in sediments.

Probably the major source of these acids are esters from epicuticular waxes of terrestrial plants (Eglinton and Hamilton, 1967). In fact, intact wax esters and also the corresponding n-alkanol moieties in the $C_{22}-C_{32}$ range have been recognised in several surficial sediment samples from this area (Grimalt and Albaigés, unpublished results). The spatial distribution of the acids defines the area of precipitation of riverine inputs, in the zone of mixing waters, following the direction of the dominant sea-water current (NE−SW).

Algal lipids on the other hand contain esterified acids in the $C_{12}-C_{20}$ range and are probably the major source of these compounds in the open sea. As shown in Figs. 10.2C and D, there is a close relationship between the distributions of n-alkanoic acids in particulate and sediment samples collected in the Balearic basin. Nevertheless, while in the suspended particulates they are basically in the esterified form, they are already hydrolysed in sediments. The free acids contain the major proportion of allochthonous acids, as would be expected from the more advanced stage of degradation of this source.

Hydrolysis of bacterial cells is known to liberate further quantities of low molecular weight fatty acids, containing such specific markers as iso- and anteiso-$C_{15}-C_{17}$ alkanoic and alkenoic acids (e.g. 15-methylhexadec-9-enoic acid). Bacterial activity may also be responsible for the depletion of n-alkenoic acids observed in samples from the open sea (Boon and DeLeuw, 1979).

Alkanes

Although structurally related to the n-alkanoic acids, *N-alkanes* are much less abundant in the marine environment (Table 10.5). The distributions of n-alkanes found in the Mediterranean are illustrated in Fig. 10.4. They are mainly characterized by an odd−even carbon number predominance in the $C_{25}-C_{35}$ range, indicating a predominantly allochthonous input from higher plant waxes (Eglinton and Hamilton, 1967). In addition, two different smooth distributions centred, respectively, at $C_{22}-C_{24}$ and $C_{28}-C_{30}$ occur in some samples. The first one (Fig. 10.4C) has been identified as detritus of microbially altered algal residues (Simoneit, 1981). This sample corresponds to that shown in Fig. 10.2B, for which a bacterial n-alkanoic acid distribution was proposed. The second distribution (Fig. 10.4D) is identified as recently 'bio-converted' short-chain alkanes from environmental petroleum contaminations (Gassmann, 1982). The former has been found in some sediments (Grimalt, 1983), while the latter is very common among dissolved hydrocarbons in open sea areas (Jullien, 1982; Hô, 1982) and could be related to the chronic pollution of the Mediterranean Sea by fossil oil products.

In contrast, with n-alkanoic acids, algal inputs are not reflected in these n-alkane distributions as would be expected for example by a predominance of n-C_{17}. This may be due to the preferential biodegradation of the shorter chain n-alkanes (Giger *et al.*, 1980) or to low levels of algal productivity. However, the occurrence of other markers confirms the evidence of autochthonous inputs. Thus, heneicosahexaene (IV), which has been isolated from algae (Lee and Loeblich, 1971), has also been detected in various sediments from the Mediterranean Spanish coast (Albaigés *et al.*, 1984a). Analyses of particulate n-alkanes from the same area confirm this evidence (Fig. 10.5).

The ratio of the two isoprenoids pristane and phytane (also shown in Fig. 10.5) appears to result from additional biogenic inputs (Keizer *et al.*, 1978), since pristane is the dominant isoprenoid in plankton (Blumer, 1963). However, pristane and phytane exhibit a more equal distribution in fossil hydrocarbons. The presence of an unresolved complex mixture (UCM) of hydrocarbons in gas chromatograms has been considered to be a strong indication of anthropogenic inputs (Farrington and Tripp, 1977).

Fig. 10.3 shows the spatial distribution of the concentrations of these unresolved hydrocarbons in the Ebro delta (as we have for the n-alkanoic terrestrial acids). The inner hatched zone indicates the area

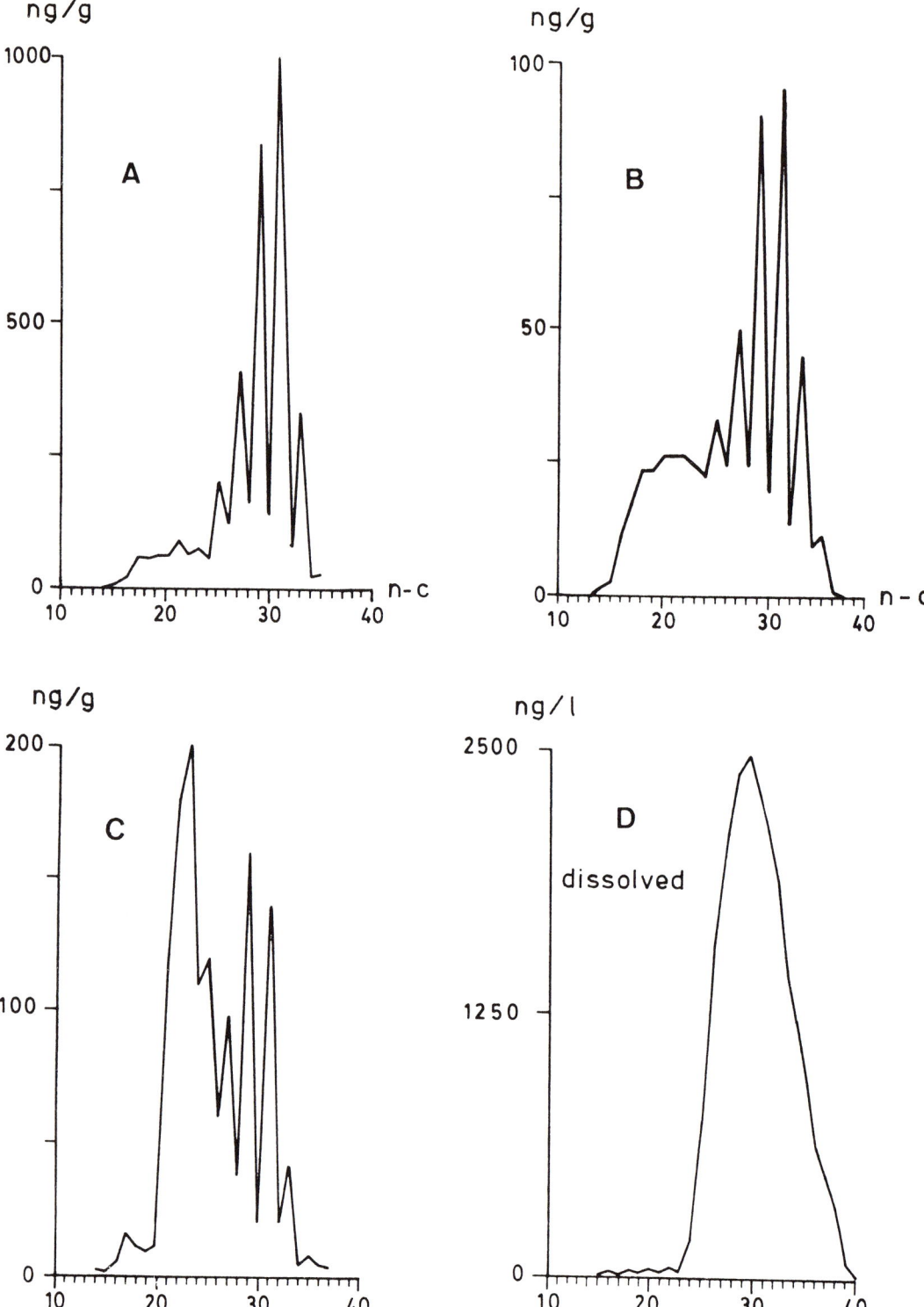

Fig. 10.4 Carbon number distribution of n-alkanes in samples from the Western Mediterranean. (A. coastal sediment; B. deep-water sediment; C. estuarine sediment; D. dissolved in water).

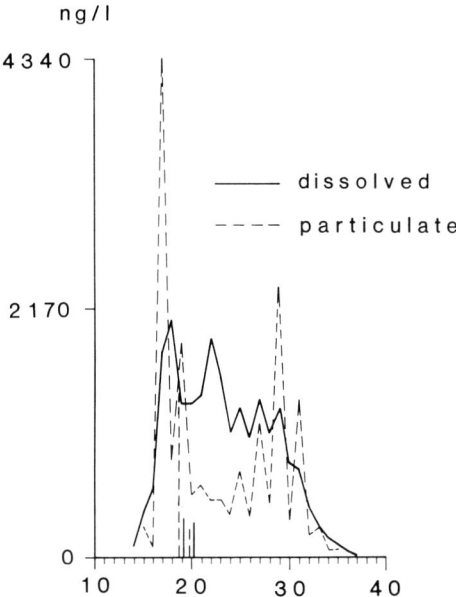

Fig. 10.5 Distribution of n-alkanes in dissolved and particulate samples collected offshore from the Spanish coast. Vertical bars indicate pristane (——) and phytane (----).

where the concentrations of UCM in sediments ranges between $5-10$ μg/g, and the outer one where the samples were only $1-2$ μg/g.

The overlapping of the patterns suggests a common origin — riverine transport, although the area of influence is slightly larger in the case of hydrocarbons. This is probably because anthropogenic hydrocarbons (oil pollutants) are preferentially concentrated in the dissolved phase and predominantly associated with the finer particles (Albaigés et al., 1983) and would therefore be expected to be transported far from the pollutant source, as is the case with measurements made in the Mediterranean (Burns and Villeneuve, 1982; Hô and Saliot, 1982) and in the Atlantic (Boehm, 1980).

The petrogenic origin of the unresolved hydrocarbon mixture can be easily confirmed from the composition of minor molecular marker compounds present in it, such as steranes (I) and $\alpha\beta$-hopanes (II). The use of these fingerprints is a good example of the way in which structural and stereochemical data can indicate the source of organic inputs into the marine environment. The availability of computerized gas chromatography-mass spectrometry is highly advisable for this method (Albaigés and Albrecht, 1979). In Fig. 10.6 the profiles corresponding to two pentacyclic triterpane series, the hopanoids (m/e 191) (II) and fernenes (m/e 243) (XVII) are presented. The former is complex containing a mixture of $\beta\beta$-hopenes, $\beta\beta$-hopanes and $\beta\beta$-hopanes, derived from the diagenetic degradation of diplopterol or polihydroxybacteriohapanols (Ourisson et al., 1979), as indicated in Fig. 10.6. The major diagenetic transformation of these hopanoids is the generation of $\alpha\beta$ and $\beta\alpha$ counterparts from the natural and less stable $\beta\beta$-isomer (Ensminger, 1977). The presence of such components in sediment extracts is invaluable for the identification of hydrocarbon inputs of bacterial or fossil origin.

Until recently, the presence of fernenes in sediments was thought to be associated with some terrestrial plant inputs (ferns) (Brassell et al., 1980), but the identification of these compounds in bacteria accords with their sedimentary occurrence (Brassell, personal communication). The predominance of the natural 9,11 fernene isomer against the more stable Δ^8 reflects the recent deposition of this type of compound.

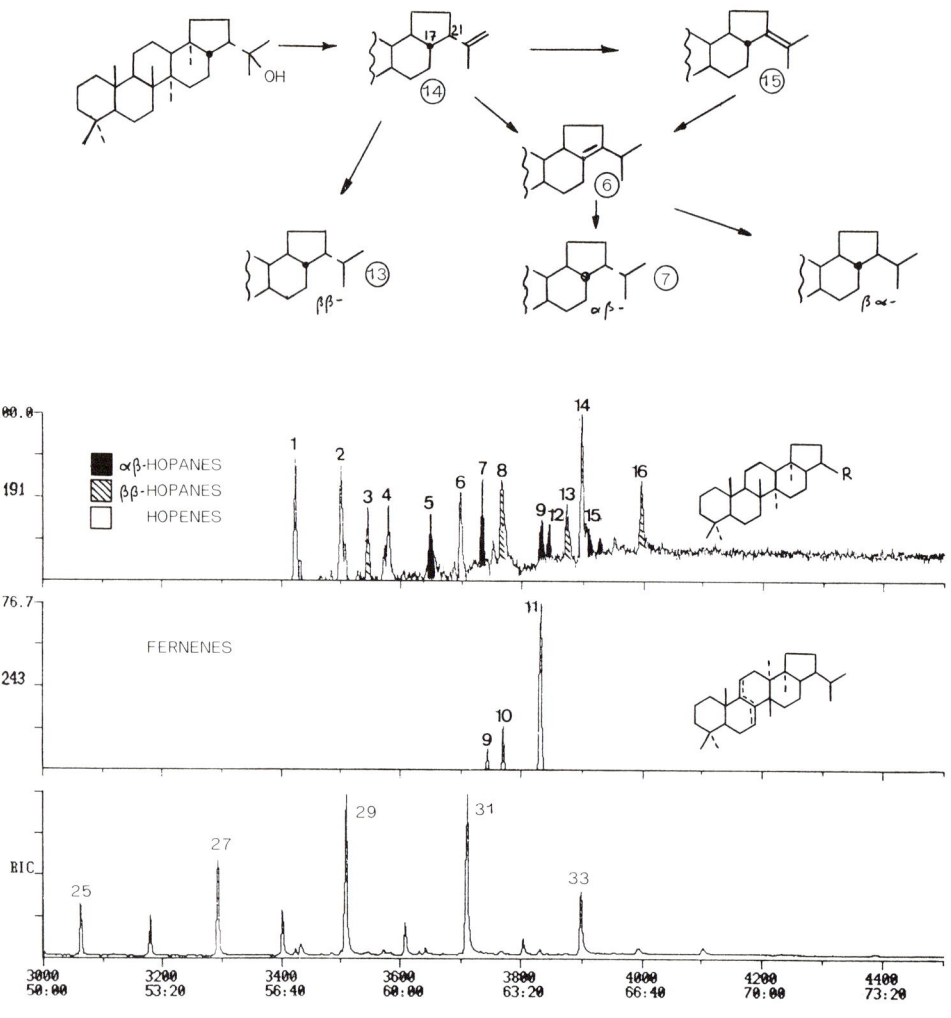

Fig. 10.6 Mass fragmentogram of the pentacyclic triterpanes, present in the hydrocarbon fraction of the inferior graph. Identifications: 1. — trisnorneohopene; 2. — trisnorhop-17(21)-ene; 3. — 17β(H)-22,29,30-trisnorhopane; 5. — 17α(H), 21β(H) -30-norhopane; 6. — neohop-13(18)-ene; 7. — 17α(H), 21β(H)-hopane; 8. — 17β(H),21β(H)-norhopane; 9,12. — 17α(H),21β(H)-homohopane; 11. — fern-7-ene; 13. — 17β(H),21β(H)-hopane; 14. — hop-22(29)-ene; 15. — hop-21-ene; 16. — 17β(H),21β(H)-homohopane.

The evidence for microbial metabolism (provided by lipid distributions) is twofold. First, the presence of specific lipids (which are only biosynthetised by bacteria) can be used as organic markers. Secondly, certain sedimentary lipids are recognisable products of the bacterial degradation of other lipids. The wide occurrence of these indicators in the marine environment has proved the important role of bacterial activity in the biogeochemical cycle of the marine organic matter.

Sterols

Sterols are excellent organic markers for they are abundant and stable in the marine environment. The variety and specificity of their structures and the extensive literature concerned with sterol compositions of marine and terrestrial organisms add to their value (Brassell, 1980). The sterol distributions in the

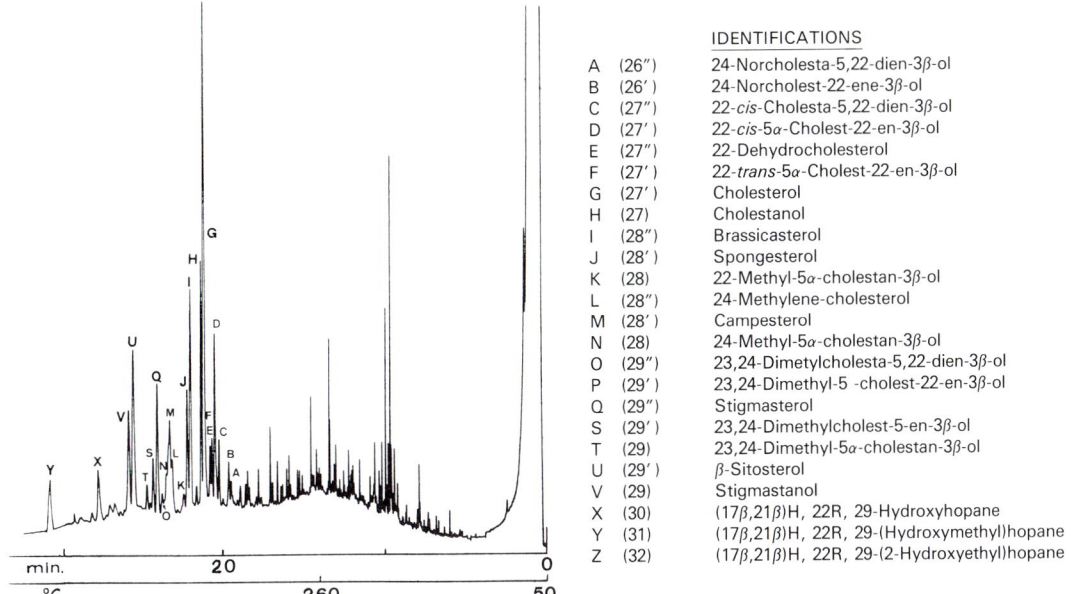

Fig. 10.7 Gas chromatogram of acetylated sterols from a Western Mediterranean coastal sediment (Spain).

marine environment are often highly complex, because of the contributions from a variety of biota (Lee et al., 1979; Brassell and Eglinton, 1983).

Sea-water sterols, either dissolved or associated with suspended matter, in the Mediterranean have until recently been poorly investigated (Saliot and Barbier, 1973). However, the application of high resolution analytical techniques has enabled analysis to be made of such complex samples. Fig. 10.7 shows a gas chromatographic profile of the sterol fraction isolated from a Mediterranean coastal sediment (Bayona and Albaigés, 1982).

Certain compounds in this profile can be identified with inputs from specific organisms. Huang and Meinschein (1976) proposed that the carbon number distribution of total sedimentary sterols could be used as an indicator of the relative input of detritus, from terrestrial and aquatic biota (in which C_{29} and C_{27} sterols, respectively, dominate). Accordingly, the distribution shown in Fig. 10.7 should reflect a dominant marine input. Both C_{27} and C_{29} Δ^5-sterols (G and U) appear also in the reduced form, the $5\alpha(H)$-stanols (H and V). As the n-alkanoic acids, these stanols may be derived from the corresponding sterols by microbially-mediated hydrogenation or may merely be the result of the selective degradation of sterols in an oxic depositional environment (Nishimura, 1981). Nevertheless, recently, Volkman et al. (1981) have suggested that there may be a contribution from autochthonous organisms.

Campesterol (M) and brassicasterol (I) have been considered, respectively, to be indicators of algal and diatom inputs. More specific marine sterols are those exhibiting multibranched side chains, such as 23,24-dimethylsterols or stanols (Volkman et al., 1981; Brassell and Eglinton, 1983). There is also a remarkable specificity of dinosterol ($4\alpha,23,24$-trimethyl-$5\alpha(H)$-cholest-22(E)-en-3β-ol) with dinoflagellates (see Table 10.4).

The sterol fraction also contains a series of components with a $\beta\beta$ hopanoid structure (compounds X-Z). These are the diagenetic products of bacteriohopanols and the precursors of the hopenes and hopanes (Fig. 10.6). The degree of preservation of such compounds depends upon depositional conditions. In any case they are indicative of bacterial inputs.

Sterols are sensitive to diagenesis and may undergo many different transformations during the early and late stages of deposition (McKenzie et al., 1982), which may afford information about paleoenvironmental conditions. In estuarine areas of the Spanish Mediterranean coast (where the prevailing conditions are slightly anoxic) the alkane + alkene fraction contains a great variety of sterenes (Δ^2—) and steradienes ($\Delta^{3,5}$—) (Albaigés et al., 1984b). However, detailed considerations are beyond the scope of this chapter.

10.2. A FEW PRELIMINARY WORDS ABOUT POLLUTION

According to an official United Nations definition, marine pollution '...is the introduction by Man of substances or of energy, into the seas, including estuaries, which can bring about deleterious effects such as: damage to biological resources, danger to human health, deterioration in the quality of sea-water from the point of view of its use and reduction of the possibilities offered in the field of leisure activities'.

This definition eliminates natural phenomena which can, locally and accidentally, modify the ecology of extensive marine zones. There is therefore no question, for example, of considering the discharge of considerable quantities of alluvium by flooding rivers as 'pollution'. Real pollution can, however, result from changes made by Man to the structure of shorelines or river banks, changes which divert or harm the natural phenomena and therefore cause ecological disorders which are sometimes very serious. But it can be even more complex: thus in the example, one can easily conceive that accidental or artificial discharge of clay-bearing earth can pollute clear coastal waters and totally lose their characteristics if the discharge takes place at the mouth of a river which naturally carries down clay or clay-bearing earth.

There is another factor which does not appear in the official definition: the question of time and where and when the pollution occurs. Certain pollutants have a short and reversible action, while others bring about profound and long-lasting changes in the seas. So, the study of pollution must take into account temporal factors, as well as the spatial effects which are usually considered. The persistence of these effects affect future ecology and the human environment of the seas, in particular that of the Mediterranean.

The multiple nature of pollution should also be stressed, for it is essential for a proper understanding of the effects of pollution. The polluting agents can be by 'substances' or by 'energy'. Experience shows that these terms cover a variety of agencies, which can be grouped into five classes: discharge of cooling water from factories (thermal pollution), discharge of urban waste water (bacterial pollution), discharge from nuclear plants (potentially leading to a radioactive pollution), discharge of industrial water (chemical pollution) and massive discharge of organic matter (which can produce profound local ecological changes). These sources of pollution are generally of land origin and persist as more or less permanent coastal pollution. On the other hand, discharge by barges, or accidental spillages at sea, or offshore drilling are generally isolated pollution events which generally occur far from the coast.

In this chapter the nature and consequences of the different types of attack on the natural environments are considered, both in relation to the ecology of the environment and to the health and well-being of man.

The preceding chapters have described the basic physical, chemical and biological processes which affect the diffusion, transport and concentration of marine pollutants. The circulation of surface water is of particular importance (Fig. 4.6). It is also important that during many months of the year, a discontinuous interface (thermocline) occurs at a depth of between 25 and 30 m depth. This creates a real barrier to dispersion. It should be remembered that the continental shelf is very narrow and in many Wester Mediterranean coasts is virtually non-existent.

TABLE 10.6. Organic Load of Domestic Sewage for the Northwest Basin

Country	D.B.O.* ton/year	Phosphorus ton/year	D.B.O., kg. per year and km of coast	P, kg. per year and km of coast
Spain	130 000	5 000	60	2.7
North-West Basin (France)	360 000	16 000	336	15.0
Italy	400 000	18 000	61	2.7

* Calculated according to the assumption that the annual organic load in domestic waste is 20 to 25 kg/person.

Approximately 100 million people live on the coasts which surround the Mediterranean and contribute to an estimated daily discharge of waste water of some 3×10^7 tons. To this must be added approximately one km^3/day of polluting discharge from inland waters and rivers for the whole Mediterranean (Table 5.2). In 1972 a group of experts established the use of the Biochemical Oxygen Demand (DBO)* and the level of phosphorus as measures of the polluting organic load to assess the polluting inflow from the various Mediterranean countries.

The highest levels occurred in the north-west basin (Italy and France) (Table 10.6). This urban and industrial inflow is, for the most part, discharged into the waters of the Mediterranean without any effective purification. We must therefore consider the origins of these pollutants, their specific dynamics in the aquatic environment, their influence on marine life and their effects on the human environment. We can then ask how far we can go with such a continuous pollution before creating an irreversible situation for Mediterranean life, and what are the necessary remedies.

10.2.1. The sources of chemical pollution in the Mediterranean

For many years it has been known that Mediterranean water and fish contain some heavy metals whose origin was routinely attributed to pollution. However, it is now realised that natural sources exist which may be related to the geological structure of the basins. We must therefore be cautious in interpreting the analytical results and not attribute them systematically to human agencies when they may arise from natural causes. Mercury from volcanic activity, or the washings from land rich in cinnabar can, for example, explain the high concentrations that are measured in the lee of or downstream from the geological sources.

With these reservations in mind it nevertheless appears that pollution largely comes directly from large industrialized and urban areas, to the waters where the highest levels of chemical or microbiological pollutants are found. Offshore water (far from the most populated areas) is generally of good quality, although some pollutants spread easily across the surface layer. The greater part of urban and industrial waste waters are discharged directly into the sea, and is a major source of pollution, although the immediate transformations of the pollutants are not completely understood.

Our knowledge of the distribution of the most important pollutants in the Mediterranean is adequate to construct a summary table and draw approximate maps of the chronically-polluted areas. The knowledge has been gained in a series of oceanographic expeditions. In particular we shall use data collected during coastal expeditions, such as the National Inventory of French Coastal Pollution (carried

* The DBO is the consumption in oxygen (expressed in mg/l) of the microorganisms present in the environment necessary to stabilise the organic substances present in this environment.

out from 1965 to 1975 by the CERBOM), the National Enquiry on the sanitary condition of the seacoast areas (carried out from 1974 to 1977 by the same laboratory), the National Observation Network of the marine environment (carried out by the CNEXO since 1975) and also surveys of the open sea by various oceanographic ships, notably: the *Jean Charcot*, the *Noroit* and the *Suroit* (for the CNEXO), the *Calypso* (for PNUE), the *Noeric II* (for the CERBOM).

Chemical pollution. Heavy metals

Mercury. The concentrations of this heavy metal are higher in the water of the Tyrrhenian Sea and Liguro-Provençal Sea (around 100 ng/l on average) than normal levels in ocean water (7 ng/l). Relatively low concentrations are found in the Atlantic current, offshore from North Africa. Areas of higher concentration appear to be associated with natural outflows from particular geological formations or from industrial sources. The very high values (up to 1700 ng/l) found three to four miles offshore from the Rhône delta (far from the much lower concentrations measured in the water of low salinity, closer to the coast) are difficult to explain.

Cadmium. The normal concentration in oceanic water is around 0.1 μg/l. This metal seems much less influenced than others by natural inflows. From relatively low levels (0.1 to 0.5 μg/l) some concentration appears (up to 2.7 – 2.8 μg/l) in marine areas close to centres of industrial activity (e.g. Naples and Massif des Maures).

Chromium. The normal concentration in ocean water is around 0.3 μg/l. The levels of this metal are very variable, with high contents predominating in the Liguro-Provençal Sea. The inflow from the rivers and certain industrial areas show marked increases of 3 to 10 μg/l.

Nickel. The distribution of nickel is similar to that for chromium, although the normal value for sea-water is higher (1.7 μg/l) with higher concentrations in the Liguro-Provençal Sea and the Gulf of Lions (between 1 and 10 μg/l on average), and very low figures for the Tyrrhenian Sea. Significant increases can be observed offshore from the mouths of certain rivers and coastal streams including branches of the Rhône, Var and Roya.

Copper. The distribution of copper is very different from that of the other metals. Offshore from the Spanish mainland coast, in the Gulf of Lions and in the north of the Tyrrhenian Sea, low values (from 1 to 5 μg/l) were measured. These figures are comparable to the normal concentration of 0.5 μg/l. Fairly high concentrations (5 – 10 μg/l) occur in certain offshore areas (in the Gulf of Valence, off Cap Creus and between Corsica and Provence). The highest levels were found around the Balearic Islands (between 20 and 110 μg/l). These high concentrations go beyond those resulting from telluric discharge. A significant, but not general, increase is found at the mouths of rivers and streams, except where there are agricultural irrigation or drainage canals and ditches (as in the south of Marseille or in the Gerace and Siderno rivers in Calabria). These concentrations rarely exceed 10 μg/l, except in the very immediate approaches to the mouth of the Grande Rhône. Some volcanic influence can be deduced from the concentrations of 7 – 19 μg/l measured in the approaches to the Aeolian Islands, as compared with 0.5 – 2 μg/l recorded further offshore from these islands.

Lead. Recent estimates of the normal content of sea-water are very low (2 ng/l), but that dissolved in river water may average 2 – 3 μg/l. Frequently values of between 0.5 and 2 μg/l have been reported for surface waters. Higher values are due to telluric discharges (around Corsica) or are a consequence of man-made pollution (as in the offshore waters off the mouth of the Rhône, the Bay of Marseille and offshore of Naples, up to 10 μg/l).

Zinc. As with lead, zinc does not follow a tidy pattern of distribution. The normal content of sea-water is 4.9 μg/l and the concentration over most of the Mediterranean is between 5 and 20 μg/l.

Certain areas, however, have higher levels (as for example the south of the Gulf of Genoa, between Corsica and Provence) which may be more than ten times that of normal waters. In coastal waters, the relatively high levels seem to have dual origin: industrial discharges, for example, in the area of Sète (to 50 µg/l), and possible volcanic influence around the Aeolian Islands and east of Sicily (30 – 45 µg/l). Higher local concentrations are difficult to interpret.

Iron. Concentration of dissolved iron in the world ocean may be as low as 2 µg/l, but our measurements are on a continuous gradient from 10 to 25 µg.l, from Gibraltar to the Tiber. Peaks may result from telluric inflow, as in the island of Elba (360 µg/l), the Balearic Islands (190 µg/l) and offshore of the Rhône, Tiber and other rivers (40-60 µg/l).

Manganese. Our analysis concerning this metal are limited (as with iron) and extend from Gibraltar to the Tiber. The concentrations are low, from less than 1 to 6.7 µg/l as compared with mean concentration for oceanic water of 0.2 µg/l.

The distributions of the different heavy metals can be related to putative input sources and to movements of the water. Increases due to input from land are obvious. The sources may be natural (rivers, streams, volcanic fall-out and undersea mineral deposits) or man-made (mining, agriculture, industry, urban effluents and polluting atmospheric fall-out).

For most of the metals, high concentrations may be related to the presence of run-off from land, but certain sources may be more specific. Volcanoes may be important sources for copper and zinc and for mercury (despite its high volatility). Areas of mining activity increase the contents of mercury, iron and, perhaps, lead.

The circulation of water controls the spread and distribution of heavy metals. The entering Atlantic water appears to contain low concentrations (close to the accepted values for the normal composition of sea-water). Inside the Mediterranean the highest levels are often found in more stable and converging areas (for example in the centre of a triangle made by Corsica, Tuscany and the Alpes Maritimes). The comparison between the concentrations of metals and the pattern of circulation in this area, shows similarities between the distributions of the metals and the two cells of circulation in the Ligurian Sea, as well as with the Ligurian current itself.

The information provided by analyses of the elements is important, but still inadequate. The nature and properties of the molecules or chemical species is decisive, and the presence of possible organometallic compounds is important. Furthermore, the extremely high concentrations of heavy metals must be temporal and local, because the chemical equilibria existing in sea-water should lead to recovery to normal levels, eventual accretion of the solid phase into new mineral deposits (as for example with fero-manganese nodules present mainly in the depths of the Pacific Ocean).

Deviations from the usual concentrations of elements in the environment cause changes in both the metabolism and the composition of organisms. Heavy metals may be concentrated to very high levels in some organisms. Contents of mercury and lead are generally higher in fish from the Mediterranean than from elsewhere and can achieve levels that may be noxious for humans (Tables 10.7 and 10.8; Fig. 10.8).

Pesticides and polychloride biphenyls

Synthetic chemicals pollutants, such as the chlorinated hydrocarbons, are more concentrated near the coasts, as would be expected from their agricultural use. They are transferred to the marine environment through the atmosphere or by drainage. The concentrations in 11 coastal samples averaged 13 mg/l, with a maximum of 38 mg/l, as compared with values of less than 0.2 to 8.6 mg/l (with an average of 2 mg/l) for 80 samples from the open sea.

TABLE 10.7. Level of Mercury in Mediterranean Animals. Average Values by Groups of Species and by Areas, in ppm (mg/Kg, wet)

Area of sampling	Analysis made by	Tuna and swordfish	Sharks and rays	Sardine and anchovy	Other fishes	Invertebrates (squid, shellfish, shrimp)
Western Mediterranean	Food Toxicol. Service Min. Agricult. (G. Curmont et al., 1972)	1.7			0.34 – 0.90	
Gulf of Lions and coast of Provence	Marseille Veterinary Labor. (G. Gilles et al., 1973 – 1974)	0.52 – 4.10	1.18	0.15 – 0.24	0.21 – 0.39	1.17
Offshore Cagnes	C.E.R.B.O.M. (M. Aubert et al., 1969)			0.46	0.31 – 0.99	2.58 (mussels)
Offshore Monaco	id. 1973		2.10	1.32	0.26 – 2.41	0.20
Offshore Cap delle Mele	id. 1973		1.30		0.32 – 1.50	0.05 – 1.82
Offshore Genoa	id. 1973		2.10		0.40 – 3.21	0.10 – 0.92
Offshore Sestri Levante	id. 1973			0.59	0.66 – 1.98	0.61 – 1.00
Offshore La Spezia	id. 1973				0.51 – 0.81	0.10 – 0.53
Offshore Viareggio	id.				0.14 – 1.73	
Offshore Livorno	id. 1973		2.40		0.29 – 1.60	0.51 – 0.85
Offshore Rosignano (Tuscany)	id. 1973				0.5 – 4.64 (grouper)	2.9 – 3.26
Offshore Cecina (Tuscany)	id. 1973				0.52 – 2.88	0.34
Offshore Porto-Vecchio	id. 1973 – 1974				0.05 – 0.25	
Offshore Bastia	id. 1973 – 1974				0.11 – 0.25	
Offshore Saint-Florent	id. 1973 – 1974				0.10 – 0.42	
Offshore Alistro	id. 1973 – 1974				0.11 – 0.60	
Cap Corse	id. 1973 – 1974		1.70		0.22 – 0.64	
40 naut. m. N Cap Corse	id. 1973		3.99 – 6.48		0.63 – 1.71	0.01 – 2.26
Ligurian Sea	Viviani		1.7 – 4.0	0.50		0.68 – 1.80

According to Marchand, Vas and Duursma (1976) the level of DDT has noticeably increased in mussels from the coast of the Gulf of Genoa, a region which specializes in the cultivation of flowers. The passage of long-lived pesticides is not without problems for the survival of certain species and for their possible human consumption. However, it seems that the concentration of these substances does not increase in the successive steps of the food chains, but, on the contrary, may decrease at the higher levels which are of interest for human consumers.

Oil products

Oil is a relatively frequent pollutant in the Western Mediterranean. It tends to be localized in harbour areas and in tanker discharge bays close to coastal refineries. Unfortunately, certain authors have provided data (based on comparisons between polluted and non-polluted samples) which seem to us to be highly exaggerated, and which do not take into account statistical constraints or rational sampling methods.

The marine zones most subject to pollution are situated seaward from oil-bases and the points of loading and discharge. They are principally in the north-western basin: close to Barcelona (for the Spanish coast) in the Gulf of Fos (for the French coast) and the Gulf of Gênes (for the Italian coast).

TABLE 10.8. Average Levels of Mercury, by Species, in the Mediterranean. All Zones Mixed

Species	Number of Samples	Average Levels of Mercury (mg/Kg $a$ppm)
Red tuna*	270	1.20
Dogfish	25	1.88
Conger eel*	10	1.30
Ray*	3	2.61
Picarel	5	0.38
Bogue	7	0.41
Sea-perch	3	0.40
Stingfish	5	1.47
Sar	3	0.58
Common bass	8	0.70
Grouper	18	0.63
Sole*	5	0.60
Hogfish	15	0.49
Whiting*	21	0.62
Bream	9	0.41
Red Mullet*	12	1.44
Becker	5	1.38
Octopus	18	0.28
Shrimp	10	0.46
Crab*	8	1.86
Coalfish	12	0.75
Chimaera	3	1.20
Prawn*	8	1.04
Cuttlefish	9	0.24
Bonito	6	0.52
White tuna*	6	0.71
Swordfish*	4	2.96
Sardina	10	0.15
Mackerel	6	0.21
Anchovy	13	0.24
Oyster	10	0.19

* The species marked with an asterisk are those which go beyond the levels authorized (0.7 ppm for tuna; 0.5 ppm for other species).

According to Lelourd (1978) the total quantity of hydrocarbons which are introduced into the Mediterranean is estimated to be between 0.5 and 1 million tons per year (0.5 g/m^{2}year).

10.2.2. Microbiological pollution

Bacterial pollution is influenced by local meteorological, oceanographical, and biological conditions. The most polluted coastlines border the areas of highest population density and industrialization, with a worsening of the situation offshore from large towns.

Along the North African coasts the polluting inflow is considerable (particularly close to urban developments) because waste is discharged directly onto the shore, or because the sewage disposal pipes are too short. Bacterial pollution is still greater in the northern part of the Western Mediterranean, along the Spanish, French and Italian coasts.

According to Ambrosiani in 1970, 90% of the localities on the Italian coast was discharging untreated domestic waste water, thus producing very high bacterial levels around certain cities (Genoa, Leghorn, Piombino, Fiumicino, Naples, Salerno). As an example, the average values of faecal coliforms for the

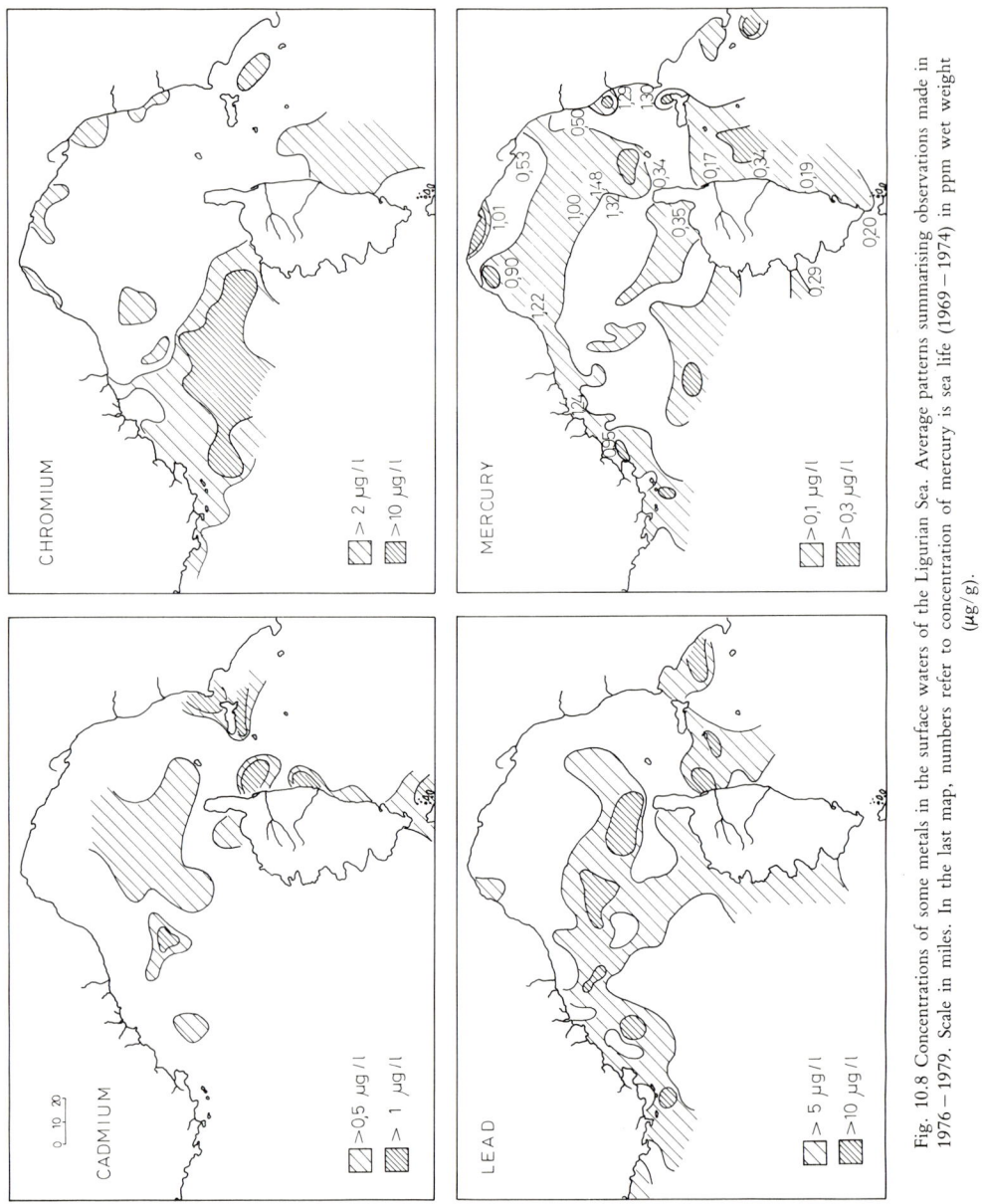

Fig. 10.8 Concentrations of some metals in the surface waters of the Ligurian Sea. Average patterns summarising observations made in 1976–1979. Scale in miles. In the last map, numbers refer to concentration of mercury is sea life (1969–1974) in ppm wet weight (µg/g).

most polluted areas of the Gulf of Naples are between 100,000 and 400,000/100 ml. (according to the data gathered by Mendia and co-workers in 1980).

The same situation occurs along the coast of Sicily. Thus, in certain points in the Messina region, Genovese found 250,000 coliforms per litre near the town of Messina, although there is no bacterial pollution in the middle of the straits. Except for certain port zones, such as Syracuse and Catania, the coast of Sicily is relatively free from pollution. The situation is the same in Sardinia (except for some ports, particularly Cagliari) where bacterial density drops rapidly within a relatively short distance.

The Spanish coast, for most of its length, largely faces open sea, and in the east has vast sandy beaches. Their contamination is generally very slight, except for the immediate proximity of some coastal towns such as Valencia, Alicante, Tarragona, and, particularly, Barcelona, where the number of *Escherichia coli* is high close to the harbour (30,000 to 50,000 faecal coliform per 100 ml).

Along the French coast, the situation has improved enormously during the past decade. However, at Languedoc and Roussillon, contamination by faecal organisms is increasing locally, especially near towns such as Sète and near rivers which carry domestic waste from the interior of the country, notably: Perpignan, Narbonne, Beziers, Montpellier and Aigues Mortes. According to the data of CERBOM, for 1975, 3000 faecal coliforms per 100 ml were found in the mouth of the Hérault and in the Grau du Roi, as well as near the industrial zones of the Etang de Berre. Bacterial pollution is even greater near Marseille, where the principal sewage discharge (between the island of Riou and the coast, as well as to the mouth of the Huveaune) results in bacterial counts of 100,000 per ml. The situation is better around Toulon, although there are some polluted spots and waste water is discharged in poor conditions. Along the Côte d'Azur, the situation is good in tourist zones (with rather low population densities) but deteriorates rapidly towards coastal towns and in areas of high population density where waste water is discharged through short outlets. In spite of increasing population, the introduction of rational sanitation techniques and, in particular, the installation of collectors which extend far from the coast, have spectacularly improved the situation. The situation in the beaches of Cannes and Nice, for example, can now be considered satisfactory.

This then is the situation concerning bacterial pollution in the Western Mediterranean. Many of the data that have been presented have not been acquired simultaneously, and different areas have been subjected to recent changes and developments. However, the general outline is still valid and can be applied to the Western Mediterranean as a whole.

Bacterial pollution is directly correlated with human population densities, all along the coastlines. The sea-water close to the coastal towns contain enteric bacteria, which, in the worst cases, extend at considerable densities for up to 6 to 7 km out to sea. The bacteria spread in the surface layers and in the direction of the dominant currents. The Ligurian current, for example, carries pollution from the Gulf of Genoa towards the westward lying coasts. Bacterial pollution always declines sharply offshore, and remains higher in the marine areas that are naturally more stable, or which have been artificially modified by restructuring of the coastlines. Micro-organism pollution does not only involve bacteria; viruses contained in run-off and in sewage can also be dangerous, and the interstitial waters on shores and beaches (commonly enriched with nutrients) may act as reservoirs for numerous bacteria and viruses. Unfortunately, no systematic information exists about the distribution of dangerous organisms in such environments.

Urban waste water contains not only micro-organisms, but also considerable quantities of organic matter. On a single stretch of the northern shore, between the rivers Arno and Ebro, it is estimated that 336 tons of organic matter are discharged per kilometer of coast, per year. The relative input is lower for the rest of the Western Mediterranean, but is still very substantial.

10.2.3. Thermal pollution

This factor is, historically, a recent pollutant. The hot water used in the refrigeration of pure thermal or nuclear thermal power plants is discharged at sea causing local increases in temperature at the outlet of as much as 7°C. Furthermore, anti-fouling substances may be used to kill marine organisms inside the water conducting systems. These types of pollution may act on the local communities, but the effects are usually restricted to a very short distance from the points of discharge. Even though the impact is limited, the increasing number of new power plants, especially in the northern coast of the basin (11 in Spain, 2 in France, about 50 in Italy, 1 in Tunisia) make it prudent to maintain vigilance in the future.

10.2.4. The consequences of pollution

Which are the dangers following pollution of the kind observed in the Mediterranean? They concern, on the one hand, the ecosystems and on the other, the health of the people in the countries bordering the sea. These effects depend upon the type of pollutants. On the basis of present knowledge, the dangers can be evaluated as follows.

Microbiological pollution

The extent of microbial populations at sea are determined by an equilibrium between input and the opposing forces of diffusion, inactivation and progressive destruction.
Fate of the bacterial discharged in the sea. Some bacteria may be dispersed in water, but most attach to mineral or organic particles. The small fraction of bacteria attached to particles of relatively large size and high density settle most rapidly. The major portion (98.5%) however, attach to small particles (of less than 20 μm) and can spread in the surface water. In these superficial layers the bacteria drift with the water and disperse. Theoretically, there should be a decrease in bacterial concentration the greater the distance from the source. However, the decrease in polluting bacteria is much faster, possibly due to the bacteriolytic actions of the marine environment. Predictions of bacterial dilution must therefore be corrected by a coefficient, expressing this bactericidal effect. The sensitivity of bacteria to marine agents must be included in mathematical expressions (Aubert *et al.*, 1981) designed to estimate the bacterial load in sea-water, according to distance from the source, the speed of currents and the sensitivity of bacteria to specific actions of sea-water. Accordingly, it is possible to choose discharge points in areas from which the polluted water could be led away, far enough out to sea, so that the self-purifying action of the environment can act effectively before the same water flows back towards other coastal areas.

Much work has been done to study the destruction of bacterial pathogens at sea. The sea is an inimical environment for terrestrial microorganisms and for those that infect terrestrial animals. Sea temperatures, high salinity and the low concentrations of essential organic substance are not ideal conditions for development. Ultra-violet radiation is also strong in the surface layers of water. However, it seems that essentially biological phenomena may be more important bacterostatic agents. These act differently according to the zone. For simplicity, we must consider three zones: the area of discharge or the estuary of a river, deeper regions, where bacteria attach to particles and an external one, in which they drift with superficial waters.

In the first, or estuarine zone, bacteriophages and *Bdellovibrio* destroy large numbers of bacteria. A number of animals also feed on discharged bacteria.

In the deep zone (i.e. the layer of sediment and the water above it) there is the antagonistic action of true marine bacteria. These exert various forms of competition, including chemical antagonism. There is also the effect of algae, some of which produce antibacterial substances.

In the open sea (where most of the bacteria disperse in the surface layers) there is filtering of water and retention of bacteria by planktonic animals and by larvae of benthic animals. Moreover, certain phytoplankton species release substances into the water which have antibacterial action. The species considered most important in this respect belong to classes of Diatoms and Crysophyceae (Haptophyceae). The effective substances have been identified in some cases: a fatty acid with 20 carbon atoms and an arabinosyl-nucleoside, in the common diatom, *Asterionella glacialis* (*japonica*), and acrylic acid, in the chrysophyceae, *Phaeocystis pouchetii*.

The marine environment thus has an inbuilt stability that results from the simultaneous, or alternating, play of several specific agents. These also regulate terrestrial bacterial pollution of terrestrial origin. The Mediterranean is no exception, and its overall biological properties have not been greatly modified in the course of the past century.

What are the consequences of microbial pollution for the marine environment and for man?

The discharge of waste water has an inimical effect on the local flora and fauna. Certain species are encouraged and others disappear, usually in rather limited areas.

Sanitary studies carried out in France and abroad have led to more precise evaluation of the dangers of bacterial contamination by bathing. It has been said that bathing in sea-water polluted by waste waters is practically without danger, even when they are macroscopically unclean.

The indication of pathogenic microorganisms in sea-water polluted in this manner is more useful to reveal coastal contamination than to indicate danger for bathers. A paradox exists between the statements of bacteriologists showing the frequency of *Salmonellae* in coastal waters polluted by sewage and those of epidemologists proving the extreme rarity of salmonellosis in those who have bathed in it. It is the same for poliomyelitis. In England a checked statistic of 150 cases among children living permanently on the coast does not permit us to think that bathing had anything to do with the transmission of the disease.

Certain muco-cutaneous conditions can, without any doubt be contracted following an immersion in water which is bacteriologically polluted; *Candida* vaginitis, conjuctivitis, or various benign conditions in the nose, ear and throat.

The risk to the consumer from marine products is quite another matter, as the introduction into the body through the digestive tract of shellfish or fish living in polluted water, which concentrates bacteria and virus at high levels can provoke very serious pathologic syndromes such as salmonellosis or viral hepatitis and toxi-infections or again infections linked to the marine germs (*Vibrio parahemolyticus* and *Clostridium botulinum*).

Some statistical data illustrate the above conclusions. An early statistical study carried out at the CERBOM (in collaboration with A. Piatier) questioned the hospital staff in coastal towns and the doctors of coastal departments concerning diseases whose origin could be attributed to bathing or to the absorption of marine products. The replies showed that a number of benign infections of the upper respiratory tracts appeared after bathing, while the ingestion of shellfish or of polluted fish had provoked, often serious, pathological conditions. These facts were confirmed by a second series of studies carried out in seven holiday camps (at the seaside and in the mountains, 1974, and in 1975, at a camping site on the Catalan coast): 'The observation during a period of 30 days of the morbidity of a population of 1,000 people showed a very low number of intestinal conditions and the high frequency of eruptive infectious illnesses. The risks which have appeared are certainly linked to overcrowding, the closeness of contacts, the high number of young children, than to the frequenting of a beach and sea-water, particularly exposed by reason of its proximity to the emergence into the sea of a contaminating river.'

To sum up, if microbiological pollution can be a danger for the consumer from marine products, it is without serious risk for the bather.

In addition, sanitary techniques, put into force during the last decade, have contributed to improve the bacteriological quality of the coastal waters. On this subject, we bring up here the ideas which came out of the final report of the experts at the International Scientific Symposium on the problems of bathing, in conjunction with the microbiological quality of the water which was held in Genoa (Italy) in 1981: 'It is established that during these last years in the Mediterranean, considerable progress has been accomplished in what concerns the supervision of discharges into the sea. Italian and French reports show that the majority of bathing areas henceforth respect the sanitary standards in force on the basis of the faecal coliform count, adopted up to the present as the means of evaluation of marine pollution.'

Chemical pollution

Dynamics of exogenous molecules. The fate of the chemical pollutants is related to their density, solubility and to the level of water in which they travel: some substances move with the surface layers, certain heavy products accumulate in marine troughs. Some compounds are decomposed by the action of organisms (biodegradation). Others concentrate inside particular organisms, and then move up the trophic ladder. If, in the process, their concentrations increase, some substances can exert toxic effects on the top consumer.

Industrial wastes. Industrial wastes are usually discharged through pipes, sometimes by barges or tanker ships directly at sea, at some distance from the coast. Besides heavy metals, the waste contains acids, used electrolythic baths, organic solvents, carbolic compounds, etc. The toxicity is variable and often unpredictable. Among the most contaminating industries are the assembly of semiconductors, dye-works, dairies, paper mills, production of plastics. They may have important effects on the local biomass, and also disrupt the mechanisms of interaction among species.

Detergents. A majority of the detergents are based on alkyl-benzene-sulfonate of sodium (ABS), and differ on the length of the hydrocarbon chain bound to the benzene core. Detergents are harmful because of their tensioactive properties. The coatings of foam can absorb and concentrate other polluting agents, they modify the transport of oxygen through the limiting films and the transmission of light. In certain cases they can have a direct, harmful effect on organisms, this can inhibit the development of plankton and reduce the self purifying capacity of the environment.

Legislation has forbidden non-biogradable products and has forced industry to manufacture products which degrade to a minimum of 80% (such as the sulphuric ester salts of secondary alcohols, whose degradation is, however, very slow). A third generation is coming out which, like soaps, has a non-ramified hydrophobic chain and a group of hydrophilic atoms at one of its extremities.

Hydrocarbons. These can be especially noxious in a closed sea. Land-based, hydrocarbon pollution originates principally in refineries, washing from heating ovens, wastes from fuel cracking plants and with accidental rupturing of pipelines, storage tanks and ships. Maritime-based pollution originates in an important part from washing of oil tanks in tankers, it is estimated that by this process 1% of the transported load is discharged into the sea in the form of crusts and heavy fuel, from accidents in tankers, from offshore oil wells, in which, besides possible accidents, chronic pollution always exists. In fact, a well delivering 1000 tons per day spills about 100 kg of oil daily.

The most frequently polluted places are the harbours, some parts of which can be reduced to the condition of abiotic zones; the beaches, particularly those affected by tankers discharges, become blackish sheets of crusts and heavy fuel, incorporating algae and floating refuse, and which stick to the sand and pebbles; the rocky shores, where hydrocarbons form a brownish and sticking fringe extended to the high water line, including masses of algae and dead coastal fauna. Oil brings about the destruction of plant and animal life in the pools left after high tides and storms. This form of pollution is severe,

TABLE 10.9. Examples of biological magnifications or accumulations (from C. M. Menzie, 1972)

Organochlorine insecticide	Range of magnification factor for 5 aquatic molluscs
Lindane (γ-BHC)	10 – 250
Heptachlor	250 – 2500
Methoxychlor	300 – 1500
Aldrin	350 – 4500
Endrin	500 – 1250
Dieldrin	700 – 1800
DDT	1200 – 9000

from the point of view of public health, and particularly detrimental to shellfish growing naturally or in cultures. In the high seas, we all know the sheets of drifting fuel, which build up the refuse, coats and kills seabirds in large numbers, and can be detrimental to the schools of pelagic fish.

An important aspect of this pollution is the eventual presence of carcinogenic, polybenzenic hydrocarbons of the benzo-3,4-pyrene type, that have been recognised in many marine organisms, especially in the plankton. They follow the food chains and in sardines fished in polluted zones, concentrations of more than 6 μg per 100 g of fish have been reported.

Hydrocarbon pollution is a very serious matter. Because of the need to increase energy sources, the petrochemical industry continues to develop, and huge amounts of oil and oil products are transported in larger and larger tankers. As a consequence immense areas of the sea are covered by a hydrocarbon film, of complex origin and composition, which has profound effects on marine life. The material is partly metabolised by marine bacteria to yield an input of organic matter to the ecosystem. This is a lesser ill, for oil destroys organisms in surface waters and even ulcerates the skins of benthic fish, making them unfit for consumption.

Pesticides and polychloride biphenyls. The list of substances used as pesticides is very long. It includes mineral products (arsenic, prussic acid, sulphur, fluorine and silica) or organic products (chlorine derivatives, phosphorus esters, nitrate compounds, and organo-metallic derivatives). They are used in large amounts, are widely distributed in the terrestrial environment and are finally introduced into the sea in appreciable quantities. Pesticides have low water solubility, in the order of the parts per billion, but can be concentrated by marine organisms, with short or long term effects (Table 10.9).

Heavy metals and inorganic compounds. Heavy metals are not destroyed and can be assimilated and concentrated in marine organisms. They can kill animals on the top of marine food chains, such as marine mammals, in which pathological conditions have been attributed to the concentration of such elements (Tables 10.7 and 8).

According to a FAO Report 'the following products are considered as pollutants to the marine environment: Aluminium, Arsenic, Beryllium, Cadmium, Fluorine hydrogenated acids, Hydrogen sulphide, Iron, Lead, Mercury, Phosphorus, Selenium, Titanium, Vanadium, Zinc...', and, the experts add: 'these substances constitute a grave problem for public health by their uninterrupted accumulation and by their considerable persistence in the marine environment. In numerous cases we are unaware of their entry into this environment. Even though Mercury and Lead are generally considered as the most dangerous pollutants, there are others on the overall table, which can be equally important, but which one can only easily discern due to their insidious toxic properties'.

Several elements are notorious for their toxicity, and the illnesses induced by them have received particular names, for example, the Minamata disease, for mercury, Itaî-itaî disease, for cadmium. Organic compounds of the metals are dangerous, and in higher animals, the metals are bound to proteins. Sometimes there is virtually no elimination by the human body and the effects of the gradual accumulation can appear only many years after commencement of poisoning, for examples in 10 years

TABLE 10.10. Experimental Food Chain (Benthic Type)

Factor of magnification of heavy metals in fishes, in relation with seawater on which the preys were grown (in the same units of weight)

In the muscle :	Copper	20
	Zinc	200
	Lead	4
	Mercury	80
In the liver :	Copper	150
	Zinc	1500
	Lead	6
	Mercury	500

for cadmium. Cadmium accumulates mostly in the liver, the kidneys and the spleen. The presence of cadmium in the kidneys has been related to cardio-vascular disorders (high blood pressure in particular) and is responsible for bone disorders. In humans, an excess of vanadium is responsible for enzymatic disorders, with an increase in the cholinesterase and a drop in amino-oxydase. It is also harmful in hepato-digestive metabolism. In animals it changes the level of cholesterol and phospholipids. Ingestion of vanidyl sulphate has induced the appearance of tumours.

Recent research has shown that an excess of titanium in man results in a perturbation of the cycle of Krebs, inhibiting the formation of pyruvic acid, which causes jaundice and anuresis.

Minamata's disease, intoxication by mercury, is characterized by nervous symptoms, with sensitivity disorders, ataxia, trembling, changes in walk, speech, hearing, vision and ideation and, also, teratogenic consequences. Victims of the Minamata disease showed concentrations of 2 – 25 mg Hg/kg dry weight in brain tissue, and up to 700 mg Hg/k in hair, the result of a consumption of about 14 mg Hg per week with polluted fish. Concentrations of less than 0.02 mg/l in blood and less than 6 mg/kg in hair have to be considered normal, and the quantity of mercury that can be tolerated in human food, according to a committee of WHO and FAO, is 0.2 mg of methylmercury and 0.3 mg total mercury per week for one person.

The fishes in the major part of the Mediterranean fall below the level of pollution set by the World Health Organisation. If such limit is exceeded, fishing has to be discontinued. The study of food chains in the laboratory enables the risk to be estimated on the basis of concentration and the likelihood of transfer. Table 10.10 provides an example from our laboratory of the transfer along food chains. Other studied food chains include: pelagic, benthic, neritic with crustaceans, and neritic with molluscs.

Secondary effects of pollution: Eutrophication

The introduction of organic matter, both dissolved and particulate, from sewage and industrial effluents (especially from paper mills and sugar factories, in lagoons, and more-or-less closed bays) leads to local fertilization and to more visible changes which are directly related to the organic matter. Oxygen content decreases, as well as transparency of water and reducing compounds such as SH_2 and ammonia appear at least below certain depths. Natural populations can be extensively decimated, to the benefit of a few specialized forms which develop in a concentric way around the almost abiotic discharge point.

Such conditions are extremely favourable for the regeneration of nutrients and to the development of planktonic primary producers. The massive bloom of microalgae can lead to a 'secondary organic pollution', if the death and decomposition of organisms follows. This may have the form of an ephemerous 'red tide'. The decomposition of the organisms and development of bacteria introduces again anoxic conditions, with sulfate-reducers, which add SH^2 very toxic for aerobiotic forms of life, and especially for the fish, which can be destroyed in mass. The lake of Nantua is a significant example.

Red tides may have pathological consequences for man, in three ways: a stinging syndrome with fever which has been reported in connection with bathing in water containing dinoflagellates; the ingestion of contaminated shellfish, a much more serious affliction which is characterized by a sharp digestive syndrome followed sometimes by shock and neurological signs, (most of which vanish without sequels, although deaths have been reported in 1 to 20% of the cases); finally, aerosols coming from water rich in certain small dinoflagellates, or cyanophyceae, which can result in respiratory syndromes.

Effects on tele-intermediaries. Interaction among species includes often pheromones, ectocrine substances, and the like, chemical messengers or 'substances synthetized by the animal and plant species in the marine environment, released in the environment and acting from a distance on the behaviour or the biological functions of the same species or other species' (Aubert, 1972). Functions of nutrition, reproduction, defence and locomotion may be affected this way. The tele-intermediary may appear as a metabolite to the receiver, or, in compounds of very low concentration, it appears as a signal in the sense of information theory.

The fragility of these mechanisms is obvious: they can be perturbed by the introduction of exogenous substances in the aquatic environment, which can destroy or modify the metabolites and the signallers. Work carried out at the CERBOM on the action of pollutants on the antibiotic action of marine microorganisms, or by the American school on the intermediaries produced by marine animals, have shown indeed that certain pollutants can perturb or modify the function of the tele-intermediaries.

Thus, pollution can have an indirect effect by modifying interspecies signals, that could lead to a progressive imbalance. Perhaps this is a real danger, although its manifestations could remain hidden for a long time.

10.2.5. Conclusion

The Mediterranean suffers at present from increased coastal urbanization and industrial development. Pollution is still moderate in comparison with other closed seas, such as the Baltic and areas of the Japanese Sea, and mostly coastal, offshore areas are relatively free of major pollutants. We do not think that the Mediterranean will be a dead sea in a few years, and prefer to look for signs of local degradation as alerting signals, to which we must react with an intensified research effort and the adoption of the technical, economical and legal measures necessary for the conservation of our sea.

The responsible authorities have been conscious of these problems, and have been active in convention and paper work. Most of the conventions have been ratified by the countries bordering the Mediterranean.

The following international conventions are applicable to the Mediterranean:
The London Convention 1954 — for the prevention of pollution of sea-water by hydrocarbons, with amendments of 1962, 1969 and 1971.
Organisation of a system of zones where the discharge of hydrocarbons is forbidden for certain ships (these zones include maritime waters for a width of 50 miles from the land).
The London Convention of 2nd November 1973 — concerns all pollution coming from maritime navigation, that is:
— prevention of pollution by hydrocarbons.
— prevention of pollution by harmful liquid substances transported in bulk.

— prevention of pollution by harmful substances transported by sea in crates, containers, in mobile tankers, in tank trucks, or rail tanks.
— prevention of pollution by ships' waste water.
— prevention of pollution by ships' refuse.

Convention of 1972 on the immersion of refuse coming from ships, aircraft, platforms or works at sea.
Brussels Convention of 29th November 1969 — on the intervention on the high seas in the case of accidents causing or possibly causing pollution by hydrocarbons.
Convention of 29th November 1969 — on the responsibility for damage due to pollution by hydrocarbons.
Convention of 18th December 1971 — set up an international fund for compensation after pollution by hydrocarbons.
Convention for the protection of the Mediterranean Sea — 1976 – Barcelona:
The framework of the convention aimed at the whole of marine pollution and is covered for the moment, by three legal protocols:
— one concerning the prevention of pollution in the Mediterranean caused by immersion operations of ships and aircraft (1976).
— one to encourage cooperation in the struggle against pollution of the Mediterranean Sea (by hydrocarbons and other harmful substances) in the event of a critical situation (1976).
— one concerning the protection of the Mediterranean Sea against land-based pollution (1980).

These international agreements make the Mediterranean Sea one of the world's most protected seas — at least by the written word.

It remains, of course, for the countries involved to respect, and to gain respect for, these conventions. They must recognize that in so doing, they not only protect an existing resource, but they also guarantee an acceptable future to their unborn citizens.

REFERENCES

Albaigés J. and Albrecht, P. (1979) Fingerprinting marine pollutant hydrocarbons by computerised gas chromatography-mass spectrometry. *Internat. J. Environ. Anal. Chem.* 6, 171 – 190.

Albaigés J., Borbón, J. and Ros, J. (1979) Source identification of tar balls from the Western Mediterranean. ICSEM Rap. 4, 103 – 109.

Albaigés J., Grimalt, J. and Bayona, J. M. (1983) Hydrocarbons in coastal sediments. I. Qualitative analysis. *Internat. J. Environ. Anal. Chem.* (in press).

Albaigés J., Algaba, J., Bayona, J. M. and Grimalt, J. (1983) New perspectives in the evaluation of anthropogenic hydrocarbons in the Western Mediterranean. *VI Journées Etud. Poll.* CIESM (in press).

Albaigés J., Grimalt, J., Bayona, J. M., Risebrough, R., deLappe, B. and Walker, W. (1984a) Dissolved, particulate and sedimentary hydrocarbons in a deltaic environment. *Adv. in Organic Geochem. 1983* (in press)

Albaigés J., Algaba, J. and Grimalt, J. (1984b) Extractable and bound lipids in some lacustrine sediments. *Adv. in Organic Geochem. 1983* (in press)

Algaba, J. (1983) Geochemistry of sterols in aquatic sediments. Tesis de Licenciatura. Universidad Autónoma de Barcelona.

Ambrosiani, P., Bisbini, P. and Marinelli, M. (1970) Enquête sur l'état sanitaire des eaux côtières de la péninsule italienne. *Rev. Intern. Oceanogr. Med.* 18(19), 225 – 234.

Arnoux, A., Bellan-Santini, D., Monod, J. L. and Tatossian, J. (1981) Polluants minéraux et organiques dans les sédiments prélevés entre la Provence at la Corse (Mission BIOMEDE 1). *V. Journées Etud. Poll.*, pp. 423 – 432, CIESM.

Arnoux, A., Chamley, H., Bellan-Santini, D., Tatossian, J. and Diana, C. (1983) Etude mineralogique et chimique de sediments profonds de la Méditerranée Occidentale. *VI. Journées Etud. Poll.*, CIESM (in press).

Aubert, M. (1972) Pollutions chimiques et chaînes trophodynamiques marines. *Rev. Intern. Oceanogr. Med.* 28, 9625.

Aubert, M. (1975) Le probléme du mercure en Méditerranée. *Rev. Intern. Oceanogr. Med.* 37 – 38, 215 – 231.

Aubert, M. and Desirotte, N. (1968) Théorie formalisée de la diffusion bacterienne. *Rev. Intern. Oceanogr. Med.* 12, 5 – 48.

Aubert, M., Gauthier, M., Aubert, J. and Bernard, P. (1981) Les systemes d'information des Micro-organismes Marins. *Edit. CERBOM* (Rev. Intern. Oceanogr. Med.) Nice, 231 pp.

Aubert, M., Revillon, P., Breittmayer, J. Ph., Flatau, G. and Aubert, J. (1979) Métaux lourds en Méditerranée. *Edit. CERBOM* (Rev. Intern. Oceanogr. Med.) Nice, 304 pp.

Aubert, M., Revillon, P., Breittmayer, J. Ph., Flatau, G. and Aubert, J. (1982) Métaux lourds en Méditerranée. *Edit. CERBOM* (Rev. Intern. Oceanogr. Med.) Nice, 23.

Bayona, J. M. (1983) Organic Geochemistry of coastal Mediterranean sediments. *Tesis de Doctorado*. Universidad Autónoma de Barcelona.
Bayona, J. M. and Albaigés, J. (1982) Evaluation of glass capillary columns for geochemical analysis. *Chromatographia* 16, 271–274.
Bayona, J., Grimalt, J. and Albaigés, J. (1983) Origen y distribución de los aportes de hidrocarburos en sedimentos del Delta del Ebro. In *Estudio Integrado del Sistema del Delta del Ebro*, Ed. M. Mariño (in press).
Bellan, G. (1967) Pollution et peuplements benthiques sur substrat meuble dans la région de Marseille. *Rev. Intern. Oceanogr. Med.* 6-7 53–87 *and* 8, 51–95.
Blumer, M., Mullin, M. M. and Thomas, D. W. (1963) Pristane in zooplankton. *Science* 140, 974.
Blumer, M. (1975) Organic compounds in Nature: limits of our knowledge. *Angew. Chem.* Internat. Ed. 14, 507–514.
Boehm, P. (1980) Evidence for the decoupling of dissolved, particulate and surface microlayer hydrocarbons in NW Atlantic continental shelf waters. *Mar. Chem.* 9, 255–281.
Boon, J. J. and DeLeuw, J. J. (1979) The analysis of wax esters, very long mid-chain ketones and sterol ethers isolated from Walvis Bay diatomaceous ooze. *Mar. Chem.* 7, 117–132.
Boussuge, C., Saliot, A. and Tissier, M. J. (1980) Utilisation des traceurs biogéochimiques aux interfaces eau de mer-sédiment profond et sédiment-eau interstitielle. In Daumas, R. (1980), pp. 191–210.
Brassell, S. C. (1980) The lipids of Deep Sea sediments: their origin and fate in the Japan Trench. Ph.D. Thesis, University of Bristol.
Brassell, S. C. and Eglinton, G. (1980) Environmental chemistry — an interdisciplinary subject. Natural and pollutant organic compounds in contemporary aquatic environments. In *Analytical Techniques in Environmental Chemistry*, Ed. J. Albaigés, pp. 1–22, Pergamon Press, Oxford.
Brassell, S. C. and Eglinton, G. (1983) Steroids and triterpenoids in Deep Sea sediments as environmental and diagenetic indicators. In *Adv. in Organic Geochemistry 1981*, Ed. M. Bjoroy, Heyden, in press.
Burns, K. A. and Villeneuve, J. P. (1982) Dissolved and particulate hydrocarbons in water from a spring sampling of the Var Estuary. *J. Toxicol. Environ. Chem.* 5, 195–203.
Burns, K. A. and Villeneuve, J. P. (1983) Vertical flux of petroleum and chlorinated hydrocarbons in the coastal Mediterranean. *VI. Journées Etud. Poll.* CIESM (in press)
Buttiaux, (1968) Le deversement en mer des eaux résiduaires. *Bull. Inst. Nat. Santé et Rech. Méd.* 23.
Cauwet, G. (1975) Optimization d'une technique de dosage du carbon organique des sédiments. *Chem. Geol.* 16, 59–63.
Cauwet, G. (1978) Organic chemistry of sea-water particulates — concepts and developments. *Oceanol. Acta* 1, 99–105.
Cenciarini, J., Fernex, F., Pucci, R., Rapin, F. and Vaissiere, R. (1981) Comparaison entre la répartition de differents polluants dans les sediments marins superficiels du plateau continental le long de la Côte d'Azur. *V. Journées Etud. Poll.* CIESM, 341–351.
Cousteau, J-Y. (1979) Rapport préliminaire de l'expedition CIESM. *IV. Journées Etud. Poll.* CIESM, 21–31.
Cranwell, P. A. (1978) Extractable and bound lipid components in a freshwater sediment. *Geochim. Cosmochim. Acta* 42, 1523–1532.
Daumas, R. A. (19%76) Variations of particular proteins and dissolved amino acids in coastal sea water. *Mar. Chem.* 4, 225–242.
Daumas, R. A. (1980) Biogéochimie de la matière organique à l'interface eau-sédiment marin. Ed. CNRS, No. 293, Paris.
Degens, E. T. (1979) Molecular nature of nitrogenous compounds in sea water and recent sediments. In *Organic Matter in Natural Waters*, Ed. D. W. Hood, pp. 77–106, University of Alaska.
Didyk, B. M., Simoneit, B. R. T., Brassell, S. C. and Eglinton, G. (1978) Organic geochemical indicators of paleoenvironmental conditions of sedimentation. *Nature* 272, 216–222.
Documento finale (1981) Balneazione e qualita microbiologiche dell acque. Gênes. *Collana CISI/2*.
Duursma, E. K. and Dawson, R. (1981) *Marine Organic Chemistry*. Elsevier.
Eglinton, G. and Hamilton, R. J. (1967) Leaf epicuticular waxes. *Science* 156, 1322–1335.
Ensminger, A. (1977) Evolution des composés polycycliques sedimentaires. Thèse Dr. Sc., Université Louis Pasteur, Strasbourg.
Faraco, F. and Ros, J. (1979) Pollution par les hydrocarbures des eaux superficielles de la Méditeranée Occidentale. *IV. Journées Etud. Poll.* CIESM, 117–121.
Farrington, J. W. and Tripp, B. W. (1977) Hydrocarbons in Western North Atlantic surface sediments. *Geochim. Cosmochim. Acta* 41, 1627–1641.
Fowler, S. W. *et al.* (1979) The role of zooplankton faecal pellets in transporting PCBs from the upper mixed layer to the benthos. *IV. Journées Etud. Poll.* CIESM, 289–291.
Gagosian, R. (1980) Transformation reactions of biogenic organic compounds at the sediment-water interface. In Daumas, R.A. (1980), pp. 211–220.
Gamrasni, (1967) Les Pesticides et l'eau. *Rev. Eau.*, 1–15.
Gassman, G. (1982) Detection of aliphatic hydrocarbons derived by recent bio-conversion from fossil fuel oil in North Sea Waters. *Mar. Poll. Bull.* 13, 309–315.
Gearing, P., Plucker, F. E. and Parker, P. L. (1977) Organic carbon stable isotopic ratios of continental margin sediments. *Mar. Chem.* 5, 251–266.
Giger, W., Schaffner, C. and Wakeham, S. G. (1980) Aliphatic and olefinic hydrocarbons in recent sediments of Greifensee. *Geochim. Cosmochim. Acta* 44, 119–129.

Gonzalez, J. M., Grimalt, J. and Albaigés, J. (1983) Amino acid composition of sediments from a deltaic environment. Submitted for publication.
Goutx, M. and Saliot, A. (1980) Relationship between dissolved and particulate fatty acids and hydrocarbons, chlorophyll A and zooplankton biomass in Villefranche Bay. *Mar. Chem.* 8, 299–318.
Grimalt, J. (1983) Organic geochemistry of deltaic systems. *Tesis Doctoral*, Universidad Autónoma de Barcelona.
Group of Experts on Marine Pollution GFCM/ICSEM (1972) Review on the state of pollution in the Mediterranean Sea. In *Marine Pollution and Sea Life*, Ed. M. Ruivo, pp. 28–32, Fishery News Ltd., England.
Helmer, R. (1977) Pollutants from land-based sources in the Mediterranean. *Ambio*, 312–316.
Henrichs, S. M. and Farrington, J. W. (1980) Amino acids in interstitial waters of marine sediments. In *Adv. in Organic Geochem. 1979*, Eds. A. G. Douglas and J. R. Maxwell, pp. 435–444, Pergamon Press.
Hô, R. (1982) La matière organique de l'air, de l'eau et du sédiment en Méditerranée Occidentale. Thèse 3 cycle. Université Pierre et Marie Curie, Paris.
Hô, R., Marty, J. C. and Saliot, A. (1982) Hydrocarbons in the Western Mediterranean Sea, 1981. *Internat. J. Environ. Anal. Chem.* 12, 81–98.
Hô, R., Marty, J. C. and Saliot, A. (1983) Les hydrocarbures à l'interface air-mer en Méditerranée Occidentale. *VI. Journées Etud. Poll.* CIESM, in press.
Horn, M., Teal, H. and Backus, R. H. (1970) Petroleum lumps on the surface of the Sea. *Science* 168, 245–246.
Horne, R. A. (1969) *Marine Chemistry*. Wiley-Interscience, New York.
Huang, W. Y. and Meinschein, W. G. (1976) Sterols as source indicators of organic material in sediments. *Geochim. Cosmochim. Acta* 40, 323–330.
ICSEM/IOC/UNEP (1979, 1981, 1983) Journées d'Etudes sur les Pollutions marines en Méditerranée. CIESM.
Jullien, D. (1982) L'interface air-mer: composants organiques, budget et processus d'évolution. Thèse 3 cycle. Université Pierre et Marie Curie, Paris.
Keizer, P. D., Dale, J. and Gordon, D. C. (1978) Hydrocarbons in surficial sediments from the Scotian shelf. *Geochim. Cosmochim. Acta* 42, 165–172.
Kranck, K. (1973) Flocculation of suspended sediments in the sea. *Nature* 246, 348–350.
Lee, C., Farrington, J. W. and Gagosian, R. B. (1979) Sterol geochemistry of sediments from the Western North Atlantic Ocean. *Geochim. Cosmochim. Acta* 43, 35–46.
Lee, R. F. and Loeblich, A. R. (1971) Distribution of 31:6 hydrocarbon and its relationship with 22:6 fatty acid in algae. *Phytochemistry* 10, 593–602.
Lelourd Ph. (1977) La pollution par les hydrocarbures en Méditerranée. *Ambio VI* 6, 318–322.
McKenzie, A. S., Brassell, S. C., Eglinton, G. and Maxwell, J. R. (1982) Chemical fossils — the geological fate of steroids. *Science* 217, 491–504.
Mendia, L. Tosti, E. and Volterra L. (1981) Studi microbiologici sulla qualita delle acque del Golfo di Napoli. In Balneazione e qualita microbiologiche dell'acque. *Gênes Collana CISI/2*, 197–207.
Menzie, C. M. (1972) *Effects of Pesticides on Fish and Wildlife in Environmental Toxicology of Pesticides*, Eds. F. Matsura, M. Bousch and T. Misato, pp. 487–500, Academic Press, New York, San Francisco, London.
Michaelis, W. (1980) Amino-acids and carbohydrates as possible indicators of sedimentation processes. In R. A. Daumas (1980), pp. 129–140.
Moore, B. (1970) The present status of disease connected with marine pollution. *Rev. Intern. Oceanogr. Med* 18,19, 193–219.
Mopper, K., Michaelis, W., Garrasi, C. and Degens, E. T. (1977) Sugars, amino acids and hydrocarbons in Black Sea sediment from DSDP Leg 42B cores. *Init. Rep. DSDP* 42, 697–705.
Mopper, K. and Degens, E. T. (1979) Organic carbon in the Ocean: nature and cycling. In *The Global Carbon Cycle*, Eds. B. Bolin *et al.*, pp. 293–316, Wiley.
Morris, B. F., Butler, J. N. and Zsolnay, A. (1975) Pelagic tar in the Mediterranean Sea. *Environ. Conservat.* 2, 275–281.
Morris, R. J. and Culkin, F. (1975) Environmental organic chemistry of Oceans. In *Environmental Chemistry*, Eds. G. Eglinton, pp. 81–108, The Chemical Society.
Nishimura, M. (1981) Geochemical information derived from the stenol into stanol conversion in early diagenesis. In *Alfred Treibs Symp.* Ed. A. Prashnowsky, pp. 93–131, Verlag Chimie.
Nissenbaum, A., Baedecker, M. J. and Kaplan, I. R. (1972) Organic geochemistry of Dead Sea sediments. *Geochim. Cosmochim. Acta* 36, 709–727.
Ostenberg, Ch. and Keckes, S. (1977) Etat de la pollution en Méditerranée. *Ambio VI* 6, 322–328.
Ourisson, G., Albrecht, P. and Rohmer, M. (1979) The hopanoids. Paleochemistry and biochemistry of a group of natural products. *Pure Appl. Chem.* 51, 709–729.
Pelet, R. and Debyser, Y. (1977) Organic geochemistry of Black Sea cores. *Geochim. Cosmochim. Acta* 41, 1575–1586.
Ryther, J. H. (1969) Photosynthesis and fish production in the Sea. *Science* 166, 72–76.
Saliot, A. and Barbier, M. (1973) Sterols from the sea-water. *Deep-Sea Res.* 20, 1077–1082.
Siezen, R. J. and Magne, T. H. (1978) Amino acids in suspended particulate matter from oceanic and coastal waters of the Pacific. *Mar. Chem.* 6, 215–231.
Simoneit, B. R. T. (1981) Utility of molecular markers and stable isotope compositions in the evaluation of sources and diagenesis of organic matter in the geosphere. In *Alfred Treibs Symp.* Ed. A. A. Prashnowsky, Verlag Chemie.
Staresinic, N. *et al.* (1981) Downward flux of particulate matter in the Mediterranean Sea. *V. Journées Etud. Poll.* CIESM, pp. 1011–1012.
Stuermer, D. H. and Harvey, G. R. (1974) Humic substances from sea-water. *Nature* 250, 480–481.

Tissot, B. P. and Welte, D. H. (1978) *Petroleum Formation and Exploration Occurrence*. Springer-Verlag.
UNEP (1977) Preliminary Report on the State of Pollution of the Mediterranean Sea. UNEP/IG 11/INF.4. United Nationals Environment Programme, Geneva (Switzerland).
Vigneaux, M. *et al.* (1980) Matières organiques et sédimentation en milieu marin. In R. A. Daumas (1980), pp. 113−128.
Volkman, J. K., Gillan, F. T. and Johns, R. B. (1981) Sources of neutral lipids in a temperate intertidal sediment. *Geochim. Cosmochim. Acta* 45, 1817−1828.
Wangersky, P. J. (1976) Particulate organic carbon in the Atlantic and Pacific Oceans. *Deep-Sea Res.* 23, 457−465.
Whelan, J. K. (1977) Amino acids in a surface sediment core of the Atlantic abyssal plain. *Geochim. Cosmochim. Acta* 41, 803−810.
Williams, P. M. (1967) Sea surface chemistry: organic carbon, nitrogen and phosphorous in surface films and subsurface waters. *Deep-Sea Res.* 14, 791−800.
Zsolnay, A. (1979) Hydrocarbons in the Mediterranean Sea. *Mar. Chem.* 7, 343−352.

Index

Abiotic factors   236, 237
Abrego   95
Absorption coefficient of water   68, 69
Abyssal zone   227 – 31
Acantharia   171
Acarina   244
Acartia clausi   172, 180, 183
Acartia danae   173, 174, 175, 185
Acartia granii   183
Acartia latisetosa   183
Acartia margalefi   183
Acartiidae   174, 183
Actinian   226
Adriatic Sea   140
Advection   130
Aegean Sea   106, 140, 142
Aetideidae   175, 179, 180
Aglaura hemistoma   176
Air, specific heat of   9
Albedo   68
Alboran Island   110
Alboran Sea   9, 10, 26 – 28, 30, 97, 110, 112, 154, 155, 171 – 73, 191, 309
Albufera de Valencia   13
Albunea carabus   208
Alcyonarians   218
Alcyonium palmatum   213, 215
Algae   194, 211, 242, 244, 246, 248 – 52, 255, 256, 262, 267, 281, 282, 289, 292
Alkanes as organic markers   329 – 32
Alkanoic acids   323, 327 – 29
Allochthonous compounds   318
Allorchestes aquilinus   244
Alosa spp. (shad)   303
Alpine belts   21
Alpine orogeny   21
Amaroucium profundum   208
Amino acids   323, 326
Ammonification   132
Amphipalaemon   208
Amphipods   171, 254, 262
Amussium cristatum   203
Anchovy   299, 302, 312
Anemonia sulcata   250
Anglerfish (*Lophius*)   305
Animals   248, 252 – 56
  benthic   292 – 93
  development of   5
  in confinement   316
  of coralligenous   270 – 71
Anomura   179
Antedon mediterranea   263
Anthozoans   213
Anticyclones   110
Antipathes fragilis   222
Aphanius fasciatus   15
Aphia minuta (transparent goby)   303
Aphroditidae   221

Appendicularia   183
Apporhais serresianus   226
Aquaculture   7, 12, 15, 313 – 16
Arbaciella elegans   206
Arca diluvii   204
Argentina sphyraena   303
Aristeomorpha foliacea (red shrimp)   306
Aristeus antennatus (pink shrimp)   300, 306, 311
Arthopyrenia halodytes   245
Ascidians   208, 213, 214, 215, 225, 252, 253, 276
Asparagopsis armata   288
Asterionella japonica   164
Asteroid   226
Astroides calycularis   271
Atherina boyeri   15
Atlantic coasts   13
Atlantic current   152
Atlantic Ocean   19 – 21, 137, 140, 198 – 99
Atmospheric pressure   8, 72
Atmospheric transmission coefficient   140
Audouin's gull (*Larus audouinii*)   304
Autochthonous compounds   318
Auxis rochei (frigate mackerel)   302

Bacterial pollution   339 – 41
Balanus   183
Balanus perforatus   250
Balearic basin   23 – 26, 30
Balearic Islands   300
Balearic Sea   111, 112, 175 – 76
Baltic Sea   8, 14
Baroclinic forces   82
Barotropic effect of wind stress   84
Barotropic motion   82
Bass (*Dicentrarchus*)   15
Bass (*Dicentrarchus labrax*)   15, 301, 314
Bathyal assemblages   219 – 27
Bathyal zone   241, 279
Bathymetry   105, 110, 111, 123
Bathypterois mediterraneus   228
Bay of Biscay   137
Belone belone   15
Benthic communities   11
Benthic populations   5
Benthos   206 – 7, 233 – 95
  primary producers   289 – 92
  succession and regression   281 – 84
Bernoulli phenomenon   123
Biogeochemical processes   324
Biological markers   317 – 34
Biological processes   131, 135
Biological productivity   144
Biotic factors   236, 237
Birds   16, 245
Bivalves   180, 204, 215, 216, 221, 224, 226, 246, 307
Black coral   222

353

Black-headed Gull (*Larus ridibundus*) 304
Black Sea 6, 7
Blenniidae 256
Blue-green algae 194, 242
Blue sail (*Velella*) 11
Blue Whiting (*Micromesistius poutassou*) 301, 305, 311
Bluefin Tuna (*Thunnus thynnus*) 302, 315
BOD (Biochemical Oxygen Demand) 133, 136, 335
Bogue (*Boops boops*) 303
Bonito (*Sarda sarda*) 302
Boops boops (bogue) 303
Bora wind 95, 144
Boundary processes 129–32
Brachiopods 218–19, 230, 231, 271, 278
Brachyura 179
Brackish environments 12–13
Branchiocerianthus norvegicus 226
Bream (*Sparus auratus*) 15, 315
Brissopsis lyrifera 226
Brittle-star 213, 214, 252
Bryozoans 216, 218, 246, 252, 256, 262, 270, 276, 277, 278
Buccinum undatum 204
Budget calculations 103–9
Budget example 107–9
Burnt almond facies 211
Bypassing model 47–49

Cadmium 336, 345, 346
Calabrian 204
Calabrian arc 27
Calcareous shelves 47
Calcium 14
Caloplacamus ramosus 208
Calothrix 242
Campecopea hirsuta 247
Candacia 187
Caneosphaera molischii 164
Caprellida 174
Capulus 217
Caranx rhonchus 302
Carbon cycle 135–37
Carboxylic acids 327
Cardium tuberculatum 204
Caryophyllia armata 220
Catalan Sea 190, 191
Caulerpa prolifera 260, 287
Caulerpa scalpelliformis 208
Cavolinia longirostris 186
Cenozoic Palaeoceanography 17–59
Central zone 176–77
Centrifugal forces 84
Centropages 181
Centropages chierchiae 172, 173, 175, 182, 185
Centropages typicus 179, 182–83
Centropages violaceus 173, 174
Centropagidae 174
Centrophorus uyatus (shark) 227
Cephalopods 224, 226, 306
Cepola 208
Ceramaster placenta (sea-star) 221

Ceramium ciliatum 249
Ceratium 168
Chaetoceros 163
Chaetognatha 173, 174, 176, 179, 181, 183, 185
Chalinura 226
Chattonella subsalsa 165
Chelophyes appendiculata 181
Chemical pollution 344–46
Chemical process 135
Chemoautotrophy 238
Chilopods 244
Chilton 221
Chlamys clavata 216
Chlamys flexuosa 204
Chlamys islandica 204
Chlamys varia 204
Chlorophyceae 244
Chlorophyll 9, 85, 152, 154–59, 162, 194
Chorizopora brongniarti 212
Chromium 336
Chrysophyceae 162
Chthamalus depressus 242
Chthamalus montagui 245
Chthamalus stellatus 245, 285
Cidaris cidaris 218
Cidaris cidaris (sea-urchin) 218, 219, 221, 223, 225, 227
Circalittoral hard bottoms 263–71
Circalittoral soft bottoms 209–17
Circalittoral zone 241, 263, 279, 292
Circulation 109–16
Cirripeds 183, 221, 230
Cladocera 171, 174, 181, 182, 183, 187
Cladocora caespitosa 207
Cladophora 247
Cladostephys hirsutus 250
Clausocalanus 172, 174
Climatic-eustatic oscillations 43–45
Climatic factors 236
Cliona 264
Cnidarians 183, 224, 225, 246, 256, 270, 276, 277, 278
Coastal detritic assemblage 210–12
Coastal upwelling 97
Coastal zones 153–54
Coasts 12–16
Coccolithophorids 162, 164, 165
Coccolithus huxleyi 164
Cockle 307
Codonella galea 183
Coelenterata 175, 176
Coho salmon (*Oncorhynchus kisutch*) 314
Coleoptera 242
Common Sole (*Solea solea*) 301, 306, 315
Common tern (*Sterna hirundo*) 304
Conservation 16
Continental margins 41–42
Continental run-off 139
Continental shelf sedimentation 45
Copepoda 171, 173, 174, 175, 178–81, 183, 184, 185, 187, 188
Copper 336
Coral assemblage 219–21
Coral reefs 6
Coralligenous 264

animals of 270–71
plants of 267–68
Corallinaceae 212, 264
Corallium rubrum (red coral) 264, 271, 291, 307
Coregonus 14
Coriolis force 72, 82, 84, 109, 110
Corsica 24, 26, 111, 308
Corycaeidae 180
Corycaeus latus 185
Coryphaena hippurus (dolphin) 303, 304
Cory's Shearwater (*Procelaria diomedea*) 305
Crab (*Pachygrapsus marmoratus*) 244, 245
Crab (*Paromola cuvieri*) 218, 220
Crabs 307
Crenimugil 15
Crevices 244
Crinoid 216
Crustaceans 185, 216, 221, 224, 225, 226, 246, 262, 270, 276, 277, 306, 307
Crysophyceae 343
Ctenocalanus vanus 185
Cumaceans 229
Curled Octopus (*Eledone cirrosa*) 300–1
Cuttlefish 306
Cuvierina columnella 173, 186
Cyanophyceae 245, 282, 289
Cyclogenesis 96
Cyclogenetic path 95–96
Cyclonic spreading centre 96
Cymodoce truncata 254
Cymodocea nodosa 260
Cyprina islandica 204
Cystoseira 249, 250, 268
Cystoseira mediterranea 285, 286, 292
Cystoseira stricta 253
Cystoseira zosteroides 271

Dardanelles 138
DDT 338
Decapods 181, 209, 211, 213, 224, 226, 278
Deep layer 130
Deep-sea coral assemblage 219–21
Deep sea fans 49–53
Deep-water formation 96, 116–22, 127, 144, 149
Deep-water movement 101
Deltas 13
Denitrification 13, 133
Dermatolithon confinis 247
Desidiopsis racovitzai 247
Desmophyllum cristagalli 220
Detergents 344
Detritic bottoms 210
Diacria quadridentata 173, 186
Diacria trispinosa 186
Diagramma mediterraneum 207
Diatoms 14, 163, 343
Diazona violacea 213, 215
Dicentrarchus (bass) 15
Dicentrarchus labrax (bass) 15, 301, 314
Dictyocysta 164
Diffusion 130

Dilution 101, 102, 105–6
Dinoflagellates 163, 164, 165
Diplodus 15
Diptera 242
Dissolved organic matter (DOM) 319
Diverging adaptations 252–56
Doliolidae 176, 180, 182
Doliolum nationalis 175
Dolphin (*Coryphaena hippurus*) 303, 304
Dromia dromia 208
Dromia vulgaris 208

Earthquakes 29
Earth's crust 4
Eastonia rugosa 206
Ebro Delta 328
Ebro fan 50, 52
Ebro River 327
Echinocyamus pusillus 211
Echinoderms 208, 211, 224, 225, 271, 277, 278, 307
Echinoid 225
Echinothrix 208
Echinus acutus (sea-urchin) 216
Echiurids 224
Ecological indicator 184
Ecological strategies 284–89
Ecophenes 242
Edaphic factors 237
Eddy diffusion 74
Eddy viscosity 75, 77
Ekman Numbers 77
Eledone cirrosa (curled octopus) 300–1
Elefsis Bay 103
Emiliania 164
Encrusting species 252
Endemics 207–9
Energy flow 293
Energy requirements 9
Engraulis encrasicholus (anchovy) 302
Enteromorpha 247
Enteropneust 225
Entophysalis 242
Equation of motion 72–73
Escherichia coli 341
Euchaetidae 179
Euchaetideidae 175
Eunice 221
Eunice floridana 221
Eunicella cavolinii 271
Eunicella singularis 271
Euphausia brevis 185
Euphausia hemigibba 185
Euphausia krohni 173, 185
Euphausiacea 173, 174, 185
Eurydice affinis 244
Eurynome 208
Euterpina acutifrons 183
Euthynnus euthynnus quadripunctatus (tuna) 302
Eutrophication 8, 145–46, 327, 346–47
Evadne nordmannii 174
Evadne spinifera 180, 181

Evadne tergestina   180
Evaporation   7, 73, 106, 137, 140–44, 150
Evaporites   199, 200, 202
Evolution   17–59

Faulting   32
Favella   183
Fernenes   331
Fertility   10
Fertility distribution   149–54
Fertilization   9
Fish and fisheries   3, 11–12, 15, 16, 144, 205, 208, 213, 226, 227, 228, 296–316
  annual catches by major species   311–15
  artisan   299, 301
  Balearic zone   312, 315
  beach gear   299
  coastal lagoon   301
  current techniques   298
  demersal resources   305–6
  development history   296–97
  disappearance of artisan gear   297
  freshwater species   314
  future   310–13
  increase in fishing power   299
  Lions zone   313
  longline   302
  management and policy   301, 307–9
  new techniques   297
  pelagic   299–303, 312
  productivity rate   297
  Sardinia zone   314
  seine gear   299
  specialized   306–7
  trawl catches   300
  trawl gear   298
  tuna traps   299, 313
  vessels and gear employed   298–99
Flow categories   90
Flow characterization   77
Flow properties   66
Fluxes and mixed-layer model   79–81
Food pellets   16
Foraminifera   171
Foraminiferan   246
Forkbeard (*Phycis blennioides*)   301
Fossarus ambiguus   247
Freshwater   8, 12, 154, 165
Frigate mackerel (*Auxis rochei*)   302
Fritillaridae   179
Funiculina quadrangularis (sea-pen)   224, 226

Gadinia garnoti   247
Gale frequency distribution   91–95
Gammarus   14
Gammarus olivi   244
Gases dissolved in sea-water   128
Gasterosteus aculeatus   15
Gastropods   180, 204, 211, 216, 221, 224, 226, 242, 245, 262, 307
Gelidium pusillum   249
General Fisheries Council for the Mediterranean (GFCM)   307
Genoa depression   91
Geochemical processes   131
Geology   4
Geopotential topography   8
Geostrophic currents   81–83
Gephyra   217
Gibraltar   7, 8, 104, 105, 106, 108, 109, 114, 122
Gibraltar arc   27, 28
Gilt sardines (*Sardinella*)   303
Glassworts   242
Global productivity   96
Glutamic acid   326
Gobiidae   256
Gobius lesueuri   216
Gravitation   72
Greater forkbeard (*Phycis blennioides*)   305
Green algae   242
Gregal   92–94
Grey mullet (*Mugil*)   15
Gryphus vitreus   218, 227, 230
Gulf of California   103
Gulf of Fos   338
Gulf of Gênes   338
Gulf of Genoa   89–91, 96, 338
Gulf of Genova   225
Gulf of Lions   1, 10, 24, 87, 96, 114, 117, 118, 142, 149, 151, 174–75, 300, 301, 336
Gulf of Marseille   153
Gulf of Mexico   12
Gulf of Naples   340
Gulf of Suez   103
Gulf Stream   81
Gurnards (*Trigla*)   301

Hake (*Merluccius merluccius*)   300, 301, 305, 308
Haloporphyrus   228
Haloptilus longicornis   173
Hard bottoms   263
Hard substrates   242, 248–52
Heat storage and radiant energy   78–79
Heavy metals   336–37, 344, 345, 346
Heliozoa   171
Hellenic arc   27
Herring gull (*Larus argentatus*)   304
Heterorhabdidae   175, 179, 180
High-frequency motions   122–24
History   198–232
Holoplankton, neritic   178
Holoplanktonic species   170
Holothurians   263
Holothuroid   213, 216
Horse mackerel (*Trachurus trachurus*)   300, 302, 312
Humectation   237
Hydrarians   254, 276
Hydrocarbon pollution   344–45
Hydrographical characteristics   172–74
Hydroids   208, 217, 252

INDEX

Hydrological factors 206
Hydrological indicator 184
Hydrozoans 262
Hypersaline environments 12

Iberian plate 27
Illumination 237
Indian Ocean 199
Indicator communities 184
Industrial wastes 344, 346
Infralittoral sands and muds 260
Infralittoral zone 240, 248−63, 281, 282
    lower 292
    upper 291
Inorganic compounds 345
Insects 242
Interface processes 129, 131
Intermediate layer 130
Intermediate water types 112−15
Ionian Sea 140
Ircinia 238
Iron 337
Isopods 213, 242, 262
Itaî-itaî disease 345

Kellya suborbicularis 212
Kelp (*Laminaria rodriguezii*) 212, 218
Kophobelemnon leuckarti (sea pen) 226

Lagoons 13−16
Lambrus massena 211
Laminaria 11
Laminaria ochroleuca 206
Laminaria rodriguezii (kelp) 212, 218
Land run-off 138−40
Langmuir circulations 80−81
Languedoc's current 174
Laridae 304
Larus argentatus (herring gull) 304
Larus audouinii (Audouin's gull) 304
Larus ridibundus (black-headed gull) 304
Lasaea adansoni 247
Late Miocene 199−202
Late Pliocene 202−3
Lead 336
Lepidochitona corrugata 246
Leptometra phalangium (sea-lily) 216
Leptometra phalangium 217
Levante 94
Levantine basin 140
Levantine intermediate water (LIW) 101, 108, 109, 112−20, 144
Levantine Sea 26, 116
Lichens 241, 242
Light extinction in water 236
Light intensity effects 290
Ligia italica 242, 245

Ligurian Sea 23, 96, 111, 117, 152, 174
Ligurio-Provincial Basin 115−16, 120
Liguro-Provençal Sea 336
Limacina trochiformis 173
Lipid molecules 322
Lipids as organic markers 326
Liquid phase processes 129−31
Liriope 181
Lithognatus mormyrus 15
Lithophaga aristata 206
Lithophyllum expansum 267
Lithophyllum tortuosum 246−47, 252, 254, 285
Lithothamniae 211, 218
Lithothamnium calcareum 211
Lithothamnium corallioides 211
Littoral pools 244
Liza 15
Local mixing 151−53
Lomentaria articulata 284
Lophelia pertusa 220
Lophius (anglerfish) 305
Lower infralittoral zone 292
Lower mediolittoral rock 245−46
Lower Sicilian 204
Lucicutia flavicornis 173
Lyriope tetraphylla 176
Lysiosquilla 208
Lysmata seticaudata 208
Lysmata ternatensis 208

Mackerel (*Scomber scombrus*) 300, 302, 303, 312
Macrura 179
Madrepora oculata 220, 221, 223
Maia 208
Maja (spider crab) 306
Malaria 16
Manganese 337
Mar Menor 13
Marine circulation 8
Marine mammals 303−4
Marine populations 5
Marine research 1−3
Marshes 12−16
Medes Islands 243, 248, 255, 261, 265, 269, 273, 278, 291
Mediolittoral rock 245−48
Mediolittoral zone 240, 243, 244−48, 291
Mediterranean 1
    ancestry of 2−3
    ancient and recent features 22
    as chemical reactor 137−44
    as concentration basin 137−44
    as living machine 9−12
    as negative concentration basin 101
    as oligotrophic ecosystem 144−46
    as scale model of real ocean 6−9
    as small ocean 100
    depth 7
    early knowledge of 2
    early research 2
    Eastern 22, 26, 101

evolution 17—22
general characteristics of 1—16
location 7
North-western 26, 49—53
physical anomalies 100—1
synthetic view of 6
Western 22—32, 101
Medusae 171, 176, 179, 181, 183, 184
Melaraphe neritoides 242, 245
Melobesiae 246
Menorca 115
Menorca Fan 52
Mercury 336, 339, 345, 346
Merluccius merluccius (hake) 300, 301, 305, 308
Meroplankton 181, 182
Mesalia brevialis 206
Mesodesma corneum 244
Mesophyllum lichenoides 267
Mesoplankton 190, 191, 193
Mesospora macrocarpa 245
Mesothuria intestinalis 226
Messinian salinity crisis 5, 6, 18, 30, 32—40, 199—202, 207
 biological response 37
 dynamic model of 37—40
 evaporite formation 37, 39
 geological evidence of 35—37
 global events leading to 33—35
 Late Miocene climatic and eustatic oscillations 34—35
 outcrops and erosional surfaces 35—37
 tectonic evolution 33—34
 world response to 40
Mestral 144
Metazoa 171
Metridiidae 175, 179, 180
Microbial metabolism 332
Microbiological pollution 339—44
Microcosmus spp. (sea quirt) 307
Micromesistius poutassou (blue whiting) 301, 305, 311
Migration 11
Minamata disease 345, 346
Miocene 199
Mistral 92, 96, 117, 120, 144
Mixed-layer model 79—81
Mixing process 78
Modiolus modiolus 204
Molecular diffusion 74, 131
Molecular scattering 68
Molecular viscosity 74
Mollia patellaria 212
Molluscs 181, 204, 205, 216, 221, 224—26, 246, 252, 256, 275, 276, 288, 307, 314, 315
Monodonta turbinata 245
Muddy bottoms 245
Muddy-detritic assemblages 213
Muggiaea atlantica 179, 181
Muggiaea kochi 179, 181
Mugil (grey mullet) 15
Mugilidae 301, 303, 315
Mullet (*Mugilidae*) 301, 303, 315
Mullus barbatus (red mullet) 300, 308, 311
Mullus surmuletus (red mullet) 301
Munida 218

Murex trunculus var. *conglobata* 204
Mussels 307, 314
Mya truncata 204
Mysella bidentata 212
Mysidacea 174
Mytilus galloprovincialis 250
Mytilus perna 206

Naples Station 2—3
Narcomedusae 179
Nassa gibberula 204
Naucrates ductor (pilot fish) 303
Navier-Stokes equation 72, 75
Navigation 1
Nemalion helminthoides 245
Nematodes 246
Nemertopsis peronea 247
Nemoderma tingitanum 246
Neogoniolithon notarisii 246
Neolampas rostellata 217, 218
Nephrops norvegicus (Norway lobster) 226, 300, 305
Nerine cirratulus 244
Neritic zones 177—82
Nickel 336
Nictemeral migrations 188—90
Nitrates 87, 153
Nitrification 132
Nitrite 159, 161, 162
Nitrobacter 132
Nitrogen cycle 132—33
Nitrogen fixation 133
Nitrosomonas 132
Nitzchia 163
Noctiluca 10
North-African coast 173
North Atlantic water (NAW) 106, 108—12, 114
Norway Lobster (*Nephrops norvegicus*) 226, 300, 305
Nucella lapidus 204
Nucleic acids 321
Nuculoma tenuis 229
Nudibranchs 256, 276
Numidian flysch 27
Nutrients 9, 13, 85, 129, 132, 145, 149, 150, 153, 154, 155, 165, 168, 169, 238, 346

Obelia 183
Oceanography 2, 8, 43, 62, 100, 127
Octopus 306
Octopus macropus 208
Octopus variabilis 208
Oculina patagonica 207
Ocypode cursor 208
Odontaster mediterraneus (sea-star) 226
Oersetgroenia digitata 213
Offshore rocky-bottom assemblages 217—19
Oikopleura 183
Oil pollution 338—39, 344—45
Oithona 183
Oithona helgolandica 183

Oithona nana 183
Oithona plumifera 173
Oithonidae 174, 183
Oligocene 198
Oligochaetes 244
Oncaea 187
Oncaea media 175
Oncaeidae 180
Oncidiella celtica 247
Oncorhynchus kisutch (coho salmon) 314
Ophelia bicornis 244
Ophidiaster ophidianus 271
Ophiopsila aranea 212
Ophiothrix quinquemaculata (brittle-star) 213, 214
Ophiuoid 216, 218
Organic carbon 318, 324, 325
Organic compounds 322, 345
Organic markers
  in Western Mediterranean 324–26
  organic matter as 321–22
  proper description of 322
  selection of 324
Organic matter 317, 346
  as organic markers 321–22
  composition of 326
  cycling 321
  degradation of 325, 326
  distribution 319
  in marine environment 318–22
  in Western Mediterranean 322–34
  source of 318
Oxygen concentrations 127, 158, 159, 161
Oxygen consumption
  in sea-water 136
  in sediments 136–37
Oxygen cycle 135–37, 145
Oxygen transport 146
Oyster 307

Pachygrapsus marmoratus (crab) 244, 245
Pacific coasts 13
Padina pavonica 250, 286
Pagrus 208
Palaemonetes (prawn) 14
Palinurus mauritanicus (spiny lobster) 218, 306
Panomya arctica 204
Paracalanidae 174
Paracalanus parvus 172, 175, 180
Paracentrotus lividus (sea-urchin) 263
Paralcyonium elegans 219
Paramuricea clavata 271
Parapenaeus longirostris (shrimp) 226, 306
Parazoanthus axinellae 271
Paromola cuvieri (crab) 218, 220
Particulate organic matter (POM) 319
Patella rustica 245
Patella safiana 206
Pecten jacobaeus 204
Pelagic ecosystem 5, 186, 301–3, 312
Pelecypods 211, 213, 216, 229
Penaeus kerathurus (prawn) 301, 306

Penilia avirrostris 174, 180, 181, 187
Pennatula phosphorea 213
Pennatula rubra (sea-pen) 220
Peraclis bispinosa 186
Peridinium 163
Perinereis cultrifera 244
Persa incolorata 176
Pesticides 337–38, 345
Peyssonelia bornetii 212
Peyssonnelia rosa-marina 212
Peyssonneliaceae 264
Phaeophyceans 268
Phalacrocorax aristotelis (shag) 305
Pheronema grayi 227
Phialidium 181
Phialidium hemisphaericum 176
Phicopomatus enigmaticus 260
Phosphate 87, 153
Phosphorus 9, 133–34, 151, 154, 170, 335
Photosynthesis 129, 131, 135, 144, 145, 158, 290
Phycis blennioides (forkbeard) 301, 305
Physiography 41–42
Phytoplankton 10, 85, 87, 135, 163, 184, 188, 239
  abundance data 169
  and water movements 168
  seasonal changes 188
  seasonal sequences 159–65
  spatial heterogeneity 165–68
  taxonomical compositions 159–65
  temporal and spatial distribution 168
  vertical organization 154–59
Pilchard (*Sardina pilchardus*) 300, 302, 309, 312, 315
Pileolaria berkeleyana 207
Pilot fish (*Naucrates ductor*) 303
Pink shrimp (*Aristeus antennatus*) 300, 306, 311
Plankton 5, 10, 15, 239, 292, 293
Planktonic indicators 183–88
Plants
  of coralligenous 267–68
  primary production by 168
Platichthys flesus flesus 204
Platidia 231
Plectonema 242
Pleistocene 202–4
Pleuromamma gracilis 173
Pleuronectes flesus 15
Pliocene 40–53, 202–3
Plocamium cartilagineum 284
Plutonaster bifrons 230
Podon 183
Podon intermedius 183
Poecillastra compressa f. *calyx* 219
Poecillastra compressa f. *placentula* 222
Pogonophora phylum 224
Polar Front Jet Stream 89
Pollicipes cornucopiae 246
Pollution 6–8, 12, 131, 165, 250, 317
  bacterial 339–41
  causes and effects 334–48
  chemical 335–39, 344–46
  consequences of 342–47
  international conventions 347–48
  marine 334

    microbiological 339–44
    monitoring activities 322–24
    oil 338–39, 344–45
    research activities 322–34
    secondary effects 346–47
    sewage 145, 335, 346
    thermal 341–42
Polychaetes 183, 207, 213, 218, 220, 221, 224, 229, 246, 252, 256, 276
Polychloride biphenyls 337–38, 345
Polydora 264
Polyprion americanus (wreckfish) 301
Pools 244
Porites 201
Porphyra leucosticta 245
Porphyra umbilicalis 245
Posidonia 210, 211, 260–63, 292, 293
Posidonia australis 208
Posidonia oceanica 208, 249, 256, 260–63, 288
Potential temperature 62, 119
Pouting (*Trisopterus minutus capelanus*) 305
Prawns 14, 224, 301, 306, 315
Precipitation 13, 138–140, 150
Predestination phase 283
Predominance phases 283
Pressure gradient 96
Primary production 168–69, 239, 289–92
Procelaria diomedea (Cory's shearwater) 305
Procellaridae 305
Proteins 319, 321, 345
Protozoa 170
Provence coast 219
Pseudocalanus elongatus 185
Pseudocucumis 208
Pseudoscorpions 244
Pseudostichopus occultatus 230
Pteria hirundo 215
Pteroides 208
Pteroides griseum (sea pen) 214
Pteropoda 174, 183, 186
Pterosagitta draco 173
Pump effect 96, 97

Quaternary 40–53, 203–6

Radiant energy and heat storage 78–79
Radiolaria 170
Raia clavata 204
Rainbow trout (*Salmo gairdnerii*) 314
Rainfall 138
Ralfsia verrucosa 245, 246
Red coral (*Corallium rubrum*) 264, 271, 291, 307
Red mullet (*Mullus barbatus*) 300, 308, 311
Red mullet (*Mullus surmuletus*) 301
Red shrimp (*Aristeomorpha foliacea*) 306
Red tides 347
Reflection coefficient 66
Regenerative processes 131
Regional depositional patterns 40–53

Regression. *See* Succession and regression
Reptantia 208
Residence times 126, 127
Respiration 135, 144, 290
Reynolds Number 77, 81
Reynolds stresses 75, 81
Rhizaxinella pyrifera 222
Rhizosolenia 163
Rhodiella tissieri 226
Rhône fan 50–52
Rhopalonema 181
Rhopalonema velatum 176
Richardson Number 77, 81
Rissoella verruculosa 245, 285
River discharges 97, 139, 141, 153–54, 192
Rock crevices 244
Rocky-bottom assemblage 217–19
Rocky substrates 245–48
Rose shrimp (*Parapenaeus longirostris*) 306
Rossby Number 77

Sagitta bipunctata 174–75, 179
Sagitta decipiens 176
Sagitta enflata 173, 174, 181
Sagitta friderici 173
Sagitta hexaptera 174
Sagitta lyra 174, 176
Sagitta planctonis 173
Sagitta setosa 174, 179, 185
Sagitta tasmanica 173
Salinity 6, 7, 9, 13–14, 73, 104–5, 119, 127, 138, 158, 176, 297
    see also Messinian salinity crisis
Salmo gairdnerii (rainbow trout) 314
Salpa democratica 175, 179
Salpa fusiformis 175
Salpae 176
Salt extraction 15
Sapphirella 183
Sarda sarda (bonito) 302
Sardina pilchardus (pilchard) 302, 315
Sardinella (gilt sardines) 303
Sardinia 24, 26, 112, 115
Sardinian Channel 111
Sarpa salpa 15
Sarsia 181
Scaphopods 216, 224
Scattering coefficient of water 68
Schottera nicaeensis 252, 284
Scleractinians 218
Sclerasterias richardi 219
Scolecithridae 175
Scomber scombrus (mackerel) 300, 302, 303, 312
Scomberomoridae 302
Scombridae 302
Scorpaena (scorpionfish) 301
Scorpaena loppei 207
Scorpionfish (*Scorpaena*) 301
Sea-air boundary (SAB) 130
Seabirds 304–5
Sea cucumber (*Stichopus regalis*) 307

Sea-grass meadows 256–63
Sea-land-air boundary (SLAB) 130, 131
Sea-land boundary (SLB) 130, 131
Sea level 6
Sea-lily (*Leptometra phalangium*) 216
Sea of Alboŕan 300
Sea-pen (*Funiculina quadrangularis*) 224, 226
Sea-pen (*Kophobelemnon leuckarti*) 226
Sea-pen (*Pennatula rubra*) 220
Sea-pen (*Pteroides griseum*) 214
Seasonal sequences 159–65
Seasonal vertical mixing 85–87
Sea squirt (*Microcosmus* spp.) 307
Sea-stars 221, 225, 226, 263
Sea-urchin (*Cidaris cidaris*) 218, 219, 221, 223, 225, 227
Sea-urchin (*Echinus acutus*) 216
Sea-urchin (*Paracentrotus lividus*) 263
Sea-urchin (*Spatangus purpureus*) 211, 212, 216
Sea-urchin (*Sphaerechinus granularis*) 264
Sea-urchin (*Strongylocentrotus lividus*) 307
Sea-water
　attenuation of solar radiation 67–71
　chemical composition 4
　chemical equilibria 127
　composition of 126–29
　compressibility 62
　density of 62
　gases dissolved 128
　major elements 127–28
　minor elements 128–29
　oxygen consumption 136
　processes controlling chemistry of 129–37
　sound propagation 63–64
　trace elements 129
Seaweeds 246, 248
Secchi disk 70
Secondary production 190–91
Sedentary species 252
Sedimentary facies 47–49
Sedimentation 41–45, 321, 325
　Continental shelf 45
　deep sea 49
Sedimentation models 45–54
Sediments 43, 136–37, 145, 325
Sepiolidae 208
Seriola dumerili 302
Serpulids 218, 221, 226
Sertularia marginata 208
Sewage pollution 145, 335, 346
Shad (*Alosa* spp.) 303
Shag (*Phalacrocorax aristotelis*) 305
Sharks 226, 227
Shelf-edge detritic assemblage 215–17
Sherki Bank 112
Shrimps 224, 226, 315
Siboglinum carpinei 224
Sicily 105, 108
Silica 134–35
Silicon cycle 134–35
Siphonaria pectinata 206
Siphonophora 171, 179, 183
Sipunculids 213
Sirocco 144

Skeletonema costatum 165
Slope facies 47–49
Soft bottom communities 256–63
Soft substrates 241–42, 244–45
Solar constant 66, 140
Solar energy 9, 66–71
Solar radiation 9, 140, 142, 143
　attenuation in sea 67–71
　flux of 66–67
Solea solea (common sole) 301, 306, 315
Solea vulgaris 15
Sound propagation in sea-water 63–64
Sparus 208
Sparus auratus (bream) 15, 315
Spatangus purpureus (sea-urchin) 211, 212, 216
Specific heat
　of air 9
　of water 9, 64
Sphaerechinus granularis (sea-urchin) 264
Sphaeroma 14
Sphaeroma serratum 244
Spider crab (*Maja*) 306
Spiny lobster (*Palinurus mauritanicus*) 218, 306
Spiratella bulimoides 186
Spiratella lesuerui 185, 186
Spirographis spallanzani 288
Spirorbis marioni 207
Spondylus gussoni 218
Sponges 211, 213, 218–20, 222, 224, 226, 238, 246, 252, 253, 256, 270, 271, 275–78, 289, 307
Sprat (*Sprattus*) 303
Sprattus (sprat) 303
Squamariaceae 212
Squid 306
Sterna hirundo (common tern) 305
Sterols as organic matter 332–34
Stichopus regalis (sea cucumber) 307
Strait of Gibraltar 109, 123, 127, 138, 150, 155, 173
Strait of Sicily 111, 112
Stratification 87, 146
Stratigraphy 4, 30–32, 42, 49
Stromatolites 202
Strongylocentrotus lividus (sea-urchin) 307
Stylocheiron suhmii 173
Submarine caves and tunnels 271–79
　dark caves 277
　semi-obscure 273–77
Subsidence 32
Sub-tropical Jet Stream 89, 90
Succession and regression 281–84
Supralittoral zone 239–44, 291
Surface layer 130
Surface water 110–12, 120, 123
Symplegma viride 208
Syngnathidae 263
Synthetic processes 131

Tele-intermediaries 347
Temora 181
Temora longicornis 173, 175, 185

Temora stylifera   174, 175, 179, 180, 182, 187
Temoridae   174
Temperature distribution   152, 153, 161, 162, 176
Terrigenous mud-shelf assemblage   213 – 15
Terrigenous sedimentation   45
Tethys, evolution of   18 – 22
Tethys Sea   208
Thalassionema nitzschioides   164
Thalassiosira rotula   163
Thalassiothrix mediterranea   164
Thaliacea   175, 179 – 82
Thenea muricata   226
Thermal change   71 – 87
Thermal gradient   73
Thermal static stability   71 – 72
Thermal structure   71 – 87
Thermal vertical structure   71 – 74
Thermic cycles   85
Thermohaline characteristics   101 – 3, 107
Thermohaline circulation   101, 122
Thunnidae   302
Thunnus thynnus (bluefin tuna)   302, 315
Thysanoessa gregaria   173, 185
Thysanopoda aequalis   173
Tidal amplitudes   122
Tidal energy   123, 124
Tidal harmonic constants   122
Tides   13, 122 – 24
Tilting   32
Tintinnida   164, 170, 183
Tintinnopsis campanula   183
Tisbe   183
Titanium   346
Topographical features and weather conditions   88
Tortonian   199
Trachimedusae   179
Trachinus   208
Trachurus trachurus (horse mackerel)   300, 302, 312
Tramontana   92, 96
Transparent Goby (*Aphia minuta*)   303
Transport processes in water   64 – 66
Triadinium polyedricum   163
Trigla (gurnard)   301
Trisopterus minutus capelanus (pouting)   305
Trophic ecology   289 – 93
Trottoir   246 – 47, 250, 252, 254
Tubularia   207
Tuna (*Euthynnus euthynnus quadripunctatus*)   302
Túnel Llarg de la Meda Petita   280
Tunicates   246, 256, 271, 276
Tunisia   111
Turbulence energy   80
Turbulent diffusion   74 – 78, 131
Turritella tricarinata f. *communis*   213
Turtles   304 – 5
Tyrrhenian Sea   28 – 31, 140, 154, 173 – 74, 336

Umbellosphaera tenuis   164
Upper infralittoral zone   291
Upper mediolittoral rock   245

Valencia   152
Valencia Fan   50, 51
Valencia Trough   23, 25, 50
Valencia Valley   50 – 52
Vanadium   346
Velella (blue sail)   11
Velocity gradient   77
Venus casina   211, 212, 216
Veretillum cynomorium   213
Vermetus   250
Verruca   231
Verrucaria   242
Verrucaria symbalana   242
Vertical exchange processes   79
Vestige indicators   185
Virgularia mirabilis   213
Volume flow   106 – 7

Water
   absorption coefficient   68, 69
   compressibility   62 – 64
   density variation   61 – 62
   light attenuation in   70
   light extinction in   236
   molecular viscosity   64
   scattering coefficient   68
   specific heat   9, 64
   surface tension   65
   transport processes in   64 – 66
   see also Deep-water; Intermediate water; Sea-water; Surface water
Water agitation   290
Water balance   138
Water budget   137, 150
Water circulation   8
Water exchange   149 – 51
Water molecules   60
Water movement   237, 273
Water properties   60 – 66
Water turbulence   168
Weather conditions   87 – 97
   and primary production   95 – 97
   and topographical features   88
Weather regimes   90 – 95
Weather systems   89 – 90
Whales   303 – 4
White corals   219 – 21, 230
Wind direction   80
Wind-driven currents   83 – 85
Wind effects   8, 73, 96, 117, 144
Wind energy   123
Wind frequency distribution   91 – 95
Wind stress   72
Winter mixing   149 – 51
Wreckfish (*Polyprion americanus*)   301

Yellow substances   68

Zinc 336
Zonation 235–36, 275
Zone typification 239–41
Zooplankton 10, 11, 170–71
  biomass 191–94
  distribution 191
  macrophagous 187
  microphagous 187
  neritic populations of 177–82
  pelagic populations of 171–77
  populations of interior waters 183
  seasonal changes of 188
  temporal changes in 190
  trophic position of 185–88
  vertical distribution of 188–90
Zostera noltii 260

KEY ENVIRONMENTS
Other Titles in the Series

AMAZONIA *Edited by*: G. Prance and T. Lovejoy

ANTARCTICA *Edited by*: N. Bonner and D. Walton

GALAPAGOS *Edited by*: R. Perry

MADAGASCAR *Edited by*: A. Jolly, P. Oberlé and R. Albignac

MALAYSIA *Edited by*: The Earl of Cranbrook

RED SEA *Edited by*: A. Edwards and S. Head

SAHARA DESERT *Edited by*: J. L. Cloudsley-Thompson

227417